Plastics

Fifth Edition

This inkstand was molded around 1850 of gutta percha, the first molded plastics material. (*Courtesy John Topham Ltd., Sidkup, Kent, England*)

Plastics
Fifth Edition

J. HARRY DuBOIS
J. Harry DuBois Company
Morris Plains, New Jersey

FREDERICK W. JOHN
Vice President—Technical Director
Revell Inc.
Venice, California

VNR **VAN NOSTRAND REINHOLD COMPANY**
New York   Cincinnati   Toronto   London   Melbourne

Van Nostrand Reinhold Company Regional Offices:
New York Cincinnati Chicago Millbrae Dallas

Van Nostrand Reinhold Company International Offices:
London Toronto Melbourne

Published by Van Nostrand Reinhold Company
450 West 33rd Street, New York, N.Y. 10001

Published simultaneously in Canada by Van Nostrand Reinhold Ltd.

15 14 13 12 11 10 9 8 7 6 5 4 3 2 1

Library of Congress Cataloging in Publication Data

DuBois, John Harry, 1903–
   Plastics.

   1. Plastics.   I. John, Frederick W., joint author.
TP1120.D8 1974   668.4   74-12450
ISBN 0-442-22179-7

PREFACE

The first edition of "Plastics" was written more than 33 years ago under the forced draft of preparing for war, with upcoming metal shortages demanding more from the plastics. It was not difficult then to ignore the confusing and complex nomenclature of the chemists and to provide an easily understood picture of the materials, methods and products. Updating revisions were made in 1943, 1945, and 1967. With each successive revision, the materials, names and processes have become more complex.

Significant changes have taken place in the industry since the last edition was published. Changes in plastics materials, processes, and machinery have been great and we have made every effort to include the most important in this edition. The oil shortage is changing materials and growth patterns.

The industry has done little to simplify the terminology and now, we have no alternative than to use some of the unpronounceable words to identify several of the materials. We also must include some electronic and thermodynamics fundamentals to assist the processing engineers. We continue to try to ignore the chemistry, rheology, thermodynamics, metallurgy and the other complicated scientific language and yet provide an understandable and workable point of entry into this increasingly complex technology.

Each molding material of contemporary or future importance has been included to help you find the essential properties that will best solve your problems. The latest processing procedures are included, along with essential machine technology, for the benefit of operating personnel, foremen, training and maintenance people. Presented here are the data that we know from our daily experience will be fundamental to the needs of our own plant people today and for the students who will want to work in the industry tomorrow. This is a practical book by practicing engineers who have met and continue to meet all of your problems on a daily basis. This book reflects more than 80 years of personal experience with the plastics and in-plant training programs.

College professors, teacher, engineers, and in-plant training personnel who use this book are urged to continue to give us the tips and benefits learned from teaching that have helped us so much in meeting the needs of this audience.

J. HARRY DuBois
Craftsman Farms, Route 10
Morris Plains, N.J. 07950

FREDERICK W. JOHN
1173 Bienveneda Ave.
Pacific Palisades, Ca. 90272

ACKNOWLEDGMENTS

The authors are indebted to the American Technical Society who published the first three editions of "Plastics." Reinhold and SPE liberally contributed photographs, diagrams, etc., from other Reinhold books: "Processing of Thermoplastic Materials" by Bernhardt, "Engineering Design for Plastics" by Baer, "Plastics Mold Engineering" by DuBois and Pribble, and Encyclopedia of Plastics Equipment" by Simonds.

Important data and information for this book were released for use by "Modern Plastics Encyclopedia," *Plastics World*, *SPE Journal*, *Machine Design*, *Materials in Design* and *Materials Engineering*. Materials and machine makers provided many special photographs, diagrams, and editorial help.

Portions of Chapter 1 were prepared under the supervision of the editors of Field Enterprises Educational Corporation for publication in "The World Book." Professor Louis F. Rahm, who initiated the Plastics Program at Princeton and Maurice Keroack, Education Director of the Plastics Education Foundation, were most helpful with advice concerning the needs of educational programs in plastics engineering. Valuable comments and suggestions were received from engineers, teachers and businessmen in the preparation of this new edition. Assistance was also rendered by Mr. M. Kaufman, author of "The First Century of Plastics," published by the Plastics Institute of London, England. Sidney Levy of the Leal Company prepared the section on environmental problems. Philip A. Heist of Revell Inc. furnished the manuscript for solid state control.

J. HARRY DuBois
FREDERICK W. JOHN

CONTENTS

Plastics

Fifth Edition

1 AN INTRODUCTION TO PLASTICS

INTRODUCTION

Plastics are nonmetallic basic engineering materials that can be formed and shaped by many methods. Plastics may be man-made synthetic resins, or they may be compositions formed from such natural resins as shellac. You see them in Fig. 1.1 as functional automobile components. Other products are squeeze bottles, paints, adhesives, bread wrappers, missile bodies, wire insulation, fabric such as nylon and rayon, pipe, shoes, baby pants, skillet lining, piano keys, telephones, etc.

Plastics is the generic name of an industry and its products; in this usage, plastics is a plural word. The industry adopted this name in the late twenties, when the formability of these compositions, while in their plastic state, was thought to be their most important feature.

The black piano keys, for example, were first made of ebony. Later on, they made a "patent" sharp by heat-forming a sheet of black "Celluloid" plastics over a wood block, thus gaining a deep black, glossy surface. Today the sharps are molded twenty at a time of black phenolic plastics, which are very durable, scratch-resistant, and not attacked by cleaning solutions. These phenolic sharps are identical in every respect, inexpensive, and never need polishing or repairing. The white keys are often called "ivories" because the early pianos had the keys covered with thin strips of ivory, glued on to the basic wood keys. Ivory, the natural plastics material, yellowed with age and became increasingly expensive

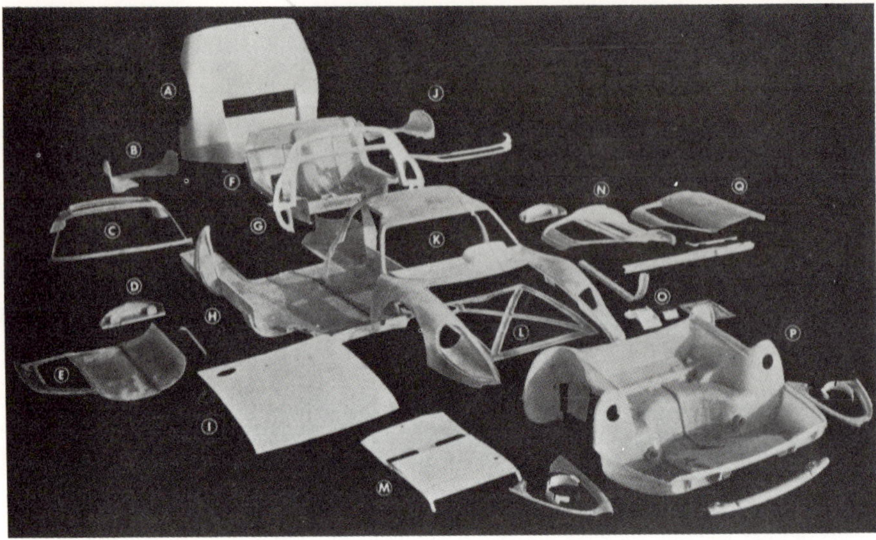

FIG. 1.1. Some of the major parts of the Carrera GTS Porsche, a car designed and tooled to take advantage of the special properties offered by polyester-glass RP. (A) rear of engine body, (B) fender support, (C) stiffener for rear of engine body, (D) door flange, (E) outer door skin, (F) bucket seat, (G) cabin bulkhead, (H) bottom, (I) lid of gas tank space, (J) fender support, (K) front end with instrument panel—roof/fender combination, (L) stiffening member of gas tank led and bottom, (M) cooling air duct, (N) inner door skin, (O) support for windshield wiper motor, (P) gas tank bottom with wheel recess, (Q) outer door skin. (*Courtesy Badische Anilin- & Soda-Fabrik AG*)

as the usage increased. "Celluloid," the first synthetic plastics, was used in replacement of ivory because it stayed white, was abundant in supply, retained its original shiny white surface, and was less costly than ivory. "Celluloid" was replaced by improved plastics for even better appearance and scratch resistance. The new piano and electronic organ keyboards take full advantage of the plastics-molding process, which produces the entire keyboard of molded plastics. Ladies' shoe heels were first made of layers of leather, then of leather-covered wood. This gave way to plastics-covered wood, and now most ladies' shoe heels are molded complete.

Today almost every manufactured product uses plastics in some form, in the end-product, or in the manufacturing process. Plastics include paints, fabrics, rubbers, insulations, foams, greases, oils, films, "glasses," structural materials, adhesives, and basic molding and fabricating materials.

KINDS OF PLASTICS

There are many kinds of chemically different plastics. Each kind may be made into hundreds of compositions or alloys. Each alloy provides some special and desirable combination of properties. For example, the basic phenolic plastics may be compared to concrete, which is a mixture of cement and sand that hardens by a chemical action in the presence of water. The cement serves as a binder for the sand, and the sand is said to be a filler which adds hardness, durability and dimensional control, thus reducing the cost of the mixture.

Phenolic plastics resin may use finely ground wood as a filler for general applications, gaining body, minimum shrinkage, strength and reduced cost. Phenolic resin and asbestos fibers are molded into missile nose cones and components, respectively, to gain high heat endurance. Pieces of heavy canvas belt duck, as shown in Fig. 1.2, bonded with phenolic resin, are used to make steel mill roll-

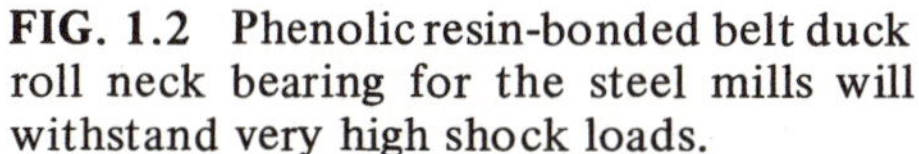
FIG. 1.2 Phenolic resin-bonded belt duck roll neck bearing for the steel mills will withstand very high shock loads.

neck bearings that will withstand the crushing and variable load of rolling steel, from billets to sheets, bars and rods. Sheets of selected paper are bonded with phenolic resin to make fine electrical insulation called "laminated plastics," because many sheets are bonded together to make the desired thickness. In like manner, mica, linen, cotton fabric, cotton linters, ferrites, powdered glass, glass fibers, glass cloth metal particles, etc., may be used as fillers with one of the synthetic resins, to produce a tailor-made product with a correct balance of special properties to serve an industrial need. Many plastics are identified by the base resin and the filler. When no filler is used, the chemical name is adopted in some short form, and a generic name for that particular chemical class of material is appropriated. Many of the plastics materials, such as the phenolics, are especially compounded or alloyed to serve a broad market with special properties at a reasonable price. In other cases, a widely used plastics compound becomes important because its properties are useful for a great variety of products, and it can be produced inexpensively from abundantly available sources of supply.

Polystyrene is one of the very lowest cost plastics, and it is made of ethylene and benzene, by-products of the natural gas, petroleum or coal distillation processes. It is a good, general-purpose material that serves well in many simple and complex operations where appearance, color, moldability and price are of the greatest importance. Many pieces of polystyrene are used in the interior of refrigerators. Styrene is sometimes used with glass fillers; its alloys are frequently synthetic rubbers. In selecting a plastics material from the many varieties, the product engineer selects the minimum-cost variety having the greatest number of desirable properties for his application. He then modifies the design wherever necessary, to compensate for the weaknesses that may be exhibited by this material. Plastics engineers have compiled property tables, similar to those used for metals and other structural materials, so that the physical properties may be used to compare the desired strength condition. Several types of plastics may be used in a single assembly to gain specific properties for each type of stress in the product. Some plastics are very hard, while others are soft and rubber-like. Foamed plastics are very light, while glass- and mica-filled plastics are heavy. Some are opaque, and others are more transparent than glass. Structural materials may not be attractive in appearance; other materials may have fine surface texture and gloss. The beauty of the finest woods may be preserved and protected by the use of suitable plastics resins and fillers, to produce a decorative laminated sheet for an enduring table-top or windowsill.

Plastics have made a substantial invasion into the furniture field and their potential is tremendous—estimated at four billion pounds by 1985. Presently, plastics are being used to simulate wood, with excellent fidelity and detail. Volume plastics furniture components are more economical to produce than wood. Foamed plastics are preferred for structural pieces because they have the feel and weight of wood. Plastics materials are used almost exclusively today for upholstering—urethane cushions and vinyl upholstering are already commodity items in the furniture industry. Cotton has been replaced by nylon, olefin and polyester fibers, with an extended life and greater economy for the buyer.

The hard plastics, such as the phenolics, ureas, melamines, alkyds and allylics, are often used because of their hardness. The melamine compounds have fine color and surface hardness, as well as good scratch-resistance, making them desirable for dinnerware. Asbestos-filled phenolic, mineral-filled urea, glass-filled allylics and polyesters provide excellent wear-resistant surfaces, which are dimensionally stable and able to withstand much surface abuse.

Many soft plastics find large markets for toys, gaskets, hose, squeeze-bottles, raincoats, upholstery, cushions and rope. The vinyl plastics are available in a wide hardness and color range, from the very soft to quite hard products, as illustrated in Fig. 1.3. Polyethylene squeeze-bottles take advantage of its softness and flexibility. The foamed plastics are produced from many of the basic resins, making either a hard, brittle, sponge-like mass, or a very soft, resilient, cushioning material. The urethane foams have excellent properties for furniture and automobile cushions. Silicone plastics having soft rubber-like properties are suitable for seal-

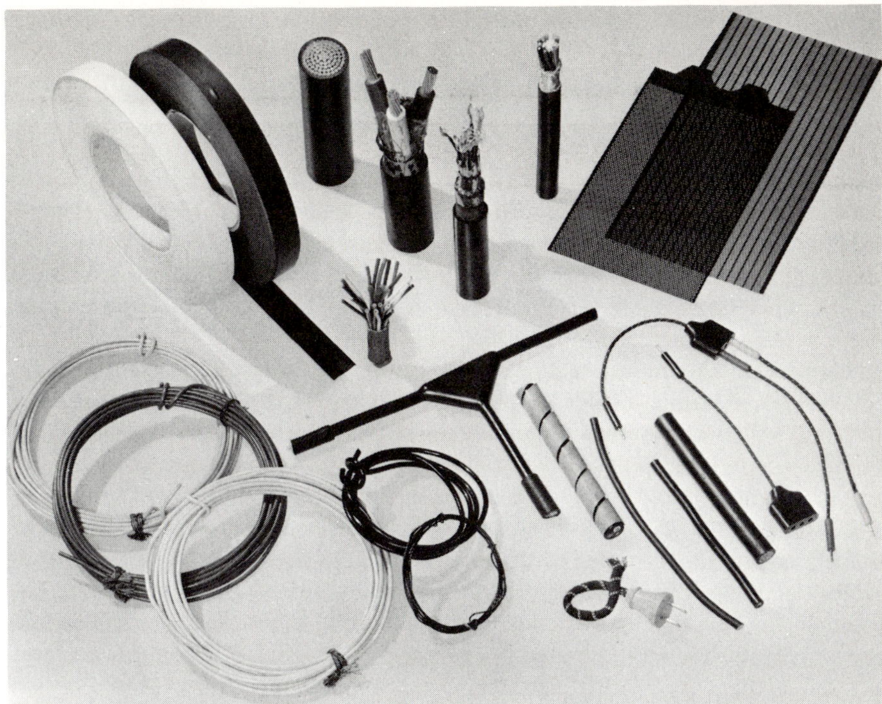

FIG. 1.3 Polyvinyl raw materials have a wide variety of applications in the electrical industry. Vinyl plastics are self-extinguishing, colorful, low in cost, and may be rigid or flexible.

ing gaskets in ovens, for high temperature electrical apparatus insulation, and for reconstructive surgery implants. Nylon, polyethylene, polypropylene and other synthetic resins may be stiff when produced in thick sections but, like the metals, when drawn into fine filaments become flexible, and may be formed into threads, fabric or rope that is soft and pliable. Nylon makes excellent rigid gears; it also makes soft and delicate hose and lingerie.

Many plastics are transparent; the acrylic resin that is used for contact lenses is more transparent than most of the glasses. Excellent optical properties may be obtained for scientific apparatus by the use of the crystal-clear acrylic plastics. The polystyrenes, vinyls, acetates, nylons, polycarbonate, polysulfones and the polyolefins can be made in transparent form.

In the decorative plastics, there is a most attractive variety of textures, colors, and designs available. Multi-colored designs may be printed on paper, simulating wood, marble, metals or geometric and artistic compositions, bonded with synthetic resins to produce the decorative laminates. These are used for the "wear" surfaces in furniture, for wall coverings and for novelty areas. The flexible plastics films may be metallized so that they gain the appearance, luster and color of gold, silver, brass or copper. They are used for nameplates, surface decorations, seat covers, "metallic ribbons," etc. These decorative metallic-colored plastics films may be formed to shape as a cover for many other colorless materials.

A variety of plastics have heat-resistant properties suitable for pothandles, carburetor sections, missile bodies, and other structural and insulating components that must withstand heat. Many of the plastics, such as polyphenylene oxide, polysulfone and melamine, can be steam-sterilized; others, such as polystyrene, may soften at temperatures below boiling water. Each plastics material has an operating-temperature limit that must not be exceeded if the original shape is to be maintained.

Most of the thermosetting materials are suitable for use in the 300° range. Recent developments include a variety of engineering thermoplastics that offer stability at working temperatures previously limited to the thermosets. Polytetrafluoroethylene, polysulfone, polycarbonate, polyphenylene oxide, polyimide, polyimide amide, polyarylsulfone, and thermoplastic polyester materials with and without glass fillers, along with other filled thermoplastics, are now doing jobs previously limited to the thermosets.

Most of the plastics are chemically resistant in some form. Polystyrene, for example, has excellent acid resistance, but styrene is attacked by gasoline and other hydrocarbons. Chlorinated polyether plastics is especially made to meet the requirement for a very high degree of chemical resistance. Some plastics, such as polyethylene, will withstand many chemicals, but will pass gas or liquid components. Vinylidene fluoride has fine chemical resistance, and does not pass components of many products.

Many plastics are compounded to gain special properties by the addition of plasticizers, stabilizers or fillers, or by alloying two or more synthetic resins. It is possible to produce a tailor-made compound for almost every type of application.

ADVANTAGES

Materials are selected for any given application by a process of elimination which finds the one material having the most desirable combination of essential properties for the application at the lowest price. In redesigning a given part, the engineer searches for the one material that will perform as well or better, but will cost less than any other material. A typical conversion to plastics is shown in Fig. 1.4. A widely recognized formula states that a sale is made by the material that offers the most "plus" properties for the price asked. The plastics are used in many applications because there are no other materials that will do the job at a reasonable price. In other cases, they are selected merely because they are as good as other materials, yet cost less. The savings in price may be the result of a low-cost material, but in most cases, the savings result from reduced processing and fabricating costs. As direct labor becomes more expensive, the automated

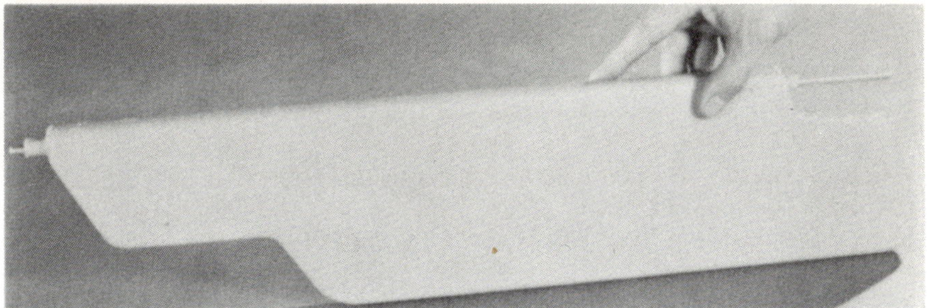

FIG. 1.4 An air-cushioned, safety automobile sun visor that eliminates all other component parts is being rotationally molded from powdered polyethylene. This design integration eliminates parts such as hard board, padding, tempered wire frames and metal tubing. (*Courtesy USI Chemicals*)

processes become an essential in overcoming the high cost of handwork, thus enabling a manufacturer to meet his competition. Use of plastics frequently eliminates the cost of finishing and painting. Plastics require no buffing after plating. Molded and extruded products come finished to dimension, ready for the assembly line, so that the final product is merely a mechanized assemblage of mass-produced parts—and the molded plastics offer a very low-cost means of producing just such pieces. All of the mass-produced products of industry use plastics as basic materials, as well as finishes and adhesives. Aluminum bicycle frames may be "welded" by an epoxy cement. The plastics materials have proven themselves superior to most other materials, with the exception of a few of the extraordinary ceramics and metallic alloys used for the space vehicles. For the missile program, the plastics contribute high thermal-endurance, a light weight with an extremely great strength, as well as fabrication ease for complex sections, plus "electrically transparent" windows for the transmission of guidance information. These "windows" must have constant electrical properties under accelerated temperature changes. Plastics are extensively used for lower-cost metal stamping and forming dies. Drill jigs and assembly fixtures made of plastics are adequate and yet less costly for many applications.

In architecture and home construction, the plastics are contributing many desirable benefits and properties, as shown in Fig. 1.5. The future will bring greatly expanded use of plastics as basic construction materials. Plastics make better and more colorful stained-glass windows. The decorative laminates of real wood are extensively used for panels, table tops, window ledges, etc. The plastics foams may be poured into voids where they will expand, filling every crevice with an excellent insulation against heat and cold. This insulation is vermin-proof, and will not rot or decay. Boards of foamed plastics are available for home construction, where they offer good structural properties, plus heat and cold insulation. Thin-walled refrigerator boxes are formed of foamed plastics. Data gained from the space vehicle construction point the way to light-weight structural

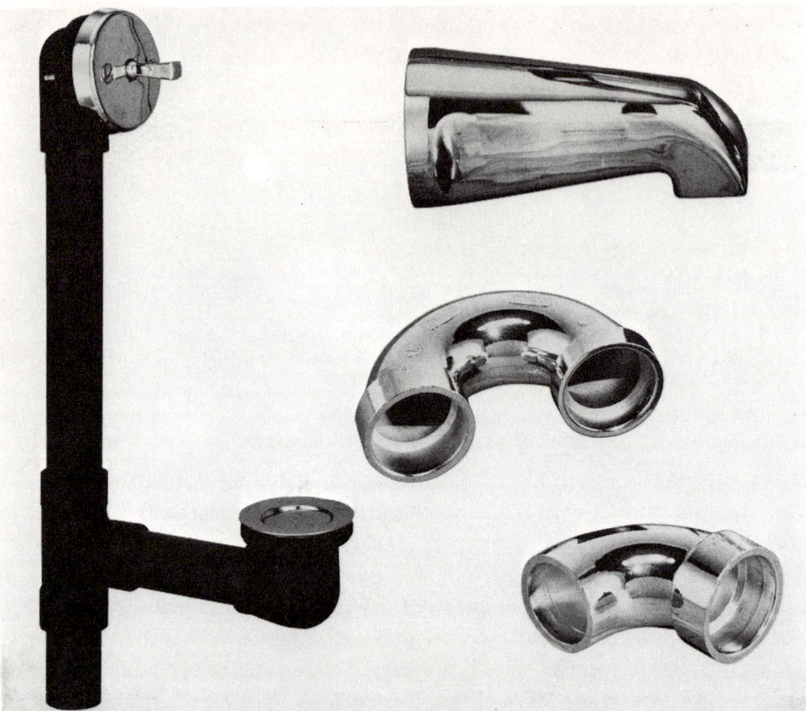

FIG. 1.5 Extruded plastics pipe with welded joints and chrome-plated plastics fixtures will greatly reduce the cost of housing and perform as well as metals. (*Courtesy Marbon Chemical Div., Borg Warner Corp.*)

sections that will take the place of steel and aluminum. Basic structural shapes for the arctic radomes have proven the capability of these materials for large, unsupported roof areas, laden with heavy snow and ice. Permanent synthetic-resin paints have virtually eliminated maintenance problems under conditions that necessitated continuous refinishing with other coating materials. The vinylidene fluoride finishes, shown in Fig. 1.6, are highly resistant to weathering and ultraviolet light.

Use of the polyester glass-reinforced plastics makes it possible to simulate any texture, such as old stone construction, etc., in a very light-weight, inexpensive structural material. Plastics sections can simulate the finest hand-carved architectural components.

In the field of medicine, the plastics are serving in many invaluable ways. Nylon and other plastics sutures have replaced the traditional materials. The dentist uses plastics adhesives for cementing the inlays in place. He uses plastics resins for his molds which consequently harden quickly by a chemical reaction, reproducing every minute detail, and also eliminating shrinkage, one cause of a troublesome fit. Many dentures are made of the plastics. A silicone resin is used to "package" the "Pacemaker," that automatic electronic timer for the heart-beat, which is inserted into the body. Plastics heart valves, intestinal sections, rivets, screws, plates, etc., are all part of the surgeon's kit today. The selected plastics is not affected by the body chemistry, and creates no internal problems of its own. Out of plastics, the prosthetic specialist makes artificial hands, legs, and other body parts, complete with finger prints, color, flexibility, etc., as shown in Fig. 1.7. The reinforced plastics are used to produce high-strength, light-weight, and correctly shaped artificial limbs.

In the fields of science and research, plastics have contributed greatly to the demand for better materials. Research work often calls for exactly controlled materials; variables cannot be tolerated. Certain plastics films will pass some gases and vapors but will not pass others, thereby making possible the fractional separation of some gases. The very complex printed circuitry, used in our micro-miniature electronic devices and computers, demands properties and fabricating

FIG. 1.6 Metal siding coated with an age and weather resistant finish of vinyl-idene fluoride is shown being placed on a new industrial building. (*Courtesy Pennsalt Chemicals Co.*)

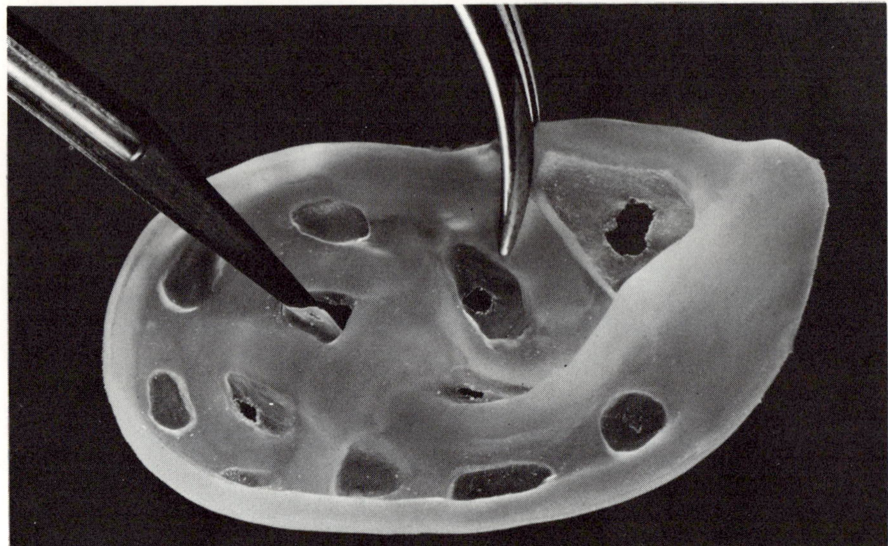

FIG. 1.7 The membrane like areas on the silicone plastics Otoplasty prosthesis may be cut out easily to permit tissue fixation as desired. Designed to serve as a supporting framework in full or partial ear reconstruction or replacement, this ear-shaped prosthesis is light weight, may be trimmed easily and exhibits tactile qualities of flesh and cartilage. (*Courtesy Dow Corning Corp.*)

methods that can be achieved only through the use of the plastics. Electron microscopes make use of minute plastics spheres for their calibrations. The ablation coatings and materials that have most effectively solved the re-entry problem of the missiles use plastics that evaporate slowly under intense heat, without increasing the internal temperatures.

ADDITIVES

The plastics industry uses the word additive to describe the products that are combined with the basic resins and polymers as extenders, or to modify their properties, or to facilitate their processing, or to achieve special color and finish.

Antioxidants are additives that retard or inhibit the oxidative degradation of the plastics materials within the intended processing and usage limits of the materials.

Antistatic agents are added to minimize the build up of electrostatic charges on plastics materials and products.

Colorants The ingredients used for coloring the plastics materials are dyes, organic or inorganic pigments, or certain specialty coloring compounds. In general, dyes are used to supply brilliant transparent colors. Organic pigments are discrete solid particles of dyes. Inorganic pigments consist of non-brilliant salts and oxides of metals. The specialty pigments include metallic flakes, pearlescent or fluorescent pigments blended with other pigments.

*Flame retardants** may be one of four general classes of product:

A. Products that produce a heavy gas that supplies an insulating or smothering effect.
B. Cooling effects by absorbing heat energy through an endothermic reaction.
C. Coatings that produce an oxygen barrier.
D. Chemical reaction producers that form less combustible products.

New regulations designed to minimize fire hazards have greatly increased the variety of flame retardants. Combustibility and the products of combustion must be limited in many applications.

Lubricants are used to improve the processibility and appearance of the plastics product.

Plasticizers are chemically and thermally stable products that are added to improve the flexibility, softness and processibility of the plastics.

Stabilizers are used to prevent the degradation of materials due to high processing temperatures or to extend their life stability under degrading environmental conditions.

Biodegradable[†] Resin additives can be alloyed with many thermoplastic resins to produce tough translucent products. When placed in contact with moist soil, they will undergo biodegradation, losing weight, strength and other properties.

Fillers

Fillers play a very important part in the manufacture of plastics compounds. They reduce cost, provide body, speed the cure or hardening, minimize shrinkage, reduce crazing, improve thermal endurance, add strength, and provide special electrical, mechanical and chemical properties. Asbestos, for example, is widely used as a filler in compounds requiring high temperature and improved dimensional stability. Mica serves as a crack-stopper in glass-bonded mica; it also improves the electrical and thermal properties of all compounds. Ferrites are added to produce magnetic materials. Glass fibers produce very high-strength compounds. Beryllium oxide-filled resins gain high thermal conductivity for heat "sinks," without loss of desirable electrical properties. Typical fillers are listed in Table 1.1.

HOW PLASTICS ARE MADE

The plastics resins that are commercially important are all derived from natural products which were in plentiful supply. Many of the plastics are by-products of

---

*Fire Retardant Chemicals Association, N.Y.
†Polycaprolactone, Union Carbide Corp.

TABLE 1.1 SOME FILLERS AND REINFORCEMENTS—AND THEIR CONTRIBUTIONS TO PLASTICS*

Filler or reinforcement	Chemical resistance	Heat resistance	Electrical insulation	Impact strength	Tensile strength	Dimensional stability	Stiffness	Hardness	Lubricity	Electrical conductivity	Thermal conductivity	Moisture resistance	Processability	Recommended for use in[a]
Alumina tabular	●	●				●								S/P
Alumina trihydrate, fine particle			●				●					●	●	P
Aluminum powder										●	●			S
Asbestos	●	●	●	●		●	●	●						S/P
Bronze							●	●		●	●			S
Calcium carbonate[b]		●				●	●	●					●	S/P
Calcium metasilicate	●	●				●	●	●				●		S
Calcium silicate		●				●	●	●						S
Carbon black[c]		●				●	●			●	●		●	S/P
Carbon fiber										●	●			S
Cellulose				●	●	●	●	●						S/P
Alpha cellulose			●		●	●								S
Coal, powdered	●											●		S
Cotton (macerated/chopped fibers)			●	●	●	●	●	●						S
Fibrous glass	●	●	●	●	●	●	●	●				●		S/P
Fir bark													●	S
Graphite	●				●	●	●	●	●	●	●			S/P
Jute			●			●								S
Kaolin	●	●				●	●	●	●			●	●	S/P
Kaolin (calcined)	●	●	●			●	●	●				●	●	S/P
Mica	●	●	●			●	●	●	●			●		S/P
Molybdenum disulphide							●	●	●			●	●	P
Nylon (macerated/chopped fibers)	●	●	●	●	●	●	●	●	●				●	S/P
Orlon	●	●	●	●	●	●	●	●				●	●	S/P
Rayon			●	●	●	●	●	●						S
Silica, amorphous			●									●	●	S/P
Sisal fibers	●			●	●	●	●	●				●		S/P
TFE—fluorocarbon						●	●	●	●					S/P
Talc	●	●	●			●	●	●	●			●	●	S/P
Wood flour			●		●	●								S

*The chart does not show differences in degrees of improvement; calcined kaolin, for example, generally gives much higher electrical resistance than kaolin. Similarly, differences in characteristics of products under one heading, such as talc (which varies greatly from one grade to another and from one type to another) also are not distinguished.
a—Symbols: P—in thermoplastics only; S—in thermosets only; S/P—in both thermoplastics and thermosets. b—In thermosets, calcium carbonate's prime function is to improve molded appearance. c—Prime functions are imparting of U-V resistance and coloring; also is used in cross-linked thermoplastics.

*Courtesy "Modern Plastics Encyclopedia."

the gasoline refineries. Principal raw material sources are coal, air, water, petroleum, natural gas, limestone, salt and sulfur. Products called "intermediates" are formed from these basic materials, and combined to make the plastics resins.

Table 1.2 shows how plastics may be tailored to meet a wide range of applications. For example, coal is destructively distilled to produce phenol, benzene and ammonia. Air is used as a source of oxygen and nitrogen. Water is converted to hydrogen and oxygen; steam is reacted with coal to make water gas, a mixture of carbon monoxide and hydrogen. Heated limestone produced calcium oxide that is again reacted with coke to make calcium carbide; another reaction with nitrogen produces calcium cyanamide. Salt and water, when electrolyzed, produce hydrogen, chlorine and sodium hydroxide.

All of these intermediates are reacted further to produce the basic raw materials for plastics. For example, carbon monoxide and hydrogen, made from steam and coke, are reacted to form methanol, which in turn is reacted with oxygen to produce formaldehyde. Phenol from the coke ovens is reacted with formaldehyde to make the phenolic resin. Phenol may be made by synthesis also. Polystyrene is made from ethylene and benzene.

Ethylene is derived from either natural gas or is obtained as a by-product from the petroleum refining process. Benzene is often obtained as a by-product from coke ovens. In combining these two intermediates, the ethylene is forced through the benzene in the presence of a suitable catalyst, thus producing ethyl benzene. Hydrogen is split from the rest of this molecule to produce styrene. This styrene is converted to polystyrene by the polymerization process, in which a number of single molecules are joined together to form larger, long-chain molecules.

The plastics materials are fabricated into end-products by several processes. The molding, extrusion and fabricating processes are similar to those developed for the metals and rubber. A very large volume of product-making is done by some form of molding. The heat hardening or thermosetting plastics become plastic under heat and pressure, and then harden by continued heat and pressure, causing condensation or polymerization of the resin. This is a one-way chemical reaction, and the plastics material thus molded cannot be resoftened or molded again. In a typical compression molding operation, a weighed load of the molding compound is preheated to the molding temperature, and then placed in a heated mold where it is compressed by the hydraulic press, and held under great pressure while the hardening reaction takes place. The hot and fully hardened molded piece is then ejected from the mold. This process is done in hand-tended presses, or in fully automatic molding machines.

The injection molding process is similar to the die casting of metals. (see Chap. 4). The cold setting or thermoplastic materials are melted in a cylinder by heat and mechanical work and, when completely melted, are pushed into the closed mold under a continuous great pressure while the plastics material hardens by cooling in the die. The cooled and hardened plastics product is then ejected by the die mechanism. The injection machines can run on a fully automatic cycle.

TABLE 1.2 HOW WE GET OUR PLASTICS. PLASTICS TECHNOLOGY'S FLOW CHART OF MAJOR PLASTIC MATERIALS.

PLASTICS TECHNOLOGY'S
Simplified Flow Chart of Plastics Resins– How They're Derived

Flow chart below illustrates representative ways in which plastic resins are currently being produced from petroleum and natural gas feedstocks.

〈 〉 Symbol denotes chemical synthesis between factor(s) within border and factor immediately preceding.

ACETAL (10)	ETHYLENE-VINYL ALCOHOL (27)	POLYALLOMER (32)	POLYURETHANE (38)
ACRYLIC (35)	FLUOROCARBON, CTFE (13)	POLYBUTYLENE (39)	POLYVINYL ACETATE (28)
ACRYLONITRILE-BUTADIENE STYRENE (19)	FLUOROCARBON, E-CTFE (14)	POLYCARBONATE (4)	POLYVINYL ALCOHOL (29)
ALKYD (15)	FLUOROCARBON, FEP (12)	POLYESTER, THERMOPLASTIC (7)	POLYVINYL BUTYRAL (30)
CELLULOSE ACETATE (23)	FLUOROCARBON, TFE (11)	POLYESTER, THERMOSETTING (16)	POLYVINYL CHLORIDE (31)
CELLULOSE ACETATE BUTYRATE (24)	IONOMER (21)	POLYETHYLENE (22)	SILICONE (6)
CELLULOSE ACETATE PROPIONATE (25)	MELAMINE MOLDING COMPOUND (9)	POLYPHENYLENE SULFIDE (5)	STYRENE ACRYLONITRILE (20)
DIALLYL PHTHALATE (37)	NYLON 6 (1)	POLYPROPYLENE (33)	STYRENE BUTADIENE (18)
EPOXY (36)	NYLON 66 (2)	POLYSULFONE (3)	UREA MOLDING COMPOUND (8)
ETHYLENE-VINYL ACETATE (26)	PHENOLIC (34)	POLYSTYRENE (17)	

Flow chart paths:

NAPHTHA[3]

BENZENE → Cyclohexane → Cyclohexanone → 〈Hydroxyl Amine〉 → Cyclohexanone Oxime → 〈Sulfuric Acid〉 → Caprolactam → **NYLON 6** (1)

BENZENE → Cyclohexane → Adipic Acid → 〈Hexamethylenediamine (from Butadiene, derived from Butane or Butylene)〉 → **NYLON 66** (2)

BENZENE → 〈Chlorine〉 → Monochlorobenzene → 〈Thionyl Chloride〉 → Dichlorodiphenyl Sulfone → 〈Bisphenol A (from Phenol + Acetone, derived from Cumene, which is derived from Propylene + Benzene)[5]〉 → **POLYSULFONE** (3)

BENZENE → 〈Chlorine〉 → Chlorobenzene → Phenol → 〈Acetone (from Isopropyl Alcohol, derived from Propylene)〉 → Bisphenol A[6] → 〈Phosgene〉 → **POLYCARBONATE** (4)

BENZENE → 〈Chlorine〉 → Dichlorobenzene → 〈Sulfur〉 → **POLYPHENYLENE SULFIDE** (5)

BENZENE → 〈Chlorosilane (from Hydrochloric Acid + Silicon Metal)〉 → Phenyl Chlorosilanes → 〈Water〉 → **SILICONE** (6)

Phenyl Chlorosilanes → 〈1,4-Butanediol (from Acetylene[7] + Formaldehyde)〉 → **THERMOPLASTIC POLYESTER** (7)

Methyl Chlorosilanes → 〈Urea + Paper Pulp〉 → **UREA MOLDING COMPOUND** (8)

Dimethyl Terephthalate → 〈Melamine (from Urea) + Paper Pulp〉 → **MELAMINE MOLDING COMPOUND** (9)

TOLUENE

XYLENES[4]

METHANE

METHANOL → 〈Hydrochloric Acid〉 → Methyl Chloride → Silicon Metal → Methyl Chlorosilanes

METHANOL → 〈Terephthalic Acid (from p-Xylene, derived from BTX Stream)〉 → Dimethyl Terephthalate

Formaldehyde → **ACETAL** (10)

METHANE → 〈Chlorine〉 → Chloroform → 〈Hydrogen Fluoride〉 → Chlorodifluoromethane → 〈Tetrafluoroethylene〉 → **TFE FLUOROCARBON** (11)

Chlorodifluoromethane → Hexafluoropropylene → 〈Tetrafluoroethylene (see TFE Fluorocarbon)〉 → **FEP FLUOROCARBON** (12)

METHANE → 〈Chlorine〉 → 〈Hydrogen Fluoride〉 → Chlorotrifluoroethane → 〈Chlorotrifluoroethylene〉 → **CTFE FLUOROCARBON** (13)

〈Chlorotrifluoroethylene (see CTFE Fluorocarbon)〉 → **E-CTFE FLUOROCARBON** (14)

ETHANE

〈Phthalic Anhydride (from o-Xylene, derived from BTX Stream) + Maleic Anhydride (from Benzene) + Propylene Glycol (from Propylene)〉 → **ALKYD** (15)

(aids in crosslinking reaction during cure) → 〈Alkyd〉 → **THERMOSETTING POLYESTER** (16)

Ethylbenzene → Styrene → **POLYSTYRENE** (17)

Styrene → 〈Butadiene (from Butane or Butylene)〉 → **STYRENE BUTADIENE** (18)

〈Acrylonitrile (from Propylene + Ammonia) + Butadiene (from Butane or Butylene)〉 → **ACRYLONITRILE-BUTADIENE STYRENE** (19)

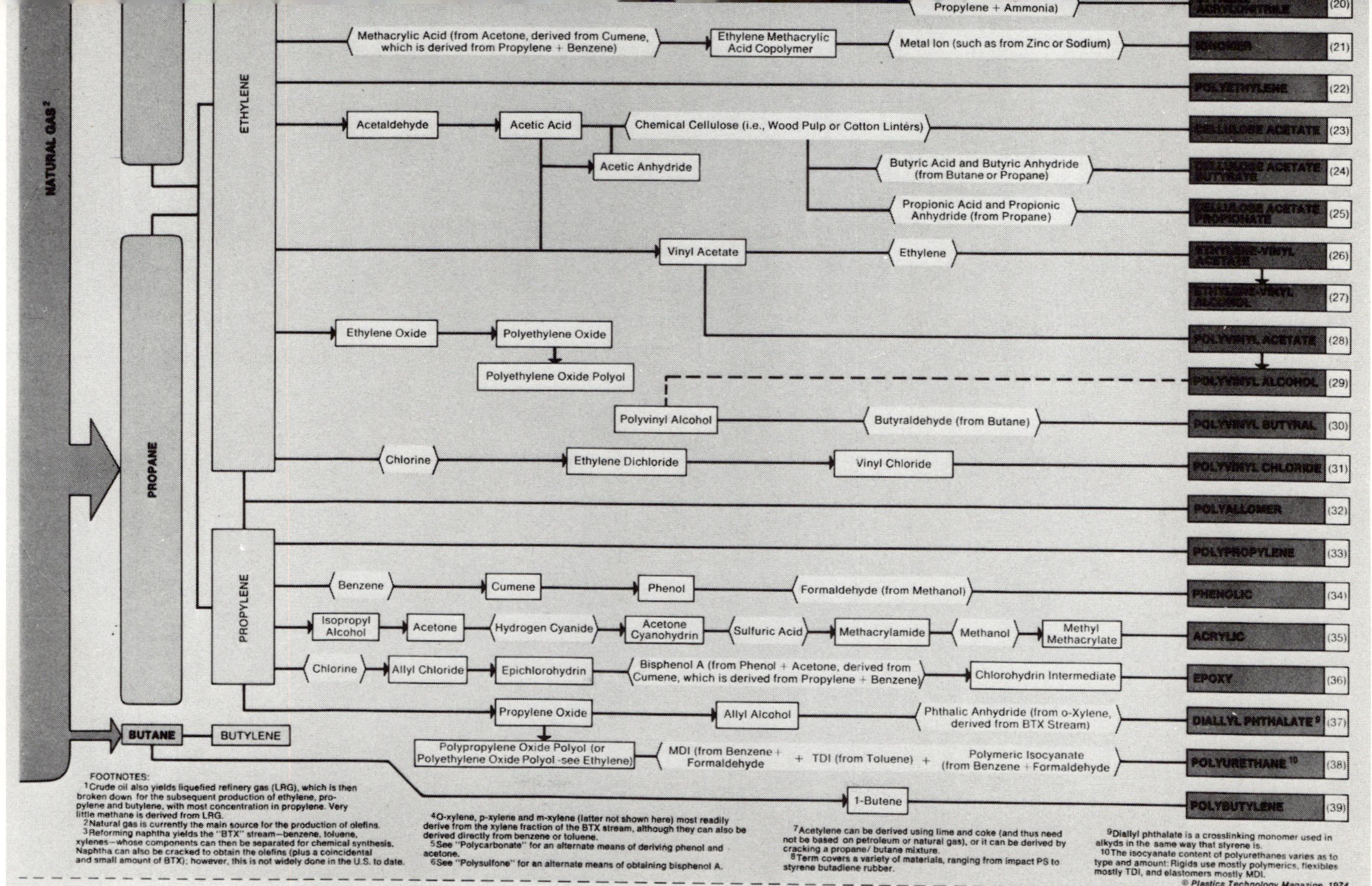

FOOTNOTES:

[1] Crude oil also yields liquefied refinery gas (LRG), which is then broken down for the subsequent production of ethylene, propylene and butylene, with most concentration in propylene. Very little methane is derived from LRG.

[2] Natural gas is currently the main source for the production of olefins.

[3] Reforming naphtha yields the "BTX" stream—benzene, toluene, xylenes—whose components can then be separated for chemical synthesis. Naphtha can also be cracked to obtain the olefins (plus a coincidental and small amount of BTX); however, this is not widely done in the U.S. to date.

[4] O-xylene, p-xylene and m-xylene (latter not shown here) most readily derive from the xylene fraction of the BTX stream, although they can also be derived directly from benzene or toluene.

[5] See "Polycarbonate" for an alternate means of deriving phenol and acetone.

[6] See "Polysulfone" for an alternate means of obtaining bisphenol A.

[7] Acetylene can be derived using lime and coke (and thus need not be based on petroleum or natural gas), or it can be derived by cracking a propane/ butane mixture.

[8] Term covers a variety of materials, ranging from impact PS to styrene butadiene rubber.

[9] Diallyl phthalate is a crosslinking monomer used in alkyds in the same way that styrene is.

[10] The isocyanate content of polyurethanes varies as to type and amount: Rigids use mostly polymerics, flexibles mostly TDI, and elastomers mostly MDI.

Reprinted from *Plastics Technology Magazine*, January, 1974.

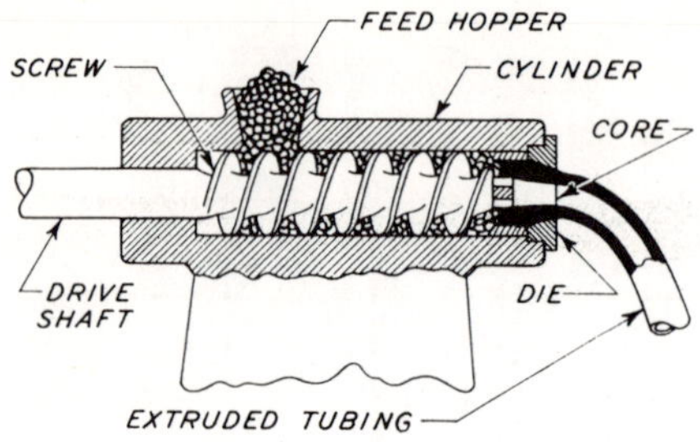

FIG. 1.8 The screw-type extrusion is used for making continuous strips or tubes of regular section.

Extrusion molding, as illustrated in Fig. 1.8, is like the extrusion of metals, except that a long screw is used to melt and plasticize the plastics material by heat and mechanical working. As the material moves forward along the screw, it is forced through a die that controls the shape of the extruded section. The extruded section is shaped in the exit die, and then hardened, by water immersion or by air. Film or sheet plastics are often made by the extrusion process.

In one blow-molding process, a tube of hot, molten plastics is extruded into the blow mold, which clamps and seals off each end, but permits air to be blown into the sealed tube of molten plastics, as shown by Fig. 1.9, where it is then expanded by this pressure to form a hollow shape.

The laminated, high-pressure plastics are produced from a pile-up of sheets of paper or cloth which has been impregnated with the desired thermosetting resin. These laminates are then subjected to high pressure and heat, which plasticize the resin and bond the sheets together while the chemical hardening takes place. The hardened resin holds them tightly-bonded together after the pressure is

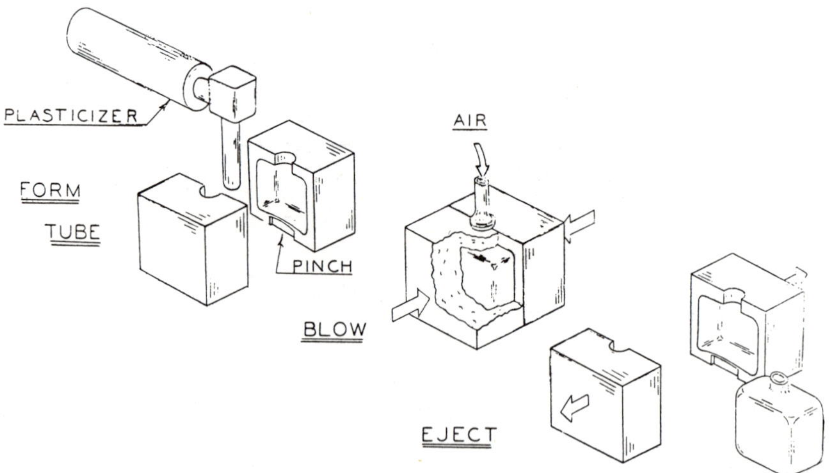

FIG. 1.9 Basic pinch-tube blow-molding process. (*Courtesy Monsanto Co., Hartford, Conn.*)

removed. Laminated and many extruded thermoplastics are fabricated into end products by conventional machining, stamping and forming processes. Welding and adhesive procedures are used to bond the thermoplastics.

Cast plastics use a hardening process without the addition of pressure. Casting of film is produced by pouring the heat-plasticized plastics on a wheel or belt from which it is stripped after it hardens by cooling. Some film is precipitated in a chemical bath that hardens it as it leaves the bath. Casting of molded products is achieved much like the casting of metals. Molten plastics compounds may be poured into simple molds for subsequent hardening by baking. Another casting process makes use of a chemical reaction, induced by a hardening catalyst that is mixed with the compound just before it is poured into the mold. The epoxy and polyester resins are used very widely for this "no-pressure" cast molding; fillers of glass fibers, mica, and other inert materials are often used to add hardness, strength and dimensional stability.

Automobile bodies, boats, missile components and many other products, such as the bucket seat shown in Fig. 1.10, are produced of reinforced plastics made by a low-pressure molding process. Many different methods are used for making the reinforced plastics. In general, the resins make use of a hardening catalyst, and are often mixed with the filler materials in the mold. Pre-mix materials, also available, permit the loading of the resin-treated fillers in the mold, where they are then shaped and cured by heat and pressure. High-strength sections are

FIG. 1.10   Reinforced polyester bucket seats are strong, lightweight, low in cost. Urethane plastics cushioning materials are upholstered with vinyl plastics. (*Courtesy Molded Fiberglass Body Co.*)

produced by the building up of the product's thickness with pre-cut sections of resin-impregnated glass or other fabrics, which are then bonded by pressure applied through a rubber bag or membrane. Heat may be added in an oven, or more heat may be supplied by the use of infrared lamps. Relatively simple molds of wood, metal or concrete may be used for the production of these high-strength reinforced plastics.

In the rotational process, a liquid or powdered plastics is introduced into a mold that spins under conditions that cause the plastics to harden on the surface of the mold; this is similar to the slush casting of metals. The fluidized bed process employs a preheated metal insert that is dipped into a bed of finely divided plastics powder which melts and fuses to the metal.

An electrostatic spray process operates on the principle that oppositely charged particles attract each other. The piece that is to be coated is grounded and attracts a charged plastics particle.

Foamed, expanded or bubble plastics are produced by many molding and extrusion procedures which employ a blowing agent in the plastics that causes foaming during the process.

The earliest known plastics-molding activities were observed in 1843 by a Dr. Montgomerie, a Malayan surgeon who reported that the vegetable material, gutta percha, a gum elastic, was being used by the Malay natives to make knife handles and other articles by first softening it in hot water, and then pressing it by hand. His report started the Gutta Percha Company, which remained active until 1930. Michael Faraday discovered that gutta percha also had good electrical properties, even when permanently immersed in water, and it became a major ocean cable insulation material. Gutta percha also replaced ivory for the making of billiard balls 30 years before Hyatt's famous search for a billiard-ball material.

The first recorded use of natural plastics in the United States was in the lantern windows (Fig. 1.11) that were converted to glass in 1740. Keratin, a natural plastics found in animal horn, hoof, hair, etc., was widely used as a fabricating material in the eighteenth century. Keratin, derived from chicken feathers, is used today for a textile material. The first plastics fabricators, called Horners or Hornsmith, did a big business in the nineteenth century. After Hyatt's invention, they became known as celluloid fabricators.

Samuel Peck* is the first known molder of plastics in the United States. He is recorded as a pioneer molder of shellac in 1852 and received his first patent in 1854. Large scale use of the shellac plastics continued until 1945.

In 1850, the plastics industry was a lusty, growing industry, making checkers, buttons, picture frames, combs, etc. The press, shown in Fig. 1.12, was used by Samuel Peck, of New Haven, Connecticut. Peck together with Halvor Halvorson of Boston was an early large-scale producer of molded daguerreotype cases. Another early maker of daguerreotype cases was Alfred P. Critchlow, who

*"History of Plastics—USA," J. Harry DuBois, Cahners, Boston, Mass.

FIG. 1.11 One of the early uses for the natural plastics material keratin from animal horn was for lantern windows. Keratin was replaced by glass in 1740. (*Courtesy Smithsonian Institution*)

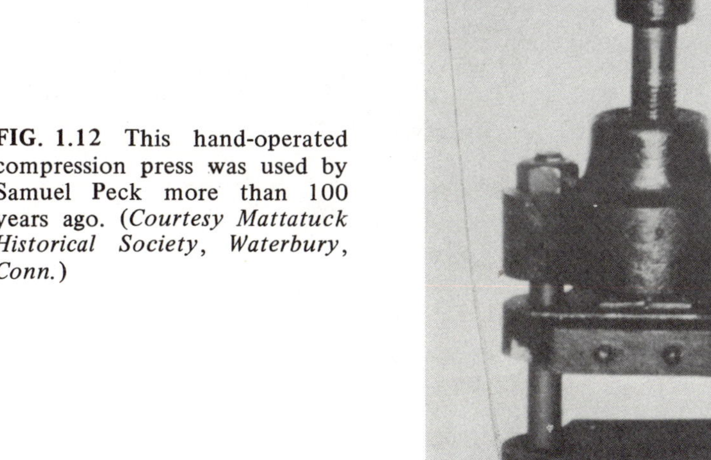

FIG. 1.12 This hand-operated compression press was used by Samuel Peck more than 100 years ago. (*Courtesy Mattatuck Historical Society, Waterbury, Conn.*)

started molding in 1853, in Florence, Massachusetts. The records show he worked two twelve-hour shifts, and this was the start of the Prophylactic Brush Company, which continues to operate in Florence as a Standard Oil Company subsidiary. A typical molded product of that era is shown in Fig. 1.13.

This early work in the development of mold-making and molded-product production, assisted by the invention of the extruder by Bewley for gutta percha in 1845, facilitated the immediate utilization of plastics resins and products as fast as they were developed by later inventors.

Many popular stories have been told about the origin of the first synthetic plastics material, "Celluloid." A careful study of the record disclosed both fact and fiction, and the following historical résumé from the record summarizes the events which led up to "Celluloid." Pelouze produced cellulose nitrate, in 1835, by dissolving cellulosic products in nitric acid. A Swiss chemist, Schönbein, in 1845, nitrated cellulose by using sulfuric acid as the catalyst. J. Cutting, of Boston, obtained patents in 1854 to cover his use of gum camphor in collodion for the improvement of photography, and this appears to be the first mention or disclosure of camphor with cellulose nitrate.

FIG. 1.13 This daguerrotype case cover was molded of shellac plastics circa 1860. Some of the finest mold making was done during that period with very primitive tools. (*Courtesy American Miniature Case Art—Floyd and Marion Rinhart, A. S. Barnes & Co.*)

An Englishman, Alexander Parkes, studied the residue left after the evaporation of the solvent of photographic collodion (cellulose nitrate in an ether-alcohol mixture), and reported that it was a "hard, horny, elastic and waterproof substance." In 1862, Parkes exhibited his "Parkesine" at the International Exhibition in London, in the form of buttons, combs, knife handles, etc., "Parkesine" contained no camphor, the all-important additive that made the "Celluloid" invention the first real synthetic plastics milestone.

John W. Hyatt, a New York printer, was attracted by the published need for the development of a substitute for gutta percha and ivory in billiard balls. After several years of study, in 1870 he patented the results of his work, and, in this patent, he described the unique and all-important action of camphor on cellulose nitrate. Thus Hyatt's invention was the start of synthetic plastics product-making in the United States.

"Celluloid" was a fabricating material which could not be molded, and the molded product field continued to be served by shellac and the bituminous cold-mold plastics, until 1909, when Dr. Baekeland announced his phenol-formaldehyde resin. This gave the world its first moldable plastics material, and added tremendous impetus to the plastics molding industry and, at the same time, put plastics resins into the paint and varnish business. The laminated plastics followed quickly.

Richard W. Seabury, founder of Boonton Rubber Company which later became Tech-Art Plastics Company, was first to see the future of the phenolic resins for molded plastics, and thus was the first molder of organic plastics. Seabury tells the story as follows:

"In 1906 I was running a small factory in Boonton, New Jersey, called 'Loando Hard Rubber Company.' About this time, the Vulcanized Rubber Company in Trenton, wanted us to try and reduce the inorganic material in our 'Loando' rubber, and we consulted a chemist—Dr. Leo H. Baekeland, of Yonkers, N.Y. He was working on a synthetic phenolic resin. Whereas Baekeland was interested in a synthetic varnish, I was interested in the resin itself to take the place of rubber in molded electrical insulation compounds.

"The first molded "Bakelite" parts, some with asbestos and some with wood flour, were made by me in 1907. Dr. Baekeland brought the resin from his Yonkers laboratory, and I made the molding powders and molded parts in our molds and presses. (These first organic-plastics parts are shown in Fig. 1.14.) My first customer for this new "Bakelite" material was the Weston Electrical Instrument Corporation, of Newark, New Jersey."

Since that time, the synthetic plastics have developed at such an accelerated pace that today it is a multi-billion-dollar industry. Each new plastics has created new markets and new methods. The injection molding machine, as illustrated in Fig. 1.15, started a total revolution in plastics materials and methods. The first automatic commercial blow-molding machine is shown in Fig. 1.16, and this touched off a major conversion from glass to plastics. The

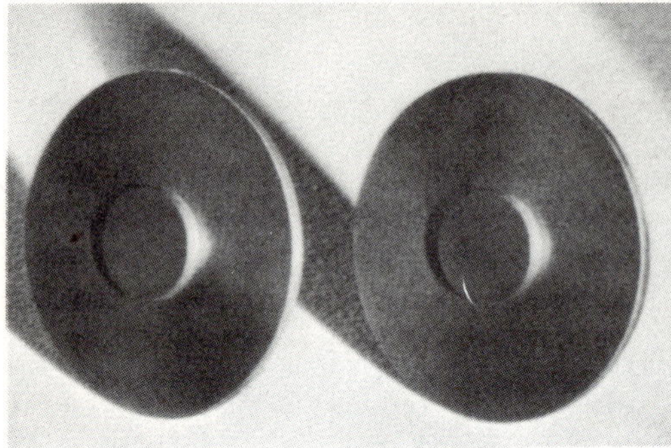

FIG. 1.14 These first molded "Bakelite" parts were used by Weston Instrument Co. (*Courtesy Union Carbide Co.*)

FIG. 1.15 Twelve of these Bucholz machines were imported in 1922 by W. D. Grote of the Grotelite Company, the first American injection molder of plastics after Hyatt.

FIG. 1.16 This first continuous plastics blow-molding machine was designed by James Bailey in the late thirties for Plax Corp.

products from the synthetic resins are so varied and versatile that they are used in some form by almost every manufacturer, and there appears to be nothing to limit their continuing growth.

While plastics have been condemned as a major ecological problem, scientific studies show that only 3% of the waste disposal problem can be attributed to plastics. Numerous studies have pointed the way to biodegradable polymers. Other studies have produced water-soluble polymeric materials. One such product, "Ecolan"* is an agricultural mulch film that degrades at a controlled rate and may be plowed under. Plastics will burn and considerable heat recovery is possible under proper conditions.

THE PLASTICS INDUSTRY

The leading plastics-producing countries are the United States, Great Britain, Italy, Germany and Japan. There is an expanding business in France, South America, Mexico, Canada and India. Growth of a plastics industry is predicated on the availability of low-cost material. A plentiful supply of petroleum, natural gas or coal is very essential to the making of the basic raw materials.

*Ecolan, Princeton Chemical Research, Inc., Princeton, N.J.

Half of the industry is located in the East, 13% is on the West Coast, 34% in the Middle West, with the balance scattered through the South. There is a fast-growing tendency to move the materials-making facilities into the oil-producing areas, where the refineries are located. The fabricating and molding plants are located in the industrial area, where many industries can be served within a reasonable distance. The petroleum product nations and the refiners are taking over the resin-making and the large-volume conversion into products.

FIG. 1.17 This plant of the Eastman Chemical Products, Inc., located in Kingsport, Tennessee, illustrates the complexity and magnitude of plastics making operations.)

All the plastics are predicated on very complex and advanced research. The principal material-producers maintain very large laboratories in their continuing effort to make better materials at lower cost. This research work requires physicists, chemists, electrical, electronic and mechanical engineers, plus a wide variety of skilled mechanics and specialists to solve the multiple problems of high polymer research, and the associated manufacturing control problems. The polymer sciences have arrived at a highly complex stage where progress is made very slowly as the result of extensive research and development work of large teams of scientists.

Much study is now being given to the synthetic inorganic plastics, since these products have very desirable high-temperature properties needed for the space vehicle and nuclear programs. These synthetic inorganics will open an entire new field when the associated material-making problems are solved.

Better adhesives, greater inertness, greater dimensional stability, lower thermal expansion and improved ultraviolet resistance are the present goals of many research workers in the field of organic plastics. Others are concentrating on the making of useful polymers from low-cost agricultural waste products. Mechanization and greater elimination of hand labor are principal goals of the conversion industries.

TABLE 1.3 APPLICATIONS OF PLASTICS IN BUILDING

Exterior

Adhesive	Illuminating panel	Shutter
Air support structure	Lighting fixture	Siding
Air vent	Louver	Sign
Cable	Moisture barrier	Skylight
Caulking	Mortar mix	Stucco
Coating-metal, wood	Paint	Sun shield
Concrete form	Panel	Swimming pool
Concrete mix	Pipe	Tape
Curtain wall	Railing	Tool shed
Door (prime and storm)	Rain system-gutter,	Topping-walk, driveway
Expansion joint	downspout, etc.	Vent stack
Facing	Roof edging, panel	Water proofing
Flashing	Safety and thermal glass	Weather stripping
Gasket	Screen	Window pane
Glazing	Sealant	Window sash (prime
Grille	Sheathing	and storm)
Hardware	Shingle	Wire insulation

Interior

Acoustical panel	Flooring	Plaster backing
Adhesive	Gasket	Plumbing fixture
Baseboard	Graphic art	Railing
Cabinet	Grille	Sealant
Ceiling	Hardware	Shower stall
Conduit	Insulation	Stair tread
Counter top	Light diffuser	Tank
Covering	Molding, trim	Tile-floor, wall,
Decorative panel	Paint	ceiling
Drawer	Paneling	Vapor barrier
Duct	Partition	Wall covering
Electrical fixture	Pipe fitting	Wire insulation

CAREER OPPORTUNITIES

The continuously expanding plastics industry offers many and varied job careers. Chemists, chemical engineers, physicists, mechanical and electrical engineers do much of the research and development work. In the resin-making field, the better jobs include chemical operators, evaporation men, millers, mixers, stillmen, control men and inspectors. Machine designers develop the processing equipment. Tool engineers design the molds and dies, and develop the production methods. There is a never ending and insatiable demand for tool makers, methods engineers, tool designers, hydraulic machinery specialists and foremen. The best plastics engineers are men with tool-making background, since the end-product is produced in molds and dies that must be accurate, long-lasting and easily maintained.

Product designers, sales engineers, schedulors, buyers, foremen and management requirements of every plastics business provide interesting and profitable opportunities for young men. A major expansion into building and construction work is envisioned by 1980, and some of these markets are listed in Table 1.3.

It is estimated that there are fifteen thousand processors of plastics throughout the United States, and there is a growing trend toward locating in the smaller communities. Plastics led all other industries in projected growth through 1980, in a recent forecast. During the seventies, plastics growth is predicted to increase from 6 to 12% per year. Most piping will be converted to plastics in this decade, with more than half of all clothes, shoes and furniture using plastics in some way. The volume of plastics in the twenty first century will exceed that of all other materials. The use of reinforced plastics for large high-volume automobile body stampings cannot be expected. The high production rates of metal stampings, combined with their lower finishing costs, give steel an edge for highest volume products.

2 THERMOSETTING MATERIALS AND COLD-MOLD PLASTICS

THERMOSETTING COMPOUNDS

The *thermosetting compounds* may be compared with concrete. Concrete is a mixture of cement and sand which has been hardened by a chemical reaction in the presence of water. The cement may be said to be a binder for the particles of sand. The sand serves as a filler, decreasing the cost, and providing body and substance to resist dimensional changes. A chemical change takes place when the cement "sets", thus producing a solid body. This change is permanent; the resulting solid substance retains the form in which it was cast and cured. By way of contrast, the *thermoplastics* materials are wax-like. They may be softened by heat and formed into the desired shape, which becomes rigid-like when it is cooled.

In a typical thermosetting material, such as a phenol-formaldehyde compound, the analogy with concrete provides an excellent means for understanding the thermosetting reaction. In this case, the binder, or "cement," is a chemically produced resin, resulting from a partial union of the phenol and formaldehyde. The phenol-formaldehyde resin is then mixed with a filler such as powdered wood, and this compound is then pressed to shape in heated dies. The combination of heat and pressure on the compound in the mold causes it to become plastic, and flow to fill out the desired contours of the molded part. Continued heat and pressure complete the chemical union of the phenol and formaldehyde. The binder has welded the particles of filler into a single mass which cannot be softened again by heat. It now resists chemicals which would have dissolved the resin before it had passed through the thermosetting reaction in the mold. A compound is the result of mixing a given resin with fillers and other additives, to produce a type of material for a specific class of work.

A major advantage which the thermosets have held over the thermoplastics has been their stability under load at elevated temperatures. *The thermosets fail slowly under such overloading over-temperature conditions*, while *the thermoplastics may fail suddenly and catastrophically*. (See Chapter 7 for the molecular structure of thermosets.)

THE PHENOLIC PLASTICS*

Development of the phenolic plastics in 1907 by Dr. Leo H. Baekeland, from phenol or carbolic acid and formaldehyde, gave industry a near-universal answer to the non-metal products market. That tremendous invention initiated the real growth of the plastics industry.

Raw Materials for Phenolic Resins

Many raw materials can be used in the making of these resins. The phenol bodies, phenol and cresol, or a mixture of several phenolic bodies, are generally used because of their availability. In the aldehyde series, formaldehyde and furfuraldehyde are most frequently used. Formaldehyde is favored because of its faster molding cycle and greater availability.

Phenol (carbolic acid) may be obtained as a by-product of the distillation of coal; large quantities are produced by this process. However, much phenol is also produced synthetically from benzene. Other phenol compounds may be used; the choice is dependent upon the ultimate characteristics desired.

The most widely used aldehyde for this resin is formaldehyde. Formaldehyde is normally a gas; in a water solution of approximately 37 per cent, it is called formalin. Large quantities of formaldehyde are produced by the oxidation of methyl or wood alcohol.

Making Phenol Formaldehyde Resin

When phenol and formaldehyde are combined in the presence of an acid catalyst, they form resins which are permanently soluble and fusible. These resins are often used in varnishes and paints, or for the two-stage process of molding resins. An alkaline catalyst, however, produces a resin which will become insoluble and infusible when the chemical reaction is completed; this is used in the one-stage process for making molding resins.

*"Bakelite," Union Carbide Corp.; "Durez," Hooker Chemical Corp.; "Fiberite," Fiberite Corp.; "Textolite," "Genal," General Electric Co.; "Plenco," Plastics Engineering Co.; "Rogers," Rogers Corp.; "RCI," Reichhold Chemicals Inc., "Plaskon," Allied Chemical Co.

One-stage Process. This is the original process used for making phenolic resins. This reaction results in the production of an insoluble and infusible resin, which is created when phenol and formaldehyde react in the presence of an alkaline catalyst. Correct molecular proportions of phenol and formaldehyde are piped to the resin kettle, and some alkaline catalyst, such as ammonia or caustic soda, is added. Heat is used to start the reaction in the resin kettle, and an agitator stirs the mixture. Heat is not needed after the reaction is started. The water which forms must be removed with considerable care. When drained from the resin kettle, the resin has the consistency of cold molasses, and cools in a clear, amber color. At this stage, a phenolic resin, which may be hardened later by heat and pressure, is called a resol.

The initial reaction between the phenol and the formaldehyde, with either an acid or a base catalyst, is called a condensation reaction. The chemists define a condensation reaction as one in which molecules join together to produce a large molecular structure, and during the process, a separation of a small amount of water, alcohol or some other substance which is left out of the new product, occurs.

The final important chemical change which causes the soluble and fusible resin to "freeze," or "cure," is called polymerization, a reaction in which a number of molecules unite to form larger molecular structures of the same composition.

Two-stage Process. The original molding compounds were produced by the one-stage process, but this process necessitated a long curing time in the mold, and the resins were very difficult to control. The first stage in the two-stage process consists of making a soluble and fusible resin called novolak, by using an acid catalyst. This resin is later converted into a thermosetting resin by neutralization and the addition of an excess of a formaldehyde-bearing compound. In this process, the phenol and formaldehyde are piped into the resin kettle, and this time an excess amount of phenol is used. An acid catalyst, such as hydrochloric or sulfuric acid, is added; heat is again used to start the condensation reaction. Since the chemical reaction is one which gives off heat, the resin kettle must be cooled during the latter part of the reaction. The removal of the water which forms need not be done as cautiously with the two-stage process; it is pulled out by means of vacuum pumps. This resin is also poured out on a hardening floor for solidification, in preparation for grinding.

Aylesworth found means to convert this novalak into a thermosetting resin by neutralizing with an alkali, and by the addition of a hexa (hexamethylenetetramine). Hexa is a compound of ammonia and formaldehyde, that supplies the extra formaldehyde required for the conversion from the soluble and fusible novolak, to the thermosetting resin, which polymerizes or "cures" when additional heat and pressure are applied. This process introduces a large gain to the molding room, since the amount of hexa added can control the speed of the curing cycle in the mold. Sufficient hexa might be added to secure a cure in the mold in a very few seconds. However, it is not yet possible to take advantage of the maximum curing speed in the mold, since ample time must be allowed for mold-closing and proper flow to insure uniformity of density.

Manufacture of Phenolic Molding Compound

The clear, amber resin which solidified on the hardening floor may be dissolved in a solvent for varnish-making, or for the impregnation of certain filler sheets used in making laminated materials. The paper-cloth sheets are often dipped in such a solution; this is called the wet process of filler impregnation. Sheets of cloth or paper, impregnated in this manner, are sometimes chopped up and used for making extra-strength molding materials.

General-purpose materials are often manufactured by a dry-process mixing. In this method, the resin is taken from the hardening floor or cooling pans, and ground into a fine powder. During this grinding, the novolak resins (two-stage process), are blended with lime, lubricant and hexa. This powdered resin is then intimately mixed with the filler and pigments (or dyes), to get a thorough blending of the components of the molding compound. Lubricants, such as vegetable wax or calcium stearate, are often added to prevent sticking in the molding operation. This compound is then ready for the final mixing. This mixing is done in a kneading and masticating machine, which thoroughly works the resin into the filler particles. Production lots are often run through a specially-designed machine called a Banbury mixer, which has rotary kneaders that work the compound thoroughly, under constant pressure. Small lots are often worked on pressure rolls, which force the component materials into an intimate mixture. Other types of fillers may be impregnated in a ball or pebble mill, which literally pounds the resin into the filler. Since an excess amount of heat would bring on the final chemical change or polymerization, the temperature during these operations must be kept low. Considerable heat is generated by the pressure and friction in the kneading process.

When the kneading operation is completed, the compound is cooled and ground into a uniform granulation so that it will flow freely, and have uniform bulk density. Molding compounds are passed over a magnetic separator to remove any particles of iron, and are then blended with previous batches. This blending is extremely important as a means of insuring uniformity, and also as compensation for the variables existing in the individual batches.

Properties of Phenolic Materials

The raw-material manufacturers offer a wide selection of standard materials with varying characteristics. Each material is designed to fill certain types of applications, at the lowest overall cost. No material has been found that will fill several types of special requirements. Fillers and resins play an important part in the special properties of the materials available, and it is therefore advisable to consider carefully all the available molding materials. Users must

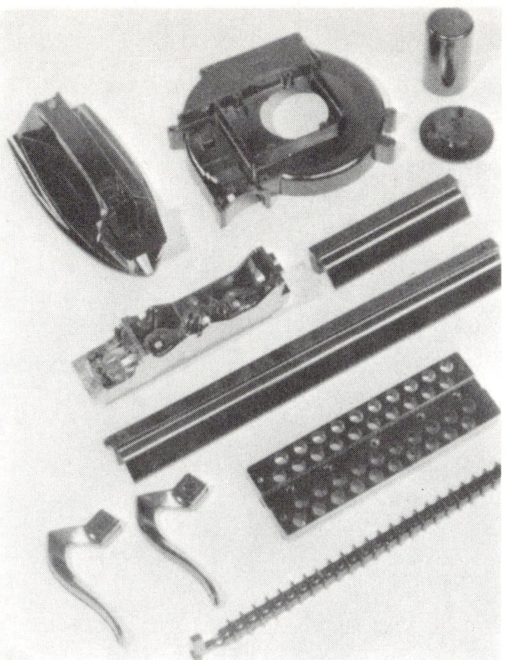

FIG. 2.1 Typical phenolic plastics pieces molded by the compression and transfer process.

select the compound that is best suited for a particular application. The phenolic molding compounds are divided into several general classes for easy selection; typical molded parts are shown in Fig. 2.1.

Widely used phenolic compounds, modified with fillers to serve the large markets, are listed in Table 2.1.

Design Considerations

When designing parts to be fabricated from phenolics and other thermosetting compounds, careful consideration of their various mechanical strengths should be made in order to use them advantageously. Phenolics do not have as great a tensile yield or ultimate strength as metals such as steel; however, the phenolic is more resistant to being pulled apart, when compared with wood, or a thermoplastic material such as polyethylene. For parts to be made from phenolics, the lower modulus of elasticity, when compared to metals, must be taken into account, and therefore a greater volume of material must be used. However, because of the lower specific gravity of the phenolics, the weight relationship is favorable.

Steel is the hardest of the above materials, followed by the melamines and

TABLE 2.1 SPECTRUM OF FILLERS USED IN PHENOLIC MOLDING MATERIALS.*

| | General Purpose | Shock Resistant | | | | Heat Resistant | Shock & Heat Res. | Rubber Modified | High Frequency Insulation |
		Improved	Better	High	Very High				
Filler	Wood flour	Flock and wood flour	Flock	Cord	Fiber glass	Asbestos	Asbestos & flock	Rubber & wood flour	Mica
Mechanical strength									
Impact, ft lb/in notch	0.24–0.32	0.34–0.50	2.00–4.00	4.00–8.00	17.0+	0.23–0.48	0.34–1.00	0.34–0.70	0.30–0.38
Flexural, psi	8000–12,000	8000–12,000	9000–14,000	9000–15,000	25,000+	6000–10,000	7000–12,000	7000–10,000	7000–11,000
Electrical tests									
Dielectric strengths, S/T, volts/mil	200–400	240–350	225–400	200–325	350+	150–300	150–250	250–350	300–450
Volume resistivity, ohm-cm	1–100×10^{11}	0.1–10×10^{11}	1–10×10^{11}	1–10×10^{11}	$1 \times 10^{13}+$	1–10×10^{11}	0.1–10×10^{11}	1×10^{6}–10^{11}	$1 \times 10^{12}+$
Dissipation factor @ 10^6 cps	0.03–0.07	0.03–0.07	0.03–0.08	0.04–0.09	0.01–0.03	0.04–0.15	0.10–0.25	0.08–0.15	0.01–0.04
Fabrication characteristics,									
Compression ratio (bulk factor)	2.1–2.7	2.3–4.4	6.0–15.0	12.0–18.0	6.0–7.0	2.0–2.5	3–12	3.5	2.1–2.7
Tableting	Good	Fair to good	Fair to poor	Poor	Fair to poor	Good	Fair to poor	Good	Good
Molding (compression)									
Temperature, °F	300–400	300–400	300–400	280–380	280–330	275–350	275–375	300–350	300–350
Pressure, psi	7000–10,000	2000–5000	2000–7000	2000–7000	3000–5000	2000–5000	2000–6000	2000–6000	2000–6000
Shrinkage, in/in	0.004–0.009	0.003–0.009	0.002–0.006	0.002–0.005	0–0.001	0.002–0.006	0.001–0.005	0.003–0.10	0.0005–0.005
Miscellaneous									
Specific gravity	1.33–1.45	1.33–1.45	1.36–1.43	1.36–1.41	1.85–1.95	1.55–1.90	1.60–1.90	1.24–1.34	1.70–1.90
Water absorption, % gain	0.3–0.8	0.3–0.8	0.5–1.75	0.8–1.75	0.5–1.0	0.2–0.4	0.2–0.5	0.5–1.0	0.01–0.07
Heat distortion, °F	290–340	290–340	290–350	290–350	600	350–450	300–400	200–280	275–350
Utility (typical applications)	Wiring device Parts, auto ignition, etc. Widest usage, lowest cost and weight	Washing machine agitators, pump housings, etc.	Heavy duty industrial parts	Heavy duty industrial parts	Heavy duty insulators	Utensil handles, appliance parts	Electrical switchgear	Power tool handles	Electronic components

*Courtesy Union Carbide Corp., Bound Brook, N.J.

phenolics. The phenolics are scratch-resistant and, because of this hardness, they are widely used for applications where there may be abrasion. For many rubbing-contact applications, compounds incorporating TFE-fluorocarbon, graphite, or molybdenum disulfide as lubricating fillers are being used. Phenolics are also low in thermal conductivity, which makes them good heat insulators, in contrast with such metals as steel or aluminum. While the heat distortion is listed at 325°F, there are some specialized compounds that will resist temperatures up to 500°F before serious distortion or warpage occurs.

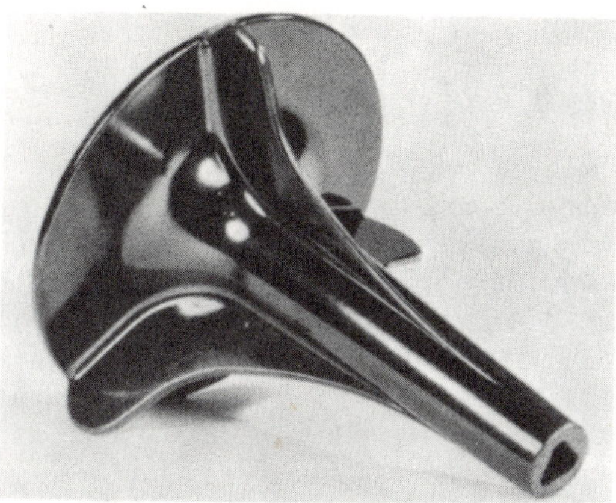

FIG. 2.2 Washing-machine agitator compression molded of phenolic compound. Injection molded polypropylene is often used for this application.

Obviously, metals are conductors of electricity, whereas the phenolics are good insulators. They are self-extinguishing after having been exposed to fire. From the information given, it may be seen that phenolics have durability that makes them attractive replacements for metals and wood in many instances.

Phenolics, because of their lower specific gravity, are cheaper than metals on a cents-per-cubic-inch basis, for appliances such as pump impellers, mixers, and dishwashers. Industrial uses include electronic parts, wiring-device parts, electrical switchgear, and automotive applications.

When properly used, phenolics have desirable attributes as structural materials. If the properties of phenolic materials are to be used to best advantage, the parts must be designed so that the loading and stresses can be accommodated by the compound selected, and the particular characteristics of the phenolic compound must always be kept in mind.

Phenolic Resin Markets

The molded product and laminated-sheet materials market were the principal original markets for the phenolics. Their versatility and low cost has greatly expanded phenolic resin usage into numerous other markets. Thermal and acoustical insulation products, consisting of various mineral fibers such as rock wool, glass fibers, etc., are bonded with phenolic resin to gain strength, resiliency and "self support." This market uses more phenolic resin than molded products. The amount of phenolic resin used for bonding plywood is almost as large. Foundry applications, where phenolics are used for bonding sand to form thin, dimensionally accurate molds ("shell" molds), and cores for metal casting, account for another large volume of phenolic resin. Other large-quantity markets include adhesives, coatings, bonded brake-linings, rubber-compound additives, particle board, and wood-residue molding.

PROCESSING PHENOLICS

Phenolic compounds are cast, compression, transfer, "injection," extrusion, fluid bed, and rotational molded. They are foamed, laminated, thermoformed, fabricated, reinforced, potted and encapsulated. In the past, "injection" molding applied only to the thermoplastics and transfer molding applied to the thermosets. Current usage describes the fully automatic molding of thermosets also as injection molding. The production of phenolic plastics pieces in modern injection molding machines has reopened many markets that had been lost to the thermoplastics. Material makers have developed a variety of new phenolic compounds for this automatic molding process. Dust free granules of phenolic molding compound that can be run in machines alongside of the thermoplastics are a boon to the molders.

ENCAPSULATED PRODUCTS

The encapsulation of semiconductor devices has become a high volume application for thermosetting resins since they can be compounded to gain adequate plasticity together with essential electrical, mechanical and environmental properties. Special encapsulation transfer molding presses have been developed for this very large electronics market. Silicone and the phenolic resins are the principal encapsulating resins. Glass beads are often added to minimize shrinkage and to improve thermal properties.

SUMMARY

The phenolic materials may be characterized as a general-utility group of plastics, which are widely used for industrial products. The several classes of phenolic materials available offer compounds which will serve for most types of applications. Many of these materials are relatively easy to mold, have excellent finish, good electrical and mechanical properties, and a low cost. Fillers play an important part in the physical properties; users will find the best clue to the value of any given material by making a study of the filler used. Phenolic materials must be properly processed and carefully selected in order to obtain the full value and fine performance which they are able to give. Future growth of the phenolic plastics will depend, to a large degree, upon the development of faster, fully automatic molding machines, that will facilitate competition with thermoplastics.

THE AMINO PLASTICS*

Melamine formaldehyde and urea formaldehyde plastics are the most important amino plastics. Both are widely used as basic materials, although many of their original markets have been converted to the equally colorful thermoplastics. The urea resins were known in 1897, but little work was done until 1920, when Fritz Pollock and Kurt Ripper studied them intensely. Carleton Ellis' work on molding compounds was patented in 1933.

In 1929, the American Cyanamid Company started manufacturing Beetle-urea plastics. Plaskon Company, Inc. was formed in 1932 as a result of the Toledo Scale Company's desire to produce a light-weight, white scale housing. Other resin makers entered this market as the volume increased. Whereas the phenolics offered only a limited range of dark colors, the amino plastics brought in all the brilliant colors of the spectrum. The makers of urea subsequently produced melamine plastics as well, since both urea and melamine are made from cyanamide. The manufacturing processes for urea compound are depicted in Fig. 2.3.

Amino Compounds

The melamine dinnerware materials, and many of the urea compounds, use alpha cellulose (purified wood cellulose), as a filler. Amino resins are water-clear; they offer the best color potential among the thermosetting materials, and are widely used for just that reason. Other fillers include cotton fabric, asbestos, minerals, wood flour, glass fibers and paper.

Urea and Melamine Applications

Very large quantities of melamine plastics are used for dinnerware, taking advantage of its hardness, attractive colors, thermal endurance, scratch resistance, stain resistance and moldability. Attractive color-printed decorative overlays, composed of clear films with identical material chemistry, are bonded to the dinnerware in a second stage of the molding process, to produce the wide variety of colorful products, as shown in Fig. 2.4. Melamine and urea materials are highly solvent and scratch resistant, making them desirable for buttons and organ-keys.

---

*"Beetle," American Cyanamid Co.; "Cymel," American Cyanamid Co.; "Plaskon," Allied Chemical Co.; "Plenco," Plastics Eng. Co.; "Fiberite," Fiberite Corp.; "Sylplast," FMC Corp.

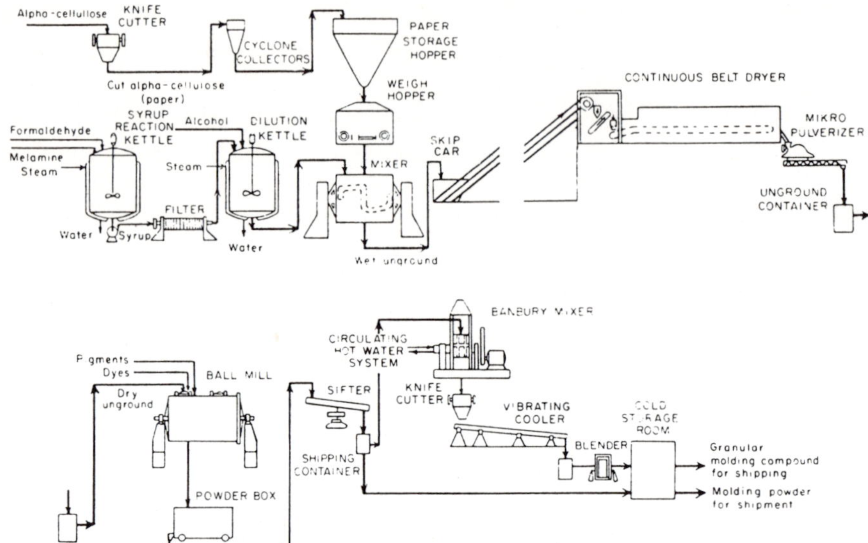

FIG. 2.3 Manufacturing processes used in production of urea molding compound. (*Courtesy Chemical Engineering*)

FIG. 2.4 This attractive dinnerware is molded with built-in decoration of melamine compound. Melamine is scratch, boil and break resistant and the brightest and lasting colors are obtainable. (*Courtesy Boonton Molding Co.*)

The melamine and urea compounds offer better arc resistance than the phenolics, and are commonly used in circuit-breaker applications, with mineral or wood-flour fillers. Melamine is the hardest of the plastics. Glass-fiber-filled melamines are often used for high-shock, electrical-arc applications. Polyester glass compounds are replacing melamines in many applications. Better dimensional stability is gained from these mineral-filled products. Urea closures are widely accepted because of their lack of cold flow, their fine colors, as well as oil, grease and stain resistance. The urea compounds are primarily selected when a colorful thermosetting compound is indicated by heat or no cold-flow product demands. Ureas have good resistance to heat and boiling distortion. The urea compounds are available in an unlimited color range, and, in addition to thermal and cold flow stability, they provide good dielectric and resistance to solvents, oils, greases, scratching and flame. Ureas do not attract dust from static. Urea compounds are low in cost. Urea compounds are extensively used for typical lighting fixtures, as shown in Fig. 2.5, because of their heat and static resistance.

The plywood and furniture industries are major users of amino resins for plywood bonding, since they meet severe exterior-use requirements, and provide boil-resistant bonds. Urea resins are also used for foundry sand cores, and for cold-setting adhesives. The laminated plastics take advantage of the color potential of the melamine resins for table tops, kitchen cabinets and abuse-resisting furniture. For these applications, the lower-cost phenolic resins are used as a core material binder, with the surface paper sheets adding an attractively

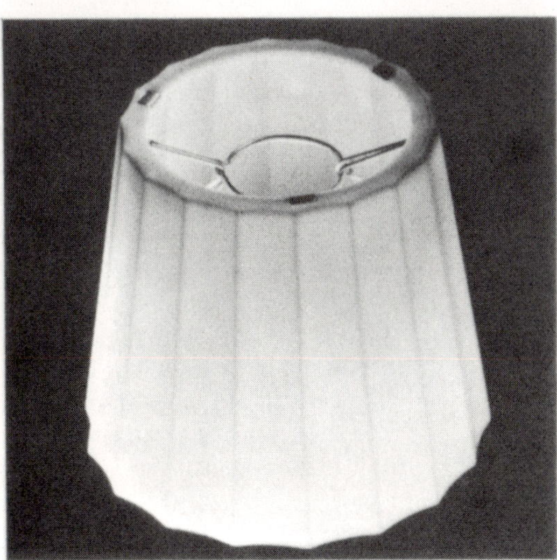

FIG. 2.5 Molded urea lampshades are colorful and translucent and will withstand the high lamp temperatures.

printed, colorful wood grain, or other desirable texture. The urea and the melamine resins are both used to increase the wet burst and tensile strength of paper and paper board.

Urea and melamine, in combination with alkyd resins, are extensively used for the production of baking enamels. The urea-alkyd enamels are best suited for indoor applications, such as ranges and refrigerators, while the melamine alkyds serve very well for automotive and other outdoor metal finishes.

ALLYL PLASTICS* DAP

The allyl plastics became available about 1955, and they have served effectively many of the sophisticated thermosetting markets because of their unique combination of desirable properties. The extraordinary environmental conditions faced by the rocket and missile components contributed to the expanded usage of diallyl phthalate compounds. Basically, these thermosetting resins will harden or cure with peroxide catalysts, to form heat and chemical-resistant products having superior electrical and dimensional control properties. Diallyl orthophthalate serves the market for compounds used to 350°F.

Fillers used with the diallyl phthalate resins include "Orlon," "Dacron," mineral and glass. The mineral fillers most commonly used are calcium silicate, silica, and treated clays.

DAP Applications

Diallyl phthalate compounds are characterized by their excellent chemical resistance to all reagents, low electrical loss, excellent weathering, very low mold shrinkage, with best dimensional stability for organic plastics. These compounds are very inert and stable, releasing no metal corroding vapors to attack inserts, and will not support galvanic corrosion in presence of moisture. DAP components, with inserts, are shown in Fig. 2.6. Whereas high temperatures and high humidity can increase the electrical conductivity of many plastics, DAP compounds retain their original resistivity under the most severe conditions. Certain formulations with glass fillers combine highest shock and arc resistance, making them particularly suitable for rugged power-circuit breakers. Orlon-filled materials give best insulation, volume and surface resistivity. The nylon-filled DAP compounds offer best abuse-resistance. The connector market is one of the principal fields of application for the DAP materials.

The DAP prepolymers are also used for improved surface laminates, plywood, hard board and particle board. They are usually applied as an overlay by means of a resin-treated nonwoven acrylic fabric. DAP resins are used to assist polymerization of the polyester plastics.

*"Acme," Acme Resin Corp.; "Durez," Hooker Chemical Corp.; "Plaskon," "Diall," Allied Chemical Corp.; "Rx," Rogers Corp.; "Dapon," FMC Corp.

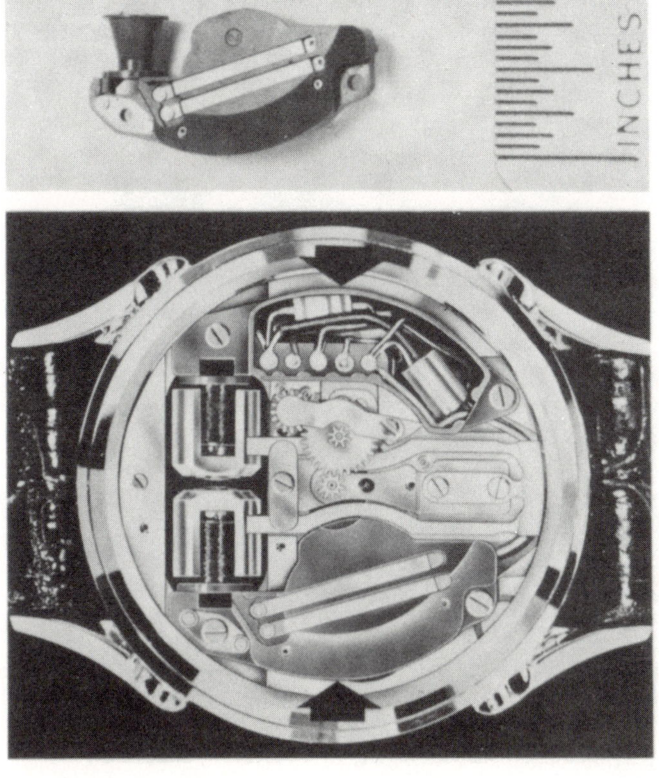

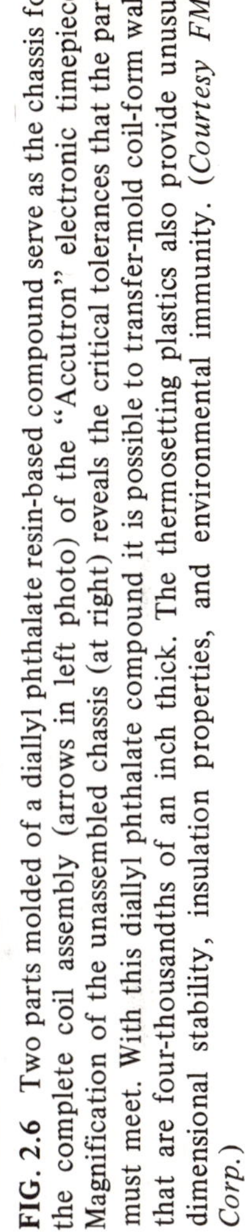

FIG. 2.6 Two parts molded of a diallyl phthalate resin-based compound serve as the chassis for the complete coil assembly (arrows in left photo) of the "Accutron" electronic timepiece. Magnification of the unassembled chassis (at right) reveals the critical tolerances that the parts must meet. With this diallyl phthalate compound it is possible to transfer-mold coil-form walls that are four-thousandths of an inch thick. The thermosetting plastics also provide unusual dimensional stability, insulation properties, and environmental immunity. (*Courtesy FMC Corp.*)

ALKYD POLYESTER PLASTICS*

The alkyd plastics were first introduced commercially in 1941, and are widely accepted for many applications. Original work on the polyesters was done by the Swedish chemist, Berzelius, in 1847. The alkyds are polyester derivatives, produced by the reaction of an alcohol with an acid. The name alkyd comes from the words ALCohol and acID. Mineral, and modified mineral fillers, are used for the basic alkyd compounds. These are characterized by their ability to flow freely under very low molding pressures. Colors are readily available. Typical volume applications include electron tube bases, automobile ignition, as shown in Fig. 2.7, television receiver insulation components, switches, etc. The

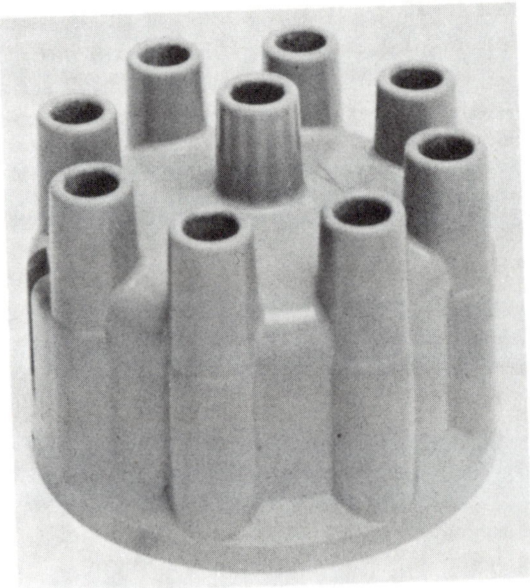

FIG. 2.7 Alkyd replaces phenolic for improved electrical performance and longer car life in distributor caps on cars and trucks. (*Courtesy Allied Chemical Corp.*)

low pressure molding capability of the alkyds is responsible for their use as an encapsulation compound for resistors, capacitors, transformers, etc. Other molding assets include short cures, freedom from volatiles, and ability to mold by compression and transfer automatically. When compounded with glass fillers, the alkyd compounds exhibit very high shock resistance, resulting in their use for switch-gear and computer components in aircraft and warships. The alkyd

*"Glaskyd," American Cyanamid Co.; "Plaskon," Allied Chemical Corp.; "Durez," Hooker Chemical Corp.; "Haysite," Haysite Div.

compounds are available in "rope" form of various diameters, to eliminate preforming and to facilitate continuous automatic molding.

Alkyd resins are used in the billion pound coating resin market and in the much smaller polyester molding compound market. The alkyds used for coating resins contain drying oils and phthalic, with other modifying acids and diols. Alkyds used for mechanical and electrical molding compounds are unsaturated polyester resins which contain maleic or fumaric acid, with modifying acids and diols. No drying oils are included.

POLYESTER PLASTICS

Unsaturated polyesters, saturated polyesters and the alkyd plastics, are each members of a family of plastics formed from alcohols and acids. The saturated polyesters are best known as film and fibers, such as "Dacron" and "Mylar." When combined with isocyanates, the saturated polyesters produce polyester urethanes. A urethane polymer base, reinforced with polyester, produces a poromeric synthetic leather.

Casting* resins are available in water white polyesters that can be cast easily in relatively large volume.

Unsaturated Polyester Plastics

The unsaturated polyesters are hardened or cured by catalysts, added, in most cases, at the time of molding. These plastics are best known for their use in the reinforced plastics field, where they are used in combination with glass fibers for the molding of automobile bodies, boats, tanks, handles, etc. (Fig. 2.8).

For the conventional compression and transfer molding processes, premixed compounds are made up of chopped glass fibers mixed with polyester resin, a pigment, and a catalyst, and extruded into rope form for easy handling, or else are used in bulk. These materials must be kept at low temperature, or used quickly after the catalyst or hardener has been added. Typical products include switchgear, trays, housings, laundry tubs, fan blades, pump housings, etc. Some products are compression molded of prepregs of polyester and glass fibers. In the prepreg-making process, a mat of fabric, nonwoven or glass-roving, is impregnated with polyester resin under controlled conditions, and the resin given a partial cure until gelation occurs. At this point, the mat, or prepreg, may be handled as a dry or slightly tacky material, which can be loaded in the compression mold to be reshaped and cured. Prepregs are used for the highest level requirements, such as aircraft and missile components.

Polyester glass laminates have taken over many of the critical laminated applications in switchgear. Low loss, high impact and dielectric strength are

*"Acra Poly," The Plastics Factory, N.Y., N.Y.

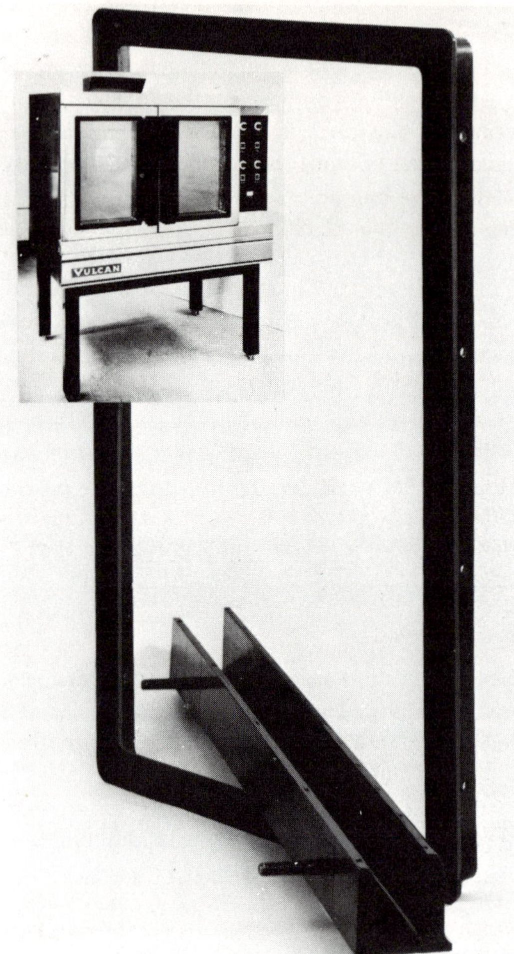

FIG. 2.8 Polyester glass window frames and handles replace aluminum and die cast metals for the doors and handles of this commercial oven with a gain in costs, maintenance and cool handles. (*Courtesy Haysite Division Synthane Taylor Corp.*)

plus factors for the polyester glass laminates. Many of the cooking bags are made of polyester.

A new family of aromatic polyesters* is said to have good physical, electrical and bearing properties at high temperatures. Their flexural strength at 500°F is reported to be 600 psi, without reinforcement. Compression molding rein-

---

*" Ekonol," Carborundum Co.

forced compounds are available. Polyester compounds are formulated and sold by many of the companies.

SILICONE PLASTICS*

The silicones are a chemical hybrid, somewhere in between the organic plastics and sand. Chemically, the silicones are organo-silicon oxide polymers. They are available in many forms; besides being excellent molding compounds for high-temperature products, the silicones produce water repellents, lubricants, polishes, rubbers, protective coatings, defoamers, and stable, high-temperature fluids. Silicones are extensively used for the reinforced plastics molds.

Silicone Molding Compounds

Two basic types of silicone molding compounds, granular and fibrous, are available. The granular compounds are free-flowing materials, filled with minerals or glass fibers. These compounds are compression or transfer molded for heat resistant parts, and may be used up to 750°F, in some cases, with 230 seconds of arc resistance. Thick sections require a long cure of one or more hours.

Mineral and short fiber-filled silicone compounds are designed for encapsulation (see Chap. 6), of electronic components, such as diodes and transistors, at very low transfer molding pressures. This encapsulation prevents damage from mechanical shock, vibration, and high "G" forces. These plastics materials have excellent properties for electronic components, with low dielectric loss at high frequencies and operational temperatures. Water has little effect on their properties.

The silicone rubbers are very useful molding materials. These remarkable elevated temperature "rubbers" have been proven to work best for high temperature gaskets in all classes of high performance engines, and for simple things like the steam iron. Silicone rubbers operate successfully from –100°F to 500°F, still maintaining their elasticity. Silicone tapes and varnishes, in combination with glass or cloth, mica and asbestos, or in unsupported form, have been responsible for considerable size reduction and improved reliability of heavy-duty electric motors and generators. Typical silicone rubber products are shown in Fig. 2.9.

Plastics molders find the R.T.V. (room temperature vulcanizing) silicone elastomers of great value for model-making, or short-run molding, molds and products. These easily molded materials are available with a pre-mixed hardener, or with the hardening catalyst packaged separately for mixing at the time of use. The R.T.V. compounds can be poured into simple molds, using a hand-made

---

*"Dow Corning," Dow Corning Corp.; "G.E. Silicone," General Electric Co.; "O-I Glass Resins," Owens-Illinois, "Dexsil," Olin Chemicals Co.

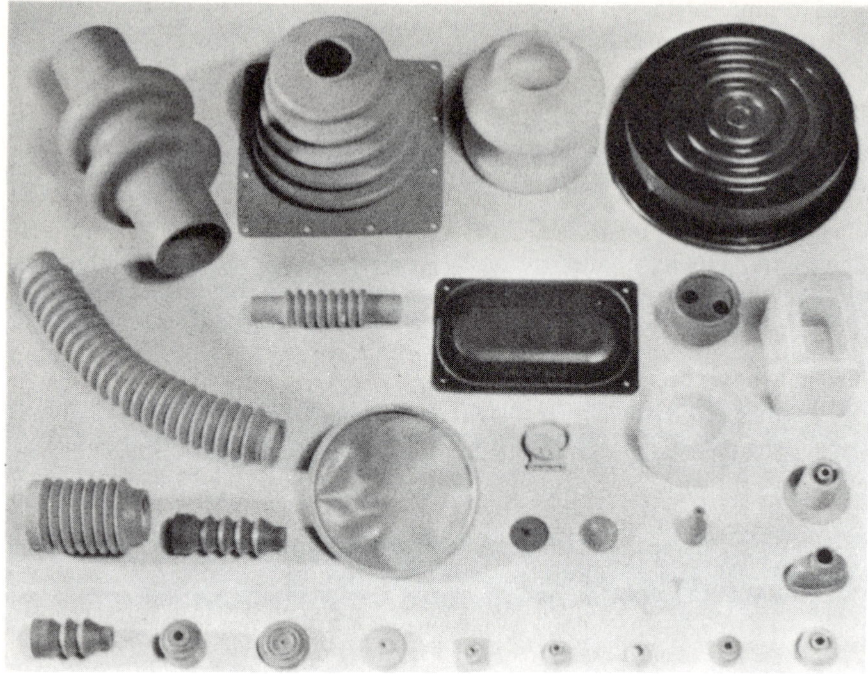

FIG. 2.9 Silicone rubber products are used in high temperature applications where natural rubber will fail. (*Courtesy Moxness Products Inc.*)

model of the end-product to shape the desired cavity. The cure can be effected either by time or heat, to speed up hardening. Such molds can be used for the molding of other castable plastics.

Silicone products perform well in a very wide variety of applications and offer highly diversified properties to the product. Transistors and other electronic components are molded of, or encapsulated in, silicone insulation. Silicone rubber coatings and sealants are used in building construction. Reaction organo-silicon compounds become coupling agents in the laminates for the formation of films permanently bonded to the substrate. Flexographic inks employ silicone additives. Concrete is made "freeze/thaw" damage resistant by the addition of silicones. Silicone fluids are available in viscosities from 0.65 to 2,500,000 centistokes. Such fluids are used in polishes, pre-shave lotions, electrical discharge machining, defoamers, paint additives, and urethane foam surfactants. Silicone fluorosilicone rubber compounds are effectively used for critical aircraft and other applications because they are unaffected by weather, age and thermal differentials. Silicone flexible molds are used for the reproduction of intricate patterns, wood grain texture, etc. Silicone release agents solve many antistick problems. Silicone products should be explored for every difficult application.

The silicone rubber products are widely used for repairs of the human body. A silicone rubber ball is used in combination with a fluorocarbon seal to replace defective human heart valves. Prosthetic devices of silicone are very successfully used in all parts of the body, and silicone rubber has had many applications in reconstructive surgery, on or near the surface of the body.

Silicone resins are used as binders for some ceramics. The silicone-ceramic compound is molded by injection, extrusion or compression and then brought up slowly in temperature until the organic silicone has been converted into an inorganic compound. Intricate ceramic products are the resultant product.

ALLYL DIGLYCOL CARBONATE*

ADC is a unique thermosetting plastics material that combines the optical properties of glass with exceptional mechanical and physical properties. This resin may be cast into transparent plates, tubes, and rods, as well as special shapes. It is not a moldable material.

The scratch or abrasion resistance of ADC is thirty to forty times better than that of the acrylic plastics, and as such is the best "window" glazing material, where scratch or abrasion damage must be at a minimum. Chemically, it resists the solvents, such as acetone, benzene, gasoline. Its temperature limit is $100°C$, with minimum distortion. Gamma radiation resistance is fifty times better than that of the acrylics.

Fabrication is done by conventional machining operations, and the sheets may be thermoformed in limited curves. Principal applications include aircraft windows, safety goggles, guards for welding, instrument windows, and watch crystals.

EPOXY PLASTICS**

Outstanding properties offered by the epoxy plastics are adhesion, toughness, chemical resistance, high bonding strength, and excellent electrical properties. Their universal value has built up many unique and special markets in fields formerly served by the phenolics, alkyds and allyl plastics. These thermosetting resins were first introduced in 1947, and are now available in many forms. The high-strength properties of epoxy adhesives is illustrated in Fig. 2.10.

The epoxy resins make use of a catalyst, or a hardening agent, mixed at the time of use, to produce the desired type and desired time of hardening. Of special interest is the complete freedom from reaction by-products or volatiles during hardening. This permits the making of very large castings and heavy,

---

*"C.R.-39," Pittsburgh Plate Glass Co.; G.R.-39, The Homalite Co.; C.R.-39, Polytech Co.; C.R.-39, Welcast Co.

**"Hysol," "Hyflo," Hysol Corp.; "Plaskon," Allied Chemical Corp.; "Scotchply," Minnesota Mining & Mfg. Co.; "U.S. Polymeric," U.S. Polymeric Inc.; "Epocast," Furane Plastics, Inc.; "Bakelite Epoxy," Union Carbide Corp.; "Epon," Shell Oil Co.

FIG. 2.10 Epoxy adhesives are being used to bond the bridge bearing plates. (*Courtesy Pittsburgh Plate Glass Co.*)

void-free sections. The shrinkage which takes place at the time of cure is also relatively small, and this facilitates stress-free parts, as well as accurate reproduction from the mold section. A variety of hardeners or curing agents are used to gain the desired cure cycle, pot life, viscosity, and environmental performance. Typical curing agents include the amines, acids, acid anhydrides, and the conventional catalyst hardeners such as dicyandiamide, and BF_3MEA.

Coatings with epoxy resin are one of the very important fields of application, since the coatings provide chemical and corrosion resistance, flexibility, waterproofing, surface hardness, wear resistance, mechanical protection, etc.

Epoxy casting resins are used for models, patterns, molds, tools, insulation, encapsulation, and potting. Epoxy mortars make fine repair materials for concrete, and are also acid resistant.

Epoxy bonded glass molding compounds for compression and transfer molding are used to produce high strength, high temperature electrical product insulation, pipe fittings, aircraft and automotive components. In the Reinforced

Plastics product field (see Chapter 10), the epoxy resin bonded glass or graphite filaments produce the very highest strength products, and are commonly used in missile components, high strength tanks, etc. Their great value in the reinforced plastics field is the result of the excellent adhesion of epoxy resins to other materials, such as fiber glass. The bond to fiber glass is so good that moisture or gas cannot travel along the bond line. Other gains result from the low rate of shrinkage, and minimum internal residual stress.

Epoxy resins are used to "weld" aluminum bicycle frames. The high rate of bond strength to other materials is the outstanding and phenomenal property of the epoxy resins. Epoxy adhesives are made in a variety of forms, with various additives. In most cases, the hardener is packaged separately to be mixed at the time of use. Some products are offered with mixed-in latent curing agents that cure under heat after their application. The automotive industry is expanding its use of epoxy, instead of welding, in applications where an inner weld might distort the exterior body surface.

EPOXY GRAPHITE COMPOUNDS*

Fiber reinforced sheets and molded products, making use of an epoxy binder and carbon filaments,* exhibit unique and desirable properties, which include: very high strength and modulus to weight ratios, exceptional fatigue resistance, low coefficient of friction, good wear resistance, and high vibration damping capacity. Additionally, thermal coefficient of expansion can be tailored from minus 1.3×10^{-6} to plus 13.0×10^{-6} in./in. °F, while thermal conductivity varies between 6 and 35 Btu-ft/hr-ft^2 °F, depending on fiber grade. Typical strength/weight and modulus/weight values for unidirectional construction run from 2.3 to 4.0×10^6 in. and 350 to 500×10^6 in., respectively, for the different fiber grades. (See Chapter 10.)

This material may be injection molded or processed by the conventional methods for reinforced plastics and offers a *very* large market potential.

THE CASEIN PLASTICS

Casein is a protein plastics material which is made from skim milk. It is little used today, having been replaced by the later materials. A limited amount is used for buttons and novelties. Casein plastics were patented by W. Kirsche and Adolph Spittler, in 1897, as a result of their search for a material suitable for white "black-boards." It was initially manufactured in the United States in 1919.

An enzyme is a complex substance which acts as a catalyst. The enzyme rennet, taken from the stomach of an unweaned calf, is used to precipitate casein from skim milk. The curd, or casein, which settles from this reaction, is

*"Magnamite," Hercules, Inc.; "Panex," Stackpole Carbon Co., Union Carbide Corp.; "Hitco," Great Lakes Carbon Co., Celanese Corp.

purified, dried and ground before mixing with suitable plasticizers and coloring dyes. This mixture is then extruded into sheets and rods, and hardened in a water solution of formaldehyde. It may be punched or formed into shapes before the final hardening with formaldehyde. Casein is highly moisture absorptive, and must not be used for products that are tightly-dimensioned.

POLYCARBORANESILOXANE*

A family of high temperature elastomeric resinous and liquid polycarborane-siloxanes retain their useful properties after prolonged exposure between 500 and 800°F. They serve markets that demand extreme heat resistance such as connectors, plotting compounds, wire insulation, gaskets, O-rings, tapes and ablative applications.

THERMOSET URETHANES**

The urethane plastics (Chapter 3) are also prepared as thermosets. Farbenfabriken Bayer AG are marketing a polyurethane duromer which is a fine celled thermosetting rigidly foamed product. It forms its own smooth, tough exterior surface during the molding operation and has a high strength-to-weight ratio.

PHOSPHONITRILIC RESIN***

Phosphonitrilic resin is a special thermosetting resin that offers better flame resistance and stability in air temperatures up to 500°F. This resin reacts with epoxy resin for added thermal stability. Processing is similar to that of the novolak phenolics.

POLYBUTADIENE PLASTICS†

Modified polybutadiene plastics materials which are processed by conventional thermosetting processes have been developed for applications demanding high heat distortion, low moisture absorption, plus good chemical and electrical properties.

*"Dexsil," Olin Chemicals Co.
**Mobay Chemical Co.
***"PH990," ElMonte Chemical Co.
†"FCR1261," "Dienite," Firestone Tire and Rubber Co.; "Ricon 1429," The Richardson Co.

P-OXYBENZOL PLASTICS*

An interesting high temperature polymer p-oxybenzol is readily formable and stable at temperatures of 500 to 1000°F. This new polymer is fabricated by a pressure sintering process. Coatings can be made by plasma spraying the powder on the desired substrate. Special values include excellent dielectric and low-loss factor, high elastic modulus, thermal conductivity, and wear resistance.

POLYBUTADIENE GLYCOL**

A relatively new high performance thermosetting resin is a 1,2-polybutadiene prepolymer with terminal primary hydroxyl groups, vinyl or carboxyl groups. The polybutadiene glycol material has high molecular weight, is soluble and rubbery. The final cure is achieved with peroxides to yield a hard, tough product. The resin is transfer and compression molded, laminated, cast, match-die molded and vacuum formed. This is particularly useful for the laminates.

HEXACHLOROBUTADIENE***

A new family of plastics which can produce ultra thin plastics films has been formed from hexachlorobutadiene. Pinhole-free films of this polymer have been deposited firmly on many metals.

POLYBENZIMIDAZOLE† (PBI)

The polybenzimidazole plastics are of special interest because of their excellent heat and oxidation stability. This thermosetting resin is infusible after the final cure at elevated temperatures. Processing is very difficult and some uses include high temperature adhesives and laminates for aerospace products. It is stable in nuclear applications and has good electrical properties.

A laminating prepreg* for aerospace vehicles has been produced in limited quantities. Fibers have been made by Celanese and DuPont.

POLYAMINOBISMALEIMIDE RESINS (PAMBS)‡

These newly developed French resins have flow properties similar to the thermosets and have greater thermal stability (500°F—long term) than many light metal

---

*"Ekonol," The Carborundum Co.
**"Hystl," Hystl Development Corporation. "Polybutadiene," Firestone Synthetic Rubber and Latex Co.
***General Electric Company.
†"Narmco," and "Imidite," Narmco Materials Division.
‡"PAMBS," Societe des Usines Chimiques Rhone-Poultec.

alloys. They are resistant to aromatic hydrocarbons, flame and radiation. The PAMBS are adaptable to many fillers and are processed like the phenolics and epoxies for molded and laminated products.

AROMIDS*

The aromatic polyamides are more heat stable than the other polymers. Chemically it is reported to be a poly-meta-phenylene-iso phtyalamide. This polymer has high heat resistance and outstanding electrical properties. These polymers cannot be described as thermosetting or thermoplastic. They decompose before they melt or soften appreciably. It has Underwriter approval for use at 220°C. Its present principal use is in paper form for electrical insulation where it provides thermal endurance and chemical resistance.

PARYLENE**

The "Parylenes" (poly-para-xylene) are used as ultra-thin films, and as pore-free protective coatings for plastics films and bottles. In one procedure, a monomer of the organic compound is heated to a vapor and condensed on a cool surface where it polymerizes into a very thin transparent application. In a typical capacitor application, the size was reduced 80 per cent, and the temperature utility increased to 340°F by the substitution of this ultra-thin dielectric. This is an excellent coating material for sensing probes.

Parylene N has unique and outstanding electrical properties. Parylene C has very outstanding barrier properties. In general, these materials may be used at temperatures up to 440°F. Unsupported films as thin as 300 Å are employed in optical and radiation applications.

POLYIMIDES***

The polyimides are characterized by their very high heat resistance, 500°F, continuous, and 900°F, short term. Polyimide chars slowly in the high temperature zone and retains some of its strength after long exposure. The polyimides compete with TFE, fluorocarbon, and glass-bonded mica in the 400 to 1000°F zone. Polyimide plastics are resistant to most acids and organic solvents, including hydraulic fluid, transformer oil, benzene, chloroform, and hexene. Polyimide is attacked by strongly alkaline solutions, hydrazine, and N_2O_4. Steam decreases its flexural strength, and the weathering resistance is not good. Arc resistance, volume resistivity, and dielectric strength are all high. Radiation resistance is exceptional when compared with many other thermoplastics.

---

*"Nomex," and "K5," E. I. Du Pont de Nemours & Co., Inc.
**"Parylene," Union Carbide Plastics Co.
***"Vespel," "Kapton," "NR150," and "Pyralin-Prepreg," E. I. Du Pont de Nemours & Co., Inc.; "Polyimide 2080," Upjohn Co., "Gemon 3010," General Electric Co.

Two compounds are available for fabricated products: the base resin (1.42 sp.gr.), and a graphite-filled composition (1.49 sp.gr.), prepared especially for high-temperature bearing applications. Molding of this material is difficult because of its rapid oxidation at the molding temperature. Polyimide resin can be molded if done in a nitrogen atmosphere. Parts can be punched from sheet or fabricated by a conventional machining process.

Typical applications include bearings, compressor valves, back-up rings, piston rings, and diamond-abrasive wheel binders.

POLYAMIDE-IMIDE*—POLYESTER-IMIDE**

These polymers have high temperature properties between the aromatic polyamides and the aromatic polyimides. The polyamide-imides are produced from an aromatic diamine and an anhydrie acid which forms a polyamic acid convertible to a polyimide by heat. The polyester-imides are used for magnet wire insulation, high temperature laminates and adhesives, decorative high temperature coatings and molded products. Polyamide-imide glass laminates have given superior performance for gear and bearing applications in helicopters. The polyester-imides are also produced for magnet wire insulation.

HYDROCARBON RESINS***

The hydrocarbon resins, introduced in 1960 by Enjay Laboratories, produce a family of compounds based on liquid butadiene styrene copolymers, such as polyethylene and elastomers. Mineral-filled compounds exhibit satisfactory electrical properties and heat stability. The coke-filled compounds offer interesting chemical resistance. Good molding qualities are achieved by the transfer and compression process.

Polybutadiene resins have desirable properties for casting, potting and encapsulation. They have been used with good success as lamination resins for the various industrial laminates.

POLYBENZOTHIAZOLES†

These thermosetting laminating resins which require a complex curing process provide higher thermal and oxidative stability than the polyimides. This resin melts and flows prior to its cure. For many purposes it is good up to 600°F.

---

*"Vespel," "Kapton," "NR150," and "Pyralin-Prepreg," E. I. Du Pont de Nemours & Co., Inc.; "Polyimide 2080," Upjohn Co., "Gemon 3010," General Electric Co.

**"Aminin," and "Aramidyn," Westinghouse Electric Co.; "AI Polymer," and "Torlon," Amoco Chemicals Corp.

***"Escorez 8115," Exxon Chemical Co.; "Buton," Enjay Chemical Co.

†Abex Corporation.

COLD-MOLD PLASTICS*

The cold-molded materials are formed to shape in the mold, and are hardened by a curing operation after removal. The original cold-mold materials were molded without any heat at all, but later developments introduced some materials which required heated molds for improved appearance and strength. Cold-mold materials were produced in this country as early as 1909, and at one time a large proportion of the electrical devices and automotive products were molded from them. A reduction in the costs of the phenolic materials, and the cost of molding, has limited the use of these products to those high-temperature and flame-resisting applications which cannot be served by the organic materials. The cold-mold materials have a price advantage for certain types of molded products, and manufacturers of these materials continue to produce a large volume. Cold mold is used today for electrical switch parts which must be subjected to open flame or arc when the circuit is broken. Cold molded products are used for cooking-utensil handles which are put in the oven or subjected to the open-gas flame, because they will not burn at these cooking temperatures. Battery boxes, wire connectors, valve wheels, and cores for high heat resistance units, are often produced from cold mold material. Typical cold mold parts are shown in Fig. 2.11.

Some of the cold-mold materials are called bitumen plastics, since many of the bitumens (mineral pitches, tars or asphalts) have been used as binders for these compounds.

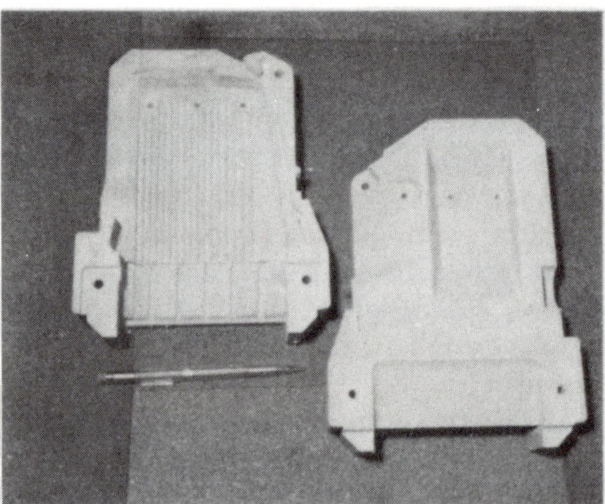

FIG. 2.11 Cold mold arc barriers produced from a cement asbestos compound. (*Courtesy Garfield Mfg. Co., Garfield, N.J.*)

*Garfield Mfg. Co., Garfield, N.J. American Insulator Co., New Freedom, Pa. Rostone Corp. Lafayette, Indiana.

Types of Cold-Mold Materials

The cold-mold materials are generally divided into two classes, organic or non-refractory, and inorganic or refractory. A refractory material is one which is difficult to burn or melt. The inorganic or refractory material is made from cement and asbestos with some clay, or it is formed from slate and limestone. The organic or nonrefractory cold-mold materials use fillers of asbestos or diatomaceous earth, and binders of pitch, asphalt, linseed oil, and sometimes synthetic resins. Gilsonite, glossy form of asphalt, frequently is used as a binder.

Organic or Nonrefractory Cold-Molded Compounds

The organic cold-molded materials are general-purpose compounds, which are often specified for wiring devices and for cookware handles. These materials will withstand temperatures up to $500°F$. The binder used for a typical formulation of this material is made from raw linseed oil, East India copal gum flour, manganese linoleate drier, stearine pitch, gilsonite and linseed oil fatty acid. For many hours, these materials are cooked and agitated, with the ingredients added at the proper time. The final quality check for this mixture is a viscosity test, which measures its ability to flow.

A typical filler formulation, used with the binder described above, is composed of Vimy asbestos, Vermont XX asbestos, sulphur, carbon black and kerosene. The asbestos, carbon black, and sulphur, are mixed in a dough mixer until dry, for about 15 minutes. The binder, which is heated to $240°F$, is then poured slowly onto the filler. Mixing is continued until control tests indicate that the mixing is completed. Three to four hours usually are required to complete the mix. Mixing is done in batches ranging from 150 to 1200 pounds. This material is then seasoned in the air for a few hours until the tackiness, which might cause sticking in the molds, is eliminated.

The post-molding cure of this organic cold-molded material is a slow process. Small pieces are baked in an oven for approximately 25 hours, while large pieces are baked for a longer period. The baking temperature generally starts at 150 to $200°F$, and the temperature is increased about $10°$ per hour until $400°F$ is attained. This temperature is held until the desired degree of hardness is reached. The baking process is complicated by the necessity for accurate control of the shrinkage and warpage. Many pieces have to be held to close dimensions, and they must also be held to a minimum amount of warpage. Some parts are packed or held flat during the baking operation. These parts, which must be held to the closest dimensions, are allowed to stand in air for three days. They are then subjected to a steam-oven bake at 150 to $200°F$, for one week before going into the regular baking operation. This slow cure permits retarded oxidation of the linseed oil throughout the molded part, and prevents the blistering, warpage and excess shrinkage which might be experienced with a fast cure.

Refractory or Inorganic Cold-Molded Materials

Refractory cold-mold products contain no organic materials; they are used primarily because of their high heat resistance, 1300°F, and their excellent resistance to high current arcs. A typical formulation for a common refractory material contains dark cement, asbestos fibers, colloidal kaolin clay, and a small amount of water. The cement, asbestos and clay are mixed until dry in a blender, for about 20 minutes. This mixture is then transferred to a dough mixer, where the water is added slowly, and where mixing is continued for half an hour. This compound is put through a fan-type grinder, which breaks the lumps and opens the fibers that were not properly mixed. The material is then ready for the molding operation.

This refractory, or inorganic cold-molded material, is cured after molding by a series of operations. The newly molded parts are first allowed to air dry for 3 hours, as a preventative against swelling during the curing process. The parts are then treated in a very humid atmosphere or fog oven for 24 hours after which they are next immersed in hot water for 24 hours, and subsequently baked for 12 to 14 hours in electric ovens, at temperatures from 250 to 500°F, to drive off the excess moisture. This material is sometimes given a wax treatment to reduce its water absorption.

The great popularity of the hot-molded materials has overshadowed the cold-molded products in recent years, and there is room for considerable research and development work on improved materials of this type. Better binders and better curing methods may be devised, which will take advantage of the automatic molding, low cold-mold die cost, and the rapid cold molding cycle.

3 THERMOPLASTICS

THERMOPLASTIC MATERIALS

The thermoplastic materials were often called injection molding materials before the reciprocating screw injection machines became available that could mold either thermoplastics or thermosets. Fillers as described in Chapters 2 and 10 are also extensively used with the thermoplastics to gain stiffness, dimensional stability and strength. Glass is the most common and effective filler for the thermoplastics. (See Chapter 7 for the molecular structure of thermoplastics and Chapter 1 about additives.) Plasticizers are added to the thermoplastics to improve their processability, flexibility, and other desirable properties. Plasticizers are generally non-volatile organic products, and are added by dry blending, hot mixing, or solution mixing with a mutual solvent. Stabilizers are added to the thermoplastics, as required, to prevent deterioration by heat, light and oxidation. Stabilizers also are antistatic, and demonstrate coloring and self-extinguishing properties. A few plastics have good ultraviolet stability; others require ultraviolet absorbers to prevent deterioration by sun and fluorescent light. Fluorescent additives are also applied to some of the plastics to obtain color attraction and customer appeal. Careful selection of the stabilizers and plasticizers is required, particularly when the plastics are used in the presence of food or drugs, or in contact with other plastics, since the plasticizer may migrate into adjacent materials. Unplasticized materials are often specified for technical products, food and drug handling, as well as for surgical implants.

As a class, the thermoplastics are tactile and colorful materials that may be produced as fabrics, molded products, films, bottles, adhesives, and coatings, by many and varied methods.

CELLULOSIC PLASTICS

Cellulose Nitrate*

"Celluloid," developed by Hyatt in 1869, was the first of the synthetic plastics. The Celanese Corporation of today grew out of this early manufacturing program. Hyatt's invention disclosed that pyroxylin could be plasticized with camphor in the presence of a mutual solvent, to produce a solid material which burned rapidly, but was nonexplosive. Camphor is a gum-like crystalline substance, obtained in Japan or Taiwan from a large evergreen tree. It is also made synthetically. The cellulose nitrate, or pyroxylin thermoplastics, were primarily fabrication materials, and were extensively used in that era for celluloid collars, mirror coverings, combs, brushes, buttons, etc. The first automobile side curtains, and later, the first bonding material for auto safety glass, was cellulose nitrate sheet. Photographic films, spectacle frames, etc., were all made of this first synthetic, prior to the development of the slow burning acetates.

Physical Properties of Cellulose Nitrate Plastics

Easy workability, excellent colors, water resistance and toughness: all are important physical characteristics that have contributed to the wide use of cellulose nitrate plastics. The ease with which it burns has been the outstanding defect that has prevented even greater use. Light and heat tend to decompose and discolor this material after long periods of time. All the cellulose plastics tend to become brittle when their moisture is removed. The cellulose nitrate plastics are little used today because of the fire hazard they present, and the availability of other low-cost materials which are suitable and better for such applications.

Cellulose Acetate**

Cellulose acetate was first introduced for photographic plates in 1912, and in 1927, it took the form of sheets, rods, and tubes for general fabrication. Molding materials followed in 1929, but the real volume use started in 1934, with the general introduction of the injection molding machine. The acetate materials were first used in World War I for airplane covering, or "dope" as it was called; the facilities that produced cellulose acetate during the war followed with acetate thread or rayon, thus starting the synthetic fabric industry.

Cellulose acetate is moldable, whereas cellulose nitrate is not. It is a tough colorful material, and was widely used at one time for many of the contemporary

*"Celluloid," Celanese Plastics Co.
**"Tenite," Eastman Chemical Products, Inc.; "Vuepak," Monsanto Co.

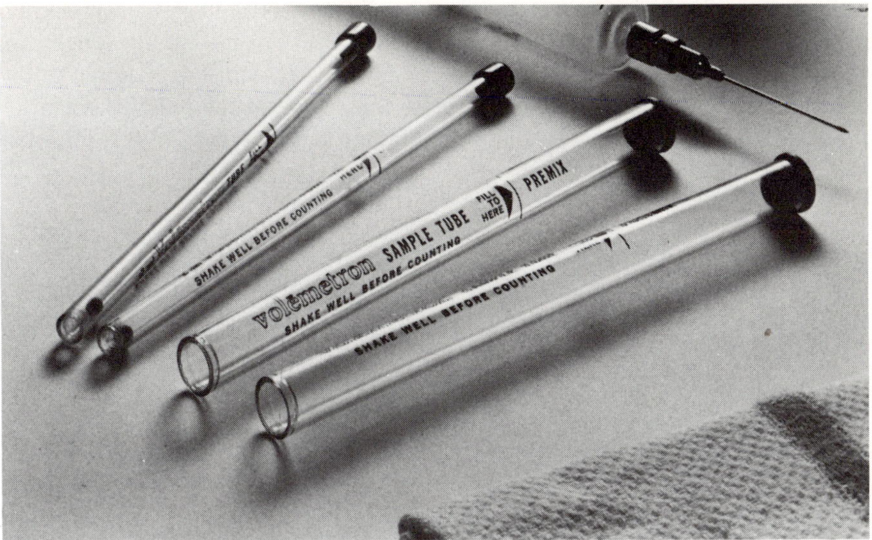

FIG. 3.1 Exactness of measure, sterility and strength are prime features of these inexpensive, disposable tubes which are extruded from cellulose acetate with excellent clarity and adequate dimensional stability. Instructions and measuring data are printed on the tube by an offset process.

products now made from other less costly thermoplastics. Its value today lies in its toughness, colorability, and fairly high impact strength. Self-extinguishing variations are available to permit its use in household appliances. Other markets include toys, beads, knobs, frames for sunglasses, as well as packaging materials, shown in Fig. 3.1 and photographic film.

Cellulose acetate can be fabricated by all of the conventional processes; it is an excellent vacuum-forming material for blister packages, etc. Some magnetic tapes are triacetate of cellulose.

Cellulose Acetate Butyrate*

The original poor moisture resistance of the acetate materials brought about the development of a cellulose acetate butyrate material, which is commonly called aceto-butyrate. It is similar to cellulose acetate, and is used for those acetate applications which require improved resistance to moisture, as well as improved dimensional stability.

In the formation of the aceto-butyrate materials, cotton linters are treated with butyric acid, in addition to the acetic acid and acetic anhydride used for the regular cellulose acetate. This facilitates the production of a molding material, using plasticizers that have better moisture and weathering resistance.

*"Tenite," Eastman Chemical Products, Inc.

There is considerable gain in dimensional stability with the aceto-butyrate materials when compared with acetate. Excellent use is made of the toughness, stability, moisture resistance, and fine colors available with this material in extruded sections. Extruded ribbons, sheets, and tubing of the material are manufactured to serve a variety of industrial applications. A typical tool application is shown in Fig. 3.2.

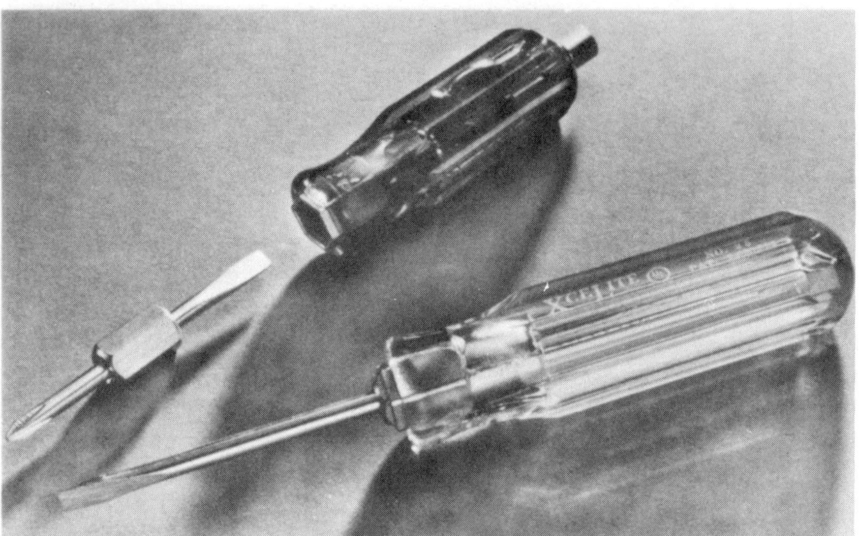

FIG. 3.2 Tough, hollow handles of cellulose acetate butyrate plastics are used for these multi-purpose screw and nut drivers. Such handles are tough, light, abuse-resistant and provide excellent electrical and thermal insulation.

The aceto-butyrate plastics may be compounded with fire-retardant additives; they also have good weatherability. All fabricating and molding processes may be used, and they provide useful machining properties. Major applications today include automobile steering wheels, knobs, lenses, street lamp globes, automobile tail and signal lamp lenses, outdoor signs, tool handles, packaging blisters, and metallized sheet and film.

Cellulose Propionate*

This cellulosic plastics material is noteworthy because of the short molding cycles that are obtainable, and its freedom from lamination during molding. Propionate compounds require less plasticizer than the other cellulosics, and may be used with a greater variety of plasticizers. It has better weathering properties and color retention than has acetate. An important feature is high

*"Forticel," Celanese Plastics Co.; "Tenite," Eastman Chemical Products, Inc.

FIG. 3.3 Reels of magnetic tape for new large-scale Honeywell computers are protected from impact during storage in transparent containers molded of tough cellulose propionate plastics.

impact resistance, and toughness. Propionate is most often used for extrusion and injection mold products, as depicted in Fig. 3.3.

Typical applications include toys, pens, automotive parts, small radio cases, toothbrushes, steering wheels, handles, and novelties.

Ethyl Cellulose*

The ethyl cellulose plastics are noteworthy for their suitability for use over a wide temperature range, and their freedom from odor. Better heat distortion temperatures are available, plus higher impact strength than are available in the other cellulosics. It may be molded, extruded and formed by all the usual thermoplastic processes.

Ethyl cellulose has been used in many military applications because of its freedom from low temperature embrittlement, and its good impact strength. Other materials with better properties have taken over some of these historical markets. Contemporary applications include decorative trim, toys, cosmetic packages, refrigerator components, tool handles, blister packages, etc.

ACRYLIC PLASTICS**

The acrylic plastics possess crystal clarity, maximum colorability, and extremely valuable optical properties. The original acrylic studies were made in Germany,

*"Ethocel," Dow Chemical Co.
**"Acrylite," American Cyanamid Co.; "Lucite," E. I. DuPont de Nemours & Co., Inc.; "Plexiglas," "Kydex," Rohm & Haas Co.; "XTPolymer," American Cyanamid Co.; "Glasflex," Glasflex Corp.

by Dr. Otto Rohm, in 1901. Acrylics were first manufactured commercially in 1931 as coating materials and for bonding safety glass. Methyl-methacrylate is the most common of the acrylic plastics; many copolymers or alloys are formed with other non-acrylic monomers.

A most valuable property of the acrylic resins is their weatherability where they maintain stability better than most other plastics. The low index of refraction, 1.49, and high degree of uniformity, makes this an excellent lens material for binoculars, camera, spectacle, contact and instrument lenses. High scratch and abrasion resistance improves its serviceability as a lens material. Many automotive tail and stop lights use ruby acrylic molded lenses. Tail lamp lenses of acrylic plastics are shown in Fig. 3.4.

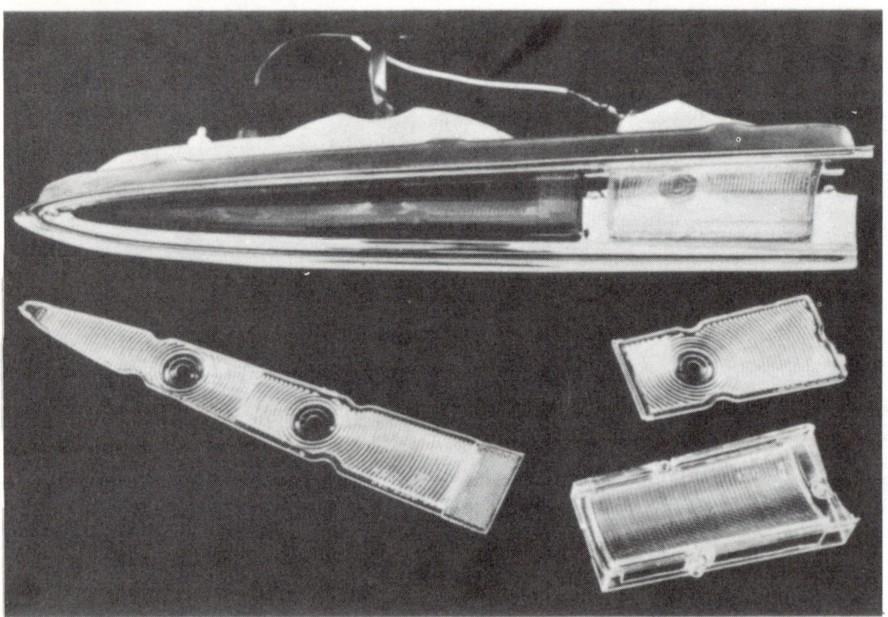

FIG. 3.4 A high-temperature acrylic compound is used for these tail and stop-light prisms.

Acrylics will withstand food oils, non-oxidizing acids, petroleum lubricants, and household alkalies. They do not react with photographic solutions, and are extensively used for this apparatus. The acrylics may not be used freely with alcohols, chlorinated hydrocarbons, esters, ketones, phenols and fluorocarbons. The acrylic plastics are slow burning, and with certain additives, may be classed as self-extinguishing. Since they burn with less smoke and toxicity than any other lighting fixture or glazing plastics, they are highly rated for these applications in building codes and insurance plans. A typical application is shown

FIG. 3.5 The transparent acrylic plastic dome is a beacon for business, especially at night when the dome is brightly illuminated. (*Courtesy Rohm & Haas Co.*)

in Fig. 3.5. Acrylic syrups are used in the making of some reinforced plastics sheets. Acrylic* and PVC alloys make an excellent thermoforming sheet that is tough and has high tear strength.

Electrical properties of acrylic plastics are good in dielectric strength, and they are also non-tracking. The combination of weatherability, high dielectric strength, and low moisture absorption has resulted in widespread use for high voltage line spacers and cable clamps. The dielectric constant is relatively high, and the loss factor is bad for high frequency work. One special acrylic type polymer has its thermal endurance elevated to 250°F, which is 50° above the conventional acrylics.

Modified acrylic compounds include a styrene acrylic copolymer, and a methyl methacrylate-alpha methyl styrene copolymer. The acrylic plastics may be injection, compression, extrusion and blow molded. Acrylic sheets are cast and extruded; these materials are readily welded, machined, and thermoformed. Specimens are frequently preserved by embedding in methacrylate casting resins. Acrylic resins are used as protective finishes for metals and for water emulsion paints, and are easy to apply as well as weatherable. Acrylic ester resins make excellent coating materials with outstanding durability, chemical and burn resistance. Leather garments are impregnated with acrylics to gain cleanability and softness.

*"Kydex," Rohm & Haas Co.

Widely used applications for the acrylic plastics include lenses, aircraft and building glazing, lighting fixtures, dishes, piano keys, knobs, dials nameplates, telephone dials, beverage dispensers, display cabinets, fountain pens, signs, skylights, packaging, and textile fibers.

Acrylic fibers are replacing glass in many fiber optics applications.

STYRENE PLASTICS

Styrene, or styrol, was first noted in 1839 by an English chemist, more than 100 years before it was produced commercially in the United States. The Germans produced styrene in 1925 for their buna-S synthetic rubber. Bakelite Corporation and Dow Chemical Company first marketed styrene resins in 1937. Chemically, styrene is vinyl benzene. It is available as a molding resin, either crystal clear or in all colors.

Preparation of Polystyrene*

Polystyrene is prepared from ethylene and benzene. Ethylene is made from natural gas or petroleum. Benzene is often obtained as a by-product from coke ovens. The ethylene is forced through the benzene in the presence of a suitable catalyst, such as aluminum chloride, to produce ethyl benzene. This ethyl benzene combination is converted into styrene by splitting hydrogen away from the rest of the molecule. This splitting off of the hydrogen is accomplished by a decomposition of the material under heat.

The styrene thus produced is a liquid which will boil at 292°F. This liquid may be converted into a clear, solid plastics material by polymerization. The polymerization of styrene into polystyrene is a process which is very difficult to control. Light, heat, dirt and agitation all tend to assist polymerization. A slow polymerization produces a long molecule chain material, while a fast polymerization will produce a short molecule chain material, with entirely different flow characteristics, as well as less strength. Both products will be identical in appearance.

Polystyrene materials are prepared in all the transparent and opaque colors. The crystal-clear and tinted materials are particularly attractive. Mottled colors can be produced. They are easily identified from all other plastics materials by the distinct metallic sound produced when they are struck or dropped. This plastics product is generally used in the pure resin form, but it can be used with glass and other fillers up to 40 per cent by volume. Excellent moldability is an important value in styrene, as shown in Fig. 3.6.

Styrene is one of the very low-cost, rigid and colorful materials, and is

---

*"Bakelite," Union Carbide Plastics Co.; "Catalin," Catalin Corp.; "Cosden," Cosden Oil & Chemical Co.; "Fostarene," Foster Grant Co.; "Dylene," Koppers Co.; "El Rex," Rexall Chemical Co.; "Lustrex," Monsanto Co.; "Shell," Shell Chemical Co.; "Solar," Solar Chemical Co.; "Styron," Dow Chemical Co.

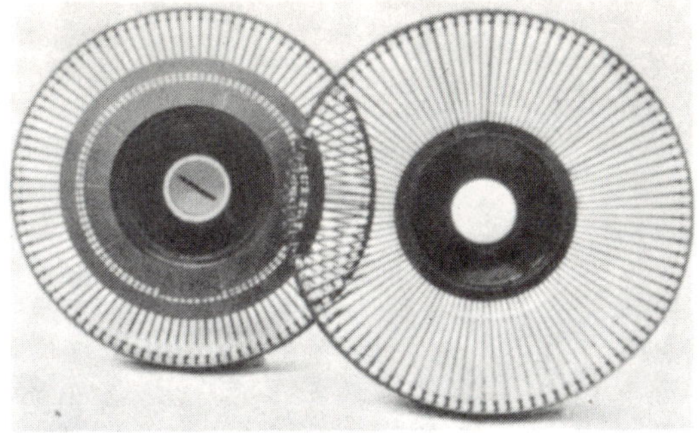

FIG. 3.6 Intricate moldings such as these slide holders are often molded of polystyrene, with or without glass fillers.

extensively used where price alone dictates. Styrene has excellent organic acid, alkali, salts, and lower alcohol resistance. Deterioration or softening occurs when used with hydrocarbons, ketones, esters and essential oils. It stress-cracks easily, and parts must be annealed for use in many applications. Styrene is an excellent capacitor dielectric.

The electrical properties of polystyrene are admirable within its temperature range. Styrene is not self-extinguishing or suitable for exposure to arc. Because of its extensive use in fluorescent lighting fixtures, it is offered with a light stabilizer to minimize long-term deterioration. A styrene-methyl-methacrylate-copolymer* provides improved light stability.

Impact styrene materials are produced by the inclusion of rubber modifiers, which give them toughness and resilience, though these modifiers sacrifice clarity. Low-temperature impact materials are widely used in refrigerator construction. Heat-resistant styrenes are available, which raise the useful range from 190 to 220°F. Glass fillers double the tensile strength, and raise the temperature limit to 220°F, also increasing the impact strength.

Typical styrene applications include jewelry, light fixtures, packages, toys, clock cases, radio cabinets, housewares, bottles, lenses, novelties, capacitor dielectrics, low-loss insulators, musical instrument reeds, medical syringes, and light-duty industrial components. Extruded sheets are widely used for packaging, thermoforming, envelope windows, place mats, signs, photographic film, and novelties.

The expanded styrene products have widely increased the market for styrene resin, which is an ideal packaging and insulation material, as depicted in Fig. 3.7. With as light a weight as 2 pounds per cubic foot, its thermal conductivity is very low, and its cushioning value high. For production purposes, pre-expanded beads are further expanded and fused in a mold. Steam is used in the pre-

---

*"Zerlon," Dow Chemical Co.

FIG. 3.7 Expandable polystyrene foam end-pads fit snugly around the ends of the set and when placed in a corrugated shipper, only the foam pieces touch the sides, providing a low-cost safe shipping package. (*Courtesy Corning Packaging Co.*)

expanders from which the molds are filled. Steam is then used for the final expansion in the mold where the foamed beads fuse to an integral foam product. Common applications include ice buckets, packaging, water coolers, wall panels and general thermal insulation applications.

Styrene Acrylonitrile (SAN)*

This inexpensive copolymer of styrene and acrylonitrile has created an important position in the market because of its better stiffness, scratch, paint-solvent, chemical and stress-crack resistance. Its heat resistance is 25°F above general-purpose styrene. SAN is not affected by food or household chemicals, and finds a considerable market for lenses, tumblers, dishes, food packages, some chemical apparatus, piano and organ keys, telephone parts, bristles, closures, batteries, and films. Modern piano keys, as shown in Fig. 3.8, are molded completely of SAN.

METHACRYLONITRILE PLASTICS

Methacrylonitrile** a vinyl with nitrile is substituted for acrylonitrile in some formulations to gain better quality, color integrity, less heat distortion and easier processing.

---

*"Bakelite," Union Carbide Plastics Co.; "Kralac," United States Rubber Co.; "Lustran, A," Monsanto Co.; "Tyril," Dow Chemical Co., "NAS," The Richardson Co.
**MAN, Sohio Chemical Company.

FIG. 3.8 Molded SAN organ keys replaced the ebonies and ivories of an earlier day. SAN is highly resistant to household and body chemicals (*Courtesy Tech Art Plastics Co.*)

BUTADIENE STYRENE POLYMER*

This thermoplastic resin is designed to serve the markets for oriented sheet or film. It combines good impact strength and crystal clarity. It has poor permeability properties and is used only in those applications where moisture transmission is permissible.

NITRILE BARRIER COPOLYMER**

The nitrile barrier resins are generally approved for the packaging of foods and beverages by FDA. Soft drinks, carbonated beverages, beer, colas, etc., may be bottled in this material. It provides an adequate barrier for gas, flavor and aroma

---

*"K-Resin," Phillips Petroleum Company.
**"Barex 210," The Standard Oil Co. (Ohio); "Lopac," The Monsanto Company; "NR-16," E. I. DuPont de Nemours & Co., Inc.

FIG. **3.9** Coca cola bottle is blow molded of nitrile barrier copolymer. (*Courtesy Monsanto Co.*)

in plastics bottles as shown in Fig. 3.9. Basically the nitrile barrier resins are rigid copolymers with a high acrylonitrile content.

ABS* Plastics

Three chemicals, acrylonitrile, butadiene and styrene, are combined to make the ABS plastics. The acrylic and styrene polymers are resinous; the butadienes are rubber-like materials. The ABS materials are superior to the ordinary styrene products. These plastics can be compounded with a high degree of hardness, or with great flexibility and toughness, and are commonly described as tough, hard and rigid. This is an unusual combination for the thermoplastics, which are often either hard and frangible like polystyrene, or flexible and tough like vinyl.

Stiffness and dimensional stability are greatly improved by the addition of glass fillers.

*"Abson," B. F. Goodrich Chemical Co.; "Cycolac," Marbon Chemical Co.; "Kralastic," United States Rubber Co.; "Lustran," Monsanto Co.; "Tybrene," Dow Chemical Co.

ABS plastics are commonly used in applications that require abuse resistance, colorability, hardness, electrical properties, moisture stability, and limited heat resistance (220°F). Certain formulations are compounded with self-extinguishing additives to be used for electrical hand tools, and other Underwriter-approved products. Adequate chemical resistance is present in the ABS materials for ordinary applications. They are little affected by water, alkalies, weak acids and inorganic salts. Alcohol and hydrocarbon solvents may affect the surfaces and cause swelling if the exposure is extended.

ABS materials are processed by thermoforming, injection, blow, rotational and extrusion molding. They may be calendered easily in various sheet-making and coating processes. ABS pipe application is illustrated in Fig. 3.10.

FIG. 3.10 Plastics plumbing "tree" for Puerto Rican low-cost housing unit weighs about 50 pounds and is easily carried to building site by plumber's helper. Cast-iron pipe and fittings would weigh more than 240 pounds. Lightweight plastic pipe and fittings for such drain-waste-vent systems have reduced plumbing costs as much as 38% and increased the number of homes completed per day by 50%.

Typical applications for ABS compounds include business machine and camera housings, telephone handsets, electrical hand tools such as drill housings, knobs, handles, cams, blowers, bearings, wheels, gears, pump impellers, grilles, deflectors, automotive trim and hardware, pipe and pipe fittings. ABS plastics may be metal-plated, and such pieces are often used to replace die castings.

VINYL* PLASTICS

Polyvinyl Chloride (PVC)

Polyvinyl chloride is one of, if not the largest, single volume plastics material in general use in the world. It is second to polyethylene on a basic resin, but ahead in the compounded, finished-product basis. It is potentially one of the lowest cost materials. The French chemist Regnault first discovered PVC in 1835, and it was initially marketed commercially in 1927. PVC has achieved this market leadership because of its good physical properties, its compounding versatility for a wide variety of applications, its low cost, and processing ease. These desirable properties include self-extinguishing characteristics, water, chemical and abrasion resistance, good strength properties, and a complete range of colors. The compounds range from soft, flexible films, to rigid, high-strength products. Plasticizers, lubricants, fillers, and stabilizers, are used to produce this versatility, and it is possible to make a compound with the right balance of properties for almost any application. PVC products will melt but will not burn, and good weatherability has been achieved by proper compounding.

Products are usually made by extrusion, injection, rotational, slush, transfer, compression and blow molding, calendering, thermoforming fluid bed coating, rotocasting, and foaming procedures. Extruded PVC is shown in Fig. 3.11.

Vinyl compounds are resistant to most acids, fats, petroleum products, salts, and they do not permit fungus growth.

Principal markets for rigid PVC include pipe, conduit, pipe fittings, wire and cable insulation, weather stripping, house siding, rain gutters, and down spouts. Flexible PVC is used for raincoats, baby pants, dolls, bottles, shoe soles, wire insulation, film, sheeting, draperies, garden hose, gaskets, closures, shower curtains, inflatable toys, novelties, and shoe welting.

Vinyl foams have become widely used as a carpet backing material.

PVC is a desirable blow-molding material because PVC bottles combine a high level of clarity approaching that of glass, with break resistance as good as polyethylene. It offers a substantial improvement over polyethylene in resistance to permeation of some atmospheric gases to the product, and also in retention of essential flavors. Its water-vapor permeability is greater than that of polyethylene. Principal PVC bottle markets are for toiletries and cosmetics, household chemicals, and food products.

Polyvinyl Acetate Plastics**

The polyvinyl acetate resins have been used with great success as an adhesive for all sorts of materials, including paper, metal, mica, glass, plastics sheets, wood

*"Dacovin," Diamond Alkali Co., "Escambia," Escambia Chemical Corp.; "Ethyl," Ethyl Corp.; " Exxon," Exxon Chemical Co.; Firestone Plastics Co.; "Geon," B. F. Goodrich Chemical Co.; "Marvinol," United States Rubber Co.; "Pliovic," The Goodyear Tire and Rubber Co.; "Opalon," Monsanto Co.; "Bakelite," Union Carbide Corp.; "Vyram," Monsanto Co.

**"Gelva," Schawinigan Resins Corp.

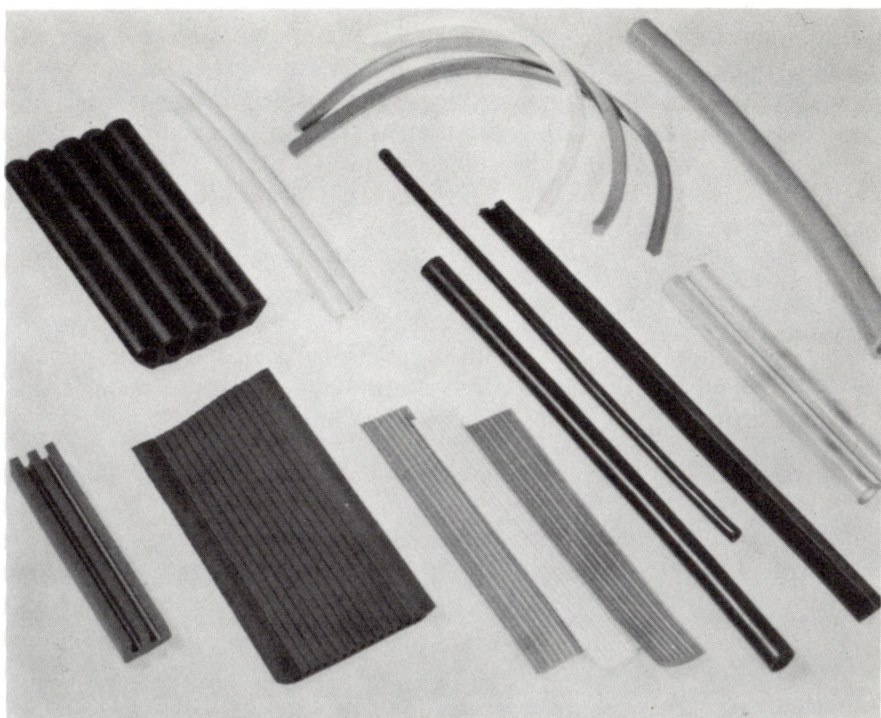

FIG. 3.11 Extruded shapes, channels, tubes and decorative strips are extruded from polyvinyl chloride plastics. These materials are also extruded to make garden hose and wire insulation and are molded to many complex shapes. (*Courtesy B. F. Goodrich Co.*)

and porcelain. These resins are unaffected by petroleum, naphtha, turpentine, mineral and vegetable oils. They dissolve in most solvents such as alcohols and ketones. There is no odor, taste or color in the polyvinyl acetate resin. Photo flashbulbs may have a coating of polyvinyl acetate on the bulb surface to eliminate the danger from glass particles, in case the lamp should be broken. The polyvinyl acetate causes the particles of glass to adhere to the resin. At the same time, the resin serves as an insulator, protecting the glass from hot particles of the burning metal. Polyvinyl acetate latex is used in large quantities for paints, and also as a sealer for cinder blocks.

Polyvinyl Aldehyde Plastics*

Vinyl acetate may be reacted with acetaldehyde, formaldehyde, or butyraldehyde, to produce materials which have commercial importance. These materials differ widely in their properties.

*Union Carbide Corp.

The reaction of vinyl acetate and formaldehyde produces a resin which is very tough, strong, and flexible. It has high temperature resistance and excellent chemical properties. This material is used for coatings and magnet wire insulation; it will withstand the abuse of modern high-speed winding and assembly operations. This plastics resin is not affected by the solvents used for the treating varnishes employed to solidify and impregnate coil forms. It will also withstand the high baking temperatures and high operating temperatures that can ruin ordinary insulating enamels.

The reaction of vinyl acetate with butyraldehyde produces a soft, flexible, very tough, rubber-like material, which is widely used as the interlayer in safety glass. An impact blow may break this glass, but the small pieces of broken glass adhere to the interlayer and prevent the damage often caused by broken particles of glass. This safety glass will actually bulge and come out of the window frame as a single entity, when subjected to a very severe impact.

Copolymers of Vinyl Acetate and Vinyl Chloride*

Vinyl chloride-acetate copolymers can be modified as required by the addition of plasticizers, stabilizers, fillers, pigments, etc., to gain the desirable balance of properties. For high-fidelity phonograph records, this material gives superior results because of the excellent reproduction from the mold, and the low noise level from the sound groove. This copolymer is extensively used for floor products because of the large amount of fillers it will bond, still retaining its good colorability and physical characteristics. Copolymer vinyl is calendered into flexible and rigid sheets for products requiring better dimensional control, heat stability and abrasion resistance.

PVC Plastisols**

Dispersion resins made by an emulsion polymerization process are commonly called plastisols or organosols. Plastisols can be used in coating machines to cover paper, cloth and metal; these coatings are hardened by an oven-bake operation, as depicted in Fig. 3.12. Plastisols are also slush and rotationally molded, foamed, extruded, and injection molded, sprayed, and then used for dip coating. Organosols make use of a liquid diluent to reduce viscosity which is subsequently removed in the fusion process.

Polyvinylidene Chloride***

Vinylidene chloride is a most important packaging and industrial material. It is tasteless, odorless, nonflammable, tough; and abrasion resistant. It has high

---

*"Opalon," Monsanto Co.; "Vinylite," Union Carbide Plastics Co.; "Pliovic," The Goodyear Tire & Rubber Co.

**Union Carbide Corp., Uniroyal, Inc.; Inmont Corp., M & T Chemicals Inc.; Witco Chem. Corp.

***"Saran," Dow Chemical Co.

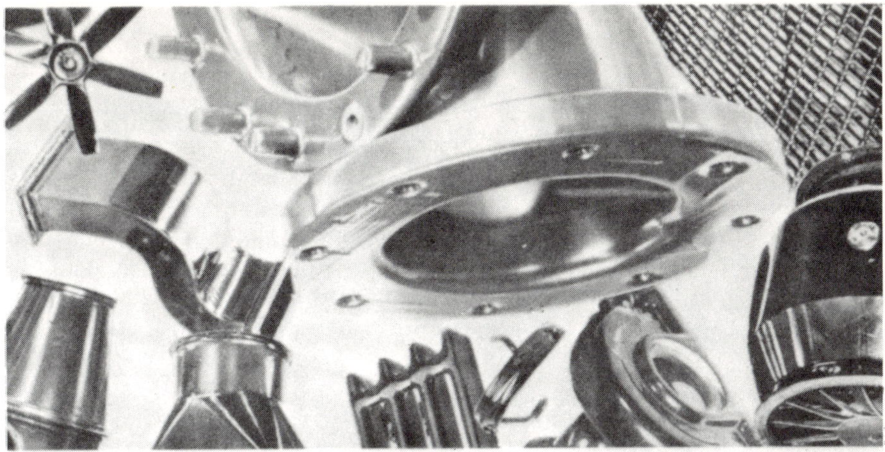

FIG. 3.12 These exceptionally durable PVC heat-fused coatings, applied as a liquid to metal parts subject to most acids, salts and exposure to 200°F or higher, give years of service where common metals are destroyed in days. Thickness of coating-$\frac{1}{32}$-$\frac{3}{16}$ in. or more takes 75 lb/in.^{2} to pull it loose. Compounded from fine particle size polyvinyl chloride resins with added plasticizers, the finished viscous product is heat-fused onto the metal base and finally cured to a smooth, tough, resilient surface. (*Courtesy Quelcor, Inc.*)

tensile strength; monofilaments are used for screens and high-strength fabrics. A major value is in its low water absorption, low vapor transmission, and impermeability to many flavorings.

Copolymers with acrylonitrile make excellent coating materials for other plastics where they add impermeability, grease, and oil resistance.

The vinylidene chloride plastics are among the most inert of all thermoplastics. Principal applications include auto seat covers, upholstery, draperies, carpeting, food packaging, film, bristles, pipe and pipe lining, filters, awnings, paper and paperboard coatings, valves, and pipe fittings.

ETHYLENE ETHYL ACRYLATE COPOLYMERS*

This flexible, rubbery vinyl compound has many of the elastomeric qualities of rubber without vulcanization. The flexibility of vinyls is gained without plasticizers and thus gains better toughness, flexibility and stability. Glass and other fillers can be added as desired. Typical applications include: vacuum cleaner hose, molded door mats, shuttle cocks and numerous other rubber-like products.

*"Flexible Bakelite," Union Carbide Corp.

POLYOLEFIN PLASTICS*

Polyethylene[†]

Polyethylene is a wax-like polymer that has almost universal use in every field. It has most desirable electrical and chemical properties, can be processed by all methods, and is well-known to the public by its large use for plastics films and squeeze bottles. It was first discovered in 1933 as a result of high-pressure reaction studies by E. W. Fawcett and R. O. Gibson, and follow-up work by M. W. Perrin, in 1935, in England. The plant, then built by ICI for polyethylene cable jacketing, supplied the essential polyethylene insulation for the radar that "won" the Battle of Britain. In 1941, DuPont and Union Carbide built plants to fill the tremendous war need for this most versatile and excellent plastics material. Polyethylene is one of the major volume plastics.

In one commercial process, ethylene gas or copolymers of ethylene, and unsaturated hydrocarbons are compressed at 15,000 to 45,000 psi, and then exposed to a catalyst of oxygen or peroxide, which causes polymerization and conversion to the white, solid polyethylene. An alternate process dissolves the component gases in a solvent at low pressure, prior to reaction with the catalyst. Polyethylene is composed of ethylene molecules joined together in long chains, and the basic characteristics of a particular formulation depend upon the number of units making up the chain, and the method by which they are combined. Short simple chains make a brittle, waxy polymer. Large molecules forming long chains produce a very tough, difficult-to-process polymer. In between is a wide variety of formulations with varying properties. The links between the crystals are shown in Fig. 3.13.

The American Society for Testing Materials has divided these materials into groups of varying density to provide a logical materials designation.

Density

	Density
Type I	.910–.925
Type II	.926–.940
Type III	.941–.965

The density of a given material depends on the shape of the molecules, as illustrated by Fig. 3.14. Closely fitting chains produce high density materials. Other chains are widely spaced and consequently are of lower density. Density is the clue to properties as shown in Table 3.1. The low density polyethylenes are

---

*"Alathon," E. I. duPont de Nemours & Co., Inc.; "Catalin," Catalin Corp.; "Dow," Dow Chemical Co.; "Dylan," Sinclair Koppers Co.; "El Rex," Rexall Chemical Co.; "Marlex," Phillips Petroleum Co.; "Fortiflex," Celanese Plastics Co.; "Grex," Allied Chemical Co.; "Monsanto," Monsanto Co.; "Petrothene," U.S. Industrial Chemicals Co.; "Plaskon," Allied Chemical Co.; "Poly Eth," Spencer Chemicals Corp.; "Shell," Shell Chemical Co.; "Tenite," Eastman Chemical Products, Inc.

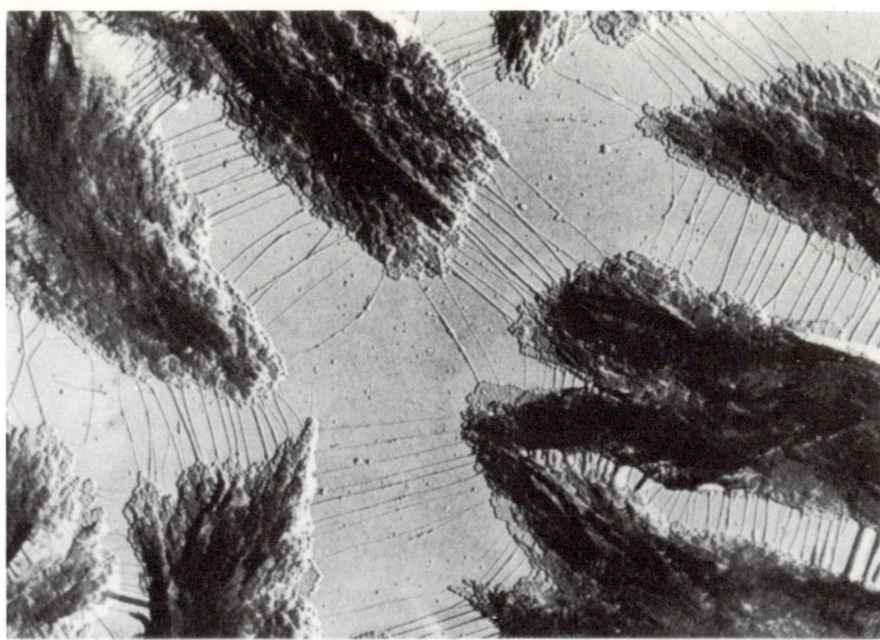

FIG. 3.13 Links in plastics crystals. The first direct observation of links between the crystals in a polymer has been made by scientists at Bell Telephone Laboratories. This electron micrograph shows a number of such intercrystalline links which measure up to 15,000 Å long and between 0.10 and 1.02 millionth of an inch thick.) H. D. Keith, F. J. Padden, and R. G. Vidimsky crystallized polyethylene from melt after blending it with the hydrocarbon, n-$C_{32}H_{66}$, which is similar to polyethylene but has a shorter molecular chain. When the polymer had cooled, the hydrocarbon—which separated the areas between crystallites—was washed away with a solvent. This technique exposed the polymer crystals and the links joining them together (*Courtesy Bell Telephone Laboratories*)

TABLE 3.1

As density increases:	
Stiffness	Increases
Yield strength	Increases
Hardness	Increases
Creep resistance	Increases
Toughness	Decreases
Softening temperature	Increases
Stress-crack resistance	Decreases
Permeability	Decreases
Gloss	Increases
Grease resistance	Increases

often called "regular" or "branched," and the high density are called "linear" polyethylene.

Another widely used measure is melt index, which is defined as a measure of the viscosity of a polymer at a specified temperature and pressure; this is an approximate function of molecular weight. Melt index will predict the performance of a given formulation when properly processed as follows:

TABLE 3.2

As melt index decreases:	
Stiffness	Increases
Tensile strength	Increases
Yield strength	Increases
Hardness	Increases
Creep resistance	Increases
Toughness	Increases
Softening temperature	Increases
Stress-crack resistance	Increases
Permeability	Decreases
Gloss	Decreases
Grease resistance	Increases

Many problems arise in finding the right polyethylene for a particular job as a result of variation in molecular distribution; Table 3.3 shows the property variation that occurs with increasing consistency of a polyethylene formulation. Materials makers and users often specify their product by its end use capability because of these complex variables.

TABLE 3.3

As molecular structure becomes more homogeneous:	
Tensile strength	Increases
Creep resistance	Increases
Toughness	Increases
Softening temperature	Increases
Stress-crack resistance	Increases

Environmental stress-cracking describes the tendency of a plastics product to crack or craze under load, when exposed to certain chemicals or conditions. Because of this potential, which is also a problem with metals, it is necessary to control molding conditions, material selection, and environmental exposure very closely. Polyethylene products, because of their extensive use as packaging materials, have been the subject of wide study. Data and procedures to meet and overcome this hazard are available, and materials makers are prepared to give

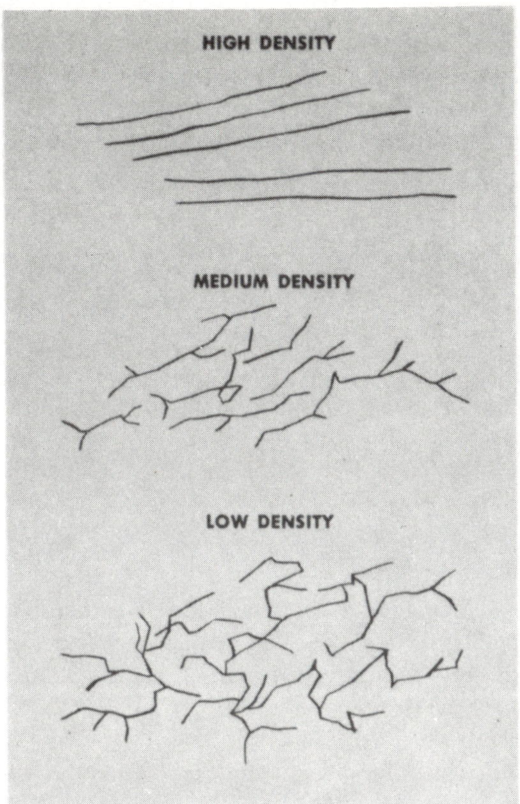

Fig. 3.14 Density depends on molecular shape.

sound advice on proper materials, molding, as well as testing procedures to avoid stress-crack problems, and their guidance must be followed closely to insure against loss. Note from Table 3.2 that the melt index gives a partial guide to stress-crack failure probability, and that a low melt index is the safest material, when molecular weight distribution is correct.

Uncolored polyethylene has poor weathering resistance, and becomes embrittled in the presence of oxygen and sunlight. Carbon black added to the natural polyethylene greatly increases its weatherability. High temperatures soften polyethylene, making it very flexible and tough, but considerably increasing its permeability. Low temperatures cause embrittlement; decrease in permeability and grease resistance are factors of importance to package material selection. Various additives such as antioxidants are used to minimize deterioration at high temperatures.

Polyethylene products are made by injection, blow, extrusion, and rotational molding. Sheets and films are easily vacuum-formed to make blister packages and

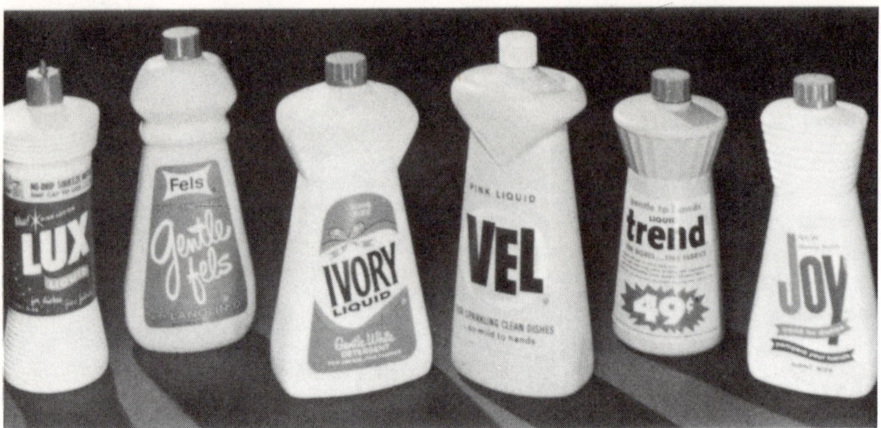

FIG. 3.15 These blow-molded polyethylene bottles are light in weight and unbreakable. They are offered in many shapes and colors.

functional products. Blow-molded bottles provide a major market for polyethylene, as shown in Fig. 3.15.

Ultra-high molecular weight polyethylene materials exhibit phenomenal impact strength and toughness, and are fabricated primarily by extrusion and machining. These materials are the toughest of the plastics, and are doing an unusual job in the textile machinery field.

Polyethylene has excellent chemical resistance and is not attacked by most acids, bases, or salts. It is attacked by strong oxidizing agents, and is affected by most solvents and hydrocarbons. The high density materials are much less affected than low density materials. Many products such as wintergreen, oxygen, carbon dioxide and essential oils pass easily through polyethylene, and prevent its use unless the surface is especially treated with a low permeability coating. The high thermal expansion rate of polyethylene may limit its use without special design compensation in many applications. High density polyethylene is a fine, low-friction material, often used for drawer slides.

The attractive electrical properties of polyethylene make it a desirable cable insulation. It offers high dielectric strength, low power factor, and low dielectric constant. Semi-conducting materials may be formulated for condenser and cable bushings.

The many large markets for polyethylene include pipe, pipe fittings, packaging films surgical implants, coatings, wire and cable insulation, disposable products, beverage cases, bottles and closures.

Many special varieties and copolymers are made to gain desirable properties—one typical product* provides increased clarity, flexibility and flow when com-

*"Zetafin," Dow Chemical Co.

FIG. 3.16 Polyethylene bags and film serve many markets. Seed orchard trees are pollinated artificially with hypodermic needles filled with pollen from selected male parent trees. Polyethylene bags protect female flowers against random, wind-blown pollen. "Super" trees for the future will result.

pared with conventional low-density polyethylene resins. Chlorinated polyethylene* resins are used as modifiers for vinyl and polyethylene sheeting. As plasticizer replacements they give permanence, freedom from volatilization and improved softness.

Another attractive class of materials, identified as ionomers,** offer resilience, glass-like transparency, exceptional low-temperature properties, grease and solvent resistance, and toughness. The term "ionomer" was coined to describe a polymer which contains both organic and inorganic compounds, linked together. The ionomers are easily colored, and will accept more than their own weight of fillers. Processing is done by all of the conventional thermoplastic processes.

*"Surlin," E. I. Du Pont de Nemours & Co.; "Bakelite," Union Carbide.
**"Tyrin," Dow Chemical Co.

The ionomers are particularly adaptable to grease-resistant coatings, for paper and other substrates. They are used also in replacement of polyethylene where its crystal clarity is desirable. Other ionomer applications include safety glasses, drum covers, skin packaging, and coatings.

Ethylene Vinyl Acetate* (EVA)

These special copolymers of polyethylene may be characterized by their extra flexibility, toughness, clarity, and high stress-crack resistance. This is accomplished without plasticizers. EVA resins are ultraviolet and ozone resistant, and their desirable properties are retained at low temperatures. They take fillers readily, and will blend with some other thermoplastics. EVA materials may be used in applications alternately served by rubber and PVC, and the processing is much simpler and faster with no processing degradation problems as experienced with PVC.

Tensile strength is double that of low-density polyethylene, and approaches that of PVC. The softening point is below polyethylene, but similar to PVC, and stress-cracking resistance is superior to both.

Typical applications include molded appliance and automotive parts, garden hose, and vending machine tubing.

Polypropylene** Plastics

The polypropylene resins are polymerized propylene gas, achieved at relatively low temperatures and pressures. These resins were first produced in 1954 by Professor Giulio Natta, at the Polytechnic Institute in Milan. Natta first discovered that it was possible to make polymer chains following a prearranged pattern. This resin is light in weight (0.90 sp. gr.), translucent in its natural state, and readily colored. It surpasses polyethylene in many fields because of its higher thermal operating range, being usable to over 250°F. Some formulations must be used carefully at temperatures below 15°F if subjected to severe impact. It has good surface hardness and fair abrasion resistance. Good dimensional stability is obtained with the unfilled resin, and this property is greatly improved by the addition of glass or other mineral fillers.

Polypropylene does not present stress-cracking problems, and it offers excellent chemical resistance at higher temperatures. It is not affected by water solutions of inorganic salts, mineral acids or bases. It can package concentrated HCL, and

---

*"Alathon," E. I Du Pont de Nemours & Co.; "Ultrathene," USI; "Bakelite," Union Carbide.

**"Avisun," Avisun Corp.; "Catalin," Catalin Corp.; "Chevron," Chevron Chemical Co.; "Dow," Dow Chemical Co.; "El Rex," Rexall Chemical Co.; "Escon," Enjay Chemical Co.; "Plaskon," Allied Chemical Co.; "Marlex," Phillips Petroleum Co.; "Moplen," Chemore Corp.; "Petrothene," U.S. Industrial Chemicals Co.; "Poly Pro," Spencer Chemicals Co.; "Pro-Fax," Hercules Powder Co.; "Shell," Shell Chemical Co.; "Tenite," Eastman Chemical Products Inc. "Calgard," Celanese Plastics Co.

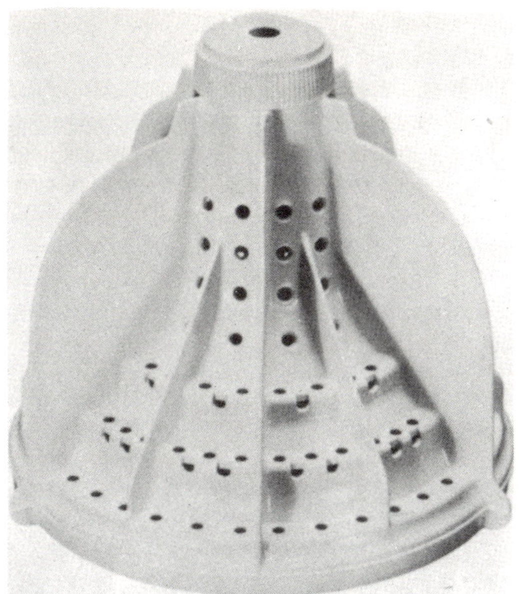

FIG. 3.17 Polypropylene is widely used for housewares and appliances because it will withstand hot water, detergents, mechanical abuse, and it is low in cost. This injection-molded washing machine agitator is a typical heavy-duty application served well by polypropylene. (*Courtesy Hercules Powder Co.*)

80 per cent solution of sulfuric acid up to 140°F. Its permeability properties are as good or better than high-density polyethylene.

Its electrical properties are all good, and its tracking resistance is noteworthy. Products are made from polypropylene by all of the conventional processes for thermoplastics.

The major markets for polypropylene take advantage of its higher use temperature. Typical applications include housewares, as shown in Fig. 3.17, dishes, appliance parts, automotive ducts and trim, washer and dishwasher components, hinges, pipe, monofilaments, bottles, packaging, radio and television cabinets, laboratory ware, rope, nets, and heavy-duty textile products.

Polybutylene* (PB)

Polybutylene has properties of considerable value for engineering applications. It has excellent creep resistance and is very resistant to crazing and stress cracks. The tensile break strength is high for a polyolefin. Polybutylene makes an excellent pipe material because its design stress-rating is 50% above that of

*Ampecet Corp., Crystal-X-Corp., Mobil Chemical Co.

polyethylene. It can be cold flared and joined by methods used for copper tubing. Other applications include industrial films, hot-fill bags and tapes.

Polyallomers*

Polypropylene ethylene polymers are produced as variations in chemical composition without change in crystalline form by a process called allomerism. The polypropylene ethylene polyallomer is similar to high-density polyethylene in many respects. It may be used safely as low as −40°F, and up to 210°F at 66 psi. Thus, it is better than polyethylene, but below polypropylene in thermal endurance. It combines rigidity and impact resistance that facilitates minimum wall section in many products. It is more moldable than high-density polyethylene, and has very high stress-crack resistance. Electrical properties are good, and its "hinging" properties are even better than polyethylene. This can be seen in Fig. 3.18.

FIG. 3.18 New binders made from sheet extruded of polyallomer offer exceptional durability and have the appearance and feel of smooth-grain leather. Hinges, stamped right in the sheet, withstand millions of flexings without breakage. One-piece construction of the binder covers and the inherent toughness of the polyallomer material serve to eliminate dog-earing.

*Eastman Chemical Products Inc.

Products are made by all thermoplastic processes, and in many cases, the procedures are less critical than those required for linear polyethylene and polypropylene. Typical applications include chemical apparatus, typewriter cases, bags, luggage shells, automotive trim, and hinged packages.

POLYAMIDE* PLASTICS

Nylon

Nylon is classed chemically as a polyamide, and is made by a complex chemical process; it is similar to the protein products created by nature in the human body. Its development started when Emil Fisher made a careful analysis of the natural animal protein. W. H. Carothers combined amines and acids to make polyamides in long chains, resulting in the production of nylon. It was first introduced in 1938 as a toothbrush bristle material, and since that time it has replaced natural bristles in many brush applications, thus causing a plastics revolution. In one manufacturing process, nylon material is heated to 450°F and forced through tiny holes to form continuous filaments. These small filaments may be stretched from four to seven times their original length, causing a molecular rearrangement which adds great tensile strength and elasticity. This high-strength property gave immediate rise to its use as a tough hosiery material, and synthetic "silk" fabric.

The molded nylon product materials are noteworthy for their toughness, abrasion resistance, strength, low friction, and heat resistance (250°F). Many special types are made to facilitate its use by injection, blow, and extrusion molding. Nylon materials resist all common organic solvents, except phenol and formic acids. It is slowly attacked by strong mineral acids and oxidizing agents. Nylon is resistant to oils, greases, and electrolytic corrosion. Electrical properties are adequate for power frequency applications where it is useful for mechanical parts. A variety of colors is available including some weather resistant products and translucent materials.

More than one hundred varieties of nylon formulation provide special-purpose products by the copolymerization of various amide-forming ingredients. Material makers have provided well-defined property manuals to facilitate proper material selection. Nylon "6/6," for example, has greater stiffness than nylon "6" and less creep. Nylon "6/10" has minimum water absorption, and is used, in one case, as a tough protective coating over other cable insulations which, in themselves, have better electrical properties. Asbestos, glass, and other fillers, are used to improve dimensional stability, minimize shrinkage, and increase stiffness of

*"Firestone," Firestone Plastics Co.; "Fosta-Nylon," Foster Grant Co.; "Nypel," Nypel Corp.; "Plaskon," "Capron," "Alpha," Allied Chemical Co.; "Zytel," "Minlon," E. I. Du Pont de Nemours & Co., Inc.; "Chemstrand," Chemstrand Co.; "Vydene R," Monsanto Chemical Co.

FIG. 3.19 This machine cuts sheets of resin-impregnated paper, linen, glass, etc. The cast nylon gear drives the large main gear. Cast nylon offers resistance to abrasive fiber glass dust, which coats gear teeth; operation without lubrication and long life are gained by the use of this cast nylon gear.

nylon compounds. A molybdenum disulfide*-filled nylon gains wear resistance, lowers friction, improves dimensional stability, heat resistance, and resistance to load deformation. Nylon is best for impact problem gears, as shown in Fig. 3.19, while acetal resin is the choice for precise small gears.

A new family of type 6 nylons called alpha nylons have a crystalline structure that adds desirable properties and processability when compared with regular nylons. A 20% increase in heat deflection temperature and stiffness is gained without loss of the impact properties exhibited by type 6 nylon. A considerable gain is reported in the molding of thin sections from alpha nylon as a result of the rapid crystallization rate from the melt phase.

Typical nylon applications include mechanical components, gears, cams, bearings, shoe heels, pipe fittings, sheets, rods, tubes, pipe, wire insulation, bottles, film, laminates, fish lines, refrigeration tubing, gasoline lines, gunstocks, reinforced hose, fabrics, and bristles. Nylon is extensively used for thin wall electrical insulation, as shown in Fig. 3.20. Many of the cooking bags are made of nylon.

*"Nylatron G.S.," The Polymer Corporation of Pennsylvania.

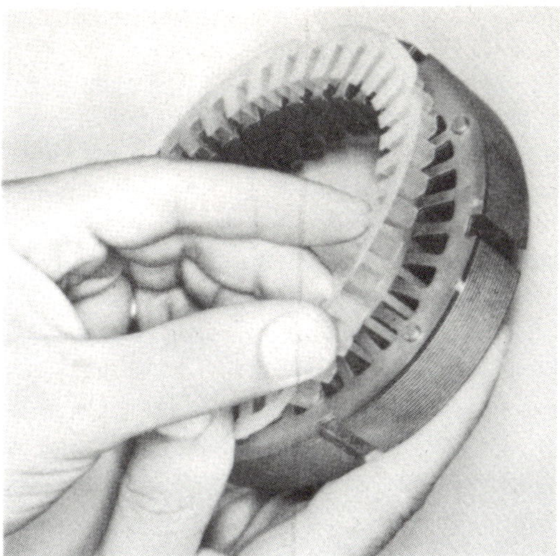

FIG. 3.20 Paper, an inexpensive material, has been replaced by nylon on the alternator which generates electricity for some of Ford Motor Company's automobiles. Although nylon is more expensive than paper in first cost, over-all costs, including assembly, are considerably lower for the nylon insulator. The life and reliability of the alternator are also increased. (*Courtesy Allied Chemical Co.*)

ACETAL* RESINS

The acetal resins are a highly crystalline, stable form of polymerized formaldehyde, and were originally designed to overcome the shortcomings of other thermoplastics, for the replacement of die castings and metal components. The acetal resins were introduced commercially in 1960 by Du Pont. The special properties achieved are rigidity, resilience, toughness, high strength and resistance to common solvents such as gasoline, kerosene, carbon tetrachloride, freon, alcohols and turpentine (below 160°F).

Acetal resins may not be used with strong acids, bases, or oxidizing agents. They are good for underground use. The acetal resins compete in cost with the nylon resins and many metals. Acetal resin has high strength per unit of volume, and low density that facilitates the volume production of light-weight, high-strength, and low-cost products.

Better-than-nylon dimensional stability is present in acetal products because of their low moisture absorption under varying humidity conditions. Rigidity and dimensional stability are greatly improved by glass fillers. The creep resistance of acetal resin is adequate, and perhaps better than many thermoplastics, at 150°F. Good holding power for self-tapping screws is an assembly asset. In the

*"Celcon," Celanese Plastics Co.; "Delrin," E. I. Du Pont de Nemours & Co., Inc.

gear application field, acetal resin is best for precise dimensional small gears, whereas nylon is best for impact problem gears.

A special-bearing material,* composed of acetal resin and TFE fluorocarbon fiber is a very useful unlubricated material for highly loaded bearings.

Acetal resin may be injection, blow, or extrusion molded. It can be heat formed, machined, welded, painted, or plated, and is available in attractive colors. Automatic screw machining processes are adaptable to mass produce acetal parts.

Very large markets for acetal resin are found in plumbing, appliance and automotive industries—a typical product is shown in Fig. 3.21. Typical applications include pumps, pump impellers, valves, valve fittings, gears, cams, hardware, screws, pipe fittings, bottles, instrument components and housings, fans, cranks, paint-sprayers, shower heads, ball retainer rings, tool handles, and dishes.

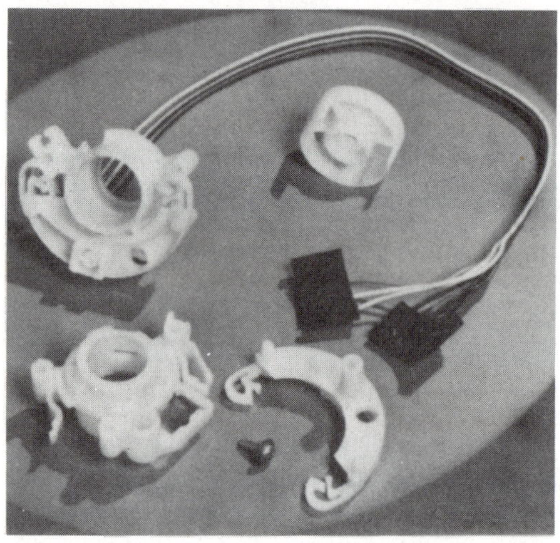

FIG. 3.21 Turn signal switch assembly of acetal plastics combines the mechanical switch with the electrical switch, eliminating a mechanical cable which previously connected them. The two complex moldings in the foreground, each incorporating two integral springs, replace several die cast zinc parts and three steel springs used in the former switch. (*Courtesy E. I. Du Pont de Nemours & Co.*)

POLYCARBONATE** PLASTICS

Polycarbonate is a most important technical thermoplastic because of its excellent heat resistance, outstanding impact strength, and good dimensional stability. Acetal resin and polycarbonate both resulted from research on pure formaldehyde. Bisphenol A, which was produced for the epoxy resins, opened the door

*"Delrin A/F," E. I. Du Pont de Nemours & Co., Inc.
**"Lexan," General Electric Co.; "Merlon," Mobay Chemical Co.

for acetal resin and polycarbonate in 1959, and Bayer, in West Germany, was first to produce and market polycarbonate, following the work done there by H. Schnell. Polycarbonate was the first of the rigid thermoplastics to offer good temperature stability up to (270°F), and impact strength.

Chemically, polycarbonate is characterized by being stable to water and to mineral or organic acids. It is insoluble in aliphatic alcohol, partially soluble in aromatic hydrocarbons. It is soluble in chlorinated hydrocarbons, and decomposes in strongly alkaline substances. This resin can be fabricated by more methods than most other plastics, including all molding methods, thermoforming, machining, hot or cold staking, and fluidized bed coating. Polycarbonate is available in many colors, as well as in transparent formulations.

Electrically, polycarbonate is quite satisfactory for most applications within its temperature range. These properties are not affected by humidity. It has high insulation resistance, low power factor, and a stable dielectric constant despite temperature or frequency changes. Excellent weathering and corona resistance is also present, indicating its potential use in some high-voltage high-altitude applications. It has satisfactory arc resistance for low-current arcs, and its tracking resistance is similar to that of the general-purpose phenolics; it is self-extinguishing. Volume resistivity is little affected by moisture or heat aging, and is good up to

FIG. 3.22 This polycarbonate coffee brewer will not crack, blister or soften. Polycarbonate resin neither stains, tarnishes, pits or corrodes and it will not add any taste or odor. (*Courtesy Chemicals Dept., General Electric Co.*)

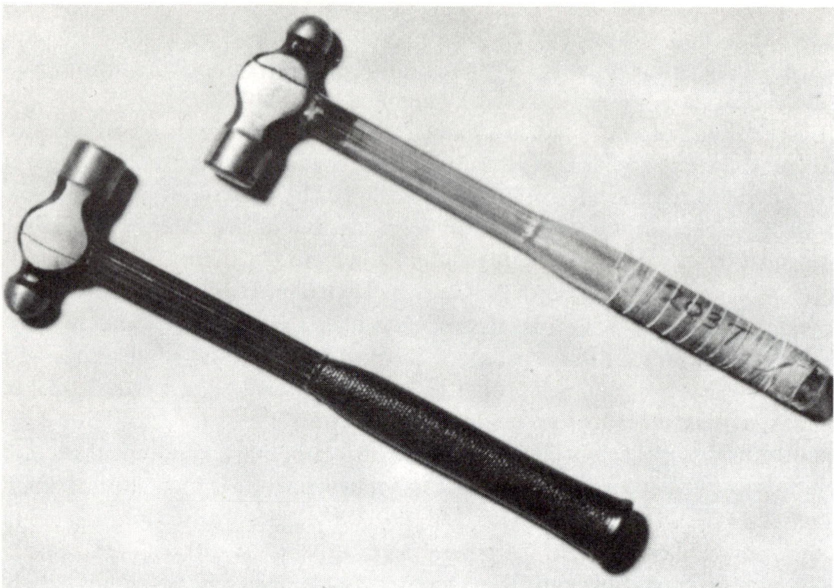

FIG. 3.23 Produced by the blow-molding process, the hollow handles of polycarbonate are anchored by forcing a solid wedge of injection-molded polycarbonate through the head opening and into the top of the handle under 1000 psi pressure. The high impact resistance of this combination was proven during destruction tests which finally split the steel head while the plug and handle of polycarbonate were unharmed. (*Courtesy Mobay Chemical Co.*)

250°F; dielectric constant and power factor are stable at power frequencies. The power factor rises fast above one kilocycle up to ten megacycles. Dielectric strength is comparable with other thermoplastics.

Polycarbonate reinforced with glass gains tremendously in tensile and flexural modulus. The heat deflection temperature is raised to 300°F, and mold shrinkage is reduced to .001 in./in. Impact strength is considerably reduced, and the molding cycle is speeded up considerably by the better thermal conductivity.

Typical applications include street-light globes to gain the advantage of heat endurance and very high impact strength, centrifuge bottles, tubes and tool housings, circuit cards, high-temperature lenses, appliance components, and hot dish handles. Shown in Figs. 3.22 and 3.23 are a blow-molded polycarbonate hammer handle, and a coffee urn.

PHENOXY PLASTICS*

The phenoxy family of compounds was introduced commercially by Union Carbide in 1962 for injection, blow and extrusion molding, and for coatings and

*"Phenoxy," Union Carbide Corp.

adhesives. Derived from epichlorohydrin and bisphenol-A, phenoxy has a good dimensional reproducibility and low shrinkage of 0.003 in./in., which is very low for any unfilled thermoplastic compound. Its creep resistance is also superior. It offers low oxygen and CO_2 permeability, crystal clarity, and because of its excellent impermeability properties, makes a better bottle material for some applications. Self-extinguishing and unusually high impact varieties are available. Phenoxy may be filled with glass or other mineral fillers to gain ever greater stability and less shrinkage. Phenoxy adhesives are excellent pipe-bonding compounds, and serve well in metal to wood laminates, as well as being an excellent bonding medium for many other metal and plastic films.

POLYPHENYLENE OXIDE PLASTICS* (PPO)

Polyphenylene oxide plastics are a family of materials that have a useful temperature range from −275°F to 375°F. They are highly resistant to hydrolytic attack, and provide excellent acid and base resistance. Repeated steam autoclaving does not deteriorate their properties. Polyphenylene oxide is nontoxic, and has very low moisture absorption. Its creep resistance is good, exhibiting only 0.75% creep after 300 hours under a 3000 psi load. Polyphenylene oxide is non-drip, self-extinguishing, poor in arc resistance, but excellent in the dissipation

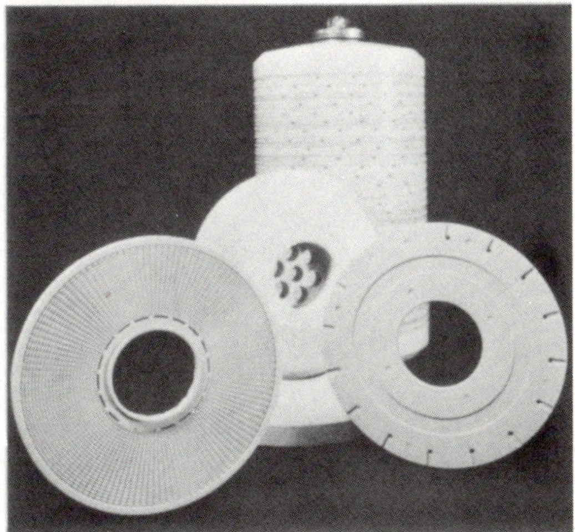

FIG. 3.24 Polyphenylene oxide thermoplastics provides autoclavable parts and chemical resistance to this micro-membrane filter that is used in contamination control. The two-piece membrane support structure shown in the foreground and the filter end plates in the center are made of polyphenyline oxide. The completely assembled unit is shown in the rear. (*Courtesy Chemical Dept., General Electric Co.*)

*Noryl®, General Electric Co.

factor where it serves nicely as a high temperature-high frequency coil form material. The low specific gravity (1.06) is substantially below that of many other technical thermoplastics. Polyphenylene oxide plastics are extruded, injection molded, thermoformed, and machined. A platable variety of PPO is available.

Applications include autoclavable surgical tools, coil forms, pump housings, valves, pipe, pipe fittings, and numerous sterilizable products. Depicted in Fig. 3.24 is a typical polyphenylene oxide application.

POLYSULFONE*

The polysulfone plastics were first offered in 1965 as an extrusion and injection molding material for high temperature and minimum creep applications. Polysulfone is one of the outstanding thermoplastics for use as a load-bearing, structural material for long-term use at temperatures up to 300°F. Chemically, polysulfone is a linkage of isopropylidene, ether, and sulfone, and it has noteworthy thermal and oxidation resistance for a thermoplastic material.

Polysulfone is resistant to mineral acids, alkalis, and salt solutions. Under moderate stress at elevated temperatures, it is resistant to most detergents, oils and alcohols. It is damaged by ketones, and chlorinated and aromatic hydrocarbons as well.

Its desirable electrical properties are comparable with polycarbonate, and its arc resistance is better than that of many thermosetting compounds. It is self-extinguishing, and contains no stabilizers or plasticizers. It is stable in its natural transparent resin form. Polysulfone plastics are available in colors, as well. Its wide molding temperature latitude, 600 to 750°F, enhances its moldability and,

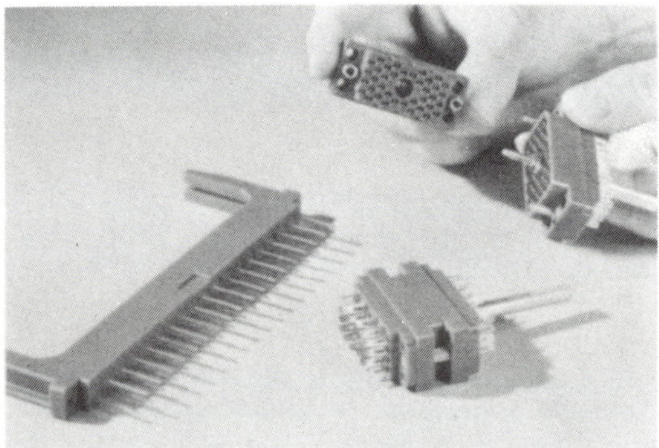

FIG. 3.25 Subminiature printed-circuit card-edge connector, at left, and rack-and-panel cable connector, at right, are precision-molded from polysulfone to gain high temperature and strength in very thin sections. (*Courtesy Union Carbide Corp.*)

*"Bakelite 1700," Union Carbide Corp.

since it does not degrade by continuous recycling, all the scrap can be used without stress-cracking or subsequent crazing.

Polysulfone is competitive with many basic thermosetting compounds, since it includes some of the useful properties that heretofore have been present only in thermosets—it also possesses the important advantage of low-cost injection molding.

Applications include hot-water pipes, under-hood automobile components, lenses, connectors, appliance housings, iron handles, dishwasher impellers, switches, circuit breakers, electronics hardware, autoclavable equipment, and wiring devices. Shown in Fig. 3.25 is a complex printed-circuit card connector, molded of polysulfone.

POLYPHENYLENE SULFIDE (PPS)*

Polyphenylene sulfide is another of the new engineering thermoplastics having high heat deflection temperature characteristics as well as generally high physical properties through a wide temperature range. It was developed by Dow Chemical Co. The resin has exceptional chemical resistance and is non-inflammable. There is no known solvent for this material below 375–400°F. Applications for polyphenylene sulfide have been in the coatings for metals to impart corrosion and non-stick resistance. It can be injection molded using reciprocating screw machines. Glass fibers are used as fillers for best physical properties.

URETHANE** OR ISOCYANATE PLASTICS

These versatile materials, introduced in 1955, are made in many forms, ranging from flexible and rigid foams to rigid solids, and highly abrasion-resistant products. Some of these can be seen in Table 3.4. They are formed by the reaction of isocyanates with hydroxyl-bearing compounds. Urethane plastics are extruded and molded.

The flexible foams are used for cushions, arm rests, upholstering, automobile seats, novelties, toys, etc. Rigid urethane foams have excellent thermal insulation properties, and serve a wide variety of such markets. A urethane jacket is used to insulate one of the moon rocket's fuel tanks, where the fuel must be kept at −423°F.

As coating resins, the urethanes provide an exceptional degree of toughness, hardness, and mar resistance, plus flexibility and chemical resistance.

Urethane elastomers, and rigid products produced by the coating, casting, and the injection molding process, are very important plastics materials because they combine metal durability and the functional properties of rubber, plus best

---

*"Ryton," and PPS, Dow Chemical Co.
**Urethanes are produced by many people.

TABLE 3.4   RELATIVE ABRASION RESISTANCE

Material	Index
Polyurethane	3
Polyester film	18
Nylon/11	24
High-density polyethylene	29
PTFE	42
Nitrile rubber	44
Nylon/101	49
Low-density polyethylene	70
High-impact PVC	122
Plasticized PVC	187
Butyl rubber	205
ABS	275
Polystyrene	325
Nylon/6	366

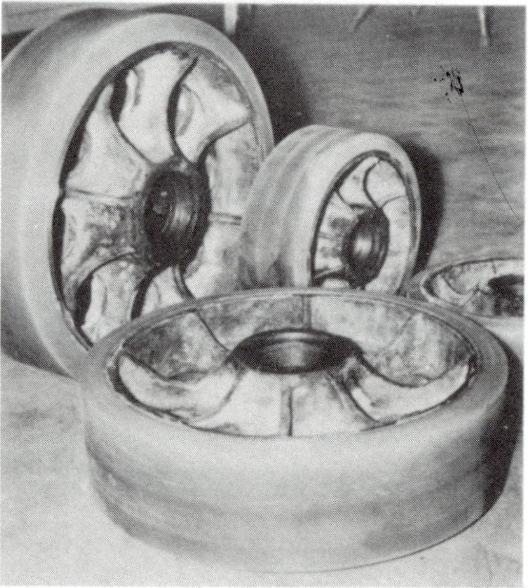

FIG. 3.26 These heavy-duty industrial wheels feature a tread cast of a urethane elastomer. Virtually wear-proof, the new material can be compounded and is pourable at room temperature without special equipment. Tread hardness may be varied from 55 to 95 Shore "A" scale simply by adjusting the amount of curing agent. Physical characteristics include high-impact resistance, high load-bearing capacity, improved resistance to elevated temperatures and excellent resistance to oils, gasolines and other hydrocarbons, dilute acids and bases and most all chemicals. (*Courtesy CPR Div., The Upjohn Co.*)

FIG. 3.27 Rigid urethane foam's *k* factor of .12 is almost twice as good as the "*k*" for the next two most commonly used siding insulating materials, styrene foam and fiberboard. In addition, urethane foam bonds permanently, is fire-resistant and lightweight, adds anti-dent strength to the aluminum panels, will not rot or mildew, is vermin-proof and muffles noise transmission. It also has good solvent resistance and low moisture permeability to minimize condensation problems between the siding and interior walls.

abrasion resistance, as listed in Table 3.4. Their desirable properties include tensile and tear strength, abrasion, fuel, ozone, oxygen and weather resistance, high elasticity and resilience, load-bearing capacity, impact resistance, vibration dampening properties, as well as hardness. Typical products include gears, bushings, shock mounts, pulleys, "O"-rings, caster wheels as shown in Fig. 3.26, industrial truck tires, cable jackets, pump impellers, drive belts, roller covers, threads, flooring tapes, and synthetic leather as well. Urethane foam-backed siding is shown in Fig. 3.27.

POLYESTER-REINFORCED URETHANE*

The synthetic leather "Corfam" was discontinued by Du Pont because of shoe marketing economics. It had porosity or breathability that made it a comfortable, long-wearing shoe material.

*"Corfam," E. I. Du Pont de Nemours & Co., Inc.

This poromeric material, with urethane impregnation or with a silicone coating, designed for heavy-duty leather applications, finds uses as an industrial material. Urethane-impregnated poromeric materials withstand heavy and continuous pressures without being reshaped or extruded. These materials make excellent gaskets, packing cups and seals, for use in the oil, aircraft, and missile industries, as well as for other premium performance applications. The silicone-coated poromeric material is also excellent for shaft seals.

These materials may be compression molded over a wide temperature band. The operation temperature range is from $-10°F$ to $300°F$.

POLYCHLOROTRIFLUOROETHYLENE (CTFE)*

This fluorocarbon resin may be molded by all the conventional techniques. Products made from CTFE are tough, have good dimensional stability and fine electrical insulation properties that are stable over a broad temperature and humidity range. Chemical resistance is excellent, and these products are inert to all common reagents. Strong alkalies, acids, organic solvents, and violent oxidizing agents do not affect CTFE over a broad temperature range. Impermeability to air, water, essential oils, etc., is good, and their weathering and ultraviolet resistance is superior.

Applications for CTFE include flexible cable jacketing, hook-up wire insulation, coil forms, tube sockets, terminal insulation, chemical ware, lining materials for handling chemicals, pipe, pipe lining, "O"-rings, as well as pharmaceutical and lubricant packaging, an example of which is shown in Fig. 3.28.

FLUOROCARBON (TFE)†

Fluorocarbon is a very reactive material, and difficult to contain. In 1930 it was first noted in Germany, and in 1938 it was isolated in the United States by Dr. J. R. Plunkett. Wartime studies on fluorine chemistry covered tetrafluoroethylene, and its amazing chemical resistance was first observed. It is truly a unique and highly sophisticated plastics material that will not melt under a soldering iron— none of the usual chemicals will react with it. TFE is completely resistant to moisture and ultraviolet; adhesives will not stick to it. TFE has a combination of properties unavailable in any other single material. It serves some of the most complex problems in missile construction, and is widely used as a non-stick surface for frying pans. Its static and dynamic friction are low, and there is no

*"Kel-F," Minnesota Mining & Manufacturing Co.; "Plaskon CTFE," Allied Chemical Co.;
†"Teflon TFE," E. I. Du Pont de Nemours & Co., Inc.; "Halon," PTFE Allied Chemical Co.; "Fluon," "TFE," ICI America.

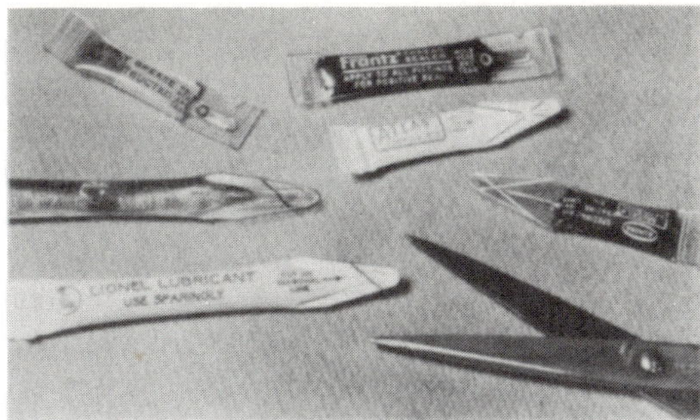

FIG. 3.28 Sealed tubes of CTFE are widely accepted for the packaging of difficult pharmaceuticals and oils. (*Courtesy Unette Corp.*)

start or stop energy required on sliding surfaces, since the static is less than the dynamic friction.

A variety of formulations is available to facilitate the various production procedures. Three grades are offered as general-purpose molding powders. A special, fine powder is available for void-free applications. Two grades are dispersed in naphtha to facilitate extrusion of thin wall tubing, wire insulation, and tapes. A fabric and metal coating is prepared in an aqueous dispersion.

Dielectric constant and loss factor remain unchanged at all frequencies up to 10,000 mc, and there is little change at elevated temperatures. Dielectric is high at 600 V/mil, and the continuous service temperature range is $-400°F$ to $500°F$, going above all other plastics except polyimides, glass-bonded mica and cold-molded products.

The mechanical properties are very ordinary, since it is a non-rigid material and has a very high expansion rate. Molding can be done only by a sintering process following preforming. Extrusion, with or without a naphtha aid, is used for wire coating and small sections. Many TFE products are formed by machining from rods or slabs. In all instances, care must be exercised to keep its temperature below $650°F$ to prevent release of an obnoxious poisonous gas.

Typical applications include high-temperature wire and cable insulation, flexible circuits, motor lead insulation, heating cables, oil well cable, chemical process apparatus and laboratory ware, food and drug handling equipment that must not be lubricated, heavy-duty slow speed applications where oil would be squeezed out, "greasing" of the commercial baking pans, frying pans, waffle irons, rolling pins, as well as for roll coating. A sheet material with two TFE-foamed surfaces is available for gaskets.

Micron-sized TFE particles are also used as fillers in thermoplastics and thermosetting compounds to produce low-friction functional products. Acetal resin, filled with TFE fibers, provides improved thermal and frictional characteristics.

FLUORINATED ETHYLENE PROPYLENE (FEP)*

This fluorocarbon formulation is available for injection molding, blow molding, extrusion and other conventional production procedures. Autoclavable laboratory ware and bottles are molded of FEP, as shown in Fig. 3.29. Its electrical properties are equal to TFE, but it has slightly less chemical resistance. Its thermal range is $-400°F$ to $400°F$.

Glass fillers may be used with the fluorocarbons to gain dimensional stability, reduce thermal expansion, and gain hardness and stiffness. Other fillers that have been used include asbestos, mica, bronze, aluminum, molybdenum, disulfide and berylia.

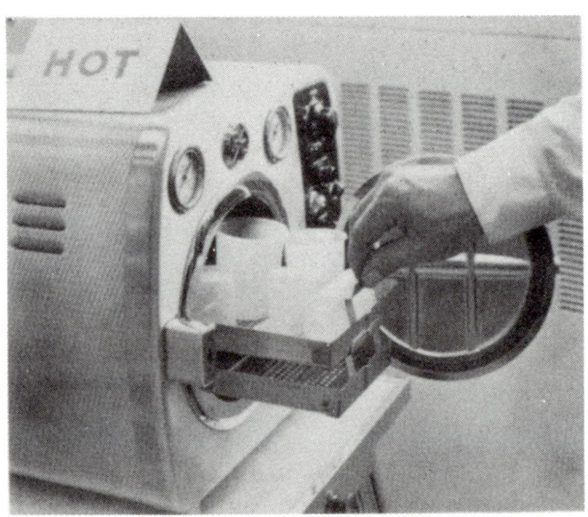

FIG. 3.29 These FEP laboratory-ware products may be autoclaved repeatedly and are highly chemical-resistant. (*Courtesy The Nalge Co.*)

ETHYLENE-TETRAFLUORETHYLENE COPOLYMER (ETFE)**

New members of the fluorocarbon family are the copolymers of TFE and CTFE. These materials are characterized by high temperature resistance, good dielectric and mechanical properties, excellent chemical resistance, easy processability and high impact strength. Their thermal range is below that of FEP, but is in the desirable range of $150°$ to $350°F$. These copolymers are available in pellet form for blow and injection molding and for extrusion. A distinct advantage lies in their relative ease of processing—when compared with the other fluoropolymers. Laboratory ware for the chemical industry, coil bobbins, wire and cable insulation and tubing are typical applications for this material. Glass fiber reinforcement may be used to enhance significantly the physical properties such as stiffness,

*"Teflon FEP," E. I. Du Pont de Nemours & Co., Inc.
**"Tefzel," E. I. Du Pont de Nemours & Co., Inc.

tensile strength, creep resistance and hardness. The coefficient of linear expansion is reduced sharply and the heat distortion temperature is increased greatly.

POLYETHYLENE CHLOROTRIFLUOROETHYLENE (ECTFE)*

ECTFE is a high molecular weight copolymer for conventional thermoplastic molding. It exhibits good impact strength and may be used at temperatures up to 350°F. With radiation treatment, it may be used up to 400°F. ECTFE is particularly useful as an insulator for high performance wire insulation and as a protective coating for metals.

PERFLUOROALKOXY TEFLON (PFA)**

This fluorocarbon resin is melt extrudable and provides high thermal performance and easy fabrication. It has better flex life and stress crack resistance making it useful as a liner for pumps, tanks, bellows, flexible tubing, etc.

POLYVINYLIDENE FLUORIDE*** CTFE (VF$_2$)

Properties of special interest in the vinylidene fluoride plastics are strength, mechanical toughness, heat stability (-80°F to 300°F), resistance to most solvents, chemicals, and ultraviolet. It also demonstrates extremely good weatherability, gamma radiation resistance, non-drip self-extinguishing characteristics, and very low permeability. It has exceptionally good electrical properties, and is available in colors as well as transparent forms. Noteworthy is the fact that no oxidative or thermal degradation occurs during continuous exposure to 300°F for one year. The "30"-year paints are based on its exceptional weather and ultraviolet resistance. Vinylidene fluoride materials may be molded by all processes, and processed by fluidized bed coatings, spray, dip, roller, slush coating and casting.

Typical applications include high-temperature valve seats, lip seals, chemical-resistant pipe and fittings, coated and lined vessels, agitators and tanks, monofilaments, coil bobbins, capacitor film, ribbon cable, protective films, and wire insulation.

CHLORINATED POLYETHER[†]

Chlorinated polyether is a highly chemical-resistant engineering material. Approximately 46% of this polymer is chlorine, and it is a highly crystalline material

---

*"Halar," Allied Chemical Co.; "Durasan film," Allied Chemical Co.
**"Teflon PFA," E. I. Du Pont de Nemours & Co., Inc.
***"Kynar," Pennsalt Chemicals Corp.; "Tetran," Pennsalt Chemical Co.; "Whitcon," Pennsalt Chemical Co.; "Plaskon CTFE," Allied Chemical Co.
†"Penton," Hercules, Inc.

which may be molded, strain free, to very close tolerances with a high degree of dimensional stability. It is one of the outstanding corrosion-resistant materials, and it has answered many of the most severe problems in the chemical equipment field. With a heat distortion temperature at 310°F at 264 psi, it retains its tensile and other strength properties for hot processing equipment. Electrical properties are above average, and quite stable under adverse conditions. Weatherability is rated as good.

Chlorinated polyether plastics are produced with and without fillers, demonstrating the usual increases in dimensional stability and improved linear expansion characteristics. Products may be made by the injection, compression, transfer or extrusion molding processes. It is very machinable and is extensively used as a fluidized bed-coating material. (Chapter 4) Water-suspension coatings may be sprayed on, and dried and hardened by subsequent fusing. In another process, the powder is sprayed on preheated metal parts, where it fuses and hardens.

Typical applications include bearing retainers, tanks and tank linings, pipe and metal pipelining, water meter components, valves, laboratory equipment, and chemical handling products, as shown in Fig. 3.30.

FIG. 3.30 Metering device fabricated from chlorinated polyether is used to dispense accurately all chemicals, detergents, and bactericides at dairies and milk processing plants. It has an orifice held to a tolerance of one-half thousandth of an inch. The metering jet is also used in hospitals, restaurants, and other public accommodations where sanitation and cleaning with corrosive chemicals are desirable and necessary. (*Courtesy Hercules Powder Co.*)

THERMOPLASTIC POLYESTER*

The thermoplastic polyester materials are injection moldable and they are used for many metal replacement applications in the appliance, electrical, hardware, chemical processing and transportation industries. These resins are used with or without fillers. See Chapter 2. Typical parts are shown in Fig. 3.31.

FIG. 3.31 Intricate part design, ease of moldability and a self-extinguishing feature, are several of the desirable factors found in new unreinforced thermoplastic polyester. (*Courtesy General Electric Co.*)

POLYMIDE (THERMOPLASTIC)**

A polyimide has been developed for injection molding purposes to provide a moldable material that is stable up to 500°F. It is used in some premium electrical, chemical and thermal environments.

---

*"Valox," General Electric Co.; "Celanex," Celanese Plastics Co.; "Tenite," Eastman Chemical Products, Inc.

**XPI-MC154 Polyimide, American Cyanamid Co.; Polyimide 2080, The Upjohn Company.

METHYLPENTENE POLYMERS (TPX)*

This thermoplastic is based on 4-Methyl-pentene-I, a monomer derived from oil cracking. With a low specific gravity of 0.83, 90% optical transmission, high temperature resistance (354°F), and excellent electrical properties, these polymers serve electrical and mechanical applications. TPX polymers are processed in conventional thermoplastics molding equipment.

POLYTEREPHTHALATE** (PTMT)

Polyterephthalate is a highly crystalline, white opaque polymer. In its form for injection molding, it has a balance of physical properties comparable to acetal and nylon resins. It is available in pellet form for injection and extrusion molding. As an engineering thermoplastic it has found use in products such as gears, bearings, furniture glides, automotive parts, machine housings, etc.

PTMT has good physical properties, chemical and abrasion resistance. The addition of glass fiber fillers results in desirable improvement of thermal and physical properties. It has a low heat deflection temperature under high loads without fillers.

Another member of this family is Poly-1.4-cyclohexylenedimethylene teraphthalate/isophthalate copolymer which is a colorless transparent material, that can be molded or extruded. Molded products are found in the hardware, automotive, appliance and consumer goods fields. Extruded sheets have excellent thermoforming characteristics and are used for blister packaging and in areas where deep draws are essential. Mold shrinkage is low—0.002 in./in.—compared with the parent material (above) which is 0.020 in./in.

POLYARYL SULFONE***

Polyaryl sulfone has the highest heat deflection temperature of the presently available engineering organic thermoplastics, with a value of 525°F at 264 psi. At room temperature, this material has a tensile strength of 13,000 psi and a flexural modulus of 395,000 psi. At 500°F these same values are remarkably in excess of 4000 psi and 250,000 psi—all this without benefit of reinforcing agents such as glass fiber or other fillers. Electrical values are also retained at elevated temperatures. Polyaryl sulfone is rated as self-extinguishing. It can be injection molded in reciprocating screw injection machines.

Applications for this premium material lie in the areas where its high temperature and arc resistance properties can be used to maximum advantage. Some of

*"TPX," Imperial Chemical Industries Ltd., ICI/Organics, Inc.
**"Tenite Polyterephthalate PTMT," Eastman Chemical Products, Inc.
***"Astrel," 3M Company.

these applications have been electrical connectors, coil bobbins, sockets, spacers, arc barriers, etc.

POLYARYL ETHER*

Polyaryl ether is another new engineering plastics that is characterized by a fairly high heat deflection temperature, 300°F at 264 psi. Other physical properties are also well retained at temperatures through 200°F. Impact strength is quite respectable at ambient temperatures, with a value of 8.0 ft lt/in. notch.

Polyaryl ether can be extruded or injection molded. The high flow properties make it an easy molding material especially for thin or intricate product sections. It is now being used for appliance components, electronic parts, housings, etc.

IONOMER PLASTICS**

Polymers composed of ionized carboxyl groups are called ionomers. They are processed like the conventional polymers. Principal usage is for film lamination and coextrusion for composite food packaging. It is limited to applications below 170°F.

THERMOPLASTIC RUBBERS

Thermoplastic "rubbers" (TPR) are being processed in conventional injection, extrusion, calendering and blow molding machines. Many are polyolefin-based "rubbers." Historically, the processors of rubber have not been included as part of the plastics industry. With the advent of the TPR compounds, the plastics "rubbers" have become another basic material for the processors of thermoplastics and natural rubbers.

Considerable nomenclature confusion exists at present in this transition period as can be seen by the materials makers definitions:

Material Maker	Trade Name	Generic Terminology
B. F. Goodrich Co.	Estanne	Thermoplastic polyurethane
Shell Chemical Co.	Kraton	Thermoplastic rubber
du Pont	Hytrel	Polyester elastomer
Mobay Chemical Co.	Texin	Urethane elastomer
Uniroyal	TPR	Thermoplastic rubber

Typical applications (Fig. 3.32) include: auto body components, gears, gaskets, seals, valves, low pressure tires, adhesives, sealants, coatings, housewares, tubing, belting, shoes, toys, sporting goods, laminates, sheeting, etc.

*"Arylon," Uniroyal, Inc.
**"Surlin," E. I. Du Pont de Nemours & Co., Inc., "Bakelite," Union Carbide Corp.

FIG. 3.32 Thermoplastic rubbers were injection molded for these tires. (*Courtesy Revell, Inc., Venice, Calif.*)

Many other synthetic rubber products are fabricated by the conventional rubber processing techniques.

ELASTOPLASTIC MATERIAL SPECIFICATIONS

A general materials specification has been prepared by the joint committee of the Society of Automotive Engineers (SAE), and the American Society for Testing and Materials (ASTM) to systemize and simplify rubber specification. It is titled "Specifications for Rubber and Synthetic Rubber Compounds for Automotive and Aeronautical Applications, and published by ASTM as Standard D735. The same classification system appears in the SAE

Handbook in Section IV, "Fabricated Materials," as SAE Standard J14. A classification based on the SAE.ASTM system also is issued by the Ordnance Department as U.S. Army Specification MIL-R-3065 and MIL-Std-417.

Table 2 in the Appendix lists the properties of the conventional synthetic rubbers, thermoplastics rubbers and the thermoplastics.

ARC EXTINGUISHING PLASTICS*

This thermoplastic offers a new answer to the continuing problem of extinguishing high voltage, high current arcs. The plasma in a high current, high voltage arc is intensely hot, generates high pressure, and is partly ionized. Bone fiber, phosphoric-acid-asbestos combinations, and glass-bonded mica are all used to minimize the damage resulting from high temperature arcs. This thermoplastic gives off an arc-extinguishing gas under flash conditions, and serves a limited group of applications. It can be injection or compression molded and extruded. Present applications include fuse tubing, lightning arrestors, circuit breakers, arc chutes, and panel boards.

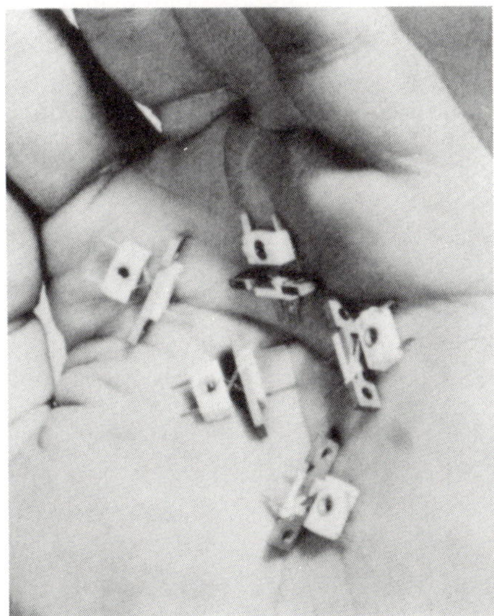
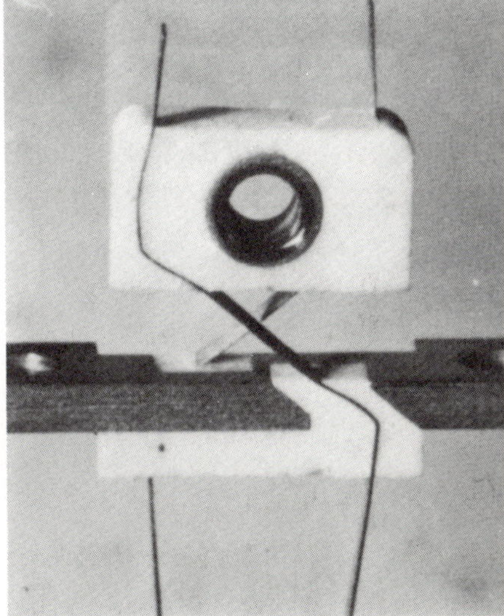

FIG. 3.33 These tiny and complex parts are used in a guidance gyro. The steel bar, beryllium copper springs and the steel insert are all molded complete in glass-bonded mica to gain absolute dimensional stability and insert tightness over the entire temperature range $-273 +750°$F. (*Courtesy Mykroy Ceramics Co.*)

*"Exarc II," Dow Chemical Co.

GLASS-BONDED MICA*

Glass-bonded mica bridges the gap between the organic plastics and the ceramics, and is often called a ceramoplastic. It is moldable with inserts like the organic plastics, and has the absolute, total dimensional stability of the ceramics, plus a useful temperature potential of $-300°F$ to $1300°F$. Other unique values are: impervious to moisture, totally resistant to radiation, does not outgas in a vacuum, is non-magnetic, is completely arc resistant and non-tracking, and its low loss factor is constant throughout great variations in temperature, pressure and frequency. It has high dielectric strength, and has a low thermal expansion coefficient that matches stainless steel. It may be used in SF_6.

Glass-bonded mica makes use of an electrical-grade glass frit as the binder, and synthetic mica flakes as the filler. Proper proportions of this compound are preformed and, after preheating to $1250°F$, the heated charge is injection molded in a transfer-type mold, heated to $500°F$. Pieces are then ejected and annealed by a slow cooling process. Sheet and rod materials are also produced which may be fabricated with much less difficulty than the ceramics. True hermetic seals may be achieved during molding. A tiny flexible component is shown in Fig. 3.33.

Typical applications include arc chutes, radiation generation equipment, vacuum tube components, memory planes, switches, connectors, hermetically sealed relay components, thermocouples, heater element connectors, electronics components, hermetic seals, and squibs and substrates.

*"Havalex," Hercules Corp.; "Mykroy," Mykroy Ceramics Co. Div. Synthane Taylor Co.; "Mycalex," Mycalex Corporation of America.

4 EXTRUSION AND INJECTION MOLDING

Extrusion molding is used to make continuous shapes such as sheets, tubes and rods. It is also used to make profile shapes as shown in Fig. 4.1. Extrusion is used to form the parison for blow molding and to plasticize materials for the injection molding process. Extrusion coating is employed to add a thin layer of plastics on a substrate such as paper. The extrusion process generally makes use of a screw that advances the material from the point of entry through a melting zone, and then forces it out through an orifice that defines the shape of the extrudate. The extrusion screw provides heat of friction in the material, and supplies the force to mold the desired shape in the die with adequate pressure to obtain the essential product density.

All compounds may require drying, blending, coloring, etc., as later described in this chapter for injection molding materials. Proper conditioning of the compound as prescribed by the materials maker is essential to quality work.

The shape of the extrudate may be considerably different than the die orifice. Final product shape is affected by die temperature distribution, material flow properties, die land length, extrusion pressure, extrusion rate, sizing dies, and external cooling systems. The extrudate may be cooled by immersion in a liquid or by air, with or without forced circulation. Tubing and profile extrudates are sized by the use of extended cooling mandrels, pressure sizing, sizing plates, and vacuum sizing.

Extended cooling mandrel procedures are used to control internal diameters

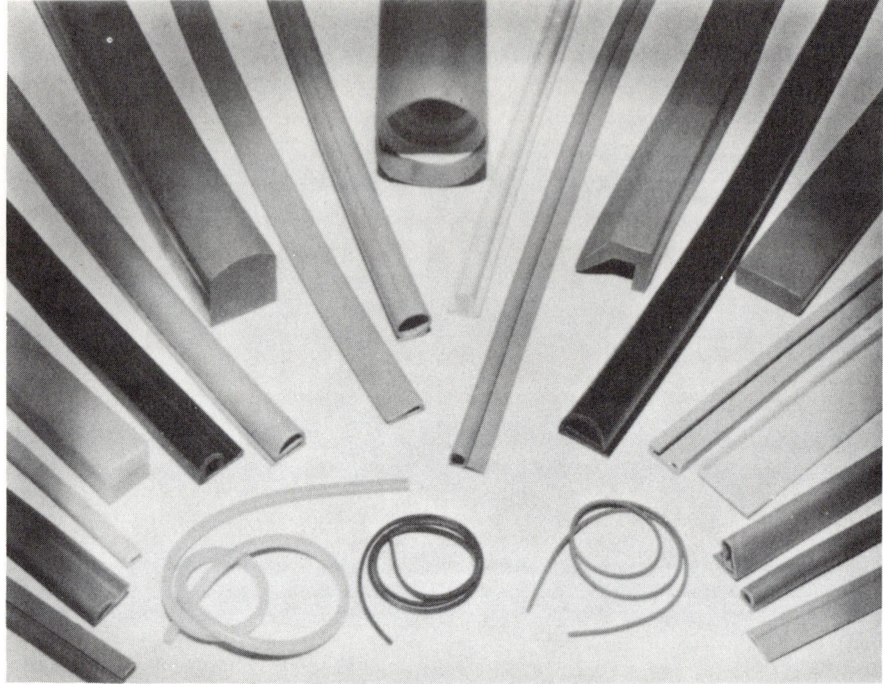

FIG. 4.1 Profile shapes extruded of vinyl plastics are typical industrial products.

by drawing the extrudate over a cooled internal mandrel while the exterior is also cooled. This "freezes" the internal dimension to that of the mandrel.

Pressure sizing is done with internal air pressure that forces the exterior of the extrudate against an external sizing sleeve.

Sizing plates are used to control the external size of flexible pipe while a cooling water spray hardens the surface to prevent dimensional change after the pipe is submerged in cooling water.

In the vacuum sizing system, the vacuum is used to expand the extrudate into a contact fit with a sizing sleeve.

Sheets are cooled and sized by open or forced air systems. Air is blown across, above, and below the extruded sheet, or released through well placed manifolds. Water cooling by immersion is effective but adds cleaning and drying problems. Air rings are often used for the air cooling of blown film. Sheets may be oriented by a stretching operation as they emerge from the extrusion die. Styrene film, which otherwise would be very brittle, may be stretched lengthwise and across the width as it comes out of the die, still in the plastic state. This orientation process greatly increases the strength and flexibility of the extrudate.

EXTRUDERS

The thermoplastics screw extruder developed from the rubber extruder, designed by Bewley in 1845. The extruder is a machine whose sole purpose is to deliver homogenous molten plastics to a die or crosshead. The basic screw extruder consists of a drive mechanism and a screw contained within a hardened steel barrel. When a plastics granulate is fed to the rotating screw, the material moves forward and, due to shearing of the plastics, frictional heat is generated to melt the material. To aid in this transition the outer surface of the barrel is fitted with electrical heaters whose temperature is controlled by thermocouples. The screw acts as a pump and also generates pressure on the melt. Figure 4.2 shows the components of a single screw extruder, and a modern extruder is depicted in Fig. 4.3.

Extruder Screws

Heat to melt the material is provided by heat through the extruder barrel, and by the shearing of the plastics in the screw flights. Ordinarily barrel heat is required to start the process, but with a well designed screw, it is used only to maintain a fixed temperature once the process is underway. In other words, the plasticizing work on the material is primarily accomplished by the screw.

A typical metering screw is shown in Fig. 4.4. The design consists of a rather long feed section with a half turn transition section to the metering section. The greatest channel depth occurs in the feed section. As the plastics granules travel forward in the feed section they are converted by shear friction and barrel heat to a plastic state. When the mass arrives at the transition zone it is converted to a viscous fluid, and enters the metering section to be pumped to the die or crosshead. The transition may consist of as many as four screw flights, to as little as one half flight or turn.

In simple terms, the output or metering section of a metering screw acts almost as the basic Archimedian screw conveying a liquid. The output of the metering section of the screw approximates the volume forwarded by each screw flight per revolution minus the losses due to backflow along the screw and the leakage due to the necessary clearance between the screw and barrel. A major discussion of extrusion theory is presented in "Processing of Thermoplastic Materials," published by Reinhold Publishing Corp. Other screw designs are shown in Fig. 4.5.

To effect best plasticity and homogenization, modern day extruder screws are built with at least a 20:1 length to diameter (L/D) ratio. A 20:1 screw with a 2-inch diameter would thus have a length of 40 inches from the rear of the feed opening to the end of the screw. Screws of 24:1 are in common use, especially in vented barrel extruders, and some machines have been built with 30:1 L/D screws.

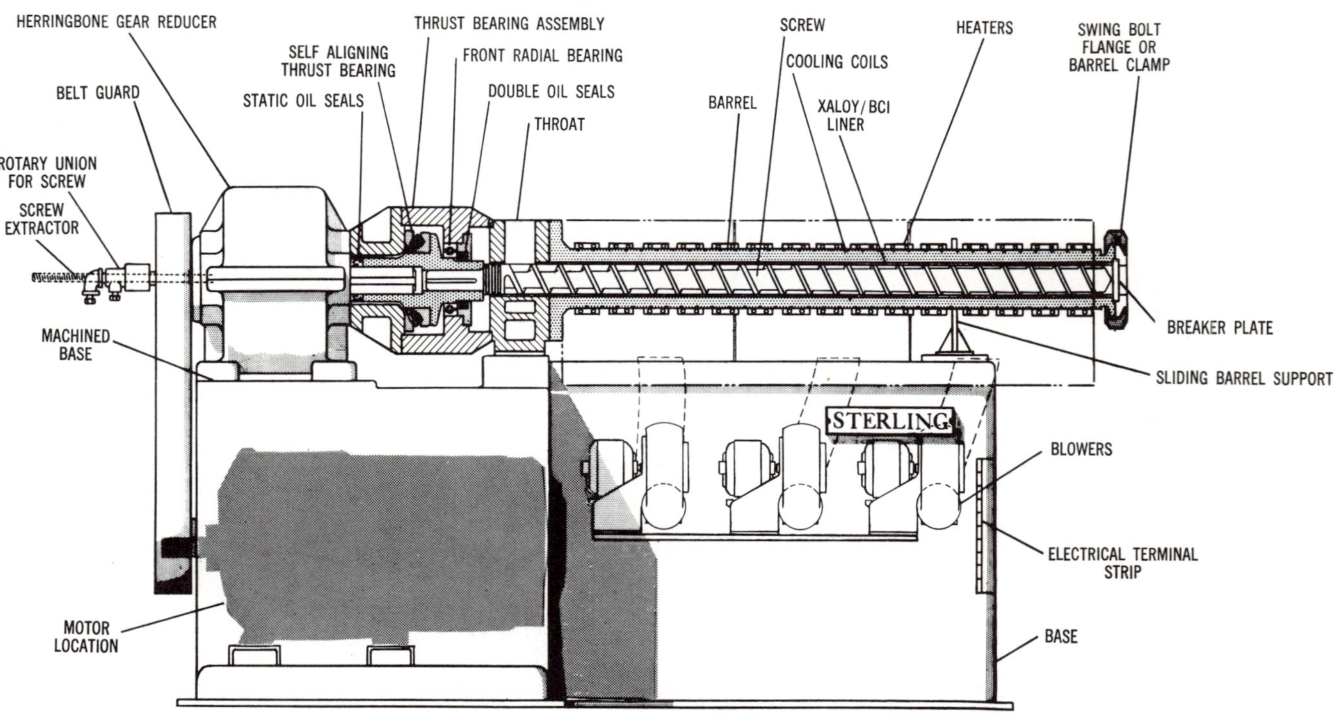

FIG. 4.2 Schematic components of a single-screw extruder. (*Courtesy Sterling Extruder Corp.*)

FIG. 4.3 Typical vented single-screw extruder. (*Courtesy Prodex Corp.*)

Extruder screws are made of steel such as AISI 4140, and screw flights are flame hardened or hard faced with "Stellite" to reduce wear. The screw surface is usually hard chrome plated to resist corrosion and wear.

Extruder Barrels

Extruder barrels are made of carbon steel with a hard wear-resistant internal surface obtained by nitriding or lining with "Xaloy," at Rockwell C 60. If corrosion resistance is required, the barrel is lined with "Xaloy-306" with a hardness of

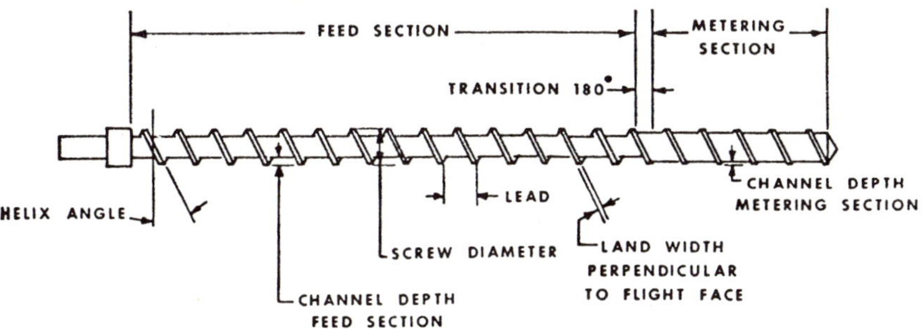

FIG. 4.4 Metering screw. (*Courtesy "Processing of Thermoplastic Materials"*)

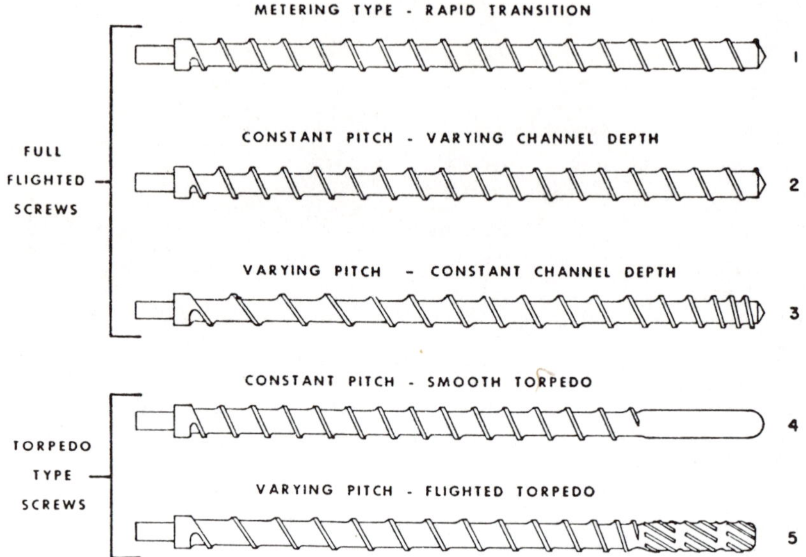

FIG. 4.5 Common extruder screws. (*Courtesy "Processing of Thermoplastic Materials"*)

RC50. "Xaloy"* is centrifugally cast inside the barrel bore to achieve a metal to metal bond required for best strength and heat transfer properties.

Barrel Heaters

The barrel is heated with electrical resistance or induction heaters. Sometimes the barrel is heated by steam or by heat transfer fluids circulated in jackets or coils installed on the barrel. By far the most widely used is the resistance heater. The heaters are arranged in zones to provide the necessary temperature gradient and all zones are separately controlled by thermocouples.

Barrel cooling to remove surplus heat caused by excess shear and frictional work on the material is generally obtained by air blowers provided for each zone. Cooling can also be accomplished by circulating cooling water or other fluid heat transfer media through barrel jackets or coils when so provided.

Mechanical Speed Variation

Screw extruders must have a variable speed drive to match the output with process or material requirements. The simplest of these is the mechanical speed variator ("Varidrive," P.I.V., etc.), whose speed variation is dependent upon the

*Xaloy Inc., New Brunswick, N.J.

pitch diameters of a pair of split sheave pulleys connected by a belt or chain. On low ratios, the machine may have one fixed pulley and one split pulley. Speed ratios of these devices range from 2:1 to 7:1, and the speed range may be doubled when used with a two-speed motor. These machines have a practical application limit to 50 hp on extruders, although they are available to 150 hp. Efficiency is relatively low at 70% in the top half of the speed range. These machines are used to drive extruders to $3\frac{1}{2}$ or 4 inch screw diameter.

Electrical Speed Variation

For larger diameter screws and for excellent speed regulation at higher speeds the magnetic eddy-current coupling drive is used. This machine consists of an ac motor of the required horsepower, driving one-half of a magnetic coupling. The other coupling half is connected to the gear box of the extruder. Excitation power of the coupling varies the speed. The speed ratio can be varied from zero to full, with constant torque available over the full range. At speeds other than full (locked coupling), the eddy-current losses generate heat which must be dissipated either by air or water-cooling. The efficiency of these units is approximately 90% at full speed. Good speed control is important for extruders.

Magnetic coupling drives are available to over 1000 hp. The advantages are low maintenance, a wide speed range, and relatively low cost. The disadvantages are low efficiency at low speed, and the need for cooling water in the higher horsepower units.

ac/dc Drives

The ac/dc drive consists of a dc motor connected to the extruder gear box, whose speed is changed by varying the field-excitation voltage of the motor. The dc voltage to the drive motor is furnished by an ac motor-driven dc generator or from a thyratron, mercury arc, or solid state rectifier.

Hydraulic Motors

Hydraulic motors are frequently used to drive extruder screws used in screw injection machines. Their advantage lies in the infinitely variable speed that can be obtained between fixed low and high-speed settings. The motor is usually attached to a gear reducer which drives the screw.

MULTISCREW EXTRUDERS

Multiscrew extruders have been developed for compounding and pelletizing purposes and for the extrusion of PVC materials. The earlier machines were fitted with twin screws but at the present time quadruple screw machines are in common use.

The advantages for the multiscrew machine are outstanding homogenization of the mass and higher output. The four-screw machine is used extensively in the

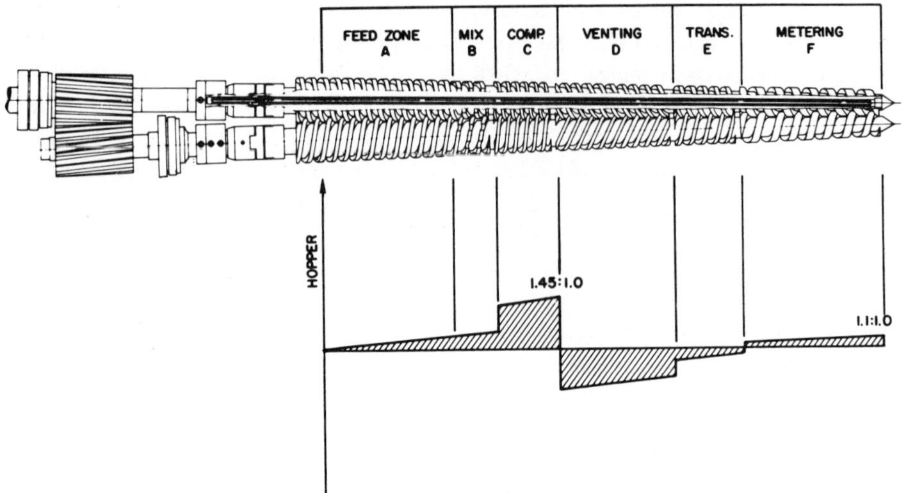

FIG. 4.6 Schematic diagram of conical twin screw geometry. Compression ratios at each zone. (*Courtesy Cincinnati Milacron.*)

production of PVC pipe where the ability to extrude rigid PVC resins from a dry powder blend at high output rates with fine size control is an advantage.

The twin screw extrusion machine is achieving greater popularity in the United States. The mechanical schematic of a conical twin screw is shown in Fig. 4.6. One of the mechanical advantages of this arrangement is more space for the drive gears and bearings than in a parallel screw machine. Numerous plasticizing advantages are also recorded. The complete machine is shown in Fig. 4.7.

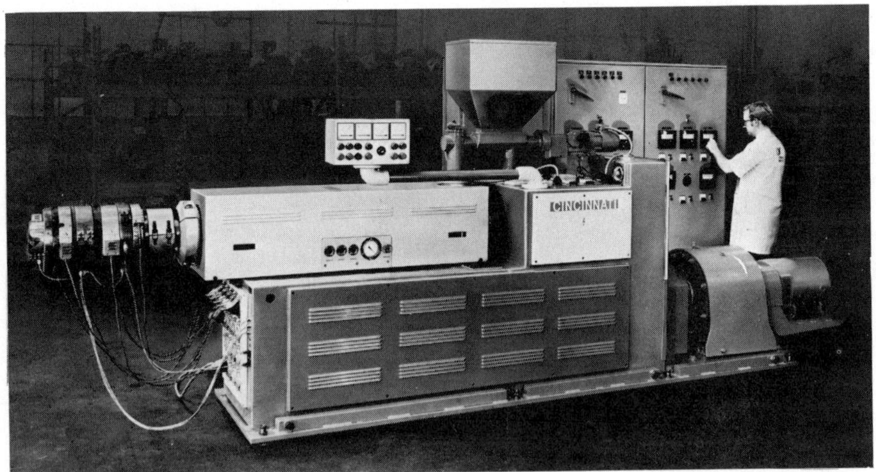

FIG. 4.7 Twin screw extrusion machine. (*Courtesy Cincinnati Milacron.*)

The twin and multiscrew machines are also used in the extrusion of PVC sheet with PVC wood grain film laminated to it. Foamed plastics profiles to substitute for wood molding shapes for door trim and molding applications are also manufactured using machines of this type.

In recent years, simultaneous extrusion of two or more dissimilar resins has become a viable process in film extrusion. The advantage of these composites lies in better chemical resistance, superior permeability characteristics and/or improved physical properties of the composite at lower cost. The process is based on feeding the output of several extruders into an especially designed die where the extrudates combine in separate layers to create the composite upon exit from the die.

Numerous markets for products made from coextruded sheet have developed in containers, packages, and disposable ware. For example, cups and plates have been made using an impact grade of polystyrene for strength, with a thin surface layer of glossy crystal styrene to enhance the surface appearance. For oil and grease resistance multilayer sheets have been used in thermoformed butter and margarine tubs. Here polyethylene has been used for the inside layer, with the outside consisting of a multilayer of ABS and polystyrene.

All kinds of combinations have been developed to take advantage of the high physical properties of one resin with a coating of a more esthetically desirable material for the surface.

Another interesting application of coextrusion is a three-layered sheet used to produce refrigerator liners. This sheet is made of ABS and high impact polystyrene with a thin polyethylene coating on the impact styrene to provide chemical resistance and barrier properties from the chemicals used for the foam-in-place low temperature insulation.

Other layer combinations provide moisture, chemical, or gas permeating properties required in a multitude of applications. The product possibilities appear to be infinite.

It is not an easy process, but one which depends on extremely careful control of the throughput of the several extruders and, of course, a most sophisticated die design.

Elastic-Melt Extrusion—"Elastirotor"

The elastic-melt extrusion principle was developed from the work of Weisenberger in Europe, on shear-melting of solids between rotating plates. The work by Bryce Maxwell in this area, conducted at Princeton University, aroused further interest, which resulted in a serious research and development program at Owens-Illinois and KPT Mfg. Company. The "Elastirotor," whose principle of operation is schematically shown in Fig. 4.8, developed from this work.

The plastics material flows from the hopper between the rotor and the outer housing (the stator), and is plasticized between the vertical faces of the rotor and stator. The resultant melt is picked up by the screw attached to the rotor, and is

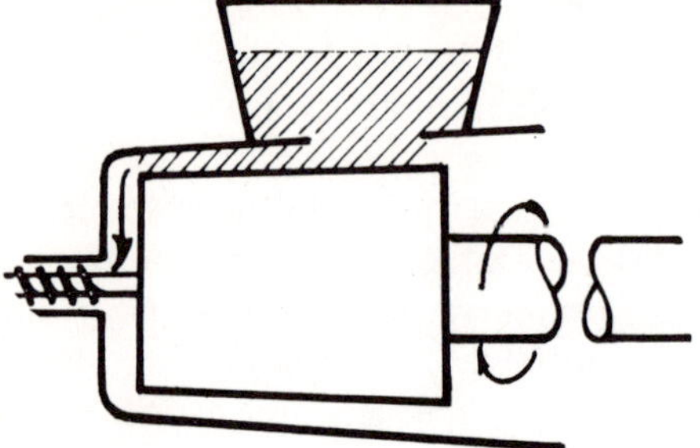

FIG. 4.8 This schematic drawing illustrates the flow of material in the elastic melt extruder.

then pumped forward through a die or cross-head as required, to make a product. The gap between the stator and rotor, as well as rotor speed, can be altered to suit material requirements. These adjustments can be made while the machine is running. (See Fig. 4.9.) The machine is provided with three heating zones, used only in the start-up, and seven cooling zones, which regulate the temperature during operation. Limited applications have been found for this unit.

Melting Plate Extruder—"Rotomeltor"

In 1963, Thomas Engel of West Germany introduced a new melt extrusion concept; Fig. 4.10 schematically depicts the principle.

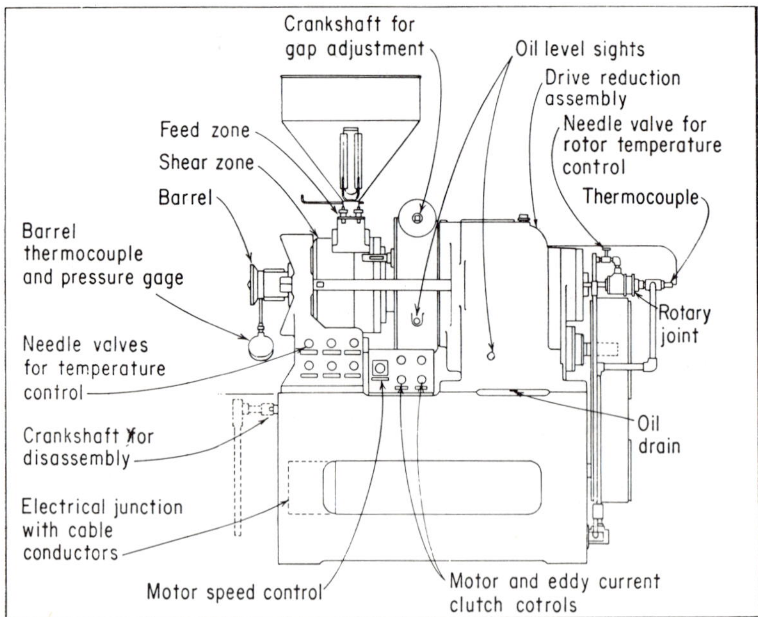

FIG. 4.9 The "Elastirotor" extruder is short in length and is cleaned very easily and quickly.

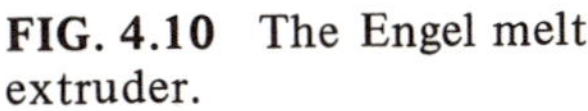

FIG. 4.10 The Engel melt
extruder.
   a. Feeder chute
   b. Melting plate
   c. Extruder barrel

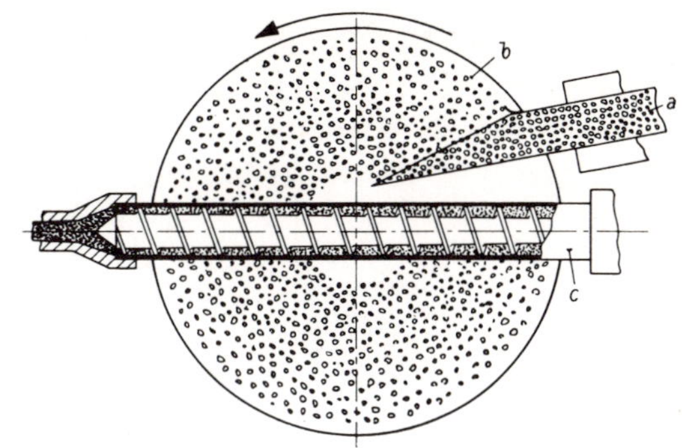

The electrically heated rotating melting plate "b" is mounted beneath a bottom opening extruder "c." Temperature of the melting plate is maintained by electric heaters and the heat level is regulated by pyrometric controllers and thermocouples. As the plate turns, plastics pellets, granulate or powder is spread on the plate by vibrating feeder "a." Melting of the plastics takes place between the end of the feeder chute and a doctor blade fixed under the extruder barrel. The melted plastics is scraped from the plate and is directed through the barrel opening into the feed end of the screw. The screw acts only as a pump, and does no plasticizing. Pressure is built up by the 2:1 compression ratio of the screw. Screw rotation is fixed at 60 rpm, and the plate rotates at 6 rpm. Output is controlled by feed rate. Figure 4.11 shows the mechanical arrangement of the machine. Advantages claimed are:

(1) Simplicity—no variable speed drives required.

(2) Ease of de-gassing and de-moisturizing the plastics melt is inherent in the machine. Gases and moisture are drawn off by suction through the cover.

(3) One screw will handle all thermoplastics.

(4) Degradation of materials is at a minimum because of low residence time on the plate (10 seconds). With particularly heat-sensitive materials inert gases such as nitrogen can be introduced into the chamber surrounding the melting plate.

(5) Color and material changes are immediate and require no machine disassembly and cleaning.

The output of a "Rotomeltor" with a 400 mm (15.8 inch) melting plate is reported as:

low density polyethylene	60.5 lb/hr
high density polyethylene	52.8 lb/hr
polystyrene	70.4 lb/hr
plasticized PVC	61.6 lb/hr
polymethylmethacrylate	56.0 lb/hr

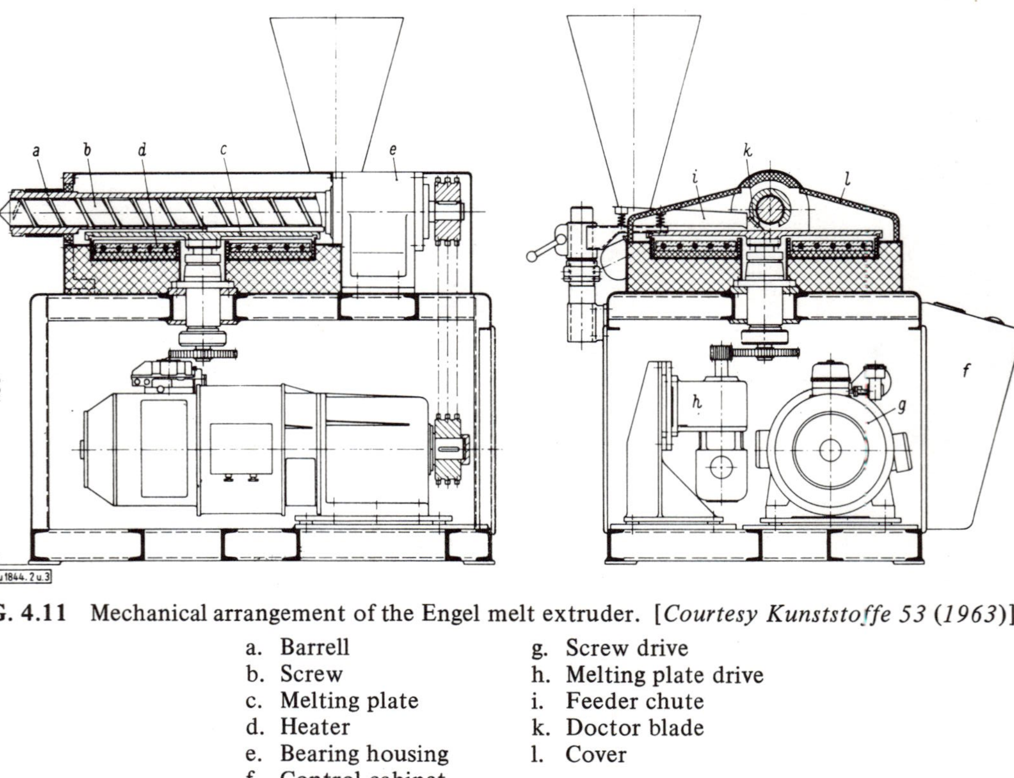

FIG. 4.11 Mechanical arrangement of the Engel melt extruder. [*Courtesy Kunststoffe 53 (1963)*]

a. Barrell
b. Screw
c. Melting plate
d. Heater
e. Bearing housing
f. Control cabinet
g. Screw drive
h. Melting plate drive
i. Feeder chute
k. Doctor blade
l. Cover

Power input for all purposes, melting, extrusion, etc., was 5.3 kw or 7.15 hp. It is estimated that the maximum practical output of machines using this principle will be in the order of 660 to 1100 lb/hr.

FOAMED EXTRUSION PRODUCTS

A practical application of the "Rotomeltor" machine lies in the manufacture of foamed products. In this process, natural polyethylene, or other resins, are fed to the melting plate. When the melt enters the extruder, an expansion agent such as ammonium carbonate (and colorants if desired) is added. The density of the foam is dependent upon the quantity of the gasifying agent added to the mass.

When the melt is extruded into a mold, immediate expansion takes place and is completed in a fraction of a second at relatively low pressure (120 psi). The molding is done in automatic machinery to produce large quantities of product at extremely fast cycles. Among such products are corks, toys, shoe soles and heels, insulated containers, etc. This "Engelit Foaming Process" is licensed by the Phillips Petroleum Company.

Dies and Specifications

The subject of extrusion dies is covered basically in "Processing of Thermoplastic Materials," by Bernhardt, and in "Plastics Mold Engineering," by DuBois and Pribble. "Plastics Mold Engineering" details the design, construction and materials for extrusion dies. Typical extrusion dies are shown in Figs. 4.12 and 4.12A.

The most important specification in the selection of an extruder is its horsepower input.

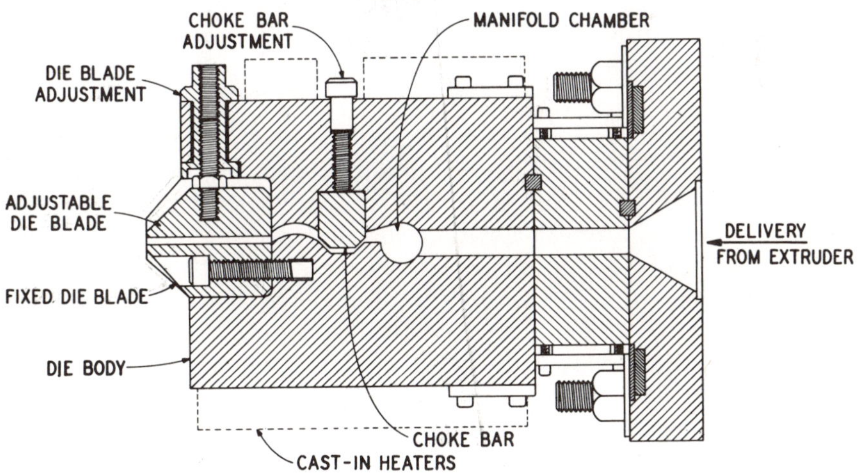

FIG. 4.12 Sheet extrusion die. (*Courtesy Phillips Petroleum Co., Bartlesville, Okla.*)

FIG. 4.12A An 84 in. tubular film die used to extrude plastics film. The giant tube is slit to provide a sheet 264 in. wide. (*Courtesy, Egan Machinery Co. Somerville, N.J.*)

In discussing extruders,* pounds per horsepower per hour is an expression of the amount of energy that is generated in the material from the extruder drive, and is an indication of machine efficiency. These data vary with the application. For example, low density polyethylene requires an extrusion melt temperature of about 300°F. (250–300 Btu/lb) and the extruder output is about 8 lbs/hp/hr. Blown film which extrudes at 350–400°F will produce at about 6 lbs/hp/hr. Cast film at 500°F is extruded at 5 lbs/hp/hr. The rates are higher for small extruders since the shear rates are lower. Large extruders often require cooling to control the melt temperature and this reduces the output per hp/hr. Another material variable which will affect the output is the melt index which varies between 0.20 to 20. The following data are representative of most currently produced com-

---

*Courtesy W. Willert, Egan Machinery Co.

mercial machines:

Screw Diameter (in)	Capacity (lb/hr)	Drive Horsepower
2½	200–300	35–60
3½	300–700	75–125
4½	600–1200	150–200
6	900–1500	250–400
8	2000–3500	400–600

INJECTION MOLDING OF THERMOPLASTICS

For the injection molding of thermoplastics, the material is melted and then pushed into the mold cavity where it solidifies by cooling, prior to removal from the mold. The Hyatt brothers patented the first plastics injection machine in 1872. The injection molding of thermosetting materials, also called fully automatic transfer molding, is described in Chapter 5; in this case, the material hardens by a chemical change. Injection molding can be done by hand, melting one charge at a time in an oven or furnace. Automatic injection molding is most often done by machines which perform each of the operations in sequence with the steps controlled by electromechanical devices. In semi-automatic molding, the press operator removes the molded pieces from the mold or press after each cycle and, by closing the gate again, starts the next cycle. In fully automatic molding, all operations are controlled by the "clock," and a machine tender checks the pieces from time to time, removing the tote boxes of finished parts. Hoppers are filled automatically with preconditioned compound.

Fully automatic molding is preferred since it delivers better uniformity of products at lowest cost. All machine functions must be perfectly adjusted and the mold operation, in most cases, degates the pieces and ejects them into a free position for delivery by gravity or other means to the tote box.

A study of Fig. 4.13 will illustrate the basic principles of injection molding. Controls are necessary to achieve the following:

(1) Proper amount of compound for each shot.
(2) Adequate plasticization of compound.
(3) Material temperature control.
(4) Clamping, filling and holding pressure.
(5) Rate of mold fill.
(6) Mold temperature.
(7) Hardening, ejection, opening, filling and mold closing time.

The quality of material to be injected on each cycle is measured by volume or by weight; exact control is essential. Each machine is equipped with its own feed devices and adjustments to meet the variables within the machine capabilities.

An inadequate volume or weight of charge will result in unfilled parts, and low

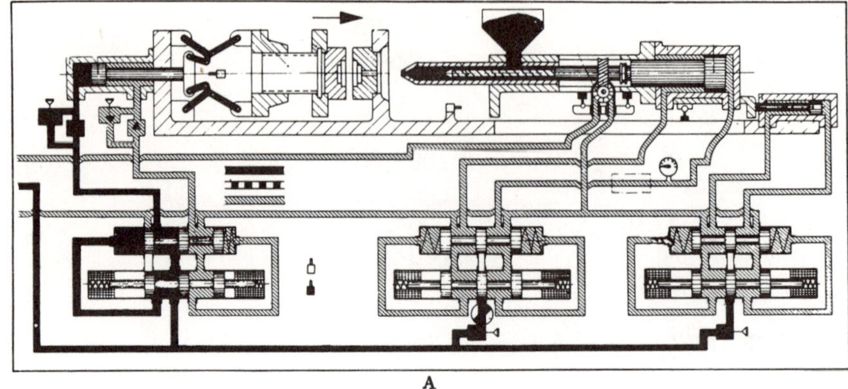

A

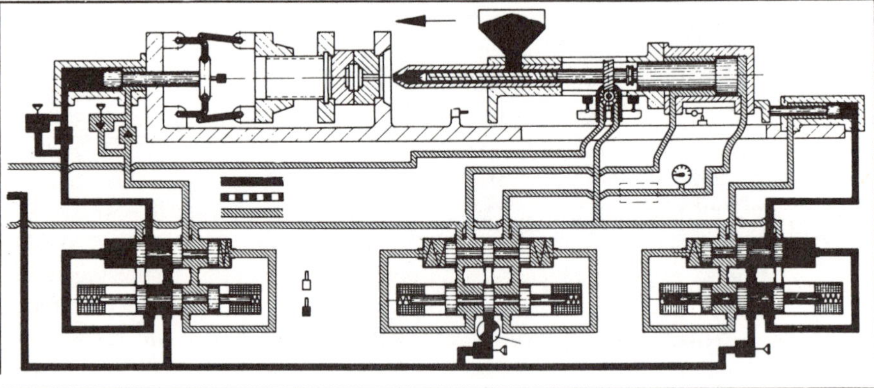

B

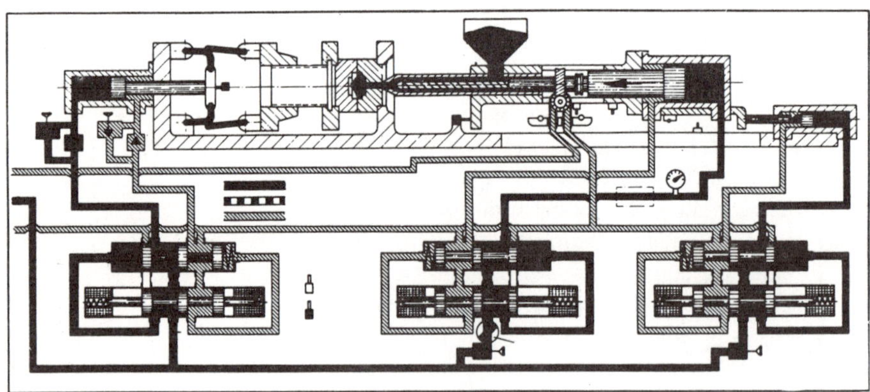

C

FIG. 4.13 Operational details of in-line reciprocating screw injection molder. A. When the mold starts to close, the screw has just finishing charging the front end of the cylinder. B. With the mold closed, the heating cylinder moves towards the sprue bushing and is ready to commence the injection cycle. C. The hydraulic cylinder then forces the screw forward under the injection cycle and fills the cavity.

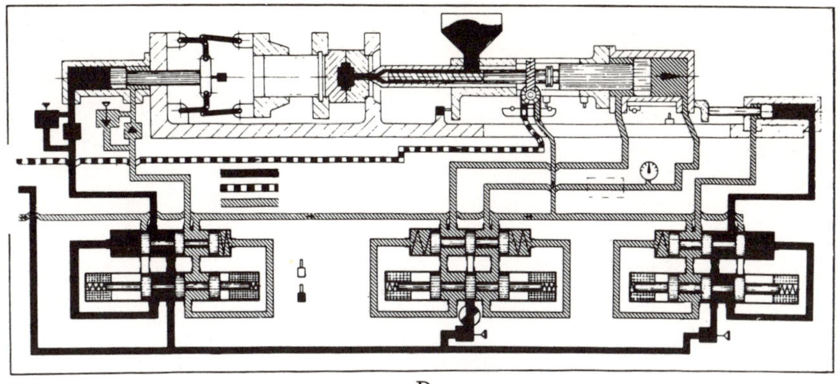

D

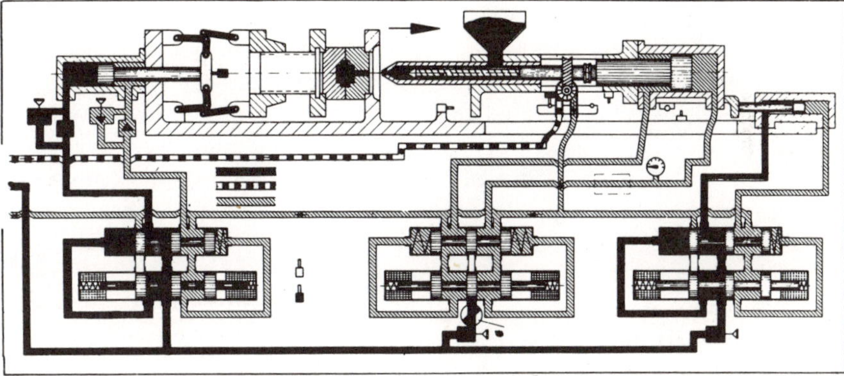

E

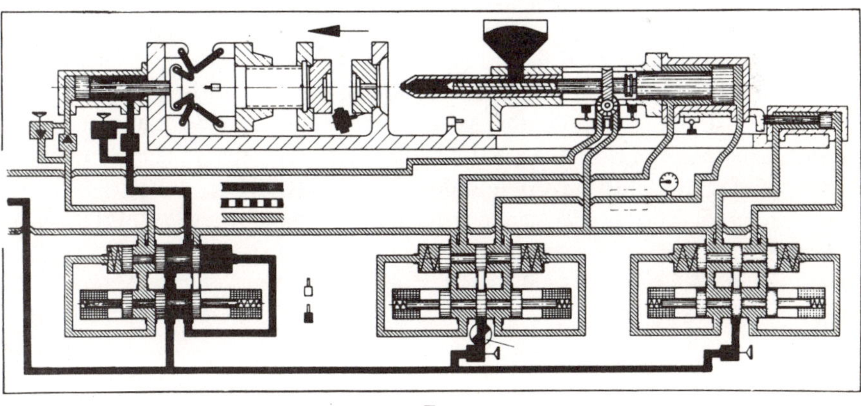

F

FIG. 4.13 (*continued*) D. Cooling and charging cycle: the mold is filled, and, after holding time, the screw starts to rotate and charge the front end of the cylinder. E. The screw is still charging the front end of the cylinder, and the cylinder carriage starts to move away from the sprue bushing. F. The finished cycle: mold opens and molded part is ejected with the sprue bushing.

density products with poor properties. Overfilling or packing the mold will cause the mold to open slightly and develop a heavy flash at the parting line where the excess material escapes. Overfill also results in pieces that stick and break during ejection, as well as oversize pieces resulting from spring back expansion after the piece is out of the mold. A very careful feed adjustment must be achieved during the start-up of the mold.

A time lag is to be expected between adjustment and achievement so all control adjustments must be made slowly with a proper waiting time between adjustments. This is especially true of feed and temperature changes which require a considerable interval between adjustments to achieve the final effect of the change and equilibrium. Any change in cycle time will change the heating time.

Drying. Moisture must be removed when present and a variety of dryers is available for this purpose. Heating without a movement of air in the dryer is useless, an air current to carry away the moisture is essential to any drying operation. Shallow pans in ovens as shown in Fig. 4.14 are often used for this purpose. Best drying is achieved by a closed air circulating system with a dehumidifier for removal of the moisture picked up by the air. Fully automatic systems are available that pump the material out of the containers, dry it, preheat it and discharge it into the press hopper in a closed system that also prevents contamination. Pure white materials, for example, will pick up specks of dirt unless completely protected in such a closed system.

Blending. Since it is necessary in many cases to feed back the sprue/runner and gates produced in each cycle, this is most effectively done concurrently with each molding cycle. In simple systems, the operator may clip the sprue and runner into small bits with pliers, and drop them in the hopper along with the virgin material. Short run programs often set up a grinder alongside of the press so that the operator can feed the sprue/runner pieces into the grinder; the grinder then is emptied from time to time, and the contents hand blended in a drum with virgin material. Volume jobs and products requiring tight material quality control make use of automatic grinder-feeder-blenders which feed a blended mix of virgin and regrind into the dryer under closely controlled conditions.

Dry Coloring. The molder can purchase the material already colored to his specification, or he may elect to color it himself. It is often desirable to stock a basic clear material and color it as required for each job. Coloring is done either by the addition of pure color pigment, or by the use of color concentrate. When screw-type injection machines are used, best color dispersion is achieved. Special dispersion nozzles are needed to obtain adequate color dispersion in plunger machines. Dry coloring is completed before the dryers are employed.

Plasticization. The injection machine has a heat exchanger in which cold granular bits of plastics are heated up to their melt temperature, and compressed to a homogeneous fluid that will flow freely under pressure to fill the mold. Many cylinders are electrically heated by means of heater bands that transfer the heat

FIG. 4.14 Drying of thermoplastic materials is often achieved by heating in shallow pans with forced dry air circulation. (*Courtesy Moxness Products Inc.*)

through the cylinder walls. The screw-type machines generate heat in the plastics by the mixing-shearing action which eliminates the need for much applied external heat after initial start-up.

There is considerable difference between the cylinder temperature and the stock temperature, since the source of heat must be above that of the material. Plastics materials transmit heat slowly and poorly, so that any change in the cylinder temperature will not be found quickly in the stock. The press operators must learn the amount of time needed to obtain a change in stock temperature by a change in the heaters. Changes must be made slowly to achieve proper temperature control, and operators need to be constantly cautioned about too much and too frequent thermal adjustment.

Uniformity from cycle to cycle is essential to high quality molded parts. Erratic time cycles change the amount of heat and plasticity. Dry materials, minimum injection pressure, uniform fill and hardening time—all are essential for good thermoplastic molding procedures.

SALVAGE OF THERMOPLASTICS MATERIALS

One large advantage of the thermoplastics is the salvage value of the scrap. Clean identical materials may be blended in the ratio of 20% regrind with 80% virgin material. It is extremely important to avoid the intermixing of various types of plastics. Materials of the same type but with melt flow differences can suffer from thermal degradation and exhibit brittleness, lamination and surface roughness. All other foreign matter must be removed from plastics scrap that is to be used again. Multi-colored scrap of the same material can be used for molding products which are to be plated or that can be run with black color concentrate to mold black parts, thus achieving 100% salvage.

INJECTION MOLDING MACHINES

An injection molding machine performs the job of plasticizing and injecting the thermoplastics materials into a mold that is held between clamping platens. Mold clamping in most present day machines is usually one of two types—mechanical (toggle), or hydraulic. Injection machine capacities range from less than an ounce to over 30 pounds, and may be horizontal or vertical. Currently used means for plasticizing are: plunger injection, preplasticizer two-stage plunger, preplasticizer two-stage screw, and reciprocating screw. A canvas of machine manufacturers shows that the reciprocating screw machines are in exclusive demand in capacity above 2 ounces.

Plunger Injection Cylinders

A typical conventional injection cylinder is schematically shown in Fig. 4.15. It consists of a barrel heated with electrical resistance heater bands, in which is

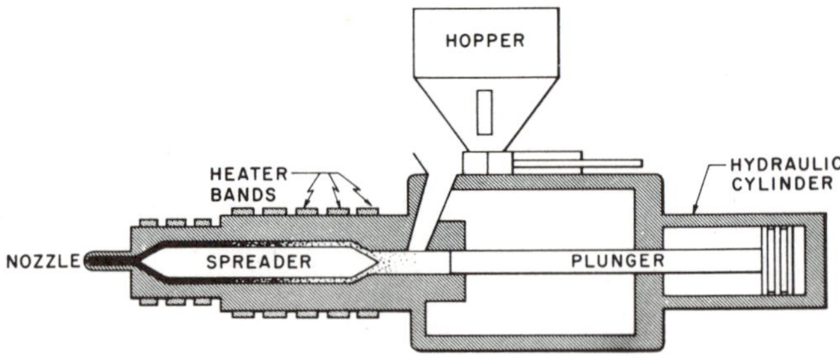

FIG. 4.15 Conventional plunger-type injection cylinder. (Reproduced with permission from the bulletin *Injection Molding Technology Fundamentals,* The Dow Chemical Co.)

fixed a spreader or torpedo. The interior of the barrel and the surface of the spreader are usually nitrided and chrome plated, thus providing wear and corrosion resistance. The spreader is also heated electrically in some machine designs. The barrel heater bands are generally arranged in zones to achieve the required temperature gradient across the cylinder.

The plunger forces the plastics granules, fed from a hopper, through the narrow passageway between the barrel and the torpedo. As the granules move forward toward the nozzle they are melted by the heat obtained from the barrel and torpedo. The granules are then fed to the cylinder, either volumetrically or by a weigh feeder, which deposits a precisely weighed charge into the cylinder. This weighed mass equals the weight of the shot which consists of the combined weights of the product, plus the sprue and runner system. Packing or underfill of the mold must be avoided to insure best dimensional control of the part, best physical properties, lowest stress level in the product and consistent product weight.

The accuracy of the charge in volumetric feeding is dependent upon the bulk density of the material, directly affected by the shape of the granule. Small granule changes can markedly alter the weight of the charge. This will result in significant changes in product quality.

There are probably a half dozen commercially used internal designs for heating cylinders other than the spreader. Among these are the reverse flow cylinder, and the melt extractor cylinder. All are attempts to deliver a well plasticized homogeneous plastic mass to the nozzle. The nozzle is the point at which the mass is injected into the mold.

Preplasticizer, Two-Stage, Plunger Injection Cylinder

Figure 4.16 shows a two-stage plunger-type preplasticizer. It should be noted that the preplasticizing cylinder, etc., shown mounted at an angle to the shooting cylinder, is identical with the plunger injection cylinder previously described. The purpose of the preplasticizer injection cylinder, of course, is to feed melted plastics material into the shooting cylinder.

The shooting cylinder is fitted with a nozzle and a plunger. The plunger is provided with a stroke control, which may be mechanical (as shown), or electrical, such as a limit switch which closes the hydraulic valve controlling the shooting plunger piston. In this way, precise shot weight can be obtained, which is the prime advantage of the system. Other advantages lie in the lower injection pressure that can be used, as well as greater shot weight and lower cylinder temperatures.

In operation, the preplasticizing cylinder forwards melted plastics through the open valve, between it and the shooting cylinder. The shooting plunger, which at this point is forward, is forced backward as the cylinder is filled. It retracts until it hits the stop, whereupon the valve closes. When the shooting cylinder plunger is made to go forward by hydraulic pressure on the piston, the plastic mass flows through the nozzle into the mold.

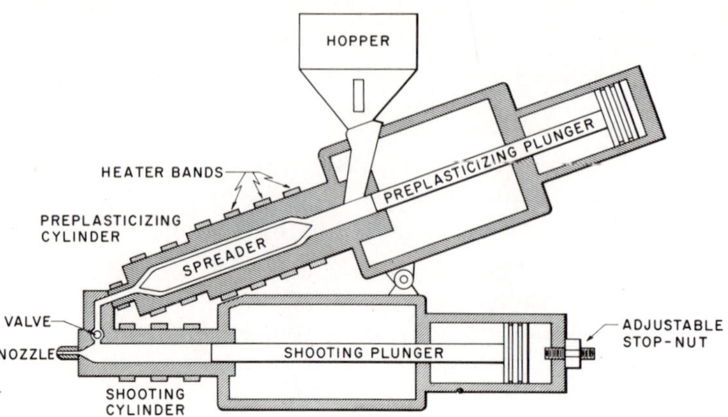

FIG. 4.16 Schematic of two-stage plunger-type injection cylinder. (Reproduced with permission from the bulletin *Injection Molding Technology Fundamentals*, The Dow Chemical Co.)

Preplasticizer, Two-Stage, Screw Injection Cylinder

The two-stage screw preplasticizer injection system is identical to the system just described, except that a screw extruder is substituted for the plunger-type injection cylinder. The advantage derived lies in the improved homogeneity of the plastic mass delivered by the screw to the shooting cylinder. Other advan-

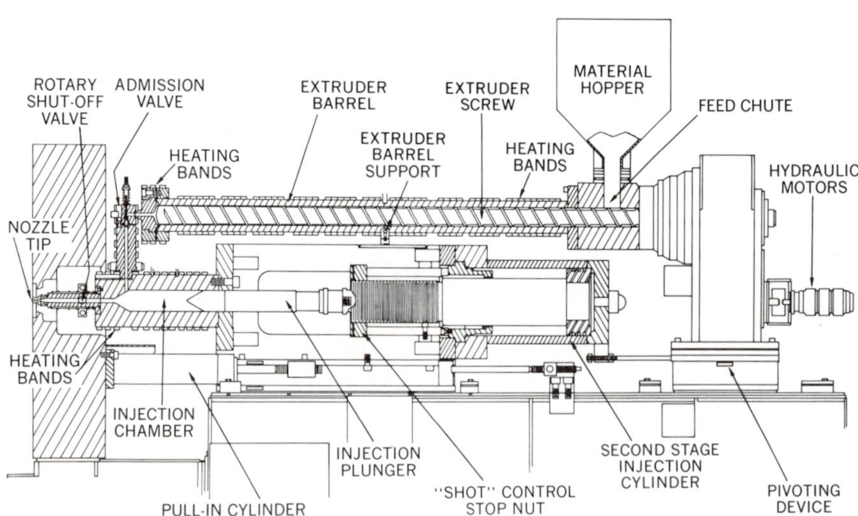

FIG. 4.17 Schematic of two-stage screw injection cylinder. (*Courtesy H.P.M. Division, Koehring Co.*)

tages are the same as the plunger version. Figure 4.17 depicts one form of this system.

The prime disadvantage with both systems lies in degradation of heat sensitive materials, which is the result of the multiple flow paths and possible hang-up areas inherent in the design.

Reciprocating Screw Injection Cylinder

The conception of the in-line reciprocating screw plasticizer and injecting unit by William H. Willert in 1952 and patented in 1956 brought a major improvement to injection molding machines. This was the first major improvement after the Gastrow torpedo in 1932. Almost all injection machines today use the reciprocating screw.

The thermoplastic resin flows from the hopper into the cylinder (Fig. 4.18).

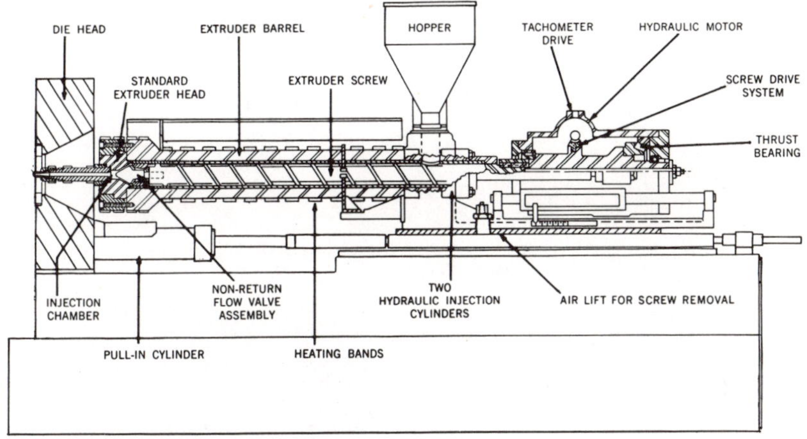

FIG. 4.18 Schematic view of reciprocating screw cylinder. (*Courtesy H.P.M. Division, Koehring Co.*)

As the screw rotates, the granules are conveyed forward and melted. The melt flows from the last flight of the screw through the nonreturn valve. As the mass collects ahead of the screw, it exerts a pressure on the end of the screw and forces it backward. Back pressure on the material can be regulated by adjustment of a relief valve in the hydraulic injection cylinder circuit. Suitable back pressure adjustment for the material being molded is very important to obtain ultimate homogeneity of the plastics being molded. Special shut-off nozzles are required when high back pressure is generated. The screw is fitted with a stroke limiting device, usually a limit switch, which stops screw rotation and feed from the screw when the required plasticized shot volume is created ahead of the screw.

During injection, the hydraulic ram moves the screw forward. The non-return valve closes as the movement starts, preventing backflow into the screw. The screw thus acquires the characteristics of a plunger and forces the plastics mass through the nozzle into the mold, as done in a conventional plunger machine.

The reciprocating screw machine has numerous advantages over the other plasticizing methods discussed. The primary advantage lies in the uniformity of plasticity obtained from the screw which produces higher quality moldings at lower temperatures and apparent pressures. A broader range of materials can be molded, including heat-sensitive materials such as rigid polyvinyl chloride. Color changes and material changes can be accomplished rapidly with small material loss while purging. A modern-day descendant of the original reciprocating screw injection machine is shown in Fig. 4.19.

FIG. 4.19 Solid state controlled 150 ton 8 oz. reciprocating screw injection machine. (*Courtesy Van Dorn Plastics Machinery Co.*)

The screw fitted with the non-return valve is used for materials such as the polyolefins, polystyrene, polycarbonate, etc. A "close-up" view of the non-return valve is shown in Fig. 4.20. Where degradation problems exist, such as with polyvinyl chloride, the non-return valve must *not* be used. The possibility of hang-up of material in the valve will lead to trouble. Special screw tips are used for PVC, such as the "smear-head," shown in Fig. 4.21 or screws whose flights extend to the tip in a tapered fashion as depicted in Fig. 4.22. The purpose of either design is to reduce the residue in the barrel after each shot to an absolute minimum.

Injection mold design is thoroughly discussed in "Plastics Mold Engineering" (published by Van Nostrand Reinhold.) A typical injection mold design is shown in Fig. 4.23. All of the machines described are fitted with instruments to indicate

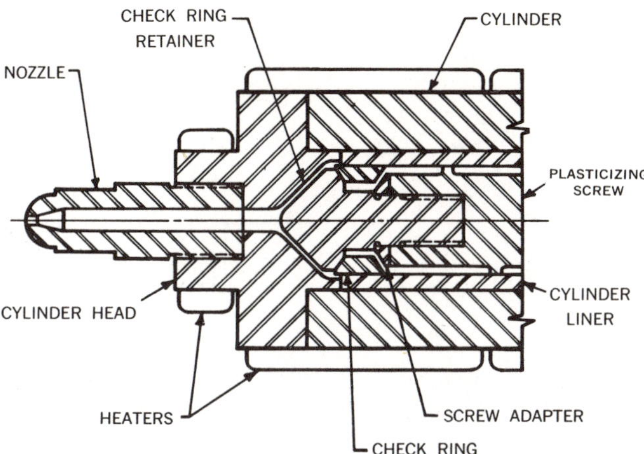

FIG. 4.20 Nonreturn valve design—supplied as standard equipment. Not recommended with heat-sensitive materials such as rigid PVC. (*Courtesy Egan Machinery Co.*)

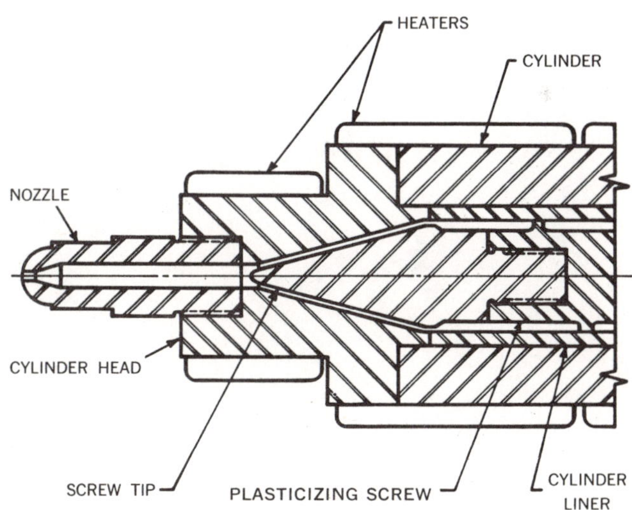

FIG. 4.21 Streamline tip design—supplied as optional equipment for running heat-sensitive materials such as PVC. Designed to mechanically clean as much of the surface in contact with the material as possible. (*Courtesy Egan Machinery Co.*)

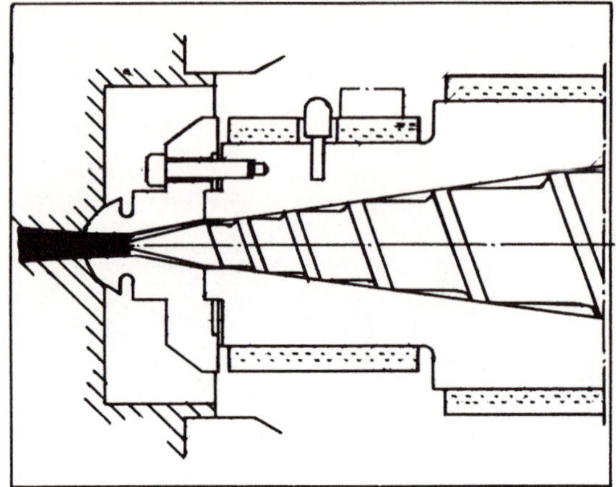

FIG. 4.22 Spiral screw tip and open type nozzle used with heat sensitive materials. (*Courtesy The New Brittain Machine Co.*)

injection pressure, to control the time elements of the injection cycle such as ram forward, dwell, cooling time, etc. and pyrometric controllers to maintain the required thermal conditions.

INJECTION PROCESS CONTROL

You have been introduced to the variables in the injection molding process. With close operator correction of variables as they occur a high percentage of quality

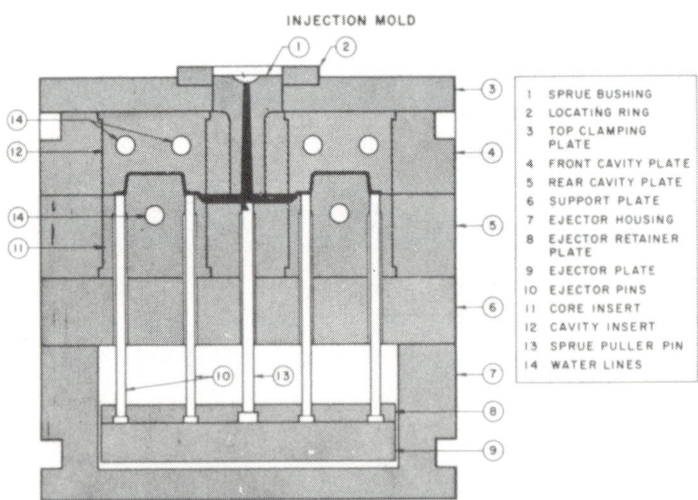

FIG. 4.23 Illustrates the various components of an average two-plate injection mold used for injection molding. (*Courtesy Dow Chemical Co., Midland, Mich.*)

parts will be obtained. Because of the scarcity of operators competent and willing to perform this function, and due to the increasingly high cost of labor, other reliable means have been found. Significant progress in automatic control of the injection process has solved many of these control problems and points the way to the future.

The injection process deals with time, pressure and temperature. To achieve *consistent* quality, each of these parameters must be controlled accurately to fill the mold with exactly the same amount of polymer during each cycle. The sub-variables which *most* cause a drift from this situation are changes in the polymer melt viscosity and the rate of injection of the polymer into the mold cavity.

Material viscosity can vary as a result of barrel temperature variation, or change in the polymer itself. To fill the mold properly, the pressure or temperature can be modified to compensate. The problem has been to observe and correct the drift soon enough to avoid poor quality production.

Equipment is now available which will automatically make the necessary adjustment as the variations occur. One way is to electronically observe the rate of injection during each cycle and compare it to the set points (go/no-go) for acceptable production. This is called Feed-Back Control. The rate of drift beyond acceptable set points (the injection pressure or flow rate) may be adjusted in small (positive or negative) increments to compensate. If a preset number of adjustment increments are exceeded, a signal or machine shutdown will occur. The feed-back control equipment used to accomplish this is based on solid state devices, and is basically useful in machines running under fully automatic control.

Another way of using feed-back control is to sense the pressure in the mold cavity during each cycle and to make injection pressure corrections to maintain acceptable product quality standards.

A more complete discussion of solid state control appears in Chapter 9.

Since the molding of thermoplastic materials is a time-pressure-temperature dependent process, proper use of this instrumentation is fundamentally important in producing quality molded products. Each variable must be carefully maintained and high quality instruments are strongly recommended. Pyrometric controllers are discussed in Chapter 9.

Mold protection is furnished by low pressure closing of the platens. If, for example, during automatic operation a molded product does not completely eject and hangs between the mold halves, the low pressure squeeze will not allow the mold to close. Usually a limit switch is so positioned to cancel the low pressure closing if nothing else prevents the mold from closing. The mold will then clamp under high pressure, and the cycle will proceed in a normal fashion. If, however, the mold is held open, a delay timer, actuated when the mold commences to close, will time out and open the mold, energizing either an audible or visual signal.

Optional equipment such as power take-off to drive core pulling cylinders, mold wipers, as shown in Fig. 4.24, and unscrewing mechanisms, are a valuable

FIG. 4.24 This press-actuated wiper removes the molded piece and the runner from the mold, assuring completely automatic operation with no possible "double-shot." (*Courtesy Bermer Tool & Die, Inc., Southbridge, Mass.*)

built-in feature, obtainable in many machines. Other important press features which must be carefully considered are plasticizing capacity, platen size, mold opening stroke and ease of maintenance of the hydraulic equipment.

A press installation cannot be considered complete without ancillary equipment such as beside-the-press sprue and runner grinders, material hopper loaders, and mold chillers and heaters, which are discussed in Chapter 9. In most molding shops, compressed air is in short supply, and self-contained vacuum hopper loaders are found to be great air savers when compared to air lift loaders.

Operator safety is of utmost importance. Most state codes in the United States require three safety devices to be part of the injection machine; an electrical interlock to prevent platen closing if gate is open, a hydraulic interlock also preventing platen closure if the gate is open and a bar which drops between the platens when the press platens are open and is automatically cam lifted when the gate is closed. These requirements will be national under the Occupational

FIG. 4.25 Many of the older plunger type injection molding machines are being modernized by replacing the obsolete plunger-plasticizer with a reciprocating screw-plasticizer as shown in this illustration. Versatility and greatly increased profitability result from this modernization procedure. (*Courtesy. Egan Machinery Co., and Engineered Plastics Machinery Co.*)

Safety and Health Act, and should be International. All injection machines should be so equipped. Failures of the electrical and the hydraulic devices have been known to occur. These failures have generally been traced to "tied down" limit switches or hydraulic valve actuators. The safety installation on the machine must be designed to prevent this.

Guards must also be provided to prevent entry into any area containing moving elements of the machine. These must be mechanically fixed. The rear guard to the molding area must be electrically interlocked to prevent operation of the machine if the guard is opened, perhaps even this should have a hydraulic interlock. Low machines must also have guards on the top of the machine to prevent operator entry into these areas as well.

The *Plastics Industry Safety Handbook* of the Society of the Plastics Industry is a reference for further details in this area.

5 COMPRESSION, TRANSFER AND INJECTION MOLDING OF THERMOSETS

One or more cavities per mold are used in the compression molding process. They are designed to provide adequate loading space above the cavity detail to contain all the unmolded thermosetting material, as shown in Fig. 5.1. As a compression mold closes, it is designed to permit the mold force or plunger to press the charge into place with a minimum of material loss resulting from some compound being pushed out of the cavity as overflow during closing. The molding compound may be in the form of loose powder, compacted pills, compact-shaped preforms, or in combinations of resin-treated paper boards and powder.

The molding process is handicapped by the fact that the mold force is designed to produce one surface of the molded part, and as such, may entrap and compress the charge with great difficulty. The shape of the force is often better adapted to pushing the compound out of the cavity than entrapping and compressing it.

To facilitate entrapment of the material in the cavity, several basic types of molds have been designed.

Figure 5.2 shows the flash-type mold. The flash-type mold is simplest and cheapest, serving many simple applications. In this case, the flash or excess material is horizontal, and the parting surfaces of the mold butt together to pinch off the flash in a thin film. This general type of cavity construction is also used for injection and transfer molding, with the addition of a gate to permit the incoming material to flow in from an external source. This type of com-

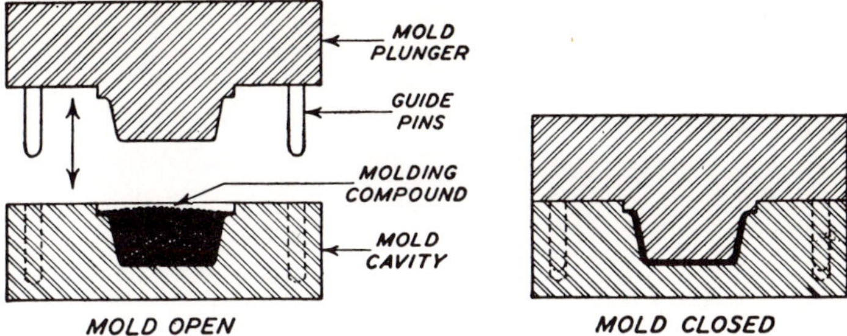

FIG. 5.1 In the compression molding process the material is placed in the mold cavity by the press operator. When mold is closed, molding compound is compressed to shape of piece and held in this form until it hardens.

pression mold must be overloaded with compounds to permit some material to excape and yet still maintain adequate product density.

Figure 5.3 shows a positive mold-parting surface design that is extensively used when the flash-type mold cannot entrap enough material to insure full product density. This requires a carefully weighed charge, since very little can escape. Any variation in the charge will affect the thickness across the parting line density, and consequently all properties.

Several compromise designs are used, as shown in Figs. 5.4, 5.5, 5.6 and 5.7.

It will be noted that the semipositive mold will entrap more material, and thus assure higher density than can be obtained with simple flash-type molds. The landed plunger mold is used in multi-cavity molds to gain loading space, reduce flash loss, and get better density in the pieces. The loading shoe type of mold provides additional loading space for powder loads, and less material is lost as flash. When the loading shoe is moved to the inner surface (compare Fig. 5.6 with 5.4), it serves the double purpose of providing added loading space, and serving as a stripper plate for removing the piece from the force.

Thermosetting materials do not become plastic until they are heated. The ma-

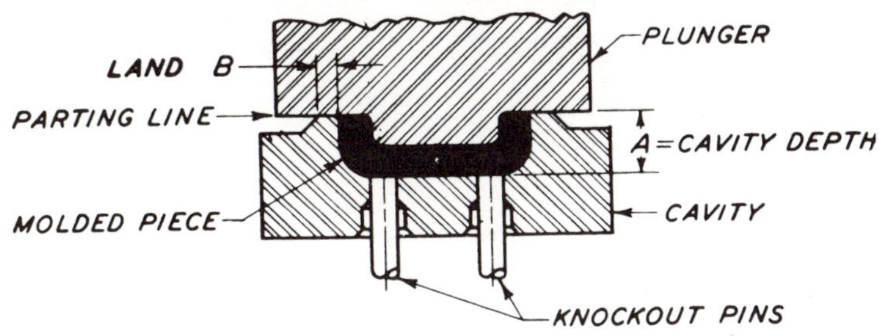

FIG. 5.2 Cross section of a simple flash mold.

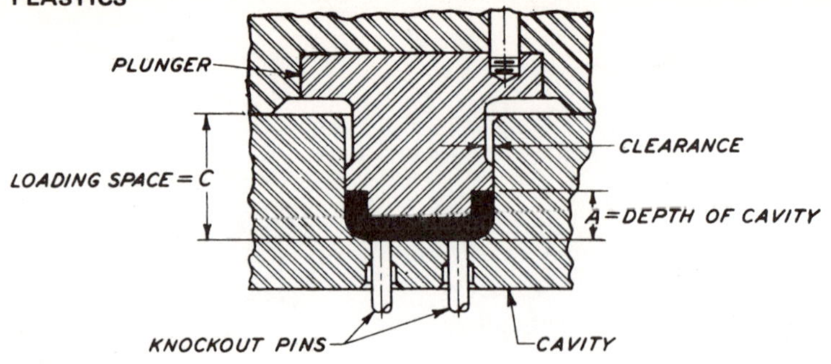

FIG. 5.3 Diagram showing construction used for a positive mold. Knock-out pins could extend through plunger instead of through cavity.

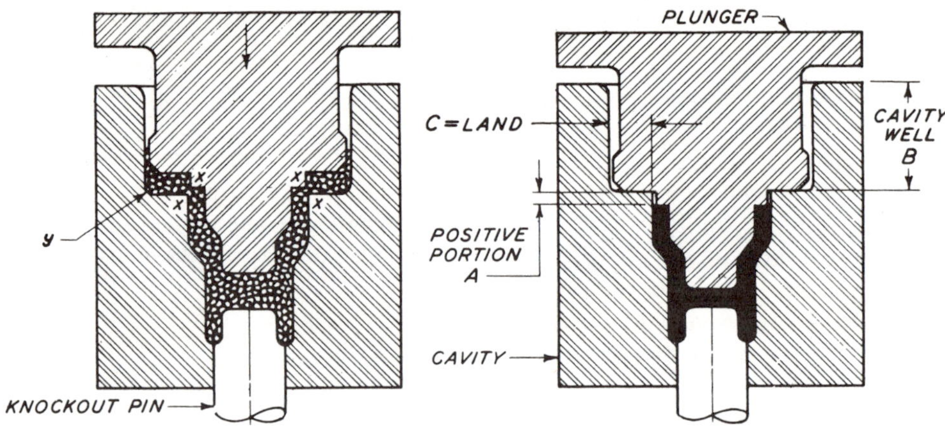

FIG. 5.4 Semipositive mold as it appears in partly closed position (left), just before it becomes positive. Compound trapped in area *y* escapes upward through overflow grooves in plunger. Mold in closed position (right). When corners at *x* pass, mold is positive, and practically all of entrapped compound is compressed.

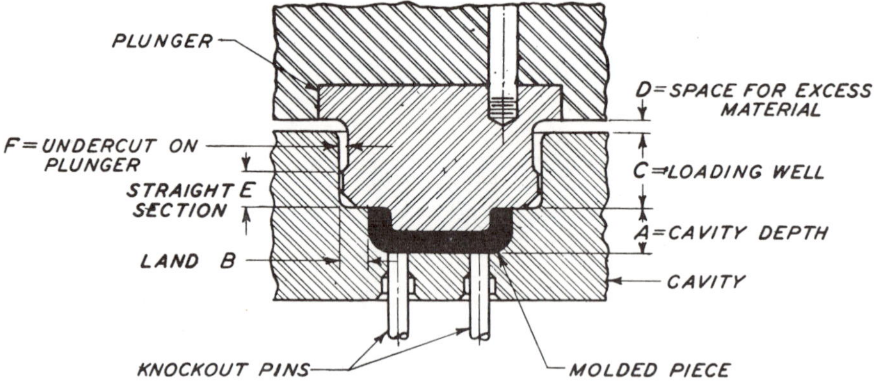

FIG. 5.5 Cross section of a landed plunger mold.

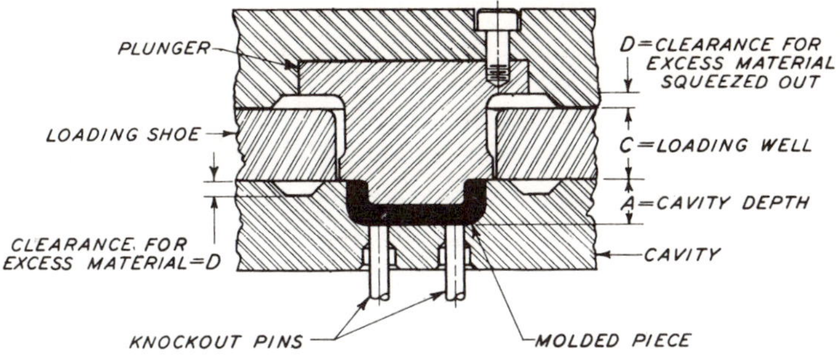

FIG. 5.6 Cross section of a loading shoe mold.

terial put into the mold varies in bulk, in heat conductivity, and in compressibility. Some materials flow easily, others with more difficulty—and as the thermosetting action initiates hardening, their plasticity and flow reduces until they become rigid. Because of the shape of the mold being used, portions of material in contact with mold surfaces will be hotter than other areas, so that it will flow easily while other portions of the charge are still cold and do not start to flow. In such cases, the areas where the material picks up much heat start to flow, and this material can be pushed out of the mold by the less fluid, colder material. Equally serious problems arise when the first-to-flow material starts to harden before the whole charge is sufficiently heated to flow and fill the cavity with uniform density. One common result is a piece which consists of an envelope of the softened charge that was in contact with the mold surfaces, enclosing an

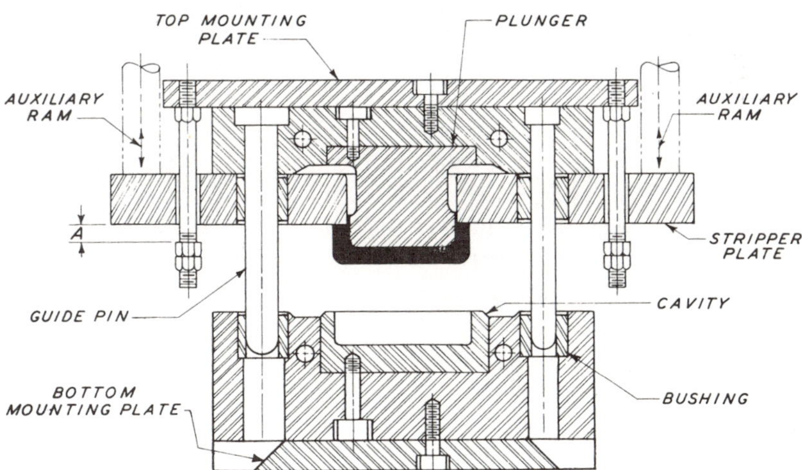

FIG. 5.7 Diagram showing construction used for a stripper plate mold. The stripper plate is pushed down by auxiliary press equipment causing molded parts to be "stripped" from plunger.

inner core of air or undensified unheated material. Such products are sure to fail since all the properties of the materials vary with the density. Other complicating factors include variable section of the part, contours that impede flow, multiple pins that must be encircled, and material welded beyond these pins. Premature hardening in thin areas may prevent complete closing of the mold.

All compression molds must make allowance for the escape of air or gas. Many flash-type molds are merely opened for "breathing" to permit the gas to escape after the compound has been densified. Failure to vent pieces properly will result in surface blisters, broken pieces, poor ejection, and poor product quality.

Several fundamental tools have been developed by the compression molders to meet these many problems and facilitate the high-speed, low-cost production of good quality parts.

The quantity of the charge must be uniformly measured; this is accomplished by volume or weight measurement. The amount of the charge must be adequate to produce a fully densified piece and some flash. Unless there is flash, the part will not be fully densified. A surplus of flash is no proof of a fully dense part.

The shape of the load introduced into the cavity can be very helpful. By compressing the charge into a preform or pill, much air will be excluded. A preform or shaped charge will fit the mold cavity contour more closely and insure better fill out. Preforms have greater heat conductivity than loose powder, and properly designed preforms minimize internal flow problems.

Proper placement and distribution of the load will help prevent blocking vent areas, and in general, provide thin sections of compound in thin product areas and thick preform where the product is thick. Annular areas gain from annular preforms since this eliminates flow and knitting problems. Preforms and pills can be preheated more easily and uniformly than loose powder.

The mold charge should, when possible, be preheated up to the temperature where it is plastic and will flow easily as the mold closes. Proper preheating of the charge will minimize many types of product failure and reduce the molding cycle. Since the plastics are poor thermal conductors, they are difficult to heat by conduction. Best results have been obtained by dielectric preheating, which insures fast preheating of the entire charge; this heat generates in all particles of the charge at a uniformly fast rate.

The molding pressure needs to be adequate to fully densify the charge and entrap it solidly within the mold. The rate of close will depend upon the amount of preheat. If a cold charge is used, the mold must close slowly so that the charge is warmed up to the plasticity state before the mold is completely closed. Too fast closing will entrap air, generate voids, exclude essential material, and make useless parts. It will also displace inserts, and break fragile mold sections. If the rate of close is too slow, the compound may become plastic and then harden in certain areas before the mold is fully closed, thus preventing complete densification of the charge and closing of the mold.

Poor compression molding procedures will produce bad parts. Electrical properties are low when the compound contains excess moisture. Drying is necessary to remove moisture. High frequency preheating will heat but *not* dry the charge. Dimensions of the part depend upon the mold dimensions, the rate of shrinkage, and the molding procedures. Improper placement and distribution of the mold charge will cause dimensional variation. Uniformity of flow and distribution of the charge will provide uniformity of shrinkage and dimension.

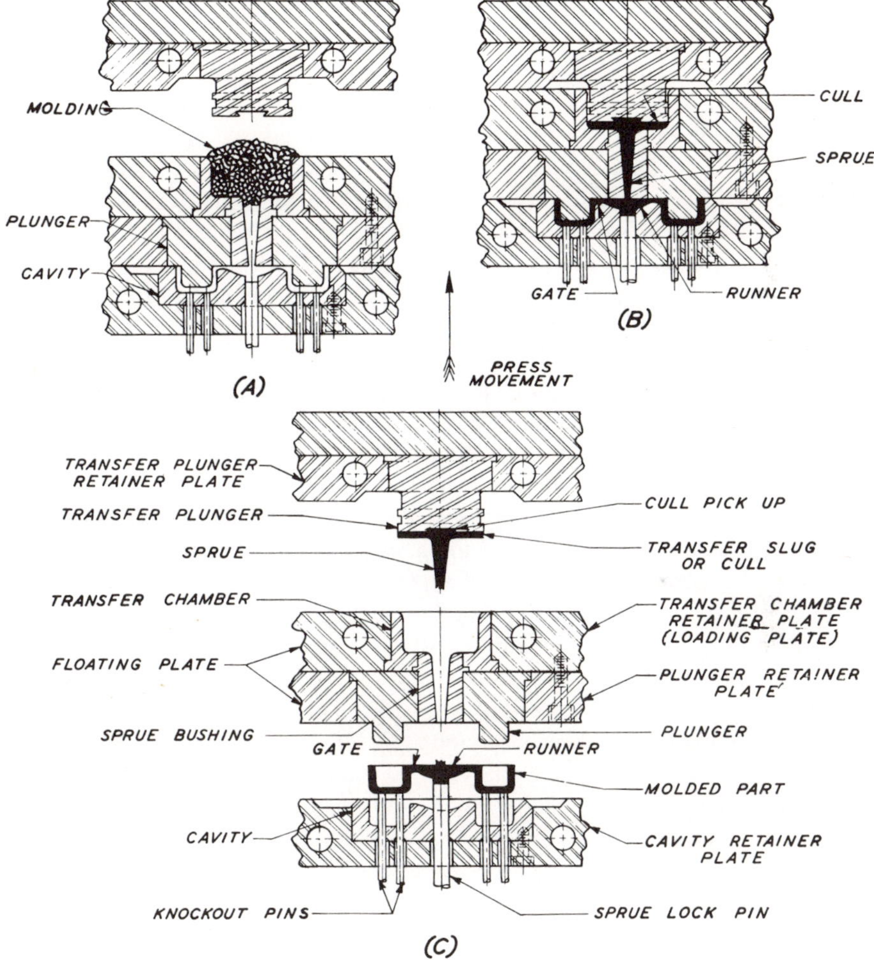

FIG. 5.8 (A) Integral transfer mold ready for pressure to be applied to compound in chamber. (B) Mold fully closed. The compound has been plasticized by heat and transferred from the chamber through the sprue, runner and gates to the cavity. (C) Mold fully open and ready for operator to remove molded parts and cull.

Non-uniformity of mold heating and mold pressures will result in nonuniform dimensions and warpage. Excess moisture affects dimensional uniformity and warpage. Preshaped charges, uniformly preheated and compacted by the correct rate of close in molds that are uniformly and adequately heated, will insure greatest dimensional uniformity.

In transfer molding, as illustrated by Fig. 5.8, the charge is plasticized outside of the mold cavity, and then forced in. Since the thermosetting materials start to harden as soon as they are plasticized, each mold load is made soft and plastic before it is placed in the transfer chamber. Then it is transferred or moved by pressure through the runner and gate channels into the mold cavity.

The transfer mold performs all of the aforementioned operations in sequence, and the molder becomes an unskilled machine operator. In the transfer molding operation, a unit charge of adequate quantity to fill all cavities, runners, vents, etc., plus a small excess, is preheated and dropped in the chamber and the press takes over. This can be manual or automatic. The rate of close can be set at the right speed to fill all cavities uniformly, pushing the air ahead and out of the vents. Since the plastic charge flows into the cavities, the placement of the charge, shape of the charge, time of gassing, etc., are eliminated as problems. The mold can be heated to maximum cure temperature to gain maximum molding speed. The cavity is clamped closed so no material can be pushed out, and density uniformity is assured with no flash at the parting lines. The initial mass is fluid so that there will be no high stresses to dislocate inserts, crush fragile mold sections, bend pins, and produce excessively thick parts.

Dimensions across the parting line are more accurately held by the transfer molding process, since the mold is closed when the compound enters. Gate location has a great effect on shrinkage and dimensional accuracy of transfer molded parts. For example, a side-gated round part may not be completely round. Transfer molded pieces will often show better surface texture because the plastics material rolls over the surface as it flows into the cavity. A compression-molded charge will precure at points of contact on the mold surface often times, producing a "frogskin" effect, instead of a smooth, lustrous finish.

The choice between compression and transfer or injection molding may be limited by strength or concentricity requirements. Annular thermosetting parts are best molded by compression when total concentricity is required. Compression molding is also preferable where maximum impact strength is desired. Transfer or injection molded pieces gain in tensile and flexural strength but may lose some impact strength.

COMPRESSION MOLDING PRESSES

Hydraulic presses are most often used for compression molding. Presses are operated by oil, water, or air pressure. Toggle systems may be used in any of these presses for the mold clamping. Individual pumping systems or a central pressure system may be used as a source of hydraulic pressure. The present trend is towards small, individual motor-driven oil pressure pumps for each press. Any press that will open and close easily will be satisfactory for this kind of molding.

Molding presses often have more than one hydraulic cylinder. The main ram or cylinder exerts the principal pressure required for molding. Push-back cylinders are often used to open the press when molding is completed. The alternative is a double-acting ram which can be driven up or down by the power source. The force available to open the press is often used to operate ejector or knock-out pins for the removal of the molded part. Air pressure cylinders or hydraulic cylinders are used in most cases to reset the ejector pins after they have been raised in ejecting the part. Side ram presses are used for some classes of work which must be split in two directions; in this case, one ram is used to keep the two halves of the mold closed, and the other to exert the compression force on the material.

HAND MOLDING

In the hand-molding process, the mold is removed from the press on each cycle and disassembled by the use of die lifts and other fixtures, as shown in Fig. 5.9. The press in this case is very simple, since it serves merely to close the mold and hold it clamped under pressure while the material cures. A work table of con-

FIG. 5.9 This hand-molding press is set up for fast operation by two operators. The mold is shown open on the adjustable height table. The mold slides in and out of the press and gets its heat while in from the steam heated platens. (*Courtesy Moxness Products Inc., Racine, Wis.*)

FIG. 5.10 Standard mold frame for hand mold. (*Courtesy Master Unit Die Products, Inc., Greenville, Mich.*)

venient height is mounted directly in front of the press to enable the molder to withdraw the mold by a sliding operation. Heated platens transfer heat to the mold from above and below while it is in the press. The work table is also heated to keep the mold hot during the open cycle while the operator removes the previously molded charge, loads the inserts and the new charge before reassembling the mold components. Hand molding is no longer done in any volume because of the high cost of hand labor and the additives thereto. Only very intricate moldings with complex side cores and insert assemblies remain as hand-molding jobs. The low mold cost of hand molds has been offset with standard mold components to run in a standard frame as semi-automatic molds, illustrated in Fig. 5.10.

COMPRESSION MOLDING SYSTEMS

Many large presses require the full-time attention of the molder. Some molding plants assign one person to each press and utilize the time during the cure to weigh material for the next charge, remove flash, make preforms, and do other post-molding finishing operations. Multiple-cavity molds are used in this system so that many pieces are produced in one cycle of press operation. This procedure is very effective if the production requirements are large enough to justify the heavy mold investment.

Another system has been developed which utilizes one operator for several presses. Small molds are used, thus minimizing mold investment. Since the operator handles several presses, the operating cost is closely competitive with multi-cavity molding. This so-called "battery" system is set up so that the operator loads and closes one press, moving along to the next one, which is just opening when he arrives. The small molds used in this system give better part uniformity, since there is less mold distortion in small molds than in large. Another gain is the fractional amount of production that is lost when one of several molds is out of production for repairs. The uniformity and quality of parts produced in small molds are generally better than that of large multiple-cavity molds.

Loading Compression Molds

The quantity of molding compound that is put in the mold is determined by volume or by weight. Loading fixtures, as shown in Fig. 5.11, holding the correct quantity of powder when filled level, are often used. These loading fixtures

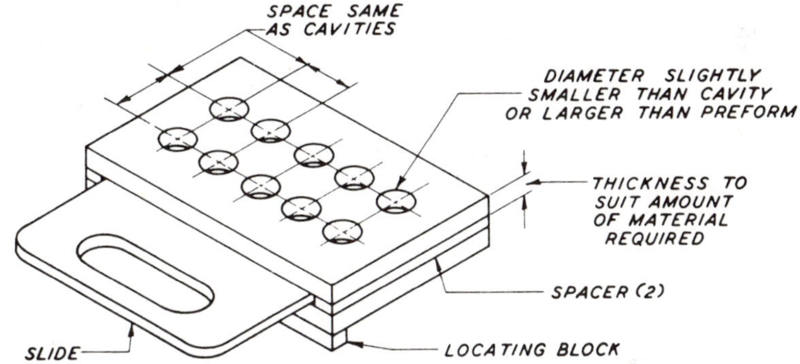

FIG. 5.11 Compound loading board made of wood or other light material.

have openings over each cavity that can drop the charge when the slide is moved. Pills of molding compound and preformed shapes are also used. Automatic preforming machines, as shown in Fig. 5.12, compress the loose powder into compact units of known volume so that one or more units are known to be a proper charge. Specially shaped preforms are used to get better distribution and flow in unsymmetrical product shapes.

The best quality and molding speed are obtained by preheating the molding compound to the maximum allowable temperature before it is put in the mold. Dielectric preheaters, making use of rectangular preforms that will stand on edge during the preheating cycle, give greatest uniformity and amount of preheat, since the heat is generated internally and there is no possibility of hardening the contact surface of the preform. Preheating ovens, hot plates, infrared lamps and live steam are also used for preheating.

FIG. 5.12 This 125-ton hydraulic preform press is used to compact plastics compounds into dense hard preforms. A rotary feed system is included for producing preforms of fines and other highly aerated materials. The vibratory feed hopper is shown above the press. These double-acting preform presses are widely used for melamines, glass-bonded mica, phenolics, allylics, with or without coarse fillers. (*Courtesy Mykroy Ceramics Co.*)

Loading Inserts in the Mold

It is frequently desirable to load metal or other inserts in a mold to become a part of the molded product. Many inserts are used to provide a strong threaded anchorage for screws. Such inserts have an external knurl that strengthens the attachment to the plastics material. The plastics materials shrink more than the metals, making a fairly tight fit. Additional details on insert design will be given in Chapter 8. Inserts may be loaded individually by hand, or special fixtures may be designed to index the inserts in relation to their holding pins, facilitating loading the entire mold at one time, as shown in Fig. 5.13. The quick loading of inserts is an important design consideration for the lowest cost molded parts. Location of inserts and their holding pins must be given careful consideration in the product and mold design, since tremendous molding pressures are generated during material flow that may "wash" the inserts off their locating pins, or break them. Baffle pins are sometimes used to protect the inserts when unusual flow conditions must be encountered. The internal pressure of the flowing material may reach 4000 or 5000 psi, a force sufficient to break large pins or thin

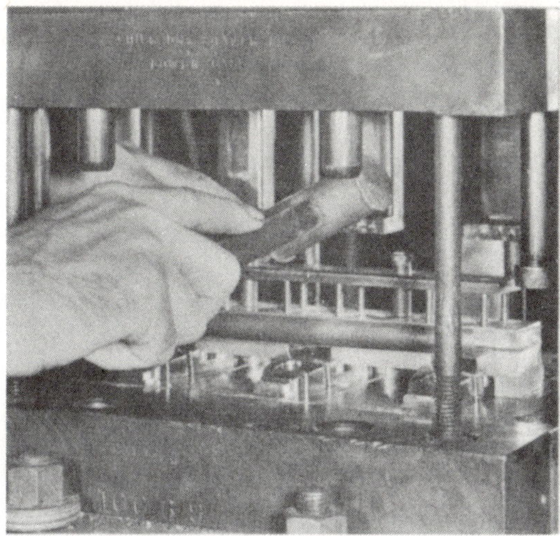

FIG. 5.13 This insert loader holds four inserts for each of the three cavities. They are released by withdrawal of the slide and are then tapped firmly into position on the mold insert holding pins.

mold sections. Fillets in corners and streamlined sections are used whenever possible, to increase the mold resistance to these stresses. Most thin-walled inserts will crush under this pressure, and often need to be retapped after molding. Tight fits eliminate compound from threads.

Curing Molded Pieces

The curing time is dependent upon the type, preheat, and volume of material used. The correct minimum amount of time required for satisfactory cure may also vary with the application, and best guidance will come from the materials maker for critical applications. Some parts which require no particular stability of shape or dimension may be taken from the mold before the cure is complete. Many materials necessitate opening the mold very slightly during the cure to allow "breathing," or to allow the entrapped gases to escape. Failure to do this may produce pieces with blisters and warped sections. Certain materials require an after-baking operation to secure maximum temperature and dimensional stability. The baking advances the cure, and drives off volatile components.

Use of Cooling Fixtures

The final dimensions of a thermosetting molded part are less than those of the mold in which it was made. This is a result of the materials having a greater coefficient of expansion than the metals. The molded piece reduces in size as it cools from mold temperature, and the final dimension will vary considerably in some materials. This is called mold shrinkage, and is generally dependent on the material type and temperature at which the piece is ejected from the mold. This shrinkage is also influenced by the molding pressure, shape of the molded piece, humidity conditions at the time of molding, dryness of compound, and air currents in the molding room at the point of ejection. Best uniformity will be achieved by long cures and chilling the mold before removal of the part.

Cooling or shrink fixtures are often used to hold dimensions with a minimum of effort. This is accomplished by the use of shapes or forms which hold the hot piece rigidly after removal from the mold, while the part cools. The form is accurately machined to the requisite shape or dimension to transmit its contour or dimension to the cooling piece. Many products require an annealing operation, such as cooling in sawdust, to minimize internal stresses.

Molding Time

The time for the cure varies with the material, preheat, cross section, desired dimensional control, pressure, and temperature. This time may vary, from a few seconds to many minutes. Tests have been devised by the material makers that will enable the molder to test the degree of cure that has been achieved. All of the properties will vary with the density of the part and the degree of cure. Molded pieces must be completely cured to provide the full anticipated properties.

FULLY AUTOMATIC COMPRESSION MOLDING

Fully automatic compression molding is extensively used, since this process produces thermosetting pieces at lowest possible cost. Fully automatic molding machines perform each of the compression molding steps automatically in sequence—one type of machine is shown in Fig. 5.14. The feed board, as shown in Fig. 5.15, moves automatically into position over the cavities after the previous load has been ejected. The feed tubes are filled by gravity during the cure cycle from the material hopper. The feed board is shown in the inverted position to view the sliding plate that moves into a position to allow the compound to drop from the feed tube into the individual cavities at the proper time. This feed board retracts, and the press closes, thus curing the parts. As it opens again, a comb plate enters, with open slots to strip the molded pieces from the knockout pins, and drop them into the discharge chute. All of this is done in a fast, fully automatic cycle. Such presses are built to include automatic insert loading where continuous volume production justifies their cost. Post-molding assembly of inserts, by fully automatic methods, is an excellent alternative.

Fully automatic compression molding machines insure following all of the essential steps that a molder may forget. Rejects are less, and mold life is extended by automatic operation. The sequence of operations follows an exact predetermined pattern. The volume and placement of the charge is controlled exactly by the loading board. The rate of close and the breathing interval is con-

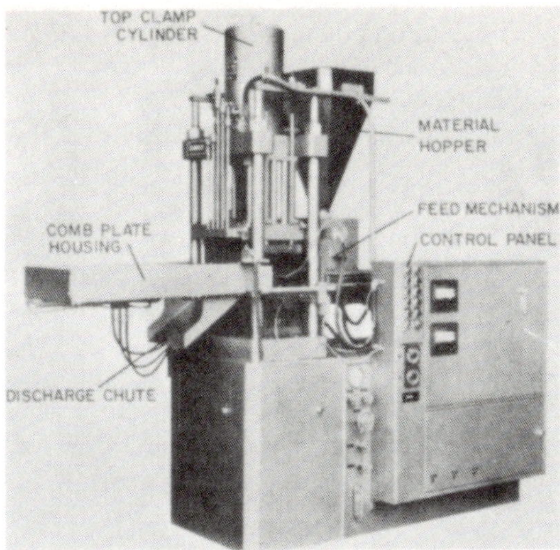

FIG. 5.14 A typical 75-ton fully automatic compression molding press.

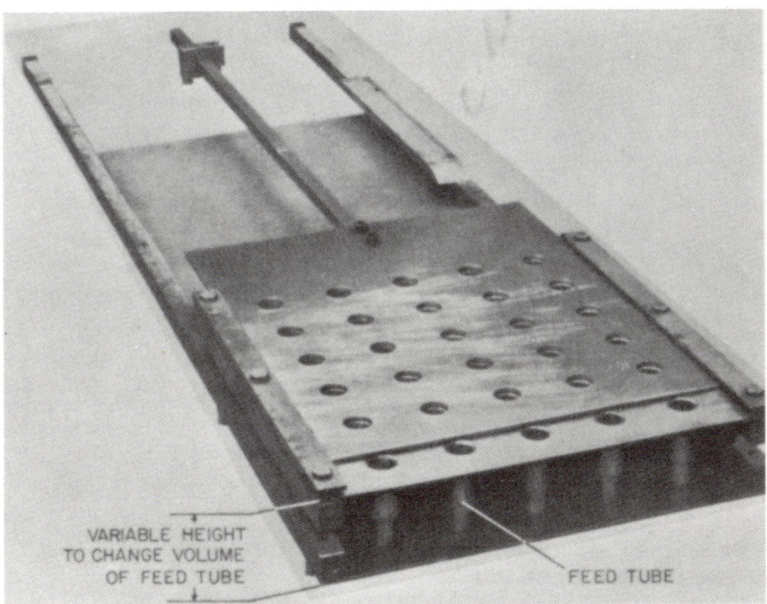

FIG. 5.15 Typical feed board from automatic press for volumetric loading of cavities.

trolled by precision timers. Mold designs include vents and material entrapment features to insure full density products. When fully automatic compression molding procedures are applicable to the product design, such parts can be made with maximum uniformity and at minimum cost. Flash-type compression molds are the least costly; fully automatic compression molds are the most expensive, and the cost of transfer molds lies somewhere in between. In all cases, the product cost and quality are the deciding factors.

Another approach to automatic compression molding makes use of an extruder to plasticize the compound and deliver the proper weight of fully plasticized compound in the mold cavity. The fully automatic unit shown in Fig. 5.16 has ten stations, any one of which may be cut out of the sequence. Each station is controlled independently to suit its time cycle and the charge weight to facilitate running a different job in each station. This is probably the lowest cost method of molding thermosets and gaining the extra advantages of compression molding.

FIG. 5.16 This 10 station automatic compression molding machine uses an extruder for plasticizing the charge. Each mold station is independently controlled. (*Courtesy Alliance Mold Co., Inc.*)

TRANSFER MOLDING

The transfer molding process may be done in compression presses by the use of integral type molds, as shown in Fig. 5.8, or in special presses, equipped with an auxiliary cylinder to operate the transfer plunger. A typical plunger mold design is shown in Fig. 5.17, and a plunger press is shown in Fig. 5.18. The pieces made in this press are shown in Fig. 5.19. Note that the plunger mold design has a transfer tube to accept the preheated charge before the plunger enters the tube, forcing the plasticized compound through the mold runners, and then into the cavities.

Applications for Transfer Molding

The compression molding process for thermosetting compounds leaves many jobs that can not be done. In the compression process, the molding compound is compressed and reduced to the fluid stage in the mold cavity. During this process, part of the material may lie in hard chunks, while other portions are flowing rapidly with great force. These internal stresses are very great, and will break large mold pins, thin mold sections, and wash inserts from their pins. The

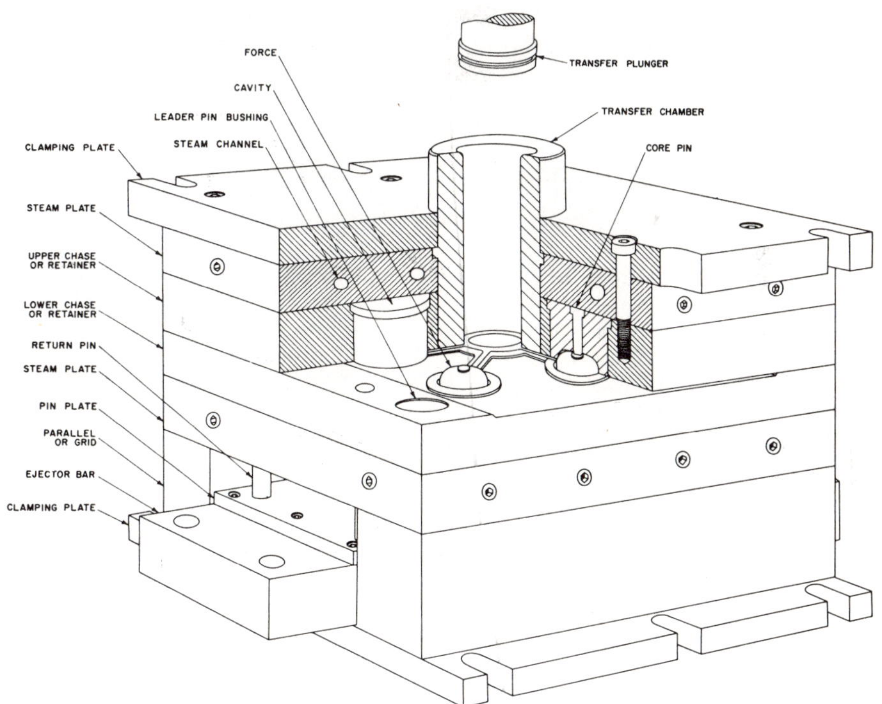

FIG. 5.17 Plunger transfer mold. (*Courtesy Wayne F. Robb.*)

FIG. 5.18 Three hundred-ton transfer plunger press equipped with top and bottom transfer plungers, for either type of molding. The 24-cavity mold for the piano sharps shown in Fig. 5.19 is installed in this press.

molding of fine holes, intricate sections, and complicated cores, is often extremely difficult with compression molds. The transfer process makes possible the molding of parts having all of these complications. Since transfer molding material flows into the mold cavity as a fluid, no large stress is built up to crack mold parts and displace the inserts. Glass and very fine wires are frequently molded in transfer molded pieces.

The fabric-reinforced materials produce a heavy flash, and are difficult to mold by the compression process. In the unmolded form, these materials are bulky and must be compressed—in some cases to one-eighth of their original volume.

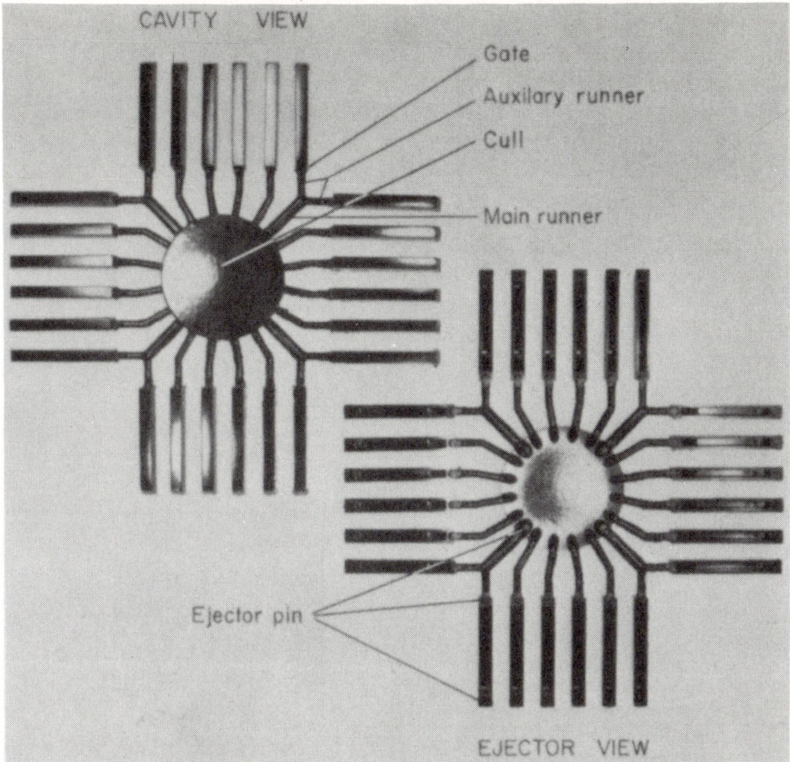

FIG. 5.19 This well-designed 24-cavity transfer mold runner and ejector system illustrates many essential design features. Note the short runners for minimum flow and runner take-off from the cull. (*Courtesy Tech Art Plastics Co., Morristown, N.J.*)

With compression molds, this necessitates large material loading space, since a considerable amount of the compound is squeezed on to the parting area, it is difficult to get a tight mold closing and a narrow parting line that will finish easily and look well. These same reinforced plastics compounds are molded easily in transfer molds, and no parting-line flash is left, thus eliminating expensive finishing. In this case the mold is closed when the compound enters the cavity. The minimum parting line is also very desirable for fine appearance products, such as flatiron handles and appliance parts. Transfer molding produces a smooth, glossy surface, that can not be attained by compression molding. For some classes of work, such as simple shapes, transfer molding may be more economical than compression molding. Encapsulation is done by transfer molding, as shown in Fig. 5.20.

FIG. 5.20 The "shuttle press" is a specially designed transfer molding press for insert molding. Using one top and two bottom mold halves, the operator achieves maximum efficiency, as one molds half is loaded while the companion half is undergoing cure. The press features a hydraulic shuttle feature; downward high-speed clamp, separate hydraulic ejection on each "wing," plus accurate control of all press functions. (*Courtesy Hull Corp., Hatboro, Penna.*)

FULLY AUTOMATIC TRANSFER MOLDING

This process is sometimes called injection molding of thermosets, since the process may be done in machines by methods that simulate injection molding of thermoplastics. By its original definition, transfer molding was applied to thermosetting materials only, since there were no thermoplastics then. When the thermoplastics came along, their molding process was called injection molding. Each process utilizes material plasticization, external to the cavity, with subsequent "transfer" or "injection" into the mold cavity.

The fully automatic transfer molding process offers many advantages, since the

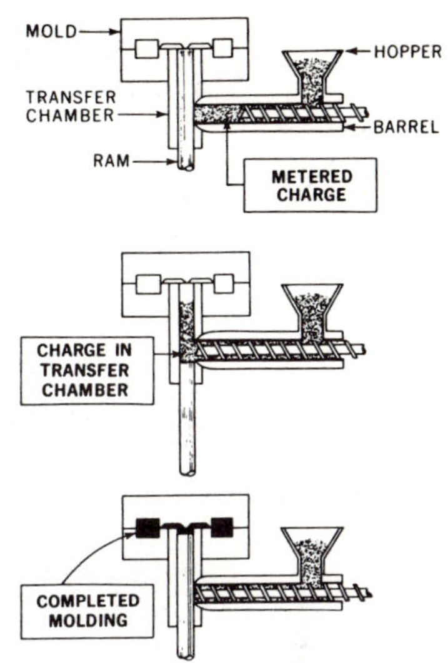

FIG. 5.21 Fully automatic transfer molding of thermosetting compounds is accomplished with machines that require no attendant. The material is heated by the screw as illustrated in position 1. The transfer plunger is then withdrawn and the screw loads the transfer chamber—position 2. The transfer ram then takes over and moves the material into the cavities (3) and at this point the screw starts to plasticize a new charge while the compound cures in the cavities. As the press opens up, combs separate parts, culls and runners, depositing them in their separate tote boxes. One operator can tend several machines since his activity is limited to emptying tote boxes and filling the hoppers.

material may be hopper loaded, with no preforming and no preheating. The machine performs all operations automatically in sequence without the help of an operator. One machine tender can take care of many such machines, only being required to fill the hoppers and empty the tote boxes of the finished parts.

Several types of automatic transfer molding machines have been developed. One type of machine makes use of an extrusion screw to preheat and feed a measured quantity of fully plasticized compound into the transfer tube. At this point the transfer plunger takes over and proceeds with the conventional transfer-molding sequence of operations, as illustrated by Fig. 5.21. This machine also has sliding combs and trays that pick off the parts, separate products and runners, and deposit each in separate barrels.

INJECTION MOLDING OF THERMOSETS

A ram type of injection or automatic transfer molding machine as shown in Fig. 5.22 feeds the material through the heating cylinder into increasingly higher temperature zones. The compound picks up heat from the cylinder by conduction and friction before being pushed into the mold cavity where the cure is completed at elevated temperature. This is similar to the older plunger type molding of thermoplastics except in this case a hot mold is used.

Invention of the reciprocating screw to plasticize and transfer the material into the mold has made possible an advanced injection system for molding the thermosets. Makers of thermosetting materials have contributed greatly to this system as described in Chapter 2.

FIG. 5.22 This fully automatic transfer press is widely used for small parts. Powder is advanced through a heating cylinder and on into the mold. This is identical with the early methods for injection of molding thermoplastics. (*Courtesy Grayhill Moldtronics*)

The standard injection molding machine can be fitted with a special cylinder and screw to mold the thermosets. Outwardly the cylinder looks much like the thermoplastic version in that it is fitted with heaters, but it is also fitted with cooling means so that control on the temperature can be held closely. This is essential so that the thermoset does not harden in the barrel. The heat from the barrel and the frictional effects from the screw changes the compound as it travels forward from granules into a semi-viscous mass. This collects in the barrel ahead of the screw thus slowly forcing the screw back, under back pressure, until the shot volume is achieved, exactly as in the thermoplastic machine. The mold at this point is in the clamped position and as the screw moves forward under hydraulic pressure it forces the viscous compound through the nozzle and runner and gate system into the mold cavities. The final cure of the thermoset is effected by the additional heat obtained from the hot mold.

Some materials may be molded in a continuous automatic transfer molding method called the cold plunger process, which is illustrated in Fig. 5.23. The material is fed in rope form at one station of a rotary table containing several molds. After the mold fills, the transfer ram is removed in a few seconds and the table rotates, bringing up a new mold for filling. The molds then proceed around the table during their period of cure until they arrive at an unloading station, where the mold is opened and parts ejected and cleaned for a repeat of the cycle.

Most machine manufacturers can furnish a suitable thermoset injection system and the injection end of a representative machine is shown in Fig. 5.24. For the thermoset molder who wishes to convert his compression molding presses to the injection process an injection module is available. This is shown in Fig. 5.25.

FIG. 5.23 The cold plunger technique for automatic molding of thermosetting materials is illustrated by this machine. It works well with polyester glass and other materials that may be transfer molded at low pressure on fast cure cycles. A series of identical molds are loaded on this turntable and filled with compound at a transfer station. After a few seconds dwell, the plunger can be withdrawn and the mold moves along around the table during the cure. At the final station, parts are ejected, the mold cleaned ready for the next shot. (*Courtesy American Cyanamid Co.*)

FIG. 5.24 Thermoset injection cylinder. (*Courtesy Van Dorn Plastics Machinery Co.*)

At this time many molding compounds specifically designed for the injection process are offered by the material manufacturers. The pelletized versions are of special value to the thermoplastic molders who also engage in thermoset molding because of the elimination of the dust problem which would contaminate their thermoplastic operation. Figure 5.26 shows a molding room in which thermosets and thermoplastics are being molded in the same shop without dust contamination.

The molder's advantages in the injection process lie in reduced cycle times, flashfree products, reduced rejects and reduced operating costs resulting from the lower skill level required by the machine operator. The recent development of the "warm" manifold system, which is similar in operation to the hot-runner mold in the injection field, has resulted in further gains because the scrap losses in the usual runner system are eliminated. The schematic of this system is shown in Fig. 5.27.

The development of the thermoset injection process has resulted in the resurgence of the use of these materials which for a long time were curtailed by the injection molded engineering thermoplastics. The principal advantages of the

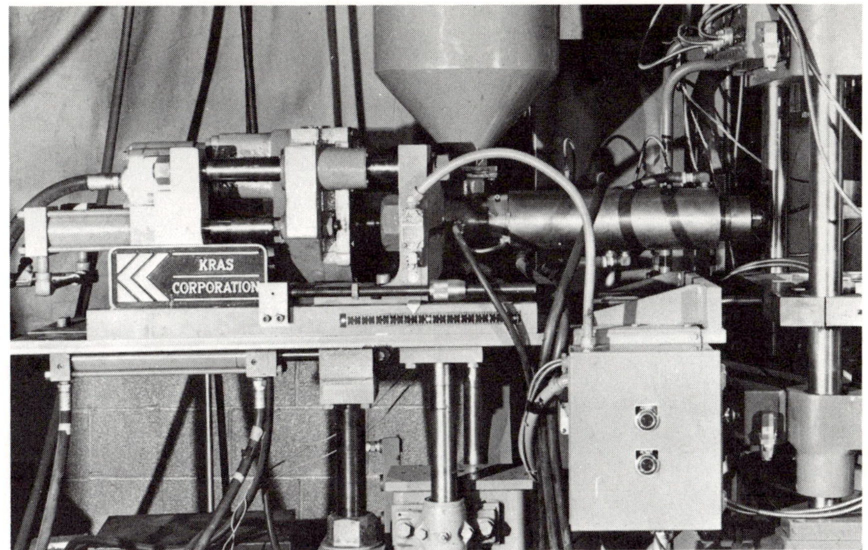

FIG. 5.25 This conversion unit will convert a conventional compression press into an injection machine for thermosets. (*Courtesy Kras Corp.*)

FIG. 5.26 Special thermosetting compounds facilitate the molding of thermosets and thermoplastics in the same room without dust contamination. (*Courtesy General Electric Co.*)

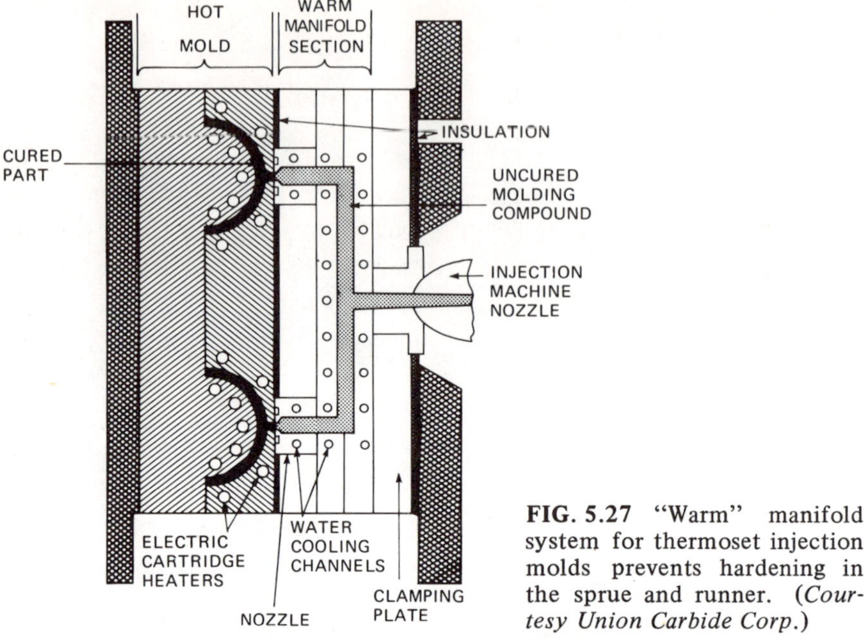

FIG. 5.27 "Warm" manifold system for thermoset injection molds prevents hardening in the sprue and runner. (*Courtesy Union Carbide Corp.*)

FIG. 5.28 This encapsulation mold makes use of an insert-product holder (front view) frame to facilitate insert loading and part removal. (*Courtesy Hull Corp.*)

thermosets, such as heat and chemical resistance, excellent flame resistant characteristics, desirable physical properties, and slow failure under very adverse conditions are all obtained at reasonable cost.

ENCAPSULATION PRESSES

Product encapsulation as described in Chapter 6 is often done by transfer molding because of the long insert loading period. Special presses have been designed for this class of molding. Molds as shown in Fig. 5.28 are designed for economical encapsulation molding. Such molds have removable insert-product holder frames to facilitate loading of inserts and removal of pieces while the alternate holder is in the press during the molding cycle. The electronics industry has made a major use of this system.

6 OTHER PLASTICS PROCESSING METHODS

A variety of special procedures are widely used in the processing of plastics in addition to the basic procedures covered in other chapters. Combinations of several processes may be used advantageously for specialty products.

CASTING

Casting resins, in the early days of the industry, were all thermosetting. Today we also have a host of thermoplastics. The original casting resin was the phenolic, developed by Baekeland in 1906, and used extensively until the mid-thirties, when the alkyd and acrylics appeared on the scene. Since that time, other thermosets such as polyester and epoxy resins were developed, and have consequently become very important. The early thermoplastics hot-melt materials were cellulose acetate, butyrate and ethyl cellulose, and we have since seen the advent of polyethylene and butyl methacrylate. Many others are under development. Acrylic-polyester resins are widely used for casting purposes, i.e., embedments and other decorative and advertising novelty products.

Casting resins are molded on a production basis in lead, plaster, rubber and glass molds, depending upon requirements. In all cases, the liquid resin is poured into the mold and the product is cured in an oven with the addition of heat, or cured exothermically by means of a catalyst. Shrinkage of the resin during cure facilitates removal of the product from the mold. Finishing operations include the removal of flash and, in closed molds, removal of the gate. Sometimes the parts are buffed or tumble polished for improved appearance.

Products made by the casting process include sheets, rods, tubes, and profile shapes. Embedments and encapsulations are also accomplished by casting. Another important use of casting resins is for tooling such as draw, bending and drop-hammer dies, as well as in the making of many kinds of fixtures.

BLOW-MOLDED PLASTICS

The extrusion blow molding, or the forming of hollow bodies of thermoplastic materials, is described simply as inflating a softened plastics tube (called parison) while it is confined between the cavity halves of a mold. When the expanding parison contacts the cold mold wall, it becomes rigid and assumes the shape of the mold cavity. The mold may then be opened, the part extracted and the process repeated.

The equipment used to accomplish this job often consists of an extruder to plasticize the resin, a cross-head with die and core to create the desired tube or parison, an air source to inflate the tube, and a press to open and close the mold.

Another method consists of pinching and thus sealing the outer edges of two heated sheets between mold halves, and simultaneously inflating them to the mold configuration. This method was used years ago in making such objects as baby rattles and toys of cellulose nitrate, and is presently popular in Europe in the production of PVC containers.

INJECTION BLOW MOLDING

The injection blow molding process has become increasingly more important in recent years. In the past the process was used to produce fine small containers, today it is commonly used to produce 32 oz. bottles and jars. Machines are being made that will produce containers up to 15 gallon capacity.

In this process the parison is created by injecting the resin around a hollow mandrel—the core rod—in a parison mold. Figure 6.1 shows the core rod and the cavity mold used in making the parison for a small bottle. Immediately after injection, the mold opens and the parison on the core rod is quickly transferred to the blowing mold shown in Fig. 6.2. Air to expand the parison to the shape of the blow mold is blown through the core rod. The finished bottle on the rod is shown in the right hand view of Fig. 6.2. When the mold halves separate, the finished bottle is blown or stripped from the core rod which now returns to the injection station to repeat the cycle. Containers molded in this fashion have accurately molded neck finishes and require no trimming or finishing of any kind. Bottle weight tolerances are also held to extremely tight tolerances and weight distribution in the bottle can be effectively controlled in this process.

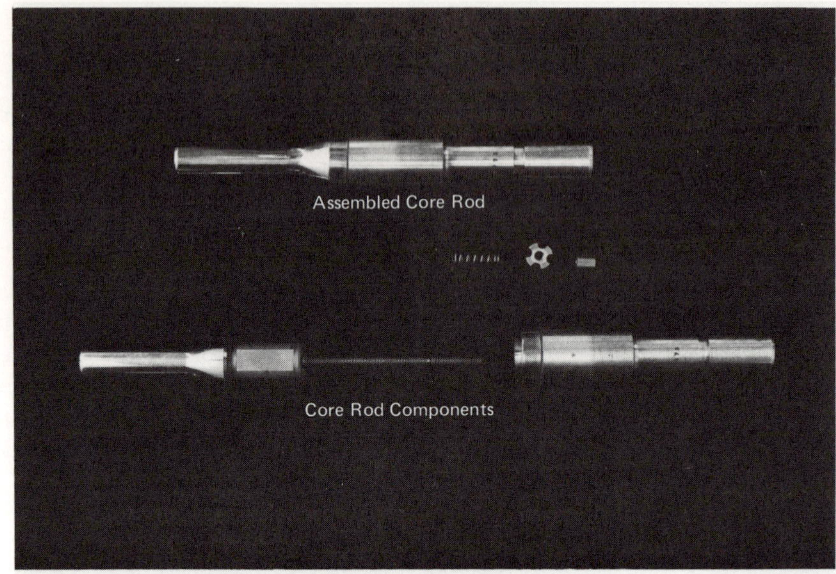

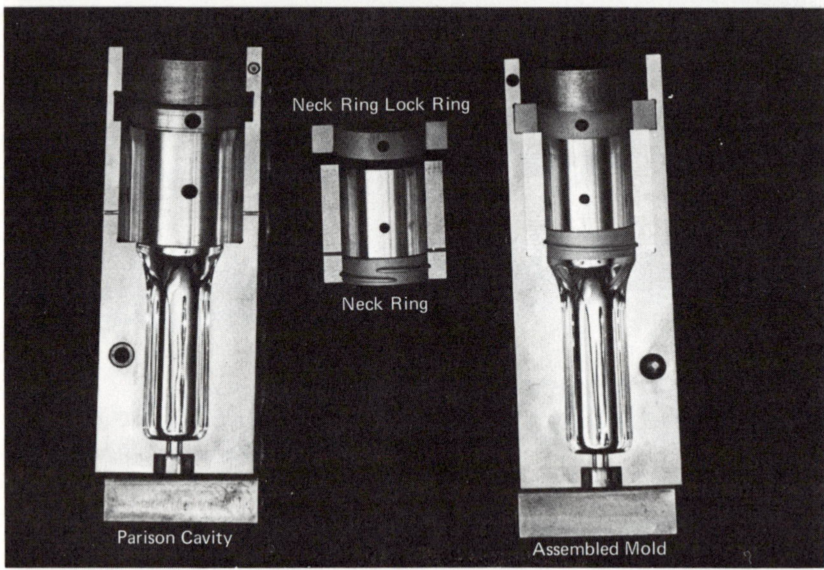

FIG. 6.1 Core rod and parison cavity. (*Courtesy Wheaton Plastics Co., Mays Landing, N.J.*)

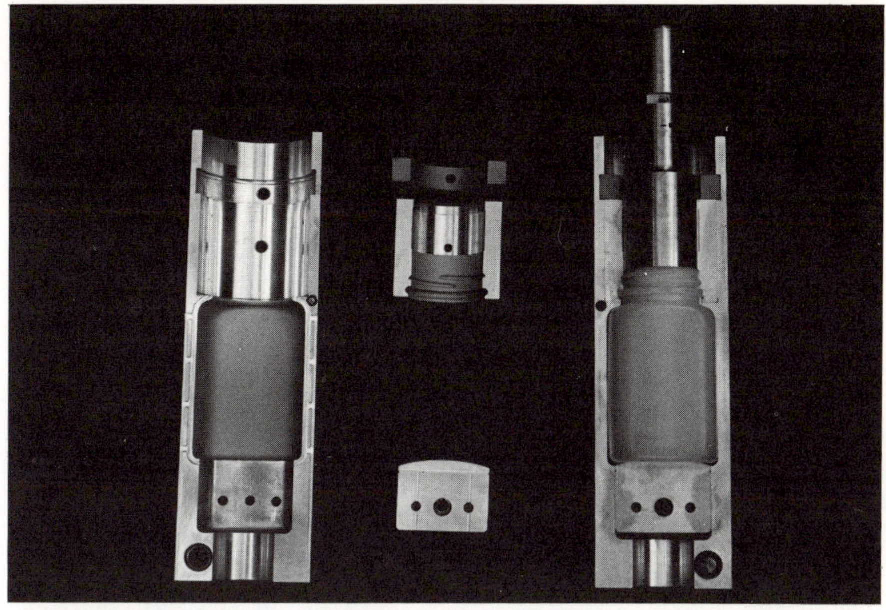

FIG. 6.2 Blow mold. (*Courtesy Wheaton Plastics Co., Mays Landing, N.J.*)

An injection blow molding machine capable of producing finished containers is shown in Fig. 6.3.

The injection blow molding process can be used with all of the thermoplastics including heat-sensitive materials such as PVC.

Extrusion Blow Process

Extrusion blow molding is of considerable commercial significance at this time. As in tube extrusion in general, the quality of the parison for blow molding is important. In order that the tube be delivered in a properly plasticized condition, screw L/D ratio of at least 20:1 is recommended. Resin and machine manufacturer's guidance should be sought in selecting the correct screw for the resin to be processed.

When the plastics melt leaves the extruder, it must be converted into tube or parison form. This is done by means of a cross-head fitted with a suitable die and core. A typical arrangement is shown in Fig. 6.4. This head design would be used for bottom blow.

Rising Mold Process

Numerous processes are used commercially to make the blow-molded products. One of the more common processes is the rising mold method, as shown in Fig. 6.5. This method also includes a bottom blow pin which is used to size the

FIG. 6.3 A commercial injection blow molding machine is shown suitable for wide range of sizes. (*Courtesy The Rainville Co. Inc., Middlesex, N.J.*)

neck finish. The neck flash needs only to be pulled off to create a bottle requiring no neck-machining operations. The molding sequence is as follows:

Position I: The mold halves C-C′ close about the parison "B" which has engaged the hardened steel blow pin "F." When the mold is closed, the parison is pinched between the blow pin and cut-off plate "D," and of course at the top pinch "G." Air is blown through the pin to inflate the tube as the assembly is simultaneously lowered to *Position II.* The tube is automatically cut with a flying knife, just prior to lowering the assembly. With some resins this cutting operation is suspended, and the tube is ruptured as the mold assembly is lowered.

Position II: The inflated bottle cools while the extrusion of the succeeding parison continues. The extrusion rate is synchronized with the rate at which the bottle cools.

Position III: Cooling completed, the mold opens, the blow pin is retracted from the neck, and the bottle is air-ejected from between the mold halves. At this time the new parison has extruded to its correct length, the mold assembly rises to *Position I*, and the cycle is repeated.

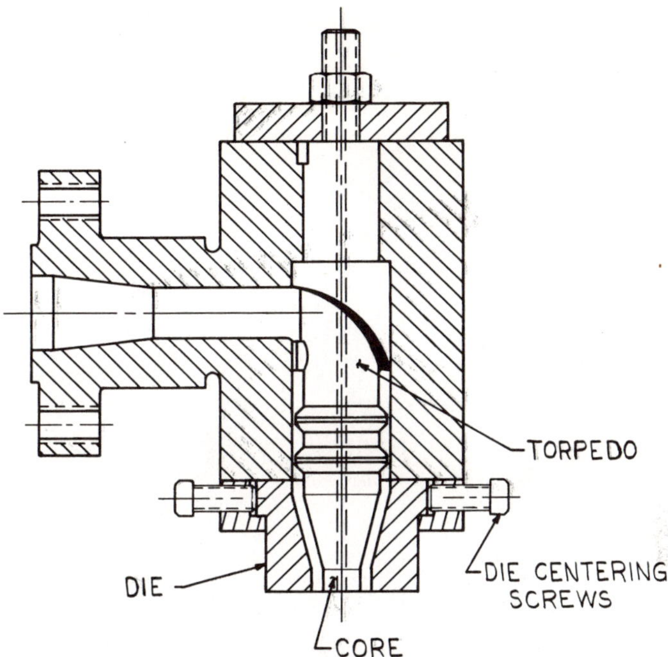

FIG. 6.4 The cross-head die is extensively used in blow molding to form the parison and this typical arrangement is used for bottom blow.

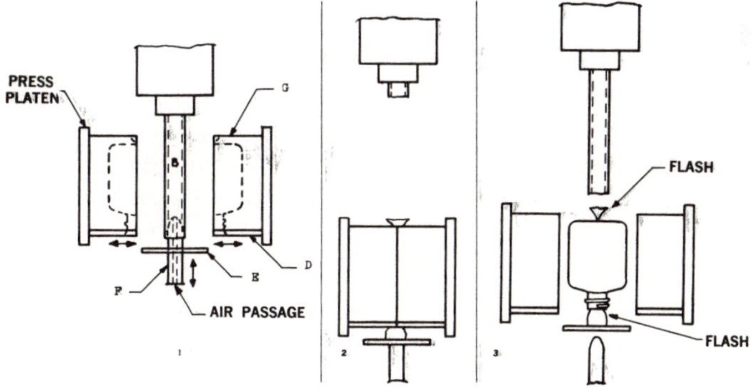

FIG. 6.5 The rising mold method is in common use on blow molding machines. In the left view, the mold is in its uppermost position ready to close on the parison. The center view shows the mold in its lowered position. Cooling of the product is completed in this position. When the mold opens as in the left hand view, the product is ejected. The mold then rises as in the first view and the process is repeated.

FIG. 6.6 Shown here is a Kautex V-8 blow molding machine using the rising mold process. (*Courtesy Kautex Machines Inc., Linden, N.J.*)

The above described method deals with single head operations. The process is also, however, commonly used with multiple heads. The rising mold method is used for producing $1/2$-ounce to one-gallon size. Beyond this, the method is impractical due to the forces of inertia created by the mold and the machine masses which must be raised and lowered quickly. The advantages of this particular method are:

(1) Continuous extrusion of the tube enables improved control of the parison and least load disturbance of the extruder, as opposed to start-stop operation.

(2) Continuous extrusion provides best utilization of extruder output.

(3) Neck finish requires no post-machining operations. A machine which operates on this principle is illustrated in Fig. 6.6.

Parison Transfer Method

Two molding methods which circumvent the disadvantage of the rising mold method in larger molded products are parison transfer, and intermittent parison

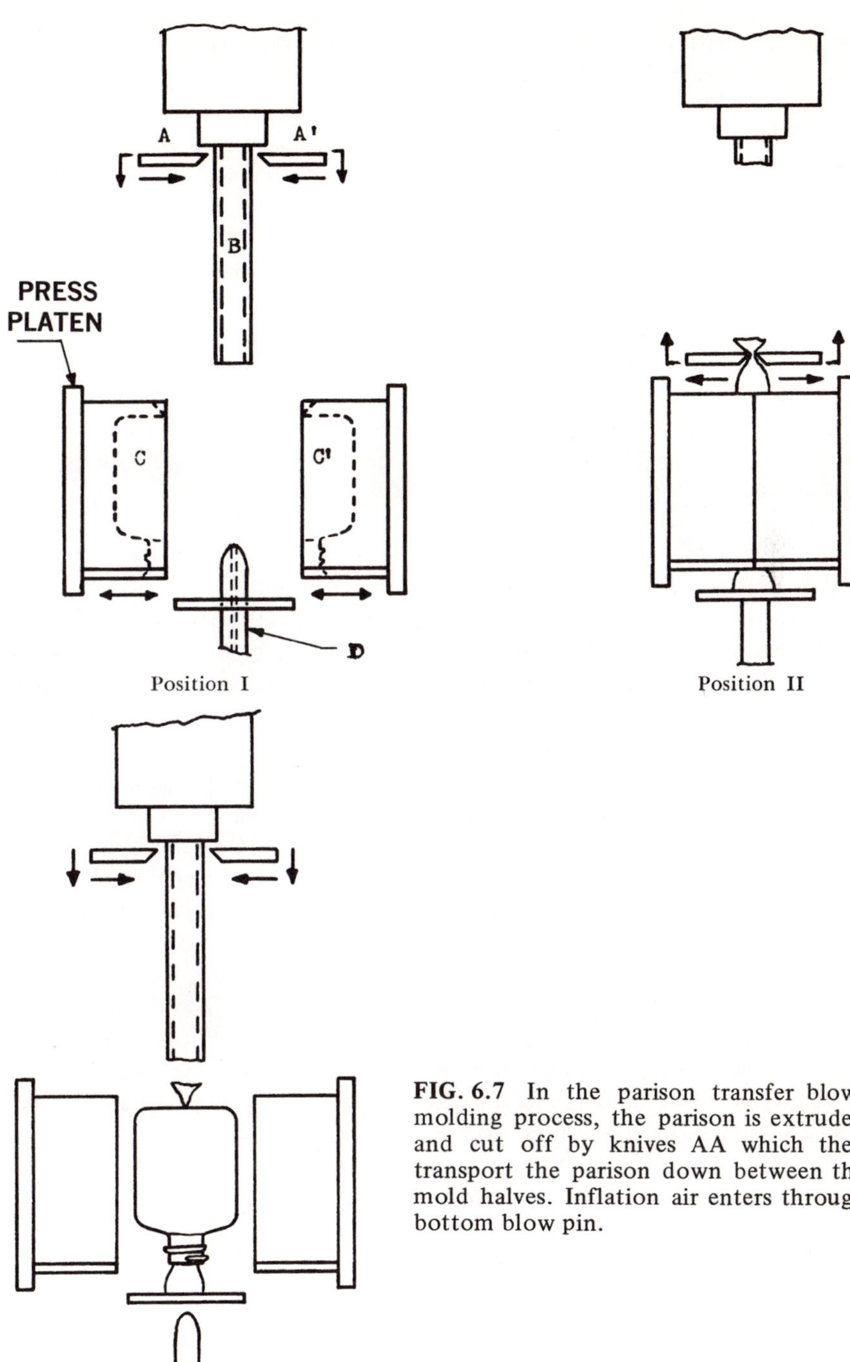

FIG. 6.7 In the parison transfer blow-molding process, the parison is extruded and cut off by knives AA which then transport the parison down between the mold halves. Inflation air enters through bottom blow pin.

extrusion. Figure 6.7 demonstrates the parison transfer method. The molding sequence is explained as follows:

Position I: The parison has extruded to proper length and the cut-off transfer knives A/A' move in, cut the parison "B," and move it down between the mold halves C/C', and over the blow pin "D." The mold halves C/C' then close to pinch the blow pin, and the process proceeds as previously described.

Position II: Depicts the end of the movement described in *Position I* above. The cut-off transfer knives now open and return to the up position, as shown in *Position I*, ready to repeat the cut-off transfer movement.

Position III: The molds have opened and the blow pin has retracted. The molded part is now extracted and the parison again moves down, as previously described in *Position I.*

The parison transfer process is sometimes used in bottles as small as one quart, with as many as five heads, and is also used in carboys as large as 5 gallons, with a single head.

Intermittent Parison Extrusion Method

As the parison becomes larger and heavier, as is required in large size blow moldings, the problem of parison stretch as it is extruded becomes important. In this case, it is best to store the melt in a plastics accumulator and to permit the extruder to operate continuously. When the mold is clear of the previous product, the plasticized resin in the accumulator is forced out through the cross-head by a hydraulically operated piston. The parison thus formed is quickly pinched by the mold to minimize "hot stretch," and expanded to form the product. Large volume containers are made by this method. The range at this time is from 5 gallons with multiple heads, to 55-gallon drums with a single head. Even larger items are in development. The machine shown in Fig. 6.8 is illustrative of an intermittent or accumulator-type blow-molding machine.

Other continuous extrusion methods, such as sliding molds, valved manifold extrusion with multiple heads and molds, are also used. The number seems infinite. However, the important aspect is an understanding of basic methods.

With the exception of injection-blow molding, the processes described are representative of the methods used in the custom molding shops, where product volume is limited to the thousands, rather than multiple millions. Change-over time in the equipment shown is at a minimum. The number of molds, heads and their dies and cores, etc., require the least time for a change-over.

Parison Programming

Parison programming of wall thickness is an important "tool" used by the blow molder. It enables the producer to increase or decrease the wall thickness to accommodate product requirements.

Programming is accomplished by the raising or lowering of a tapered core, relative to a fixed tapered die, so that the parison wall thickness is increased or

FIG. 6.8 Large blow molding machine. (*Courtesy Beloit Plastics Machinery Div.*)

decreased as the bottle walls require. Schematically this is shown in Fig. 6.9. For example, in a neck down-blowing machine, if the neck and shoulder section must be thick, the core moves forward to provide a heavier tube at the beginning of the tube extrusion. If the body then needs a thin wall for "squeezability,"

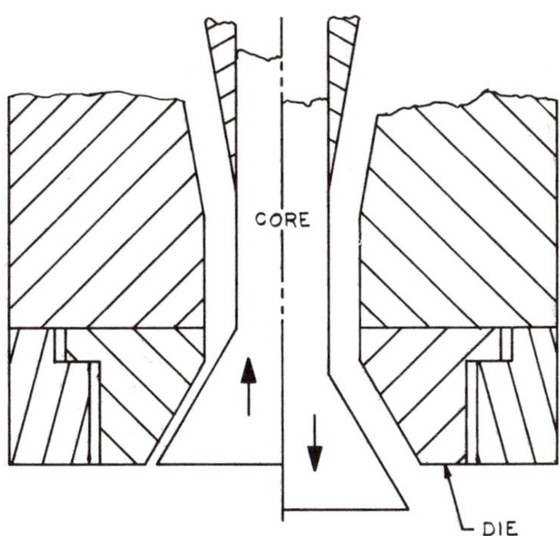

FIG. 6.9 Schematic presentation of parison wall control.

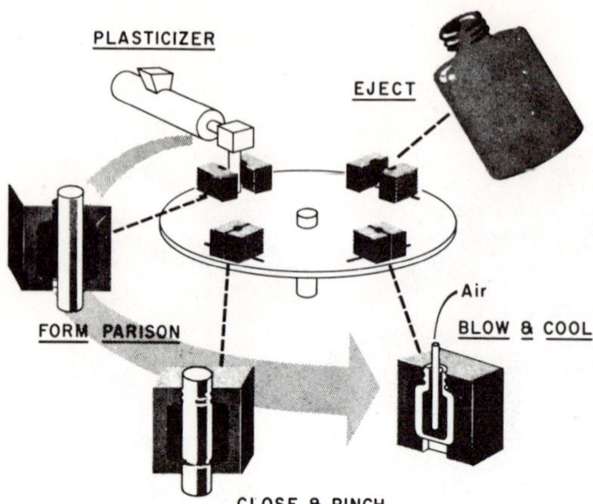

FIG. 6.10 The rotary blow-molding process makes use of multiple molds which clamp in sequence on the parison so that no time is lost in the process. (*Courtesy Dow Chemical Co.*)

the core moves backward to reduce the parison wall thickness. And again, if the base needs thickness, the core moves forward to provide this. In a proper programming system, any combination of values can be obtained.

In very high volume production the continuous tube and rotary process methods have been favored. Two of these are: the rotary process, schematically shown in Fig. 6.10, and the continuous tube process, represented in Fig. 6.11. In both procedures many molds are required. The output is high, and costs are low. The changeover time from one product to another is large, and to justify the cost of the changeover, the product run must be even larger.

In the rotary process, the blowing air is introduced through a blowing tube inserted through the neck. The continuous process utilizes a hollow needle which pierces the parison wall and injects the air to inflate the product.

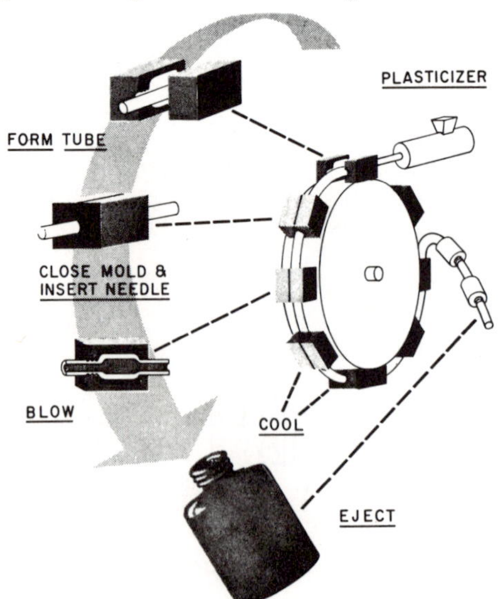

FIG. 6.11 In the continuous tube-blow molding process, the parison is a continuous extrusion that is clamped by the mold halves in sequence as the mold table rotates. Inflation air is applied through a needle that pierces the tube in the blow mold beyond the finish. (*Courtesy Dow Chemical Co.*)

The bottle obtained from either method usually requires neck-machining as a secondary operation. This is generally done in a completely automatic machine.

CELLULAR PLASTICS

Cellular plastics, also referred to as expanded or foamed plastics, are growing members of the plastics family. They are used in the making of insulation, as core materials for load-bearing structures, as cushioning materials used in product protection during shipping and for furniture, bedding and upholstering. Among those plastics which are commercially produced in cellular form are: acrylonitrile-butadiene-styrene, cellulose-acetate, epoxy, phenol-formaldehyde, polyethylene, polypropylene, polystyrene, silicone, styrene-acrylonitrile, urea-formaldehyde, urethane and vinyl.

FIG. 6.12 Radio frequency heated expanded-polystyrene expansion—machine and products, (*Courtesy Creative Packaging, Inc., Indianapolis, Ind.*)

Cellular plastics are available in two types: closed cell, where each individual cell is a completely enclosed void, and open cell, where the cells are interconnecting as in a sponge. Foamed plastics can be made flexible, semi-rigid, as well as rigid, in densities ranging from .1 to 60 lb/cu/ft. Obtainable forms are block, sheet, slabs, molded products, and extruded shapes. These plastics can also be sprayed, and adhere to walls, for example, as insulation, may be foamed in place between walls, or used as a core in mechanical structures.

The plastics are foamed by additives to the resin called blowing or foaming agents. Among these are solvents which cause foaming through volatilization and inorganic compounds which decompose during the foaming cycle, thus creating gases to make the foam. The blowing action in either case is initiated by heat, which is added to the resin (blowing agent mix) during the foaming cycle. The heat may be added by the process, such as during extrusion, or by steam lances which introduce live steam into polystyrene preexpanded beads for "in-place" foaming. Another could be through the exothermic reaction caused by a catalyst mixed with the resin-blowing agent mix. Specific details on procedure are obtained from the resin manufacturer. The "Engelit Process" is described in Chapter 4.

Still another method of polystyrene foaming process has been recently developed. This procedure, called "Dri-electric," utilizes radio frequency heat instead of steam to expand the polystyrene beads. Fast-operating cycles are claimed, with a lack of moisture in the product as in the steam expansion process. This permits immediate packaging. Low cost plastics molds are used. Another feature lies in the ability to include decorative substrates such as paper or plastics inlays, which are impossible to include in the steam lance method. Typical products and the molding machine are shown in Fig. 6.12.

CELLULAR PLASTICS MATERIALS

The urethane foam materials in recent years have taken a greater share of the market because of their unique properties.

The urethanes can be produced in products ranging from cellular, with densities as low as $\frac{1}{2}$ pound per cubic foot; to non-cellular, with a density of 80 pounds per cubic foot.

The major markets for the flexible foams (density, 1.0 to 6 pounds/cubic foot) are in bedding, automotive and furniture applications. The rigid foams enjoy their greatest use in the construction industry as insulation materials; obtained by spraying, pouring in place or as cut slab stock. Figure 6.13 shows the application of a sprayed in place urethane insulation.

Urethane foam moldings are produced by injection molding or casting. The Duromer process, patented by the Phillips Petroleum Co., is predicated on machinery based on a two-part foam system with metering and injection equipment which can accurately and repeatably deliver the correct volume of foaming

FIG. 6.13 Close-up view showing application of sprayed-in-place rigid, closed-cell, fire-retardant urethane insulation. Window casements and underside of roof slabs were insulated with 1 in. thickness at a rate of 2 lb/cu/ft. (*Courtesy Hooker Chemical Corp.*)

mixture to the molds in rapid succession. The molding cycle is fast and products with densities ranging from 30 to 40 lbs. per cubic foot can be produced. Mold pressures are from 25 to 50 psi, and low cost cast aluminum or epoxy molds can be used.

The casting of urethane foam requires equipment to process, meter, mix and disperse the chemicals to produce the foam. This apparatus is quite intricate and its' specifications are best obtained from the manufacturers of such machinery.

In addition to this processing equipment, the casting plant will also require conveyors to carry the molds through the various operations such as mold filling, clamping and through the ovens that heat the molds and cure the cast product. The conveyor will also carry the molds from the oven to unloading stations where the finished product is removed and the mold treated with a suitable antistick release in preparation for the next pass through the cycle.

INJECTION MOLDING OF STRUCTURAL FOAM

A structural foam is a plastics product having a rigid cellular core and solid integral skins.

The stiffness of a structure is a function of its thickness or depth. The strength of the structural element in bending varies as the cube of its' depth. Therefore, to achieve stiffness one increases the depth (or thickness) of the product.

Molding solid thick sections in thermoplastics is impractical because of molding difficulties and cost. It has been demonstrated, for example, in the aerospace field, that if two skins were firmly cemented to a slab of foam, a very stiff, yet light weight structure, would be obtained. The same results are derived in one continuous process with injection molded foamed thermoplastics.

In the foam molding process, the solid skin is formed when the injected foaming mass is chilled by cool mold surfaces. This skin, immediately creates an insulating barrier which still permits the central mass to complete its expansion to form the foamed core. Because of the heat insulating properties of the foam molded products, the molding cycles will be found to be quite slow when compared with those obtained in normal injection molding practice.

All thermoplastics can be foam-molded. Typical of these are polystyrene, polyethylene, polypropylene, polycarbonate, acylonitrite butadiene styrene (ABS), vinyl, and many others. The plastics used in this process range from factory pre-blended resins and foaming agents, to materials that are dry blended with the foaming agent in a machine mounted hopper blender. The foaming agents are usually based in inorganic solids which, when decomposed by the heat in the plasticizing cylinder, evolve carbon monoxide or nitrogen gases. To control the process, a shut-off nozzle must be used on the injection cylinder.

The basic processes currently practiced fall into two categories; (1) high and (2) low pressure molding. Either process is accomplished with the use of reciprocating screw injection molding machines.

The high pressure method is known as the USM process. In this process a mold is completely filled with the foamable plastics mass at normal injection molding pressures from 5,000 psi upward. The mold, known as an expansion mold, is then allowed to slowly open with controlled low resisting pressure to permit the injected mass to expand to a set fixed height.

The advantages of this method are excellent detail in the solid skins, product density is closely controlled, relatively fast molding cycles, and perhaps most importantly a savings of about 50% on weight in comparison to a solid plastics molding.

The disadvantages lie in the special molding machine required to allow platen retraction for the controlled expansion of the mold and, of course, the more expensive expansion mold which is generally constructed of steel to resist the initial high injection pressure. On closer inspection it would appear that almost any hydraulically clamped molding machine could be modified easily to permit platen retraction for mold expansion.

The low pressure processes are generally based on using a standard molding machine and usually with low cost molds, cast in aluminum or kirksite, since the injection pressures are in the order of 100 to 500 psi during foam expansion. Perhaps the only addition required to the standard machine would be a means for increasing the injection speed (an accumulator with suitable valving and controls, is one method).

It is most important to inject the mass as rapidly as possible (less than two seconds). Since the decomposition of the chemical blowing agent is spontaneous as it enters the mold the initial foam would be collapsed by the following mass if the injection were slow. An example of a plaque used for mounting a "half hull" ship model molded using this process is shown in Fig. 6.14.

In practice, the expandable melt is plasticated in the barrel of the screw injection machine during the latter half of the molding cycle and briefly held under pressure by the shut-off nozzle. After the previously molded part is ejected and the mold again clamped, the screw moves forward at high speed with

FIG. 6.14 Low pressure foamed injection molding was used to mold the simulated wood frame. (*Courtesy Revell, Inc.*)

simultaneous opening of the shut-off nozzle to inject a controlled volume into the mold. This "short-shot" immediately expands to fill the mold and create the finished product.

The plastics used in this process are, as in the high pressure process, pre-blends or resins that are mixed with suitable foaming agents in a machine mounted hopper blender.

There are at least two variations from the process just described, but all are based on injecting rapidly a "short-shot" into a mold at low pressure. These are the Union Carbide process, and the molding methods developed in Europe.

The Union Carbide (patented) process is based on using direct gas injection into the molten plastics mass. The gas most generally used is nitrogen. The machinery is not an ordinary screw injection machine but is specially built, consisting of an extruder, an accumulator, a ram cylinder, and a mold clamping press. The nitrogen gas used for the expansion is introduced into the plasticized plastics mass in the extruder cylinder. The process has enjoyed success in the large product area such as pallets, crates, furniture and industrial parts.

The European approach to foam molding has been to develop special machinery specific to the process. These consist of presses with very large platen area and plasticizing equipment capable of extremely rapid mold injection rates. The plastics used are resin/foaming agent blends as previously described.

The advantage in low pressure (short-shot) foam molding is the ability of the custom molder to use his standard molding equipment with only minor changes (if necessary) to increase the injection speed. The basic advantage, as in the high pressure method, lies in the significant reduction in density of the product.

The disadvantage in the low pressure process lies in the poorer surface detail obtained. However, for most commercial purposes this has not been a deterrent when cost and physical properties are considered.

Another interesting sandwich molding process has been developed by the German furniture manufacturer Robert Hanning in collaboration with Siemag, a German machine manufacturer.

The machine consists of two screw plasticizing cylinders extruding material into two short transfer chambers. Connecting these two chambers is a special nozzle shown in Fig. 6.15, which is fed with foamed core material or unfoamed skin material from the transfer cylinders. The center detail is a composite shut-off nozzle, one inside the other, and operated by separate hydraulic cylinders.

The sequence during injection is: (1) Both nozzles are in the forward closed position. (2) The outer nozzle is opened and a small amount of unfoamed skin material is injected. (3) The inner nozzle is opened and both foamed and unfoamed material is simultaneously injected to partially fill the mold. (4) Both nozzles are now closed. The foaming action now causes the skin material to be forced against the mold surfaces and as expansion of the core is completed the mold is filled. Conventional injection molds are used. The more expensive expansion type molds with vertical lands are not necessary in practicing this process.

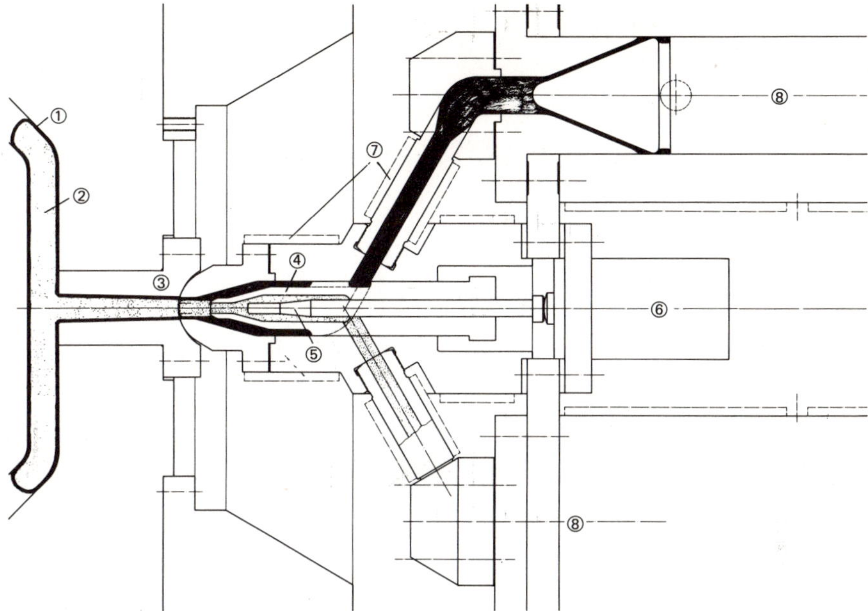

FIG. 6.15 Hanning/Semag process of sandwich molding. (*Courtesy Europlastics & Hanning /Semag*)

Nozzle block, heart of Hanning system: 1—Solid skin of molding; 2—foam core; 3—sprue bush; 4—outer shut-off nozzle; 5—inner shut-off nozzle; 6—nozzle hydraulic actuation; 7—heater band; 9—injection ram.

EXTRUSION OF STRUCTURAL FOAM

The extrusion of structural profiles consisting of a dense homogenous skin with a cellular core was developed in France by Ugine Kuhlmann et Cie of Paris. The method is called the Celuka Process.

The process uses a conventional extruder which plasticizes a thermoplastic compounded or dry-blended with a blowing agent as previously described. This foamable mass is extruded through a cooled profile die head which is fitted with a torpedo (core) having a shape similar to but suitably smaller than the die. This causes the plastics mass to extrude as a fairly dense hollow shape in the form established by the die. The rate of extrusion and the temperature of the die fixes the thickness of the skin. The strength of the resultant skin prevents outward expansion of the melt on the outside of the profile and the remaining plastic mass expands inwardly to form the cellular core. The process can be used with any of the thermoplastics, the choice is based on the economic and end use factors. The density of these extruded products can range from 25 to 50% of the solid polymer.

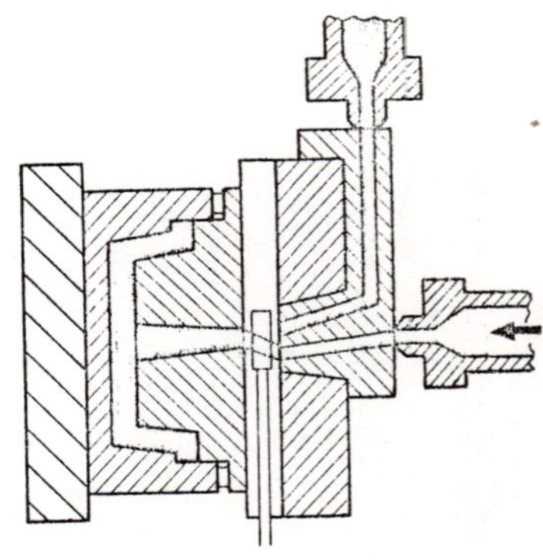

1

The sequence of operations for producing a sandwich molding with an expanded core starts with the injection of the skin material which partly fills the mold.

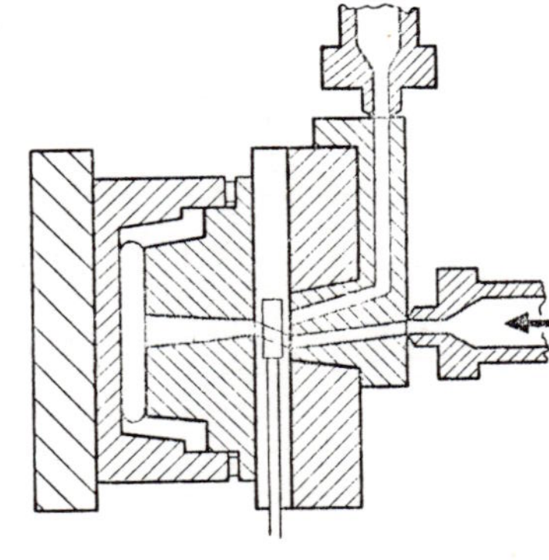

2

The flow-control valve then shuts off the flow of skin material and allows the core material to flow into the mold from the second injection unit. The core material pushes the skin material ahead into the remainder of the cavity without displacing it from the mold surface.

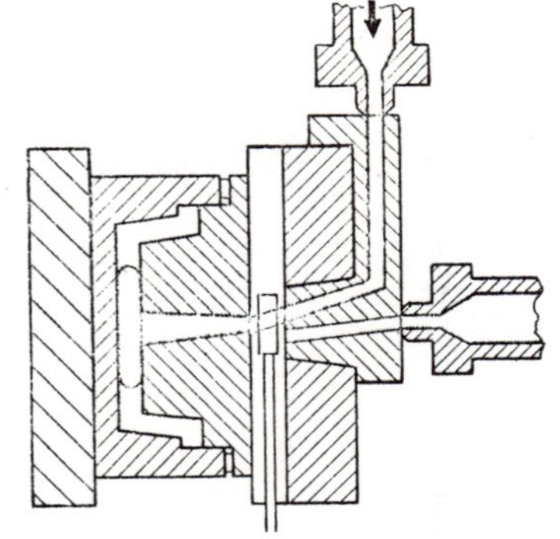

3

Given the right combination of melt and mold temperatures, injection rate, and other variables, the core material forces the skin material to the edges of the mold.

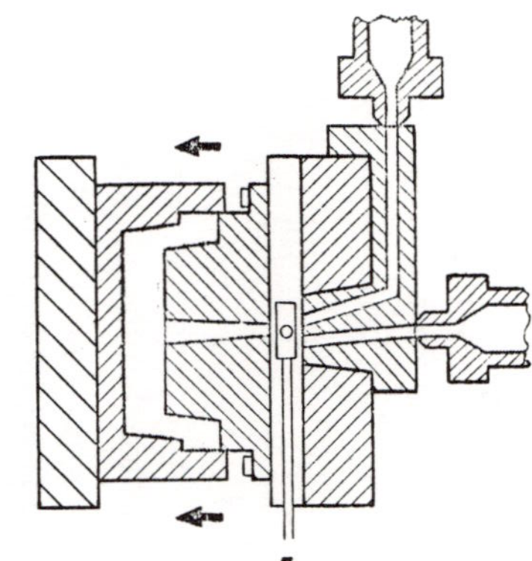

4

After the mold has been filled, the flow of the core material is shut off and a small amount of the skin material is injected to clear the sprue. This is necessary to prevent the core material from appearing at the surface when the sprue is eventually cut off from the molding, and to clear the core material from the valve and sprue. Otherwise the core material would mix with the skin material of the next molding and mar its surface.

5

The valve then shuts off flow to the mold, which is held for a few seconds at full clamping pressure before one half of it is moved a short distance away from the other to allow the core material to expand. The last phase of this cycle is omitted to produce moldings with solid cores.

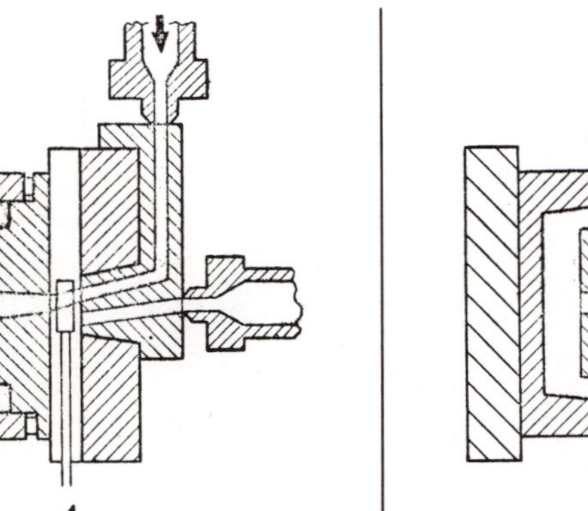

FIG. 6.16 Molding "sandwich" product with foam core. (*Courtesy Machine Design*)

The Celuka process has been extensively used to make all manner of finish "lumber" for housing, such as trim moldings, door and window frames, thresholds, window sills, mullions and a multitude of other architectural shapes.

These shapes can be handled by the carpenter just as he would their wooden counter part. They can be sawed, nailed, or used with screw fastenings. Suitable finishes can be molded in or they can be painted on-site.

All of the processes described have contributed to the use of many millions of pounds of plastics in the furniture industry. In fact, it has been predicted that by 1980 over 3 billion pounds of plastics will be used in the construction of furniture alone, thus far exceeding the use of wood in that field.

The "sandwich molding process" was developed in the early 1970's and is used for molding combinations of plastics materials as well as for the production of foamed core products. It is a two-stage injection molding process that permits the use of two different thermoplastics in the same product as, for example, in acetal skin with a foamed polystyrene core. The process was developed by Imperial Chemical Industries Ltd., England.

Figure 6.16 (1 through 5) schematically shows the sequence of events in producing a sandwich molded product. It should be noted that the last phase of the sequence is the same as in expansion molding previously discussed. The difference lies in the fact that in this new process the skin material can be different than the core.

Typical combinations of materials that have been successfully used are:

Acetal–polystyrene	Polypropylene–filled polypropylene
Acetal–SAN	Nylon–filled nylon
Acetal–ABS	Flexible PVC–ABS or EVA
Rigid PVC–flexible PVC	PTFE nylon–glass filled nylon
EVA–Polypropylene	

The process can be used to produce foamed core products, products with soft outer skins, and flexible cores or moldings having an expensive wear resistant outer skins with a cheaper material in the solid core. It is, of course, obvious that the same material can also be used in both skin and core to produce the conventional form of foamed core product.

Injection molding machines are commercially available to practice this method, or conventional injection molding as required. The elevated side view is depicted to show the angular relationship of the two injection cylinders (Fig. 6-17). An interesting ancillary development to the sandwich molding process resulted from ICI's efforts to minimize mold cost in the experimental products. The molds were made of a metal shell .150 to .200-inch thick sprayed on the model and backed with a reinforced concrete bolster complete with cooling channels. The core is made by building up this cavity with wax of the required product thickness and again spraying this to obtain a metal shell. Once again this is reinforced with concrete. To absorb the compressive loads on the parting face steel

FIG. 6.17 This injection machine is used for producing sandwich molded products using an inner and an outer material. Note the angular relationship between the two cylinders. (*Courtesy G. K. N. Windsor Ltd. G.B., London, England*)

blocks are embedded in both core and cavity halves. The sprayed metal shell is a 70:30 zinc eutectic alloy with a melting point of 280°C. Resulting mold cost is reported to be 10% of a steel mold to produce the same product. The manufacturing life of this tooling is not available. License to practice this technology is obtained from ICI Ltd.

MACHINING THERMOPLASTICS

The plastics materials may be fabricated by conventional machining operations. Twist drills such as are used for soft metals are commonly used. Better results are obtainable by the use of special drills, coolants and techniques that remove the shavings and chips. Figure 6.18 shows modified drill designs that will give best results with many of the thermoplastics. Larger holes make use of hollow end mills or fly cutters ground with zero rake. CO_2 jets are often used as a coolant to prevent melting of the thermoplastics materials during drilling and other machining operations. Coolants must be selected carefully to make sure that they will not craze the plastics; kerosene for example can ruin polystyrene.

Lathe bits should have a zero or negative rake and be held below center in all turning operations. Cutting tools should have no rake and should be cut by a scraping action. Circular saws are best when hollow ground or set to prevent binding. Band saw work must be done at speeds that do not melt the material, and clean chips are good evidence of proper cutting.

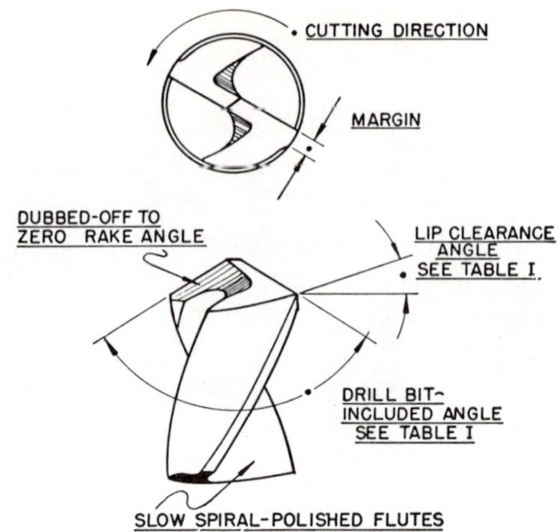

Selection of Drills for Plexiglas

	Shallow	Medium	Deep
Ratio of Hole Depth to Diam.	Less than 1½:1	1½:1 up to 3:1	Greater than 3:1
Chip removal	No problem	Continuous ribbon cleared by flutes without clogging	Material removed in form of powder or minute chips
Drill Bit— Included Angle	55-60°	Depends on size of flute	140°
Drill Bit—Flutes	Wide, polished, slow spiral	Wide, polished, slow spiral	Wide, polished, slow spiral
Drill Bit— Rake Angle	0°	0°	0°
Drill Bit—Lip Clearance Angle	15-20°	12-15°	12-15°
Coolant	None	None	(See text)
Rate of Feed	—	As necessary, to cut continuous chip	Slow—approx. 2½″/minute so that powders, not shavings are formed

FIG. 6.18 (*Courtesy Rohm & Haas Co., Philadelphia, Penna.*)

Many of the thermoplastics can be cut with dinking and blanking dies; some of these operations are helped by warming the stock. Tapping must be done slowly and tapped holes may need reinforcement with metallic inserts. In assembly operations it is important to remember that the thermoplastics will cold-flow under pressure. Glass-reinforced plastics will dull cutting tools rapidly and carbide cutters are recommended for long runs.

The laminated and reinforced plastics are widely used as fabricating materials as presented in Chapter 10.

STATIC AND ROTATIONAL POWDER MOLDING

The molding of plastics powders is a process whereby finely ground plastics is heated in a mold, causing the powder to melt, thus creating a wall of plastics on the inner surface of the mold. The product is strain-free, unlike pressure molded products. In polyethylene, this is especially significant if the product is used to contain oxidizing acids.

There are four processes used in the production of powder molded plastics products: static (Engel), single-axis rotational (such as Heisler), multi-axis rotational molding and centrifugal casting, and also a single-axis rotational process.

Static Powder Molding Process

The static powder molding process consists of filling a simple mold representing the exterior shape of the product with a finely ground powder of polyethylene, as seen in Fig. 6.19. This mold with a heat insulating cover is placed in a circulating air oven, at 400 to 700°F, for a period of time sufficient to melt and

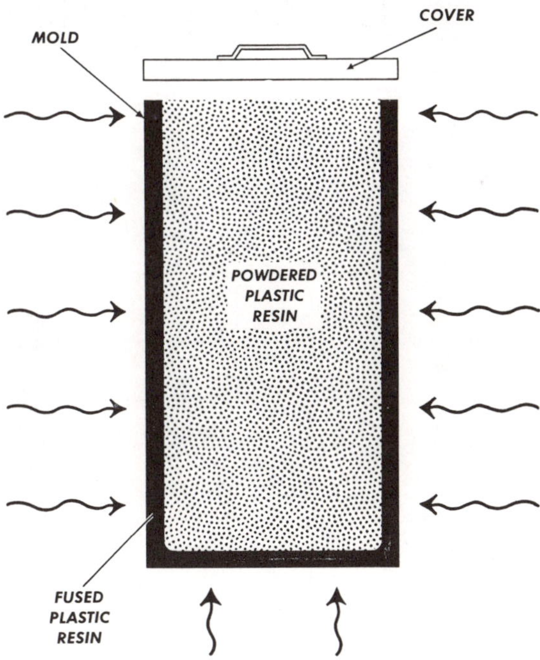

FIG. 6.19 Molding process schematic for static powder molding. (*Courtesy Agile Div., Nalge Co., Rochester, N.Y.*)

fuse the material adjacent to the mold surface, thus forming the required wall thickness. The mold is then removed from the oven and the unmelted powder dumped back into the raw material bin. The inner wall at this point is extremely rough due to unmelted powder particles adhering to the melted wall. The mold is then returned to the oven to melt the powder particles, and smooth the inner wall.

When the smoothing phase is completed, the mold is taken from the oven and allowed to cool at room temperature. As cooling takes place, the part shrinks away from the mold surfaces and can then be removed from the mold. Under some circumstances, the outer mold surfaces are sprayed with water to accelerate cooling. From this description, it is seen that the molding cycle consists of:

(1) Fill the mold with powder.
(2) Heat the filled mold for a specific time to build the required wall thickness.
(3) Dump the excess powder.
(4) Return the mold to the oven to smooth the inner wall.

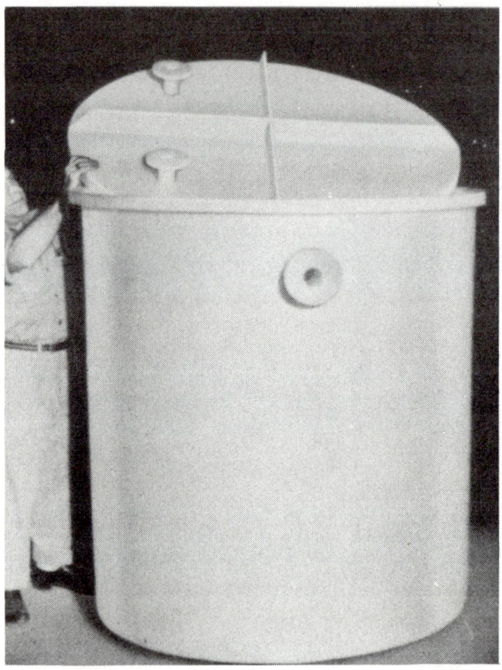

FIG. 6.20 Static powder-molded 500 gallon tank. (*Courtesy Agile Div., Nalge Co., Rochester, N.Y.*)

FIG. 6.21 Static powder-molded products. (*Courtesy Agile Div., Nalge Co., Rochester, N.Y.*)

(5) Remove the mold to cool the product.

(6) Remove the product.

This is a time- and labor-consuming process, and is obviously expensive. It has been the least costly method for producing modest quantities of molded products at lowest mold cost; at this time it is still a rational process for making large size molded products in any quantity. The process is limited to making open-end containers. Figure 6.20, for example, illustrates such an item, a 500-gallon chemical tank with fabricated cover. Figure 6.21 shows other articles made by static sinter molding.

Heisler Process

In the Heisler process, a hot mold is clamped into a machine and filled with powdered polyethylene. The machine causes the mold to rotate about a vertical axis, and simultaneously permits the mold to slowly tip toward the horizontal. The excess powder is thus completely dumped from the mold as it reaches a horizontal position.

If the resulting wall is too thin, the mold may be reheated and the process repeated. When the desired wall is obtained, the mold is again reheated to smooth the inner wall.

Closed-end containers can be made in a two-step operation. A second mold, representing the "head" end of the container, proceeds through the same process just described. It is again partially, or completely filled with the powder, and attached to the first, still hot mold. The process is again repeated, and the two molded parts fuse together. Again it can be seen that quite a bit of labor is involved in this process, especially in powder handling, even though the dumping operation is mechanized.

Rotational Molding

Rotational molding of polyethylene evolved from the vinyl plastisol rotational molding process, which has been in use for many years. In rotational molding, the mold, or molds, are charged with a predetermined weight of powder to produce the desired part. The closed molds are then caused to simultaneously rotate in two perpendicular planes while in a hot oven. The temperature will be from 500 to 900°F, depending on material and product. This causes the powder to melt and uniformly deposit on the mold surface. When the melting phase is completed, the molds are moved while still rotating from the oven into a cooling chamber, where the molds and contents are cooled by moving air, water fog, or water spray. Internal cooling to speed the cycle by means of nitrogen, CO_2 or water fog, is currently being investigated as a means of further lowering the cost.

FIG. 6.22 Rotational powder-molding machine. (*Courtesy McNeill Machine Co., Akron, Ohio*)

Following the cooling cycle the molds are opened, molded parts removed, reloaded with powder, and the cycle repeated. The cost-saving aspects of this process are important when compared with the labor involved in other powder molding practices. Further important advantages lie in the uniform wall thickness, and minimal material waste. Closed- or open-end products can be produced. The product is also essentially strain free.

Figure 6.22 depicts a three-spindle rotational molding machine. In this machine the oven section is heated with hot-air convection heaters. Other heating media are direct gas, molten salt spray, and heated liquids such as ethylene glycol. The cooling chamber is provided with water fog nozzles or water spray nozzles and forced air cooling. These can be used individually or in sequence, as required by the resin or product. Each spindle has independent rotational speed control at the oven position. This permits compensation for the requirements of oddly shaped products, or for the molding characteristics of different resins. A rotationally molded part is shown in Fig. 6.23 depicting a 20-gallon water tank.

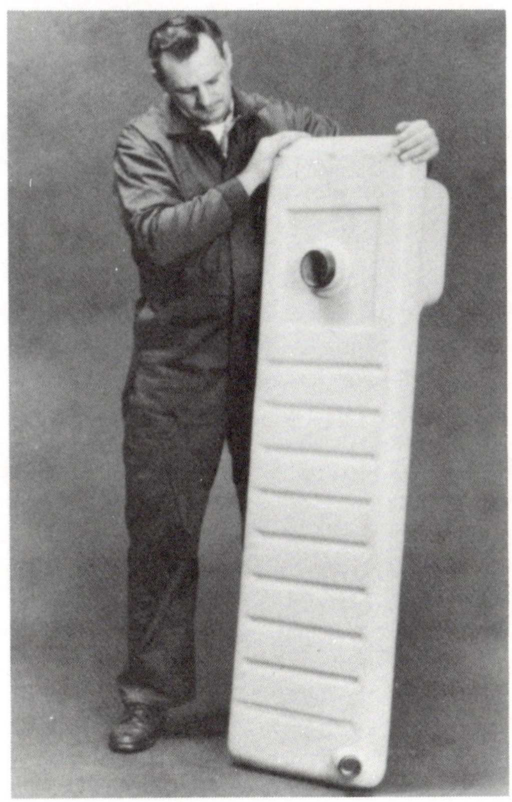

FIG. 6.23 Rotational powder-molded water tank. (*Courtesy U.S.I. Chemicals*)

FIG. 6.24 Large 3000 lb. capacity rotational molding machine. (*Courtesy McNeil-Femco-McNeil Corp.*)

Figure 6.24 illustrates a very large three-arm rotational molding machine capable of handling 3000 pounds of molds and material per arm. Parts to 9 feet in diameter and 15-$\frac{1}{2}$ feet long can be molded in this machine. Molds 10 feet in diameter by 6 feet in length may be mounted on the straight side arms.

Centrifugal Casting

The centrifugal casting process is generally used for making large-size tube forms. The equipment consists of rotating a heated tube which is uniformly charged along its length with powdered plastics, thereby creating a molded tube of the desired wall thickness. Previously described processes melt and coat the inner wall of the tube. When the melting is completed, the heat source is removed while the tube mold continues to rotate, thus maintaining uniform wall thickness of the tube throughout cooling. Upon completion of the cooling cycle, the tube, which has shrunk away from the mold, is removed, and the process repeated.

A typical centrifugal casting machine, with guards removed to show drive, is shown in Fig. 6.25. Length and diameter are limited only by machine size. Current production sizes range from 6 to 30 inches in diameter, and up to 96 inches in length.

The apparent advantages in powder molding lie in the ability to quickly produce prototypes at minimum mold and product cost. In most cases the production economics are attractive even when large numbers are involved, especially in large or heavy weight items.

FIG. 6.25 Centrifugal casting machine used to produce large diameter tubes. (*Courtesy Agile Div., Nalge Co., Rochester, N.Y.*)

In large powder-molded polyethylene tanks containing liquids of high specific gravity, the sidewalls will deform when the tank is filled. This sets up stresses in the tank, which should be restrained. There are several ways in which this is done. One is to build a restraining outer tank of reinforced plywood; another is with an outer casing of structural fiber glass, as shown in Fig. 6.26.

The molds used in all four processes, due to the low pressures involved, are relatively inexpensive cavity forms. Commonly used materials are sheet steel, spun aluminum, machined- or pressure-cast aluminum, electroformed copper and nickel and others.

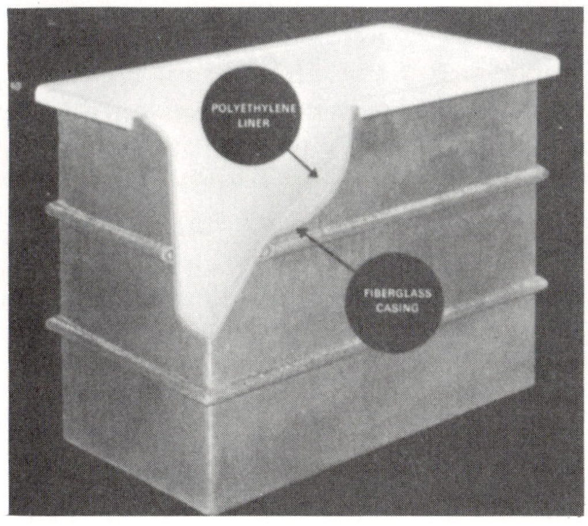

FIG. 6.26 Reinforced-polyester tank for supporting polyethylene inner liner. (*Courtesy American Agile Co.*)

Materials most generally used in powder molding at this time are low, medium and high density polyethylene. Aceto-butyrate, polystyrene, polycarbonate, vinyl and vinyl copolymers, are also in production. Many other resins are being evaluated, and it is certain that good processing procedures will be developed.

Important elements in the powder-molding process are close attention to time and temperature to avoid degradation. Of equal importance is the quality of the powder grind. The grind must provide particles of relatively uniform size, with no fluff, tails or shreds. This will result in a powder of high-bulk density which

TABLE 6.1*

POLYETHYLENE POWDERED RESINS—Low, Medium and High Density ASTM Types I, II, and III						
MARLEX Resin Numbers	**TR-905**	**TR-906**	**TR-915**	**TR-916**	**TR-925**	**TR-951**
SPECIAL CHARACTERISTICS	Uniform Coating and resin flow	Uniform coating, excellent flow	High flow, uniform distribution with very little dusting	Excellent mold release	Increased stiffness	Excellent impact strength for a powdered resin, excellent stiffness abrasion resistance and flow properties
SUGGESTED APPLICATIONS	Rotational molding, fluidized bed coating	Rotational molding, powder molding and fluidized bed coating	Carpet backing, rotational molding	Rotational molding	Industrial rotational molding applications	Rotational molding
Density, gms./cc. ASTM D1505-63T	0.914	0.916	0.924	0.925	0.932	0.950
Melt Index, gms./10 min. ASTM D1238-62T	12	22	22	9.0	5.0	7.0
Environmental Stress Cracking Resistance, hrs. @ F_{50} ASTM D1693-60T	<1	<1	<1	<1	<1	<1
Tensile Yield Strength, psi — 20″/min. ASTM D638-61T	1550	1400	1900	1700	2500	3800
Die "C" of — 2″/min. ASTM D412-62T	1300	1200	1600	1400	2300	3500
Elongation, % — 20″/min. ASTM D638-61T	80	135	50	50	35	10
Die "C" of — 2″/min. ASTM D412-62T	210	155	75	100	190	90
Vicat Softening Point, °F. ASTM D1525-58T	193	186	203	199	221	252
Brittleness Temperature, °F. ASTM D746-57T	—65	—73	—67	—127	—180	—180
Flexural Modulus, psi ASTM D790-63	36M	41M	48M	57M	97M	165M
Hardness, Shore D ASTM D1706-61	50	47	53	50	60	67
Mesh Size	35	50	16	35	35	35
Bulk Density, lbs./cu. ft.	19	26	24	25	21	21.5
Pourability, gms./min.	200	280	300	250	260	185

*Courtesy Phillips Petroleum Co., Chemical Dept.

will produce parts of highest physical and chemical resistant properties. The properties of typical powdered resins offered by one supplier are shown in Table 6.1.

PLASTISOL MOLDING

Plastisol molding deals with the conversion of a liquid dispersion of finely divided polyvinyl chloride resins and plasticizers, to a solid form. This is accomplished by the addition of heat to the dispersion during slush, rotational, dip or low pressure injection and cavity molding. In all processes, fusion is accomplished at temperatures from 300 to 400°F, depending on the plastisol.

Vinyl plastisols can be compounded to obtain a great variety of chemical or heat-resistant properties. A broad spectrum of physical values is used, ranging from those of soft rubber to a high degree of hardness and modulus, as exhibited by the vinyl copolymers. Fillers, such as calcium carbonate and others, may be used to alter the physical characteristics or reduce the cost of the compound. A wide range of colors is available, including fluorescent and phosphorescent tints. Specific formulations can best be established by consulting with the manufacturers of vinyl resins.

In slush molding, a one- or two-piece hollow mold is used. The preheated mold is filled with liquid plastisol, and vibrated or spun to remove entrapped air bubbles from the surface; it is then heated for the time necessary to create the required wall thickness. At this point the mold is dumped to remove the remaining and usable liquid plastisol. Final fusion is obtained by heating the mold to 350 to 400°F. When fusion is complete, the mold is cooled by water spray or water fog. The product is removed from the mold and the cycle repeated. The most common slush molded products are such doll parts as heads, arms, legs, bodies, syringe bulbs, etc.

The single-pour process described often results in surface imperfections. A double-pour procedure circumvents this problem. In this system, room temperature molds are filled with the liquid plastisol, vibrated for a short time, and then dumped. The film remaining in the cavity is virtually bubble-free, depending on the flow characteristics of the plastisol. The mold is then heated to partially fuse the film. When this is done, the mold is again filled with the plastisol, and the cycle proceeds as in single-pour slush molding, previously described.

Rotational molding of plastisols is done in the same machinery described in rotational molding of thermoplastic powders.

The molds are charged with a measured volume of the liquid plastisol. When rotation commences and heat is added to the molds in the oven at 300 to 450°F, the liquid distributes itself about the mold wall, and fuses to a solid form. Upon completion of fusion, the molds, while still rotating, are cooled with a water spray in the cooling station. When cooling is finished, they are moved from the cooling station and opened. The product is removed. The molds are again charged and the cycle repeated.

As in rotational powder molding, the operation is clean and economical, since there are no dumping losses in time or plastisol. Product quality is at a high level, and it possesses excellent uniformity in wall thickness. The rotational process is used to make hollow objects such as balls of all kinds (handballs, basketballs, and footballs), toys, industrial parts, and automotive components such as arm rests and sun visors.

The dip-mold process consists of dipping a male mold into a plastisol solution, withdrawing it at a controlled rate, and then fusing the mass in a heated oven. The product is then stripped from the form and trimmed if necessary. This process is usually highly mechanized, and consists of a conveyor to transport the many mold forms through the preheating oven, the dip tank, fusing oven, cooling station and last, the stripping station. Stripping is usually done by means of compressed air blowing the parts from the form.

Product thickness is a function of the amount of preheat the mold forms contain, and also the nature of the plastisol and its temperature. For products in the order of .06 to .09-inch thick, the form may be preheated to approximately 300°F. For greater thickness, higher preheat temperatures are employed. The reason should be evident when one considers the process mechanics. When a hot form is submerged in the liquid plastisol, fusion begins to take place immediately, and a wall of solid plastics commences to form. The more heat energy available, the greater the thickness of the wall. Greater thicknesses may be obtained by reheating and re-dipping.

As the wall commences to fuse on the form, the form is slowly withdrawn from the tank. The uniformity of the wall depends in part on the withdrawal rate. The rate may range from 4 to 8 in./min.

Products made in dip molding are gloves, boots, automotive components such as spark plug covers, electric insulation of tool handles, etc., and toys. Sometimes the product is used as the "mold" itself, as in the case of wire racks used in plating industry, and the common household dish drain rack.

Cavity molding is accomplished by pouring, or by low pressure injecting the liquid dispersion into a cavity until it is filled. Filling is generally indicated by the overflow of the plastisol through suitably placed vents. The mold is heated at this time to fuse the material enclosed. The cross-sectional mass of material will dictate the fusing or curing time. When fusion is completed, the mold is thoroughly cooled before removing the molded part. This process is used to encapsulate electronic devices, conductors for electrical switch-gear, or for mechanical products requiring close dimensional tolerances.

Molds used in the systems just described are machined, cast, or electroformed. The materials are steel, aluminum, brass, or copper. In all cases the quality of the material, workmanship, and finish in the mold will be reflected in the quality of the product.

WELDING OF PLASTICS

Welding of plastics materials is limited to the thermoplastics. In all methods, a melting of the plastics must be achieved at the interfaces of the joint to create a weld. The welding processes in common commercial use at this time are:
(1) Hot gas.
(2) Fusion.
(3) Friction.
(4) High frequency (dielectric).
(5) Ultrasonic.

Hot Gas Welding

Hot gas welding consists of directing a stream of hot gas into the joint area to cause a melting of the plastics which will then fuse, thus uniting the two elements. Usually a filler rod (of the same plastics) is melted by the hot gas stream to properly fill the joint, as shown in Fig. 6.27. The hot gas may be air as used in

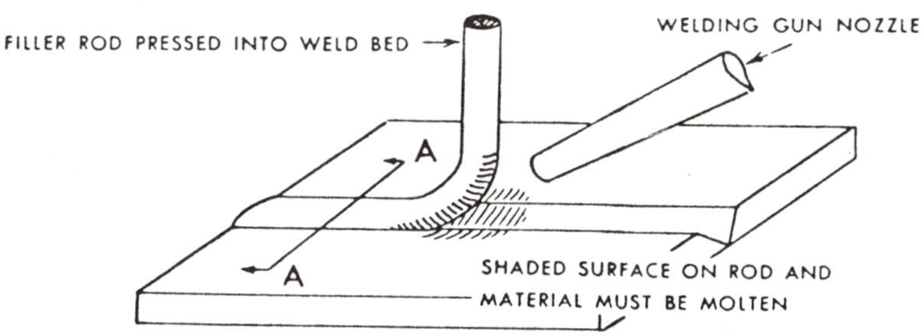

FIG. 6.27 Heating position of gun in welding. (From "Welding of Plastics," J. A. Neumann and F. J. Bockhoff, Van Nostrand Reinhold Publishing Corp., 1959)

welding polyvinyl chloride resins or an inert gas such as nitrogen when welding the polyolefins. Nitrogen will avoid oxidation of the polyolefins during welding. It is thus seen that the hot gas welding of plastics may be directly compared to the welding of metals with an oxyacetylene flame.

There are two types of welding guns used in plastics welding. Figure 6.28 depicts an electrical heated gun. A gas-heated gun is shown in Fig. 6.29. This gun has particular value in field use where electric power may not be readily available. The exiting hot gas temperature from either gun will be from 400 to 800°F, depending on its adjustment.

As in metal welding, joint preparation is important so that maximum joint strength values may be achieved. The joint area must be clean. Figure 6.30

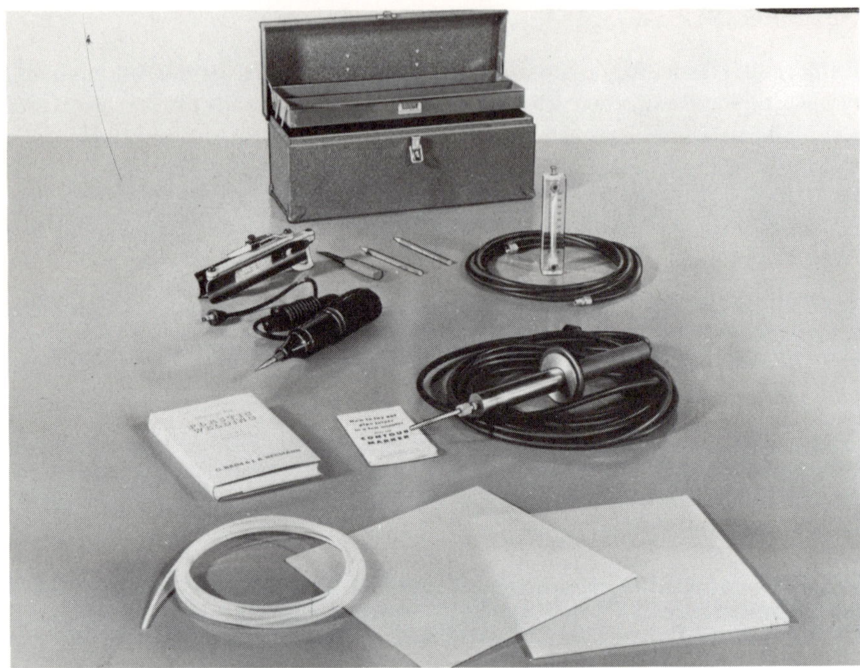

FIG. 6.28 Electrically heated welding apparatus for thermoplastics. (*Courtesy Nalgene Piping Systems, Div. of Nalge Co.*)

FIG. 6.29 Propane gas-heated welding apparatus for thermoplastics. (*Courtesy Nalgene Piping Systems, Div. of Nalge Co.*)

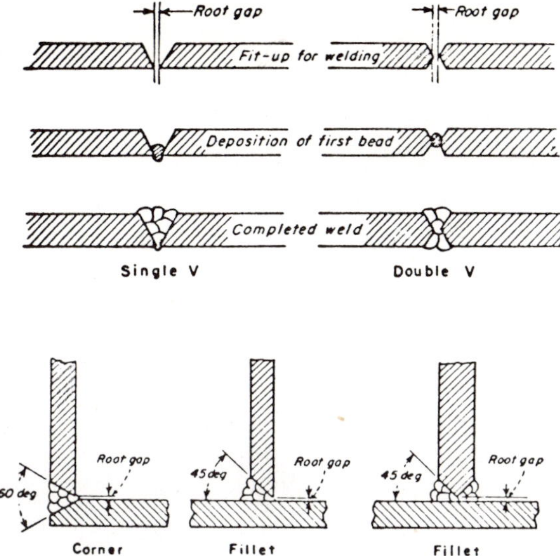

FIG. 6.30 Fit-up for welding. (From Huscher, J. L., "Handbook of New Engineering Materials," p. 45, published by *Materials in Design Engineering*, Reinhold Publishing Corp., N.Y., 1957). (Courtesy "Welding of Plastics," Van Nostrand Reinhold, 1959)

illustrates single-V and double-V joints. The included angles should be approximately 60°. The proper arrangement for perpendicular joints is also shown.

The subject of hot gas welding has been thoroughly covered by J. Alex Neumann and his associate, F. J. Bockhoff in their book "Welding of Plastics" (Van Nostrand Reinhold). It is recommended to those who wish to pursue the subject in detail.

Fusion Welding

Fusion, or hot tool welding, in its simplest form is accomplished with an ordinary electrically heated hot plate. The two plastic elements to be welded are placed on the hot surface to melt the surfaces to be joined. When this is accomplished, the surfaces are quickly placed together with a slight wringing motion to rupture the surfaces which may be oxidized from the direct heat causing the melt. The whole operation may be performed in ten seconds or less, depending on the material and mass. The two elements must be firmly pressed together to assure perfect contact, and held until the joint cools to achieve maximum strength of the weld.

To prevent degradation of the plastics, overheating must be avoided. To aid in minimizing sticking of the plastics to the surfaces of the hot plate, the plate should be chrome-plated. In some applications, "Teflon"-coated plate surfaces have been found useful in reducing this problem. In all circumstances the residue

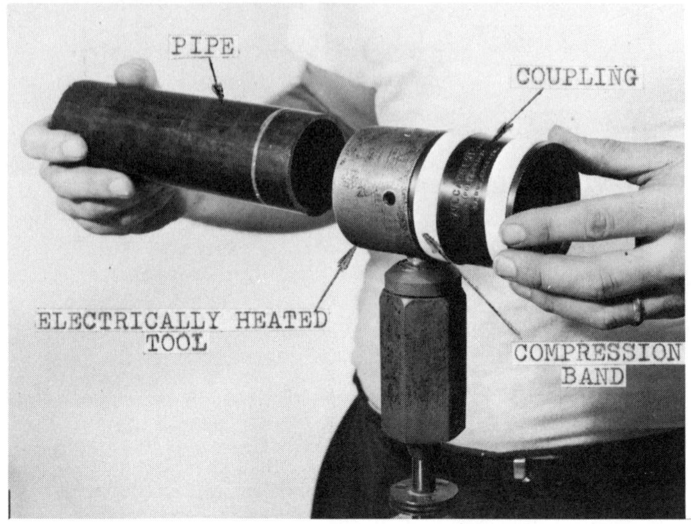

a

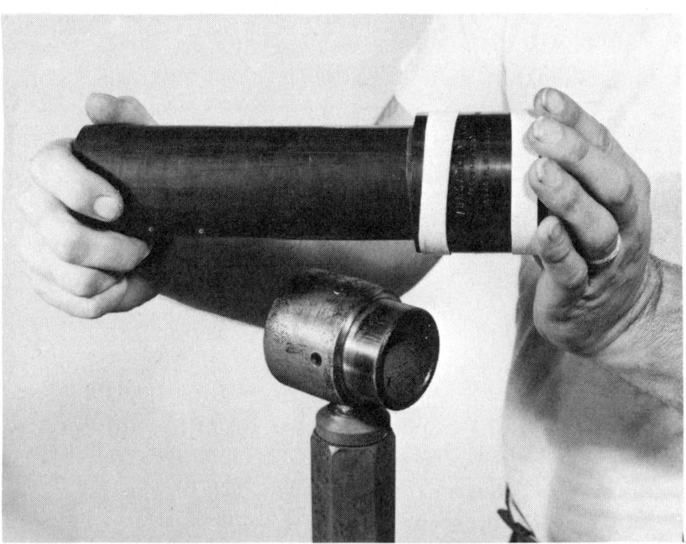

b

FIG. 6.31 Electrically heated "Polyfusion"® tool for welding polyethylene pipe and fittings. (*Courtesy Nalgene Piping Systems, Div. of Nalge Co.*)

remaining on the plate must be removed before proceeding with succeeding joints.

The use of release compounds such as stearates, waxes, silicone oils, etc., to prevent sticking to the hot plate, is not recommended, since their presence in the joint due to carry-over will invariably result in low physical values of the weld.

Commercial use of the process described is generally limited to objects of relatively small size, but variations of the basic concept are sometimes used to butt weld sheets and strips of considerably length. In this case the heating element is two-sided and vertical so that the ends of each sheet may be heated simultaneously while in a horizontal position. When melting is achieved, the ends are pressed together and held in a jig until they are cool. A variation of this process, called *"Polyfusion®,"* is used in making pipe and fitting joints in polyolefin waste-drainage systems used in the chemical industry. Figures 6.31, a and b, show an electrically heated tool which includes both male and female sections for melting the coupling socket I.D. and the pipe O.D. When melting of the plastics has taken place, both elements are quickly removed from the tool and inserted into one another with a simultaneous slight twisting motion. The purpose of the compression band is to apply a constant compression force on the joint as the weld cools. The band may be later removed if desired. The force exerted by the band will help develop optimum physical values in the joint.

The tool must be cleaned of oxidized residue before proceeding with succeeding joints. The time required for each weld will be less than thirty seconds, depending on pipe size. Figures 6.31 a and b show the welding of a pipe and coupling joint. This process is applicable to joining polyethylene fittings of all kinds—elbows, tees, "y's," etc., to pipe and to one another.

Friction Welding

Friction or spin welding of thermoplastics consists of rotating one of the two elements to be welded against a second stationary element. The heat generated by the friction between the two causes a melting of the surfaces, resulting in a weld when the rotation ceases. The job is usually done in machines such as modified lathes or drill presses, and the operation, including securing the elements into the machine, can be accomplished at high production rates, depending on the size and mass of the parts. The speed of rotation, pressure between elements and time of rotation is easily established by trial. Complete fusion is obtained when flash extrudes from between the pieces being welded. The cycle must be long enough to ensure complete fusion of the parts and when this is achieved, rotation must be stopped instantaneously to guarantee highest physical values in the joint. This is necessary to prevent shear stresses in the joint which result when the two melt surfaces cool as rotation ceases slowly.

It should be obvious that this process is limited to objects having a circular configuration. Typical weldings are dual-colored knobs, molded hemispheres, insulated drink cups, injection-molded bottle halves, etc. Most thermoplastics can be spin-welded.

High Frequency Welding

Dielectric or high frequency electronic welding is used in joining those thermoplastics which have high dielectric loss characteristics. Among these are cellulose acetate, ABS, polyvinyl chloride, and others. Table 6.2 illustrates the effect of high frequencies on various thermoplastics and the resultant weldability.

TABLE 6.2   ELECTRONIC SEALABILITY*

Material	Remarks	Material	Remarks
Acetate	Excellent	Nylon	Excellent
ABS	Excellent	"Orlon"	Poor
"Dacron"	Poor	Polyesters	Poor
"Dynel"	Poor	Polyethylene	No effect
"Dylene"	Good	Polypropylene	No effect
Epoxy	Excellent	Polystyrene	No effect
Fluorocarbon	No effect	Polyurethane sheet	Good
Foam, vinyl	Excellent	Polyvinyl acetate	Good
Foam, urethane		Rubber	Good
Polyether	Excellent	Saran	Excellent
Polyester	Good	Vinyl film	
"Kel-F"	Poor	and sheeting	Excellent
"Mylar"	Poor	Vinyl, semi-rigid	Good

*Under ordinary conditions using standard equipment. Some "poor" sealability materials can become "good" or "excellent" under special sealing conditions. Courtesy Thermatron Electronics Div.

Welding is accomplished as a result of the molecular disturbance created by the high frequency energy to which the plastics are subjected. In other words, the high frequency field causes the molecules in the plastic to vibrate and "rub" against the other at such a rate to create frictional heat sufficient to melt the interfaces. The effect—a weld at the interfaces. The device used to accomplish this is essentially a radio transmitter operated at frequencies between 27 and 40 Mc, as stipulated by the Federal Communications Commission. The equipment must be either shielded or else contained in a shielded room to prevent transmission of its signal. The energy obtained from the "transmitter" is directed to electrodes of the welding apparatus, rather than an antenna as in a radio transmitter.

The essential variables in dielectric welding are heat, time, and pressure on the elements to be sealed. The generated heat is a function of the applied frequency. These welding variables for a particular plastic are best established by trial.

FIG. 6.32 Electronic sealing machine used for manufacturing chaise lounge pads. (*Courtesy Thermatron, Electronics Div.*)

Among the commonly H.F. sealed plastics are vinyl film products such as toys, beach balls, children's ware, mattress pads, diaper bags, rainwear, etc. Shown in Fig. 6.32 is a machine used to dielectrically assemble chaise lounge pads. The press consists of a strong yoke mounting a fixed lower platen, and an upper platen which is raised and lowered by means of dual-air cylinders. The cycle is push-button actuated and controlled by means of timers. The high frequency generator is enclosed in the cabinet shown behind the press.

Induction Welding. Some films and products are welded by the use of electrical induction heaters which generate heat in a wire that is loaded in the interface. In this case, the wire is sealed in the weld.

Ultrasonic Welding

In ultrasonic welding, the molecules of the plastics to be welded are sufficiently disturbed by the applied ultra high-frequency mechanical energy to create frictional heat, thereby causing the plastics to melt. The components are thus joined quickly and firmly. The weldability is a direct function of the modulus of elasticity and the thermal characteristics of the plastics.

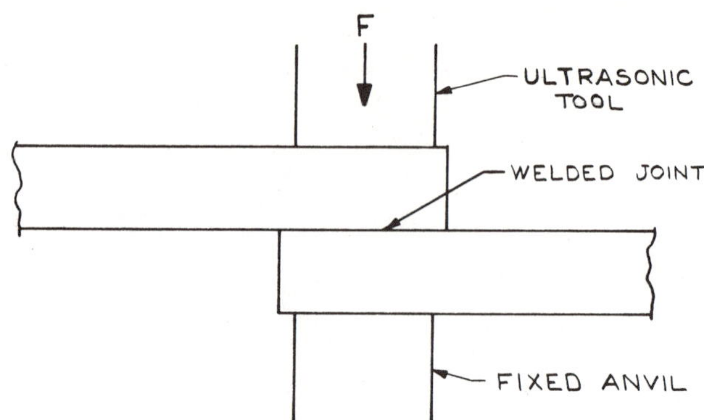

FIG. 6.33 Ultrasonic contact welding.

The "machinery" used in ultrasonic welding consists of an electronic device which generates electrical energy at 20/50 KC/sec., and a transducer (either magnetostrictive or piezoelectric) to convert the electrical energy to mechanical energy. Much discussion regarding the relative merits of the transducer systems exists at this time. Needless to say, the results obtained in a particular welding job will determine the best transducer system for the application. Both should be tried if necessary.

Two methods are used in applying the ultrasonic force—contact and remote, as depicted in Fig. 6.33 and Fig. 6.34. The ultrasonic force "F" is transmitted from the transducer through a tool or "horn" to the objects to be welded. The amplitude of motion of this horn is from 0.0005 to 0.005 inch depending on the design. The tools are generally made of titanium. In experimental work,

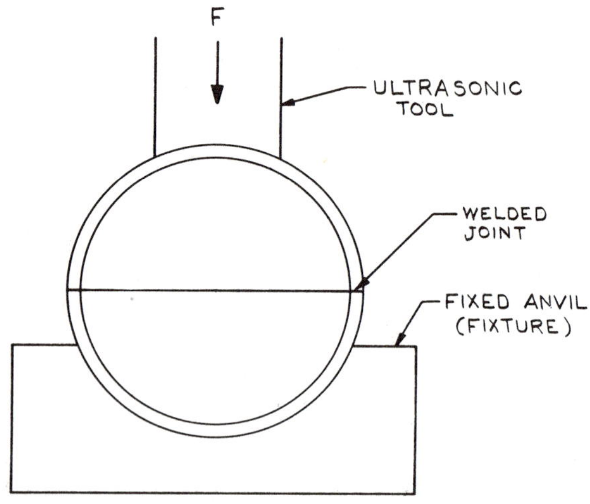

FIG. 6.34 Ultrasonic remote welding.

aluminum is often used for the tool. In any case, the horn design is most critical for each application and should be worked out by a competent acoustical engineer. Help can be obtained from the equipment manufacturer. Joining time in ultrasonic welding is quite short; in the order of 0.2 to 5.0 seconds, depending on the material and area of the joint. Tool pressures vary from 10-100 psi gauge.

Contact welding is most generally used for welding thin, or less rigid thermoplastics, such as films or sheets of polyethylene, plasticized vinyl, and others having equally low stiffness. It should be observed that the energy is directly applied to the joint area to effect the weld. In remote welding, on the other hand, the sonic force is applied at some point away from the joint—in some applications as far as 8 or 10 inches. The energy travels through the plastic to the joint area, the heat is generated, the surfaces melt, and when they cool, are welded. It is important in either case that pressure be maintained on the joint until it is cooled to obtain good physical values.

Joint design is extremely important in contact welding. In general, the initial area should be minimal to induce the spread of a complete weld. Basic parameters are shown in Fig. 6.35 and Fig. 6.36. The equipment manufacturer should be consulted for advice in design of joint configuration.

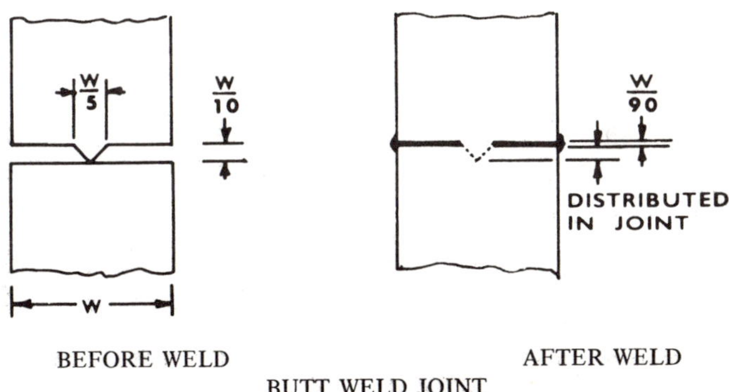

FIG. 6.35 Joint design parameters for ultrasonic welding. (*Courtesy Branson Sonic Power*)

Remote ultrasonic welding will have increased commercial usage. An interesting current application is a disposable filter used in biological research. It is shown in Fig. 6.37. An important feature lies in the fact that the assembly includes both horizontal and vertical welds.

Other interesting applications lie in securing metallic inserts, staking metallic inlays into plastics, or upsetting molded studs to hold plastic assemblies together

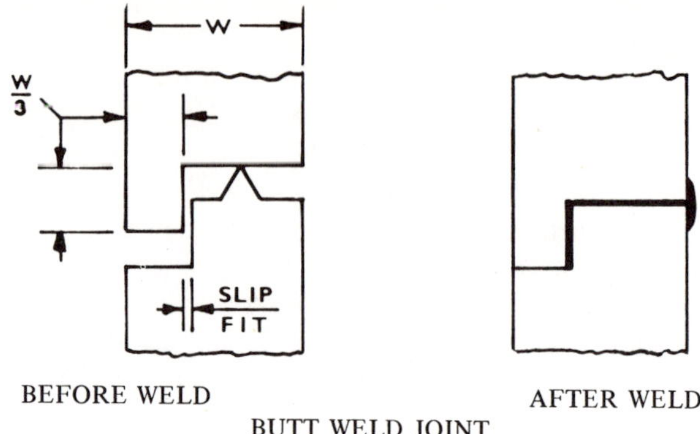

FIG. 6.36 Joint design parameters for ultrasonic welding. (*Courtesy Branson Sonic Power*)

at far greater rates than by other means. Other applications not completely explored are welding partially cured assemblages of thermosetting resins, and cementing cured thermosets together with an uncured resin such as an adhesive, polymerized by the ultrasonic energy.

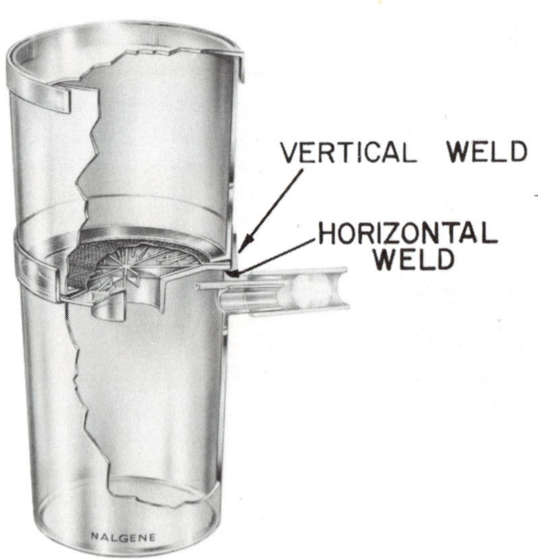

FIG. 6.37 Ultrasonically assembled disposable filter. (*Courtesy Nalge Co., Div. Ritter Pfaudler Corp.*)

SURFACE TREATMENT OF PLASTICS PRIOR TO DECORATION

Some plastics, notably the polyolefins and acetals, are highly resistant to bonding other media to themselves. To overcome this deficiency, several treatments are available. The most common is flame treatment; electronic treatments such as corona discharge, plasma discharge and chemical treatments are also widely used.

Flame treatment of bottles, film, etc., consists of passing the object through an oxidizing gas flame. The momentarily impinging flame on the plastic causes an oxidation of the surface which makes it receptive to adhesion of inks, enamels, and other media used in decorating the product. The process is usually completely mechanized and performed immediately before printing or decorating the product. On irregular-shaped objects, the product is oftentimes flamed with a hand torch which an operator passes over the surface to be treated. Flame treating in several variations is patented by Kritchever and Kreidl.

The "corona" discharge process is used to treat films and molded products. When film is treated, it passes over an insulated metal cylinder beneath conductors charged with a high voltage. When the electron discharge, or corona, existing between the charged conductors and the drum, strikes the intervening film surface, oxidation occurs which makes it receptive to coatings. Molded products are handled in a similar manner and often by fully automatic machinery.

The plasma process consists of directing air at low pressure through an electrical discharge and then expanding it into a closed chamber (also at reduced pressure) containing the object to be treated. While passing through the discharge chamber, the air components, nitrogen and oxygen, are partially disassociated into their component atoms. In the atomic state, these elements react with the surfaces of the plastic to alter their physical-chemical characteristics in such a manner that excellent adhesion is afforted surface coatings. The process is adaptable to individual or batch processing of products, including film which may be unreeled in the vacuum chamber for treatment. The "Aerovac" process described is patented by Aerovac Research Laboratories, Inc., of Princeton, New Jersey.

The chemical treatment of acetal resins consists of subjecting the product to a short acid dip resulting in an etched surface which makes it receptive to paint. Great care must be exercised in choosing the proper ink, paint or other decorating material so that the container components or the permeation of its contents do not cause swelling, softening or loosening of the decorative surface.

Screen Process Printing

The process of printing through a screen consists of forcing ink or paint through the interstices of a stencilled screen of the required image with a squeegee. The screen material is often nylon mesh or other synthetic material. Metallic screens such as stainless steel are also in common use.

To create the stencil, the art copy positive is light-exposed to a photo-sensitive stencil film secured to the screen. When this is immersed in a developer bath, the exposed areas wash out, and the clear or unexposed areas harden. In this way, the required open and closed areas are obtained. After fixing in a hardener and drying, the screen is ready for use.

Screen printing of plastics products is accomplished manually to produce small production lots at rates of from 200 to 600 pieces per hour. Automatic screening of large volume products, such as detergent bottles, is accomplished at rates of from 2000 to 3000 impressions per hour. Multicolor screening will of course require a screen for each color, with adequate drying time for each color before the next color is applied. If colors are not mixed in the design, split screens can be used which will allow a simultaneous printing of several colors. Imprint drying is generally accomplished in a circulating air oven, and may require up to 15 minutes of drying time.

Hot Stamping

Hot stamping by the roll leaf process is one of the original methods used for decorating plastics materials. The leaf or foil consists of a thermoplastic color coat applied to an acetate or cellophane carrier film. When a heated die is pressed against the foil carrier film, the color coat is released and adheres to the product placed beneath it. Die temperature, die pressure and dwell time are important factors in achieving optimum results, and must be precisely controlled. Product accuracy is also essential to successful printing.

A broad range of colors, including genuine gold and silver, is available. Costs of obtaining silver and gold effects are reduced by using aluminized polyester foils. The gold effect is obtained by applying a colored overcoat to the aluminum imprint. The roll leaf, in any case, must be chosen to suit the plastics to be imprinted.

The machinery for applying a roll leaf imprint is relatively simple. It generally consists of an air-operated cycle-controlled press with a built-in foil holder and feeder. The bottom or fixed plate is fitted with a product holding or locating fixture. The heated printing die is attached to the reciprocating pressure head of the machine. The foil passes under the die between the product to be decorated. When the heated die descends and presses the foil against the part the color coating is released and adheres to the product. The operation may be manual, semi-automatic or fully automatic.

MOLDED-IN LETTERING

Lettering in the molded product is used to identify and sometimes decorate the product. It is produced either by raised or depressed letters.

Raised lettering is by far the cheaper since it is cut or stamped into the cavity of the mold, thus appearing raised on the product. It may be used as molded, or the faces of the letters coated with paint using a roller technique.

Depressed letters on the other hand require raising steel on the cavity surface to create the depressed letters on the product. This either calls for a hobbed or EDM cavity with the hob engraved with the depressed letters or cutting the cavity in the solid and engraving each letter. The latter is an extremely expensive task. The depressed letters can be filled with paint, the surface wiped. The result is very attractive, however at a high cost.

Lettering can also be achieved by protoetching the mold. The practical depth is from .003 to .005 inch and this is clearly legible. Raised or depressed lettering usually does not exceed .030 inch. (See also Fig. 7.78.)

HEAT TRANSFER DECORATION

Heat transfer printing has become an important one-pass method of single and multi-color decoration of films, molded products and blown containers. One such process is known as "Therimage®,"* developed by the Dennison Manufacturing Co.

The structure of the printing stock is shown in Fig. 6.38. It consists of a paper carrier, a fusible release coat, and a thermoplastic ink image. The method of

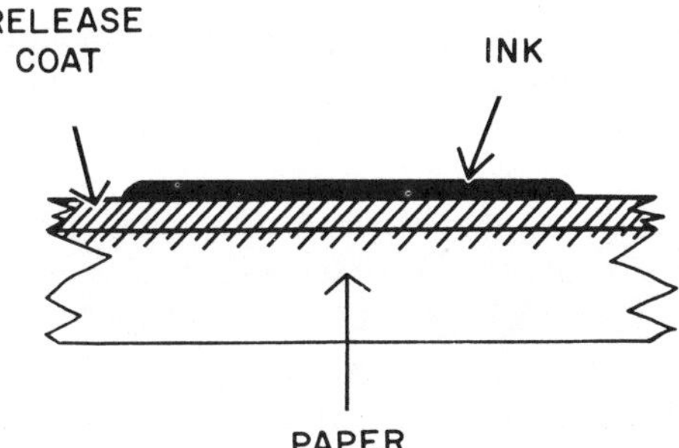

FIG. 6.38 Structure of heat transfer printing stock. (*Courtesy Dennison Mfg. Co.*)

transferring the image is schematically shown in Fig. 6.39. The ink becomes tacky when heated by the hot platen. When it is then pressed against the receiving surface, the ink bonds to the surface. Some of the fusible release coat carries over with the ink, thus providing the image with a glossy protective coating.

In actual practice, the process is accomplished by pre-heating the label carrier paper and effecting the ink transfer by means of a heated rubber roll. The rubber roll insures a full line contact with the variable product surface. The rubber roll is designed to suit the product configuration, whether round, oval, concave or

*®Dennison Mfg. Co.

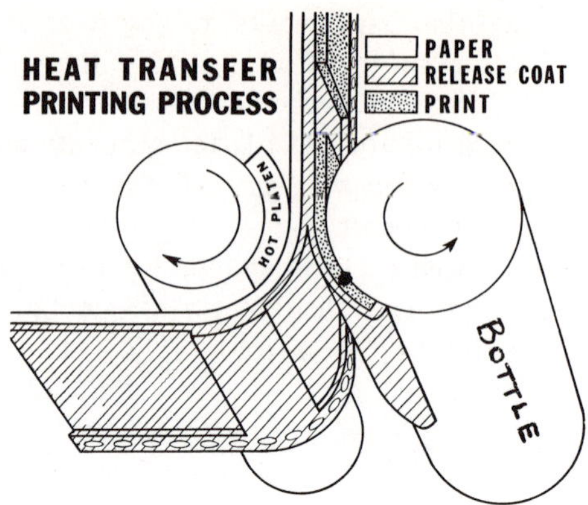

FIG. 6.39 Rolling method for transferring. (*Courtesy Dennison Mfg. Co.*)

convex in shape. A machine which includes these abilities is shown in Fig. 6.40. This machine is designed to decorate only one side of the product per pass. Two passes are required to decorate both sides. A cylindrical bottle can be decorated all around in one pass. Another machine is available which will imprint all four sides of a square container in one pass.

An alternate process for imprinting in this manner is known as "Electrocal," and was developed by Noble and Westbrook.

In-mold Decoration

Most high quality melamine dinnerware is decorated by this method, as are a host of "housewares," such as bread boxes, canisters, wastebaskets and hardware knobs and handles. The cost of the overlay is relatively inexpensive, especially when a multiplicity of colors is required in the decoration. The result is one of the most durable and permanent decorations; in fact, the decoration becomes an integral part of the product.

In decorating such thermosetting plastics, a two-stage process is utilized. The mold is loaded with the resin in the usual manner and closed. After a partial cure it is opened, and the overlay is placed in position. The mold is again closed and the curing cycle proceeds to its completion. The "foil" or overlay consists of a printed cellulose sheet covered with partially cured clear melamine formaldehyde resin face. During the molding cycle the overlay is fused to the product; when the cure is completed, it is part of the molding.

For decorating thermoplastic moldings, a single stage process may be used. The overlay is placed in the mold cavity prior to the injection of the polymer. The overlay, in all cases, is printed on a film .003 to .005 inch thick, of the same polymer. Thus a styrene overlay film will be used with a styrene product, and a polypropylene film is used when decorating a polypropylene product. The ink used to imprint the film must have heat resistance to prevent color shift due to

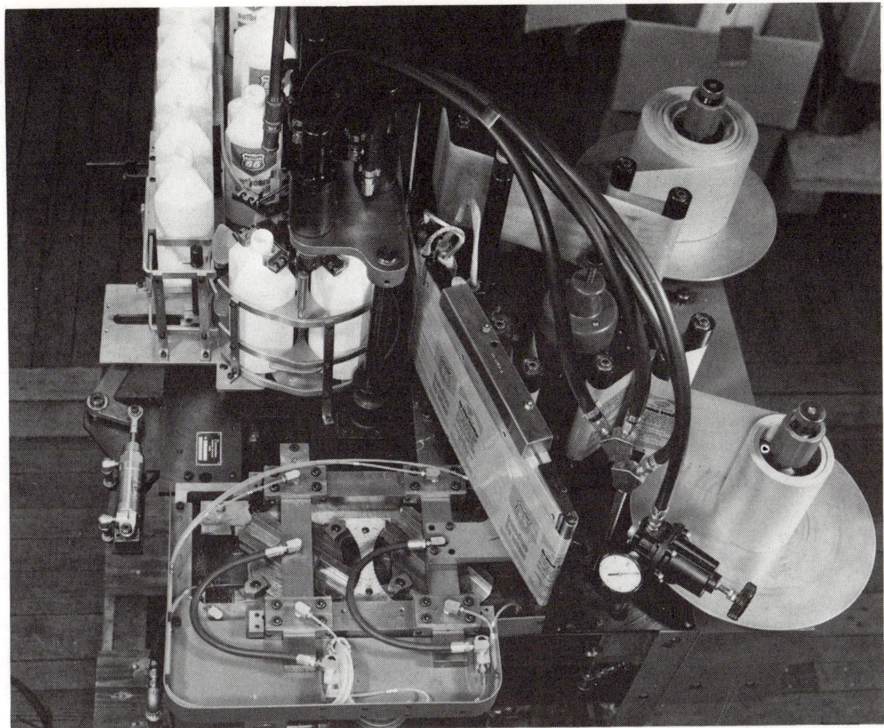

FIG. 6.40 "Therimage" application machine. (*Courtesy Dennison Mfg. Co.*)

the heat absorbed during the molding cycle. Pigments must also be carefully chosen to provide light fastness in the finished decoration.

The film is held in place in the mold by its inherent static charge. To guarantee that no shifting takes place during molding, an additional charge is generally induced by passing the wand of an electronic static charging unit over the foil after it has been placed in proper position. Even though the film is securely held in the mold, if the injection molding gate is improperly placed, the foil will be wrinkled or even washed away. It is most important that the proper gate location be established in the mold design if a part is to be decorated with this process. Inked areas must not be in the vicinity of the gate since the impingement of the hot polymer will melt the film and destroy the printing. Printed areas should be $5/16$ inch or at least $1/4$ inch away from the gates. Coated metal foils have also been successfully used in this process to decorate many molded products.

A similar procedure may be used for decorating blow-molded products. It is possible to transfer ink from a printed paper matrix that is positioned in the mold and held by vacuum. The ink transfers from the paper to the hot plastic material as it strikes the matrix. Printed plastics film may be bonded to the blow-

FIG. 6.41 Products decorated by the "in-mold" decoration process. (*Courtesy Kaumagraph Co.*)

molded product in a similar operation. Figure 6.41 shows a group of molded products decorated by the in-mold decoration process.

Numerous methods may be used to print plastics products. A novel method to imprint not only flat but irregular surfaces as well as peripheral products is shown in Fig. 6.42.

The system uses a debossed metal plate made by acid etching the required decoration or design into the plate. The etched area is filled with a pigment by means of a moving brush in the first pass of the head. On the return stroke a doctor blade removes all of the pigment from the plate, leaving the debossed area filled. The soft silicone rubber transfer pad is pressed against the debossed design and picks up the pigment. The transfer pad is then pressed against the product and the pigment is transferred. Multicolor printing units are also available. Manual production rates approach 2200 pieces per hour. Automatic systems produce to 20,000 pieces per hour, depending upon the nature of the product.

Painting

Painting of plastics products is done to achieve decorative effects by painting the molded surface. Many examples of this are to be seen, such as electric iron handles, television cabinets and tube masks, escutcheons of all kinds, toys, etc. Oftentimes a product is molded in one color only—or of multi-colored scrap—to obtain lowest cost, and then spray-painted with colors to satisfy market requirements.

All plastics can be decorated by painting, although it may be difficult and bothersome to do so. All require special consideration of the resin-solvent system of the paint to secure proper adhesion, abrasion and chemical resistance,

FIG. 6.42 Printing process for soft and irregular surfaces. (*Courtesy Dependable Machine Co., Cedar Grove, N.J.*)

adequate covering, freedom from deleterious effects on the base resin such as crazing, etc. It is recommended that advice from the materials and paint suppliers be obtained during product design to assure success. Mold lubricants must be carefully chosen and used sparingly to assure proper adhesion; if at all possible, it is better to avoid them.

Paints for plastics are formulated to meet three factors: the plasticizer system of the plastics, the heat distortion point of the plastics, and the plastics' resistance to a solvent system for the paint. The paint must be formulated so that the plasticizer does not migrate into the coating and soften it. Heat-distortion point dictates whether an air-dry or oven-dry finishing system can be used. The solvent system must be so designed that it etches the surface for good adhesion but causes no crazing. Thus, depending on the thermoplastic, an air-dry lacquer or low temperature baking enamel may be chosen; more lustrous finishes are obtained with the enamel. In most instances a higher temperature baking enamel or an epoxy paint is used on thermosetting compounds.

Paint is applied by manual or automatic spraying and dipping, just as for solid products such as knobs. Roller coating is used to create wood-grained surfaces.

If two-tone effects are to be achieved, close fitting masks are used to obtain a sharp demarcation between adjacent colored surfaces. Accurately molded parts are essential in roller coating and masked products.

Three-dimensional effects such as may be seen in most automobile horn buttons are achieved by multicolor painting intaglio designs on the undersurface of clear plastics products.

TWO-COLOR MOLDING

The most common examples of this technique are typewriter and business machine keys as shown in Fig. 6.43A.

Two-color molding is an injection molding process where two colors are successively molded to create their portion of the finished product.

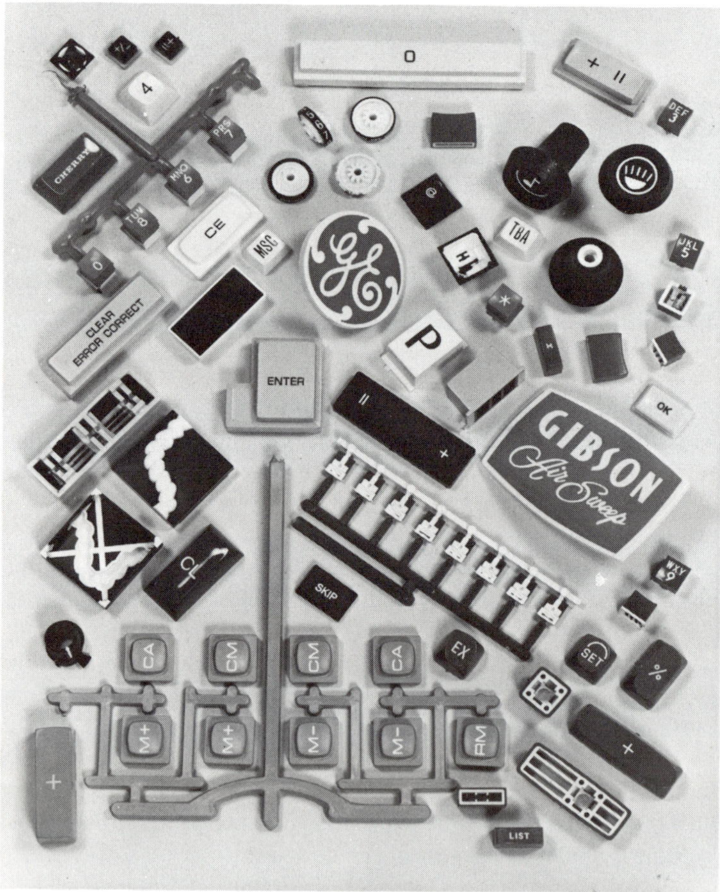

FIG. 6.43A Two color or double shot molded products. (*Courtesy Master Unit Die Products, Inc.*)

FIG. 6.43B Two-color molding is done in machines such as this. After the first color has been injected, it rotates into position for the second color. Each mold has the appropriate details to produce the desired end effect. (*Courtesy Trueblood, Inc., Tipp City, Ohio*)

The shell is molded first and then is transferred to another cavity where the second color is injected to complete the product.

Shuttle presses Fig. 6.43B are used for this kind of molding which are fitted with two injection cylinders to provide the two colors. The injection molds must be accurately made so that the transferred elements fit and seal properly. Proper runner and gating systems are also critical. Standard mold frames* are available to index the cavities correctly for each color shot.

*Master Unit Die Products, Inc.

ELECTROPLATING OF PLASTICS

Plastics, though non-conductors of electricity, can be electroplated if the surface to be plated is properly conditioned and sensitized. There is nothing about the equipment used that is not standard in any electroplating operation, except that the conditioning, sensitizing, activating, and electroless tanks are all lined with polypropylene. Specific information on plating procedures will be supplied by the resin manufacturers and suppliers to the plating industry.

Plating is done on many plastics including phenolic, urea, acetal, ABS, polycarbonate, etc. The purpose is to provide a hard wear and corrosion-resistant surface, as opposed to the somewhat delicate surface obtained in vacuum metalizing. Physical properties such as tensile strength, heat deflection temperature, flexural strength, etc., are also enhanced in the plated plastics. Metal die castings require buffing in most cases after plating. The plastics ordinarily do not require this extra expensive operation. Attractive texture contrasts may also be achieved by the use of plated plastics.

Many automotive applications exist today which include shift knobs, instrument cluster panels, instrument knobs, and door knobs. All provide the high-quality appearance of metal with the light weight and corrosion resistance of the

FIG. 6.44 Chrome-plated ABS plastics replaces zinc in these marine search lights to gain corrosion resistance light weight and lower cost. (*Courtesy Marbon Chemical Co.*)

plastics. Many appliance and hardware uses of plated plastics are also in evidence today—these include knobs, bezels, speaker grilles, nameplates, etc. Figure 6.44 illustrates an ABS search-light with chrome-plated surfaces giving minimum weight and greater corrosion resistance.

Product design of parts to be plated is particularly important if a successful product is to be created. Basic plastics design practices to achieve a good molding in the unplated product must be adhered to, and it is strongly recommended that the design be reviewed by the plater to avoid later difficulties in plating. The proper choice of resin for products to be plated is of basic importance and the resin manufacturer should be consulted while the product is in design stage.

VACUUM METALIZING OF PLASTICS

Vacuum metalizing of plastics is a process whereby a bright thin film of metal is deposited on the surface of a plastics molded product or film while being subjected to a high vacuum. The deposited metal may be gold, silver, or most generally, aluminum. Deposition is the result of vaporization of small clips of the metal to be deposited attached to an electrically heated filament. When electrical energy is applied to the filament, the clips melt and coat the filament. Increased energy causes vaporization of this coating, and plating of the product takes place. The deposition is accomplished at high vacuum, in the order of $1/2$ micron. The thickness of the resulting coating is about 5 millionths of an inch thick—5 micro-inches.

The metalizing process starts with dipping or spraying a lacquer base coat on the parts. This minimizes surface defects and enhances the adhesion of the metal coating. The coating is oven dried. The lacquered parts are then secured to a rack, as shown in Fig. 6.45. The rack is fitted with filaments, to which are fastened clips of the metal for plating. These are shown in the geometric center of the rack. It should be noted that the part holding fixtures are on axles which are chain-driven to rotate the parts during the plating cycle. This is necessary since the vaporized metal travels in a line-of-sight fashion, and deposits only on what it "sees." To get full surface plating, the product must therefore be rotated. The rack is next placed in the vacuum chamber and when full vacuum is achieved, power is applied to the filaments and the plating proceeds as previously described. When deposition is completed the rack is removed from the chamber. Parts are unracked and dipped or sprayed with top-coat lacquer to protect the tender plating from abrasion. If color tones such as gold, copper, brass, etc., are desired, they may be included in this coating.

Films of cellulosics (acetate and butyrate) mylar, etc., used for decorative, electronic or industrial purposes, are handled in essentially the same way. Entire rolls are unreeled and rewound during the deposition process to metalize the desired surface. The metalized roll is subsequently run through an automatic coating machine to apply the protective abrasion-resistant coating.

FIG. 6.45 A 90-in. vacuum plating tank being loaded with planetary fixture. (*Courtesy Revell, Inc.*)

Encapsulation

The purposes of encapsulations are to protect the enclosed components from physical damage, chemical attack and/or to enhance the overall electrical properties of the components. Both kinds of plastics are used—thermosets and thermoplastics. Among the thermosetting materials are epoxies, polyurethanes, polyesters, and silicones. The thermoplastics in common use are polyethylene, polypropylene, polystyrene, polyvinyl chloride, polycarbonates, fluorocarbons, glass-bonded mica, and others. There are many methods used in encapsulation. Among these are hot-melt pour, centrifugal casting, foaming, dip coating, fluidized bed coating, hand brush and spatula coating, injection molding, compression molding, and transfer molding.

Hot melts and "fluidized bed" encapsulation are often used for short runs. For mass production, transfer, compression, and injection molding are favored. In any case, in the product to be encapsulated, temperature resistance, delicacy of components, etc., dictate the method to be used.

Figure 6.46 depicts various products that have been encapsulated by low pressure transfer molding procedures. Special automatic presses as shown in Fig. 5.20 have been developed for this work.

FIG. 6.46 Low pressure-transfer molded (encapsulated) products. (*Courtesy Stokes Equipment Div., Pennsalt Chemical Corp.*)

THERMOFORMING OF SHEET PLASTICS

This is a process whereby flat sheet material is converted to a three-dimensional shape. The process requires that the sheet material be brought to a thermoplastic state by the addition of heat and then forced into a mold, pneumatically or by mechanical means. Containers of various kinds are, for example, produced from oriented styrene sheet by a mechanical punch-press method, but the majority of thermoformed products are made using a pneumatic forming process, which may be pure vacuum forming, pressure forming or a combination of these processes. The various methods are schematically shown in Fig. 6.47 A through K. Sheet heating is accomplished with heaters above the sheet or with simultaneous heating above and below the sheet to effect a more rapid, uniform heating before forming.

There are machines in current use which are capable of the conversion of the thermoplastics pellet to the finished thermoformed product. These consist of a sheet extruder, sheet cooling equipment, a reheat oven to bring the sheet to the proper thermoforming state, an automatic thermoforming press and a die cutting machine to cut the finished product from the sheet web. The waste material from the web is immediately reground automatically and pneumatically con-

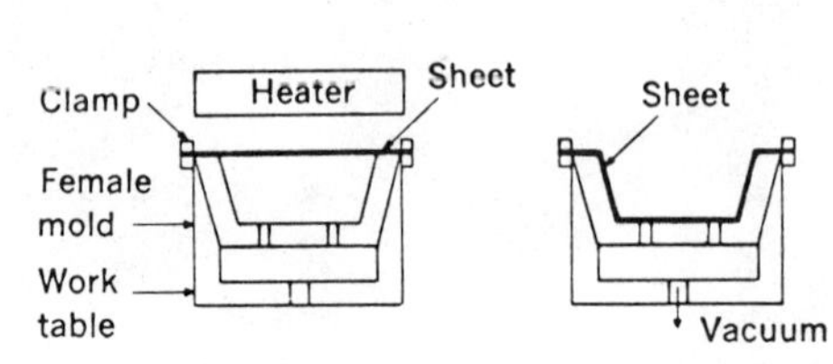

A. Straight vacuum forming in a female mold is recommended for low-profile parts where deep draw is not a requirement.

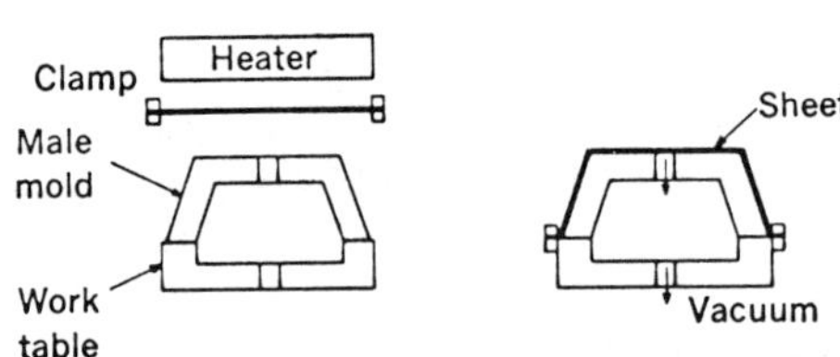

B. Drape forming over a male mold usually results in better material distribution and depth-to-diameter draw ratios.

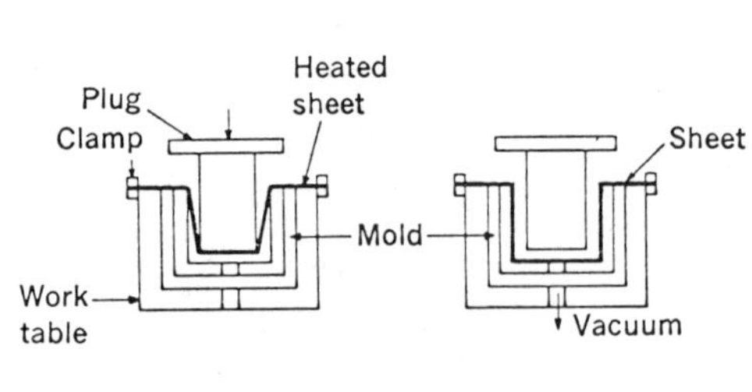

C. Thinning of material in deep-mold cavities can be overcome by use of plug assists designed for fast penetration.

FIG. 6.47 Thermoforming methods. (*Courtesy Mobay Chemical Co.*)

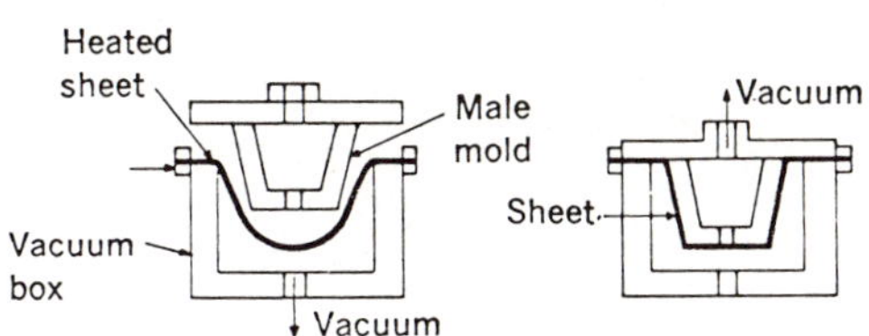

D. Vacuum forming with **snap-back** can reduce starting sheet size, aids material distribution, minimizes chill marks.

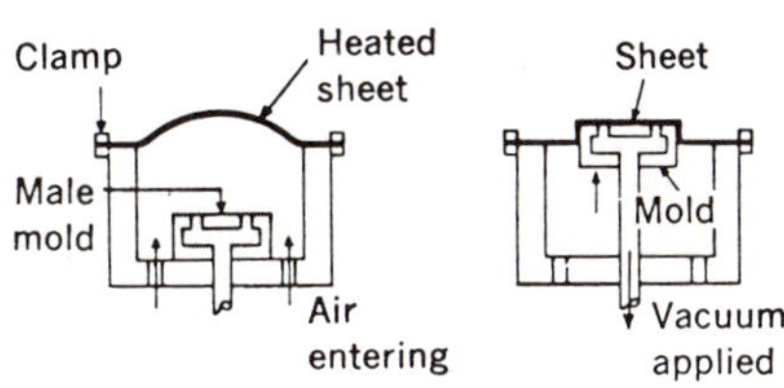

E. Air-slip forming is similar to vacuum snap-back except that heated sheet is billowed up and mold rises to meet it.

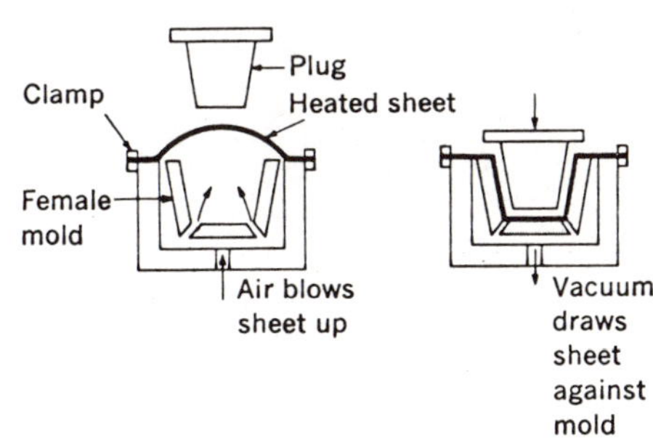

F. Forming with billow plug is often used to produce thin-wall items with depth-to-diameter-draw ratios up to 1.5:1.

FIG. 6.47 *(Continued)*

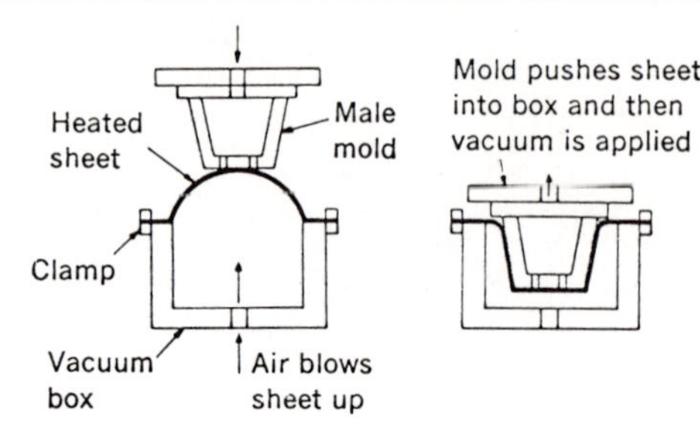

G. Forming with billow snap-back is recommended for any parts requiring a uniform, controllable wall thickness.

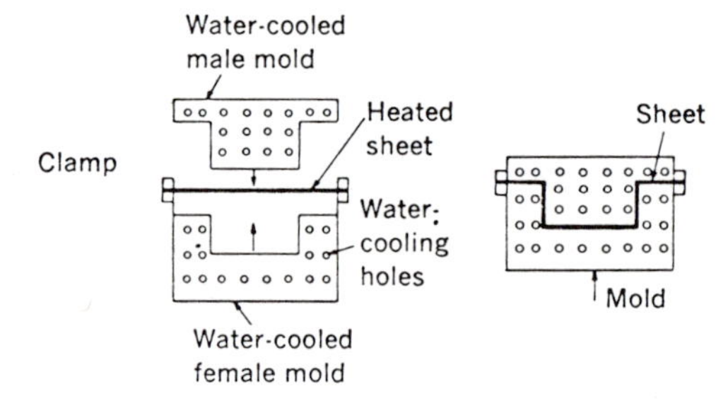

H. Matched metal-die pressure forming, although expensive, is a highly positive thermoforming technique.

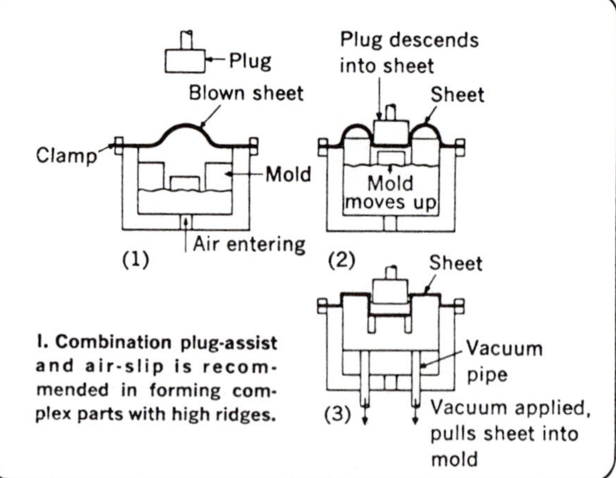

I. Combination plug-assist and air-slip is recommended in forming complex parts with high ridges.

FIG. 6.47 *(Continued)*

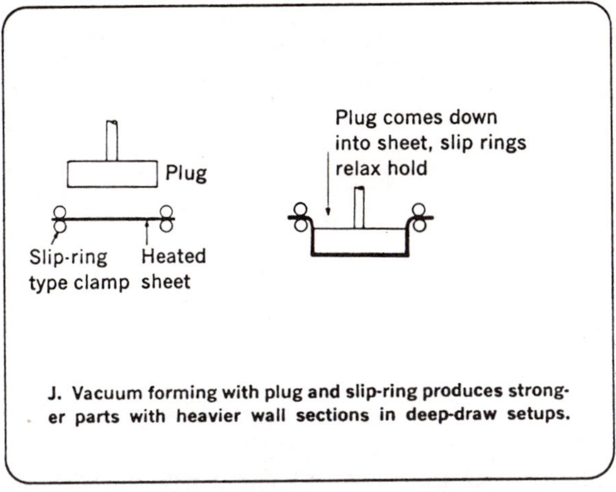

J. Vacuum forming with plug and slip-ring produces stronger parts with heavier wall sections in deep-draw setups.

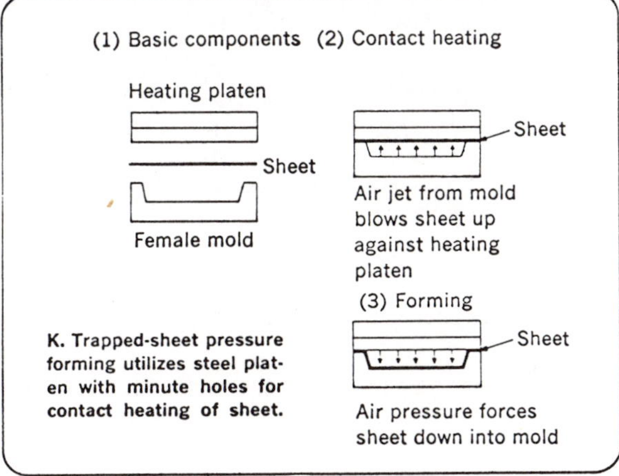

K. Trapped-sheet pressure forming utilizes steel platen with minute holes for contact heating of sheet.

FIG. 6.47 (*Continued*)

veyed to the raw material bin to be mixed with vigin pellets for another trip through the process. Products such as refrigerator door liners and the like are made in this way.

At the other end of the spectrum is the work-horse of the industry; the hand fed machine. In a straight vacuum forming cycle cut sheets of plastics are placed into the clamping frame. By simultaneously putting pressure on a pair of safety switches, the clamp closes and the heater moves forward to bring the sheet to a thermoforming state. When this is achieved the heater retracts and vacuum is applied to the mold causing the sheet to assume the shape of the mold. After the cooling period, vacuum is released and the molded part is removed from the

FIG. 6.48 Hull of a two-man submarine is thermomolded from ABS sheet. (*Courtesy Marbon Div., Borg-Warner Corp.*)

mold. After the safety switches are again energized the whole process proceeds automatically.

An interesting thermoformed product is shown in Fig. 6.48. The hull of this two-man sub is thermoformed of $^5/_{16}$ thick ABS and is about 15 feet long, 4 feet wide and 5 feet high.

THERMOFORMING DRINKING CUPS

A direct method of thermoforming drink cups, cheese containers, yogurt cups, etc., can be made in the machine shown in Fig. 6.49a. This machine, fitted with a reciprocating screw injection unit, injects a precise amount of the resin into an open mold. The mold is closed to create a preform disc. This disc is indexed to the next station where the container is formed by the plug-assist and draw method. At the third station the product is ejected. As shown in Fig. 6.49b, the rate of production is in excess of 1200 pieces per hour with no waste, as would be experienced in the same product made from extruded sheet. This "Mona-former" process lends itself to producing containers with reinforced lips which is very important in this type of product.

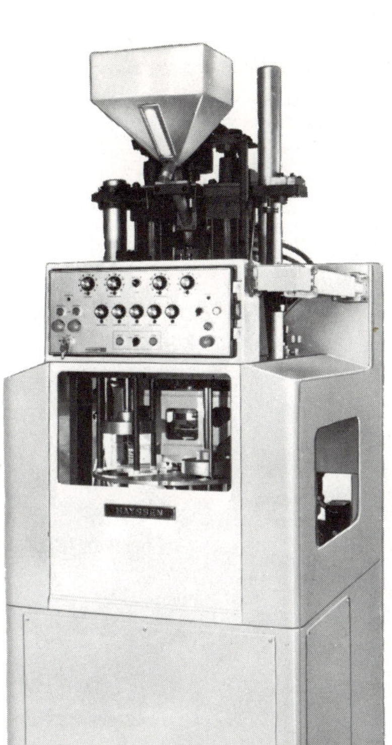

FIG. 6.49A Direct method thermoform machine. (*Courtesy Haysen Mfg. Co.*)

REPAIR OF THERMOPLASTICS PRODUCTS

Scratches and loss of gloss may often be helped by the use of automobile body buffing and rubbing compounds. Some spray type household cleaners will serve to remove lightly embedded surface dirt. Some will ruin the finish. A hot air gun will be helpful in softening the surface sufficiently to remove more deeply embedded particles. Some deep cuts, etc., may be filled with a paste made from the product's compound obtained by dissolving it in methyl ethyl ketone or methylene chloride. After the solvent has evaporated the surface may be sanded and buffed as required.

Hot melt guns are very useful in repairing many surface cuts and depressions. A hot air gun with welding rods cut from the material of the product may produce effective weld—similar to procedures used for the metals. A scab plate is

FIG. 6.49B Twelve-ounce cups produced in direct method thermoform machine. (*Courtesy Haysen Mfg. Co.*)

often adhered to the underside of a fractured area to reinforce the piece prior to one of the foregoing repair procedures.

FLUIDIZED BED COATING PROCESS

The fluidized bed process of coating metallic objects with plastics is unique in that both thermoplastics and thermosetting materials can be used for the coating. The obvious reason for the coating is to enhance the substrate with improved characteristics. Among these are corrosion resistance, improved electrical properties, colorful appearance and wear resistance. The properties of various fluidized bed coatings illustrating these advantages are shown in Table 6.3.

The "moving" or fluid mass of resin is created by air flowing through a porous plate fixed into the bottom of a tank. When the resin is placed above the plate it becomes buoyant and acquires the apparent qualities of a fluid and any object immersed therein is completely "wetted" by the resin particles.

The coating process consists of dipping a heated metal object into the "moving" mass of powdered resin. When the resin particles touch the surface of the hot ob-

TABLE 6.3 PROPERTIES OF FLUIDIZED-BED COATINGS[a]

	Vinyl	Cellulosic	Epoxy	Nylon	Polyethylene	Chlorinated Polyether
Chemical resistance						
Exterior durability	E	E	F	F	F	F
Salt spray	E	E	VG	G	G	E
Water (salt, fresh)	E	VG	G	F	VG	E
Solvents						
Alcohols	E	F	E	G	E	E
Gasoline	E	G	E	E	VG	VG
Hydrocarbons	G	G	E	E	VG	E
Esters, ketones	P	P	F	VG	G	VG
Chlorinated	P	P	E	E	F	F
Salts	E	VG	E	VG	E	E
Ammonia	E	P	P	G	E	E
Alkalies	E	F	VG	G	VG	E
Mineral acids						
Dilute[b]	E	G	E	F	E	E
Concentrated[c]	G	P	G	P	VG	E
Oxidizing acids						
Dilute[b]	E	P	G	P	VG	E
Concentrated[c]	G	P	P	P	P	G
Organic acids						
Acetic, formic, etc.	F	P	F	P	VG	E
Oleic, stearic, etc.	E	F	E	VG	VG	E
Mechanical and physical properties						
Abrasion resistance	G	VG	VG	E	F	VG
Flexibility	E	G	F[e]	G	E	F
Impact resistance	E	E	G	VG	F	G
Max. service temp., °F	200+	180	350+	180[d]	160	250
Dielectric strength	VG	VG	E	G	E	VG
Decorative properties						
Color range	E	E	P	F	G	P
Color retention	VG	E	F	VG	VG	G
Initial gloss	VG	E	G	G	VG	G
Gloss retention	G	E	P	—	—	—

[a] These data are intended only as a preliminary selection guide. Final selection should be made after consulting with coating formulator and after suitable testing. Data are based on Corvel fusion bond coatings as supplied by The Polymer Corp.

Key: E = excellent, VG = very good, G = good, F = fair, P = poor.

[b] Dilute = 10%.

[c] Conc. = over 30%.

[d] Up to 300° F in nonoxidizing environment.

[e] Ranges from good to poor, depending upon composition.

Courtesy "Modern Plastics Encyclopedia."

TABLE 6.4 SELECTING THE RIGHT ADHESIVE*

	Acetal	Acrylics	Cellulose acetate	Cellulose acetate butyrate	Cellulose nitrate	Ceramics	Glass	Metal	Nylon	Phenolic	Poly-carbonate	Styrene	Urea	Vinyl, flexible	Vinyl, rigid	Wood
Acetal	E,R	B,E,R	R,E	R,E	R,E	E,R	E,R	E,R	E,R	E,R	E,R	E,R	B,E,R	S,B,R	B	E,R
Acrylics	B,E	M,S,B	B,E	B,E	B,E	B,E,R	R,E,B	E,R,B	R,E,B	E,R,B	B,R,S	S,R,B	E,R			B,R,E
Cellulose acetate	R,E	E,B	S,B	B	B,S	B	R	R,E	R,E	R,E	B,R	R	E,R	R	E	B
Cellulose acetate butyrate	R,E	B	B	B	B	E,R	E,R	E,R	R,E	E,R	B,R	E,R	E,R	R	E,R	B
Cellulose nitrate	R,E	R,E,S	S,B	S,B	S	B,R	B,R	R,E	R,E	R,E	R,E	E,B,R	E,R	R	R,E	B
Nylon	E,R	R,E	R	R	R	E,R,S	R,E	R,E	S,B	S	R,E	R	R,E	R	R	R,E
Phenolic	E,R	E,R	R,E	R,E	R,E	E,R	E,R	E,R	S,E	S,E	E	R,E	R	R,E	R,E	E,R
Polycarbonate	E,R	M,R,E	E,R	E,R	E,R	E,R	E,R	E,R	E,R	E,R	S,E,R	E,B	E,R	R,E	E,R	E,R,B
Polyester	E,R	E	E	E	E	E	E	E	E	E	E	E	E	E	E	E
Styrene	E,R	R,B	R	R	R	R,B	R,B	B,R	R,B	R,B	R,B	R	R,B	R,B	R,B	R,B
Vinyl, flexible	B	S,B,R	B	B,R	B,R	B,R	B,R	R,B,E	R,E	R,E	B	R,E	E,R	S,R,B	S	B,R
Vinyl, rigid	B	B,R	R,B	B	B,R	B,R	R,B	R,E,B	B,R	E,R	R,B	B,R	E,R	B,R,S	S,B	R,B

Abbreviation and definitions:

M—Monomeric cements, based on a specific plastic which must be catalyzed to produce a strong bond.

S—Solvent cements, which dissolve the plastic to provide molecular interlocking, then evaporate. Normally they require close fitting joints, produce strong bonds.

B—Bodied adhesives contain a thermoplastic or thermosetting resin, and solvents, sometimes plasticizers, which dry by evaporation. The bodied adhesives can compensate for substantial variations in mating surfaces while still providing strong joints.

R—Elastomeric adhesives, based on natural or synthetic rubbers. Some contain the rubber in solvent or water suspension or solution. They may be cured at room or elevated temperatures to provide extremely strong joints.

E—100% reactive adhesives which depend upon catalytic action to join the two materials with an interlayer of thermosetting resin. The various dry and liquid epoxies, the polyesters, and the phenolics are being considered in this category.

Polyethylene can be bonded only by some of the rubber based cements with extremely low joint strengths.

Fluorocarbons can be bonded to themselves and to other materials only if pretreated or etched. Normally, epoxy adhesives are used. Joint strength is moderate.

*Courtesy "Modern Plastics Encyclopedia."

ject, they melt and fuse to the metal surface. Residence time and residual heat in the object determines the thickness of the coating. Post-heating is often used to completely smooth and fuse or cure the resin coating. Cooling is accomplished by water immersion or cooling in air. The item to be coated must always be degreased and sometimes primed to obtain maximum adhesion of the coating.

The advantages of the process lie in the ability to build uniform coating thicknesses from .005 to .080 inch on many substrates such as aluminum, carbon steel, brass, expanded metal, etc. Products coated by this method include valve bodies used in the chemical industry, electrical transformer casings, covers, and laminations; steel pipe, appliance parts, window sash, hardware, etc.

ADHESIVES FOR PLASTICS

The joining of plastics elements depends on the use of monomers, solvents and cements such as the epoxies and elastomerics. Elastomerics and epoxies are usually used in cementing thermosets and the monomers and solvents for thermoplastics. Tremendous bond strength may be obtained. In any case, the joining surfaces must be correct to obtain the ultimate in joint strength. Depending on the product requirements, the lap joint, V joint, as well as tongue and groove joint, should be considered in comparison to the simple lap joint, to obtain the required joint strength. Joint fit must be close and accurate when using solvent or monomer cements with thermoplastics. Greater latitude exists when using the epoxies in joining thermosetting materials. Since so many variables are involved it is suggested that the material supplier and adhesive manufacturer be consulted for specific advice. Table 6.4 presents a general approach to the problem. Amide-Imide bonds 90 to 700°F.

Plastics adhesives are extensively used today in place of rivets, welding, soldering, brazing, etc. Aluminum bicycle frames for example are "welded" with epoxy adhesives. Many metallic components of automobiles are adhesive bonded in place with superior results and reduced cost.

ELECTROSTATIC POWDER COATING

Electrostatic spraying of plastics powders has developed in recent years. This method of solventless coating has been of especial interest to manufacturers faced with the increasingly stringent requirements of solvent-pollution laws.

The spray gun used in this process looks like, and is used like, a paint spray gun. The system consists of a container for the powder with a feeder to a pump that conveys the powder to the gun through a tube. The gun is fitted with an electrode which imparts a high voltage charge to the powder as it passes through the gun. The powder is emitted from the gun in a form of a spray which is attracted to the grounded object to be coated. Ordinarily the parts are not preheated prior to spraying but are fused in an oven after spraying. The resultant films are usually 2 to 4 mils thick. Thicker films, 20 mils or better, can be obtained by preheating the substrate prior to spraying.

Two hundred mesh is the usual particle size of the powder used in this process. By decreasing the particle size to 400 mesh, thinner coatings in the order of 0.5 to 1.0 mil, are possible.

ELECTROSTATIC–BED COATING

Another method of powder coating using the fluidized bed is obtained by installing electrodes in the bottom of the fluidizing tank. With high voltage applied to the electrodes, the bed becomes electrostatically charged and the particles are attracted to the grounded product as in electrostatic spraying.

This process is used to coat coiled stock, wire mesh and other products of simple configuration. The electrostatic spraying method is used to coat furniture, wire racks, automotive body parts, machine frames and other such products having intricate configurations.

FORGING OF THERMOPLASTICS

The process is misnamed, since it is not truly a forging process but rather a re-forming, cold or warm, or as the technically oriented call it, solid-phase forming. However, the usual appelation is forging, and this is what we shall call it during this discussion.

In the forging process a preheated blank of the plastics is placed in a male/female die and re-formed under pressure to its final shape. The preferable pressure media is a hydraulic press but mechanical toggle presses are used as well. The obvious preference for the hydraulic press lies in the ability to control easily the rate of pressure application. Forming pressures range from 1,000 to 10,000 psi depending on material and temperature.

Dies are generally made of a machineable prehardened steel such as is commonly used in the construction of injection molds. A smooth machine finish is suitable with no time consuming polishing necessary. Means for cooling the die have also been found unnecessary.

The most commonly used plastics in the forging process are those which are very difficult to produce using the injection process, or at any rate impractical to injection mold in the heavy sections required in the product.

Ultrahigh-molecular weight polyethylene, glass reinforced UHMW polyethylene, and polyvinylidene fluoride are among the "difficult" materials that are forged. The former for its physical and chemical resistance properties and the latter for its exceptional chemical inertness. Another material used in large volume is polypropylene.

The blank referred to earlier is prepared by cutting from an extruded bar or by compression molding. It has been found that it is best to preheat the blank to a

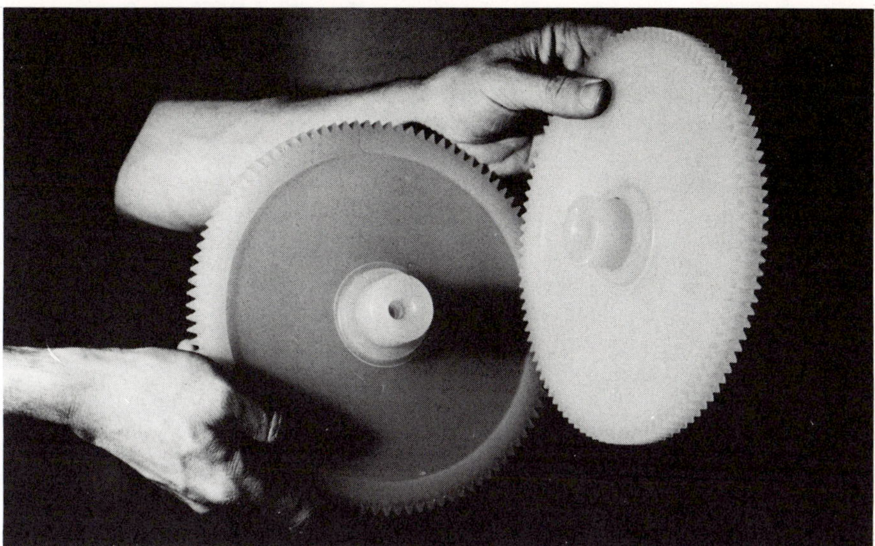

FIG. 6.50 This heavy sectioned spur gear pf 8.5 in. diameter was forged from polypropylene. (*Courtesy Mayfair Molded Products*)

temperature somewhat below its melt temperature before forging. A properly preheated blank will form at lower pressure and the finished piece will have a much reduced tendency to spring back as it would if a cold blank were used. In complex shapes it has been found that the latter method results in a more easily forged product.

The forging process lends itself to the manufacture of thick section plastics products of otherwise difficult to process plastics. The re-orientation of molecules obtained during forging enhances the physical properties of the end product considerably. The illustration shown in Fig. 6.50 is a fine example of this process.

PLASTICS TOOLING

Many plastics resins and compounds are used for tooling plastics and metal dies, jigs and fixtures as shown in Fig. 6.51. Molds for the reinforced plastics, RTV silicone rubbers, cast phenolics, etc., are often made of plastics. For this purpose, it is only necessary to produce a pattern of some material such as wood, plaster or concrete that will define the desired shape and dimension. This pattern is then coated with a parting compound that prevents sticking. The resin is then poured or laid in place for curing by the applicable process.

Fillers of glass fibers, fabrics or beads, powdered metals, mica, etc., are added

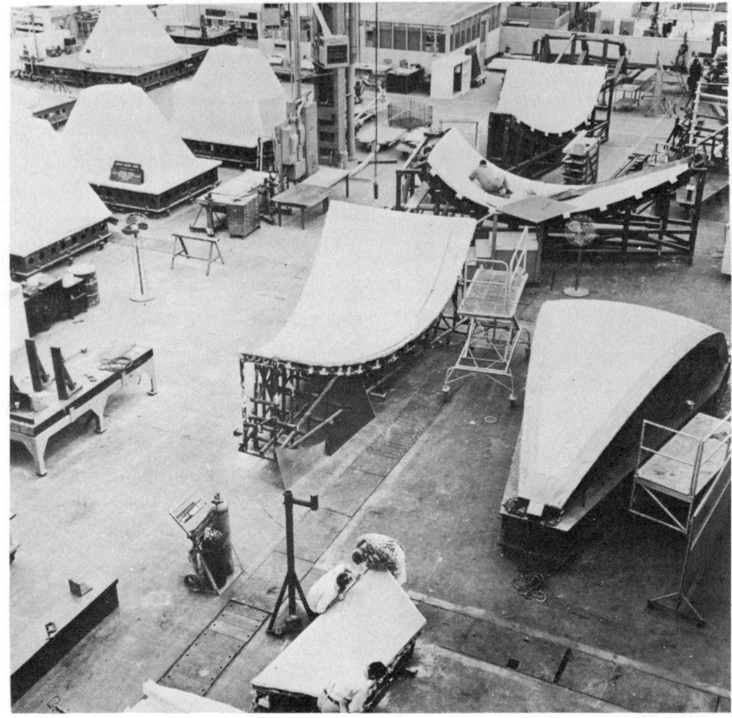

FIG. 6.51 Tooling for reinforced plastics molding. The molds shown here were used for making the Saturn space vehicles. They were made from a two component, white colored epoxy laminating resin system. (*Courtesy Fruane Plastics, Inc.*)

for stability, heat conductivity, etc. A variety of epoxy resins and compounds are available for this work; some are available for tools that will perform at temperatures up to 300°F.

Drill jigs are often made from these materials, especially when the surface to be drilled has a complex curvature. In this application, the drill jig bushings are cast in place. Trial molds for injection and transfer molding have been made successfully from metallic particle filled epoxy resins.

CALENDERING

Calendering is a rolling process used for coating webs such as fabrics, paper or plastics with another plastics material (Fig. 6.52). In essence, it is a continuous "extrusion" process between a pair of mating rolls that establish the thickness and surface characteristics of the applied coating.

The first calender was designed and built for rubber in 1836 by Edwin M.

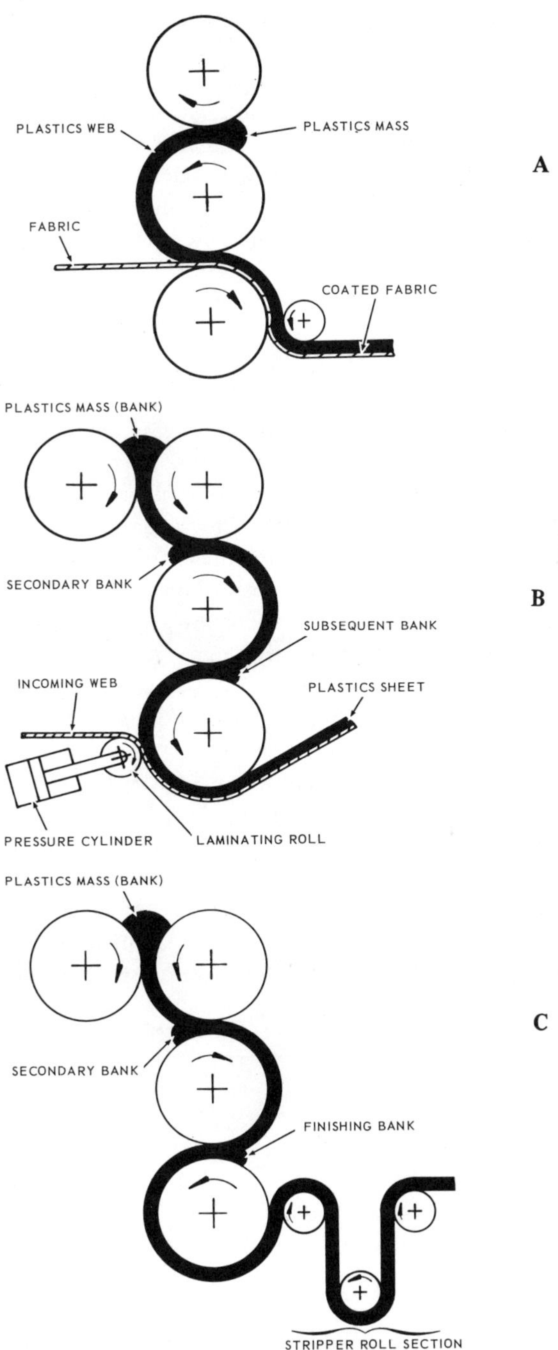

FIG. 6.52 Various calendering processes. (*Courtesy Adamson United Co.*)

Chaffee and was used to waterproof fabrics. Vinyls are used extensively today for this same purpose. Calendering has become a highly technical process with fine controls and wide product latitude. "Frictioning" formerly was used to drive a high viscosity material into the interstices of the fabric; today, the desired effect is achieved by placing a thin coating of plastics on the fabric.

A basic type of calender coating is illustrated in Fig. 6.52A wherein elastic plastics are applied to a prepared web of nonsimilar material. Note that the plastics sheet has been formed in the prior nip and is applied by the two coating rolls. Fig. 6.52B illustrates the common method for plastics in which a laminating roll, hydraulically operated, pushes the web against the plastics formed by the calender.

The most widely used calendered products are light gage polyvinyl chloride film or heavier gage sheeting. These applications include hospital sheets, printer's blankets, raincoats, tablecloths, draperies, counter tops, flooring and myriad other items. They may be heat-sealed, heat-shrunk for blister packaging or used plain as wrapping materials. They serve as mulches, swimming pool liners, reservoir films, vapor barriers, and so forth. They may be soft, stiff or filled.

They can be homopolymer, copolymer, filled or plasticized compounds. Approximately one half of all PVC resins and 40% of all rubber are calendered. Some polyethylene, some ABS and a few like products are calendered.

Plastics films generally are produced in gage ranges from 0.002 inch to a maximum of 0.200 inch. The most commonly used calender arrangement is shown in Fig. 6.52C. Because of gage and product variations, temperature control must be adjustable. Embossing units may be included in the calendering operation. Other supplemental units include laminating units (floor tile), festoon or accumulators and valley printers.

A single calender may produce a range of gages that permit material consumption of 700 to 20,000 lb/hr. Material preparation is done in-line with calendering. Premixed pellets and other prepared compounds are never used. Calender product quality is dependent on compound formula. Most preparation is done in high shear, intensive batch-type mixers—the rest is done with screw-type or continuous mixers.

7 MATERIAL SELECTION AND PRODUCT DESIGN

MATERIAL SELECTION

The large increase in the number of sophisticated plastics, and the fine points of their capabilities, place a greater responsibility on the product designer who must specify the materials that are suitable for the application. Subsequent economic analyses will determine which of these materials will be selected for production.

In general, it must be remembered that selection of the proper material requires analysis of the good points and the weak points of each material considered for the job. No one material will possess all of the qualities desired and none of the weak qualities. Undesirable characteristics must be compensated for in the product design.

There is a workable elimination approach to the selection of the right plastics material which will narrow the field to a limited choice. In this case, the final material selection problem becomes one of field testing under actual use conditions to prove the endurance and stability of the product.

Elasticity. If the product requires flexibility, the choice is limited to one of the conventional synthetic rubbers, the thermoplastic rubbers, EVA, ionomer, urethane/polyester, polyethylene, vinyl, polypropylene, fluorocarbon, silicone, polyurethane, plastisols, acetal, nylon, or some of the rigid plastics that have limited flexibility in thin section; e.g., thin laminations are quite flexible. See Table 3.1.

Temperature. Thermal considerations will quickly eliminate many materials. For products operating above 450°F, the silicones, polyimides, polyarylsulfone, hydrocarbon resins, cold mold, glass-bonded mica or phospho-asbestos plastics may be required. A few of the organic resin-bonded inorganic fibers such as TFE-bonded ceramic wool perform well in this field. Epoxy, diallyl phthalate and phenolic bonded glass fibers may be satisfactory in the 450 to 550°F range. A limited group of *ablation* material is made for re-entry use.

Between 250 and 450°F, glass or mineral-filled phenolics, melamine, alkyd, silicone, nylon, polyphenylene oxide, polysulfone, polycarbonate, methylpentene, fluorocarbon, polypropylene and diallyl phthalate must be evaluated. Household heating devices use melamine and phenolic handles, as depicted in Fig. 7.1. The addition of glass fillers to the thermoplastics can raise the useful temperature range as much as 100°F and at the same time shorten the molding cycle. In the 0 to 212°F range, a broad selection of materials is available. Low-temperature considerations may eliminate many of the thermoplastics. Polyphenylene oxide can be used at temperatures as low as -275°F. Glass-bonded mica, TFE, and FEP are serviceable down to -400°F. Thermosetting materials exhibit minimum embrittlement at low temperatures.

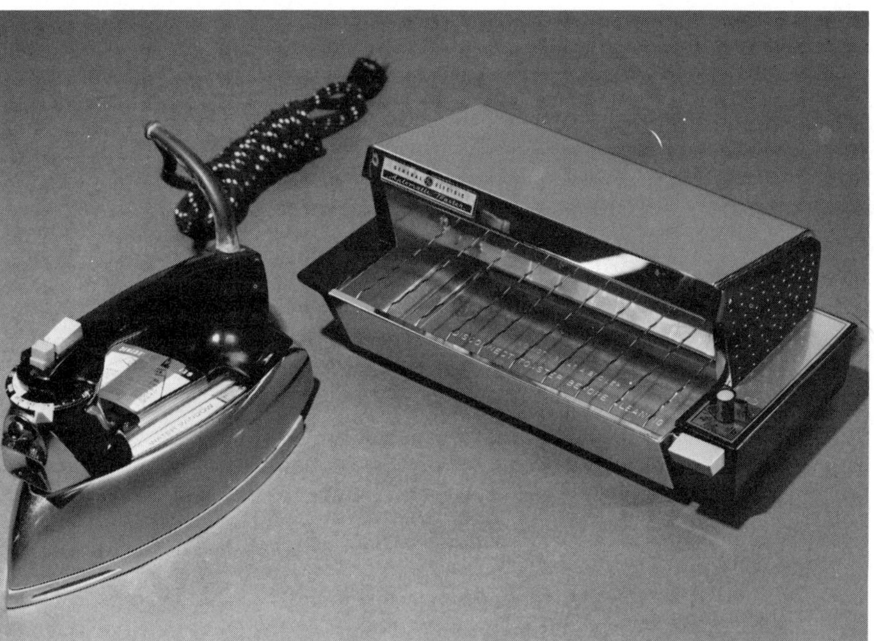

FIG. 7.1 Household appliances such as irons and broilers use phenolic compounds for handles.

A major advantage which the thermosets have held over the thermoplastics has been their stability under load at elevated temperatures. The thermosets fail slowly under such overloading at above-temperature conditions, while the thermoplastics fail suddenly and completely.

Flame Resistance. The underwriters' ruling on the use of self-extinguishing plastics for contact-carrying members and many other components introduces critical material selection problems. All thermosets are self-extinguishing. Nylon, polyphenylene oxide, polysulfone, polyarylsulfone, polycarbonate, vinyl, chlorinated polyether, chlorotrifluoroethylene, vinylidene fluoride and fluorocarbon are thermoplastics that may be suitable for applications requiring self-extinguishing properties. Cellulose acetate and ABS are also available with these properties. Glass reinforcement improves these materials considerably.

The Underwriter Laboratories SE–O rating has been met by many contemporary modified thermoplastics. Some include long glass fibers and asbestos fillers, as well as flames-retardant additives. Some of the flame retardants are also smoke repressive—a fast growing product design consideration.

Impact. Although impact strength of plastics material is widely reported, the properties have no particular design values and can be used only to compare relative response of materials. Even this comparison is not completely valid because it does not solely reflect the capacity of the material to withstand shock loading, but can pick up discriminatory response to notch sensitivity. A better value is impact tensile, but unfortunately this property is not generally reported. The impact value, with this limitation, can broadly separate those which can withstand shock loading versus those which are poor in this response. Therefore, only broad generalizations can be obtained on these values. Comparative tests on sections of similar size which are molded in accordance with the proposed product must be tested to determine the impact performance of a plastics material. The laminated plastics, glass-filled epoxy, melamine and phenolic are outstanding in impact strength. Polycarbonate, urethane, and ultra high molecular weight PE are outstanding impact strength thermoplastics. The impact strength of some plastics films is outstanding, as shown in Fig. 7.2.

Arc Resistance. Electrical Devices often require arc resistance, as a high-current, high-temperature arc will ruin many plastics.

Polyarylsulfone (Chapter 3) is one of the better thermoplastics for use in arc-resistant applications. Many of the plastics burn momentarily or longer under an electrical arc or flash-over. This burning may leave a conducting carbon path or "track" on the surface that will carry the current when voltage is reapplied. Materials are called non-tracking if this cannot occur.

Some special arc resisting plastics are available, such as Ex-Arc II. The more serious cases may require cold mold, phosphoasbestos, glass-bonded mica or mineral-filled fluorocarbon products. Lesser arcing problems, as shown in Fig. 7.3, may be solved by the use of polysulfone, polyester glass, D.A.P.-glass, alkyd, melamine, urea or phenolics. With low-current arcs, general-purpose phenolic

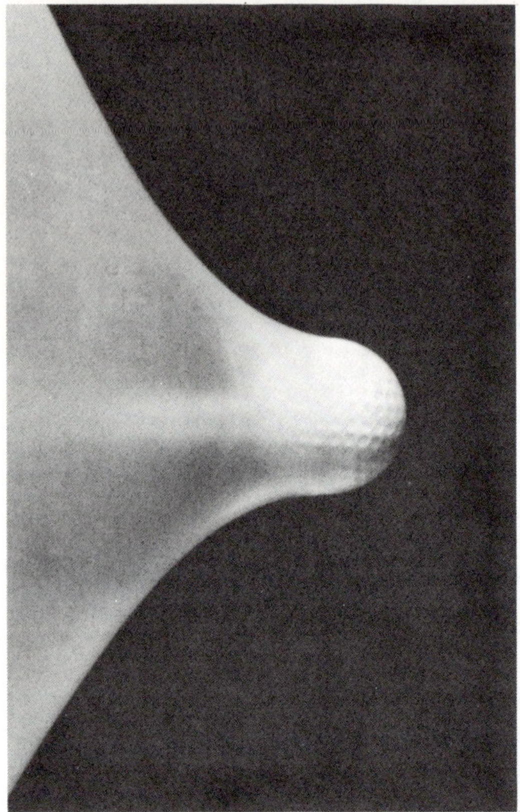

FIG. 7.2 A golf ball does an "about face" after speeding into film of polyurethane only 2.8 mils thick. This impact strength, combined with tensile strength of 5800 pounds per square inch at 73°F., and all around toughness, qualifies polyurethane film for rugged packaging duty. (*Courtesy B. F. Goodrich Chemical Co.*)

and glass filled nylon or polycarbonate, acetal and urea may be used very satisfactorily. A coating of fluorocarbon film will improve arc resistance in some cases. All circuit breaker problems must be scrutinized with respect to product performance under short-circuit conditions and mechanical shock.

The most popular materials today for Navy circuit breakers are the alkyd-glass and diallylphthalate-glass compounds.

Radiation. In general, plastics are superior to elastomers in radiation resistance but are inferior to metals, glass-bonded mica, and ceramics. The plastics materials which will respond satisfactorily in the range of 10^{10} power and 10^{11} erg per gram are glass and asbestos-filled phenolics, certain epoxies, polyurethane, polystyrene, mineral-filled polyesters, silicone and furane resins. The next group of resins in order of radiation resistance includes polyethylene, melamine, urea formaldehyde resins, unfilled phenolic and silicone resins. Those materials which have poor radiation resistance include methyl methacrylate, unfilled polyesters,

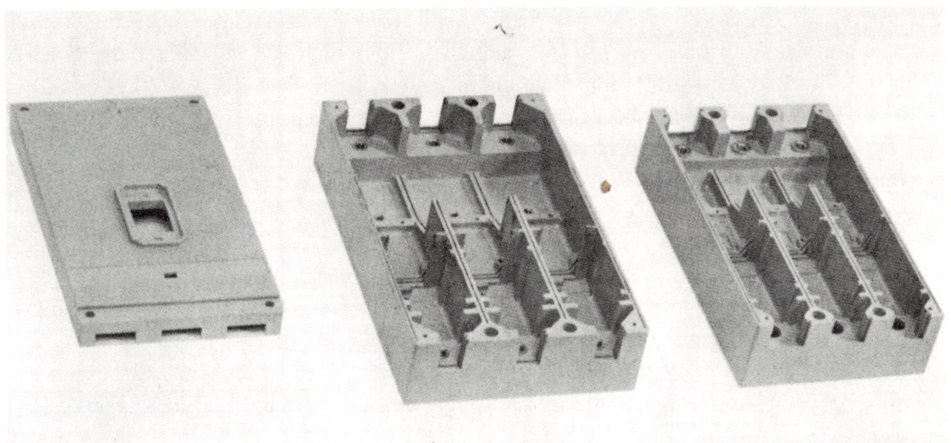

FIG. 7.3 Melamine, polyester and diallyl phtbalate resins are often used for the making of complex circuit breaker components. Urea compounds are widely used for the small household breakers.

cellulosics, polyamides and fluorocarbons. Glass-bonded mica is completley unaffected by prolonged radiation exposure. Polyimides resist ionizing radiation.

Color. Urea, melamine, polycarbonate, polyphenylene oxide, polysulfone, polypropylene, diallyl phthalate, and the phenolics are needed in the temperature range above 200°F and up to the materials limitation for good color stability. Most thermoplastics will be suitable below this range, as indicated in the tables.

Bearings, Cams, Gears. Heavy-duty gears, bearings and cams are produced very successfully from fabric-filled phenolic laminates. A few special moldable phenolic compounds, polyester, polyurethane and polyethylene exhibit interesting properties for bearings and cams. In high load products, consideration must be given to nylon and acetal resin. These resins, compounded with glass fillers, graphite or molybdenum disulfide, exhibit very desirable properties. An FEP-fluorocarbon-acetal compound has shown attractive bearing properties. Poly-imide materials are suitable for high temperature bearings. Nylon is best for impact problem gears while acetal resin is best for precise dimensional small gears.

Transparency. Maximum transparency is available in acrylic, polyolefin, iono-mer and styrene compounds. Many other thermoplastics may have adequate transparency; polyethylene with improved transparency is now produced. Poly-carbonate, polysulfone, and polyphenylene oxide provide high temperature transparency. ADC gives transparency and maximum scratch resistance.

Applied Stress. Many thermoplastics will craze or crack under certain environmental conditions, and products which are highly stressed mechanically must be checked very carefully. Polypropylene, ionomer, chlorinated polyether, phenoxy, EVA and linear polyethylene offer greater freedom from stress crazing than some other thermoplastics. Low melt index, and low density polyethylene may withstand applied or environmental stresses. Solvents may crack parts held under stress. Thermosets are generally preferable for parts under continuous loads.

The several thermoplastic polyesters are replacing some of the older plastics in load bearing electrical applications. Nylon, acetal and the phenolics are giving way to these thermoplastic polyesters. The reinforced thermoplastic polyesters have an adequate distortion temperature for many applications and also exhibit adequate tensile strength, volume resistivity, dielectric strength, arc and chemical resistance, plus superior dielectric properties when wet.

Moisture. Deteriorating effects of moisture are well known. For high moisture applications, polyphenylene oxide, polysulfone, acrylic, butyrate, diallyl phthalate, glass-bonded mica, mineral-filled phenolic, chlorotrifluoroethylene, vinylidene chloride, chlorinated polyether, vinylidene fluoride and the fluorocarbons should be satisfactory. Diallyl phthalate, polysulfone and polyphenylene oxide have performed well with moisture/steam on one side and air on the other—a troublesome combination and they also will withstand repeated steam autoclaving. Long term studies of the effect of water have disclosed that chlorinated polyether gives outstanding performance. Impact styrene plus 25% graphite and high density polyethylene with 15% graphite give long-term performance in water.

Chemical. The chemical resistance of plastics is well known, and Table 7.1 will serve as an excellent initial guide. Most material makers have developed long term data for commonly used chemicals. Great care must be exercised in this selection, and environmental conditions are very pertinent. Two materials which do not attack a plastic material when used separately may be troublesome when used in combination, or diluted with water. Chlorinated polyether is formulated particularly for products requiring good chemical resistance. Other materials exhibiting good chemical resistance are all of the fluorocarbon plastics, methylpentene, hydrocarbon-coke, polyolefins, some phenolic and the diallyl phthalate compounds. Additives such as fillers, plasticizers, stabilizers, colorants and catalysts can decrease the chemical resistance of unfilled resins. Temperature is also highly important. Careful tests must be made under actual use conditions in final selection studies.

Certain cosmetics will affect plastics, and tests are necessary in most cases with new formulations. Most materials manufacturers provide reliable data.

Surface Wear. Hardness is not necessarily the proper index for scratch resistance. In general, the thermosets have the best abrasion resistance. Acrylic, ABS and SAN also have good fingernail scratch resistance. Again, tests simulating

TABLE 7.1   EFFECTS ON PLASTICS IN DIFFERENT CORROSIVE ENVIRONMENTS AT TEMPERATURES OF 77°F AND 200°F*

PLASTIC	Aromatic solvents		Aliphatic solvents		Chlorinated solvents		Weak bases and salts		Strong bases		Strong acids		Strong oxidants		Esters and ketones	
TEMPERATURE °F	77	200	77	200	77	200	77	200	77	200	77	200	77	200	77	200
Acetal	1	2	1	2	1	2	1	1	1	2	5	5	5	5	1	2
Acrylic	5	5	2	3	5	5	1	3	2	5	4	5	5	5	5	5
Acrylonitrile-butadiene-styrene (ABS)	4	5	1	5	5	5	1	3	1	3	2	5	4	5	5	5
Cellulose Acetate (CA)	1	3	1	2	1	4	1	3	3	5	3	5	3	5	5	5
Cellulose Acetate Butyrate (CAB)	4	5	1	3	4	5	2	4	3	5	3	5	3	5	5	5
Cellulose Acetate Propionate (CAP)	4	5	1	3	4	5	1	2	3	5	3	5	3	5	5	5
Chlorinated Polyether (CP)	1	4	1	2	1	4	1	1	1	1	1	1	5	5	1	4
Epoxy (glass fiber filled)	1	2	1	2	1	3	1	1	2	3	2	3	4	4	2	3
Ethylene-vinyl Acetate (EVA)	5	5	4	5	5	5	1	5	1	5	1	5	1	5	4	5
Furan (asbestos filled)	1	1	1	1	1	1	2	2	2	2	1	1	5	5	1	1
Ionomer	2	4	1	4	4	4	1	4	1	4	2	4	1	5	1	4
Melamine	1	1	1	1	1	1	2	3	2	3	2	3	2	3	1	2
Phenolic (asbestos filled)	1	1	1	1	1	1	1	3	4	5	1	1	4	5	3	3
Phenoxy	4	5	2	5	5	5	1	5	1	5	1	5	4	5	5	5
Polyallomer	4	5	4	5	4	5	1	1	1	1	1	1	3	5	1	1
Polyamide	1	1	1	1	1	2	1	2	2	3	5	5	5	5	1	1
Polybenzimidazole (PBI)	1	1	1	1	1	1	1	1	1	2	1	2	1	3	1	1
Polycarbonate (PC)	1	1	1	1	5	5	1	5	5	5	1	1	1	1	5	5
Polychlorotrifluoroethylene (CTFE)	1	1	1	1	3	4	1	1	1	1	1	1	1	1	1	1
Polyester (glass fiber filled)	1	3	1	2	2	4	2	3	3	5	2	2	2	3	3	3
Polyethylene (PE)	4	5	4	5	4	5	1	1	1	1	1	1	1	1	4	5
Polyimide	1	1	1	2	1	1	2	3	4	5	3	4	2	4	1	1
Poly-para-xylene (PPX)	1	2	1	1	1	2	1	1	1	1	1	1	2	3	1	2
Polyphenylene Oxide (PPO)	4	5	2	3	4	5	1	1	1	1	1	2	1	2	2	3
Polypropylene (PP)	2	4	1	4	2	4	1	1	1	1	1	3	1	4	1	3
Polysulfone	4	4	1	1	5	5	1	1	1	1	1	1	1	1	3	4
Polystyrene (PS)	4	5	4	5	5	5	1	5	1	5	4	5	4	5	4	5
Polytetrafluoroethylene (TFE)	1	1	1	1	1	1	1	1	1	1	1	1	1	1	1	1
Polyurethane (PU)	3	4	1	4	4	5	1	1	3	4	1	4	1	4	2	3
Polyvinyl Chloride (PVC)	4	5	1	5	5	5	1	5	1	5	1	5	2	5	4	5
Polyvinyl Dichloride (PVDC)	4	5	1	2	5	5	1	1	1	1	1	3	1	2	4	5
Polyvinylidene Fluoride (PVF$_2$)	1	2	1	1	1	1	1	1	1	1	1	2	1	2	2	4
Silicone	4	4	2	3	4	5	1	2	4	5	3	4	4	5	2	4
Styrene Acrylonitrile (SAN)	4	5	3	4	3	5	1	3	1	3	1	3	3	4	4	5
Urea	1	3	1	3	1	3	2	3	2	3	4	5	2	3	1	2

Code: 1. no effect or inert, 2. slight effect, 3. mild effect, 4. softening or swollen effect and 5. severe deterioration.

*Courtesy *Plastics World.*

actual conditions are necessary to get the best answer. When abrasive wear is the problem, ultra high molecular weight polyethylene, urethane, high density polyethylene, nylon 11 and polyester film are good performers. ADC combines transparency with best scratch resistance.

Permeability. Many plastics rate very poorly in permeability properties. Polyethylene will pass wintergreen, hydrocarbons, essential oils and many other chemicals. It is used in certain cases for the separation of gases since it will pass one and block another. Chlorotrifluoroethylene and vinylidene fluoride, vinylidene chloride, polypropylene, EVA and phenoxy merit special study.

Electrical. Electrical considerations will eliminate many plastics, and published data are reasonably comparable on similar sections. Final tests must be made on the actual section under field environmental conditions to make sure that the design is adequate. High-frequency, high-temperature applications are the most difficult to solve. High-altitude, high-voltage applications with the included ozone problems are often solved only by use of *glass-bonded mica*, which matches the thermal expansion rate of steel and prevents corona gap formation. Vacuum impregnation with epoxy resin may be suitable for some products. All organic plastics except polyimide can give off vapors that may cause contact failure when used in a vacuum. Many special dielectrics and encapsulant foams are available, as shown in Fig. 7.4.

Dimensional Stability. If *absolute* dimensional stability is mandatory, then the organic plastics are eliminated in favor of the ceramics and glass-bonded mica, as shown in Fig. 7.5. Several organic plastics are available with very good dimensional stability, and they are suitable where some age and environmental dimen-

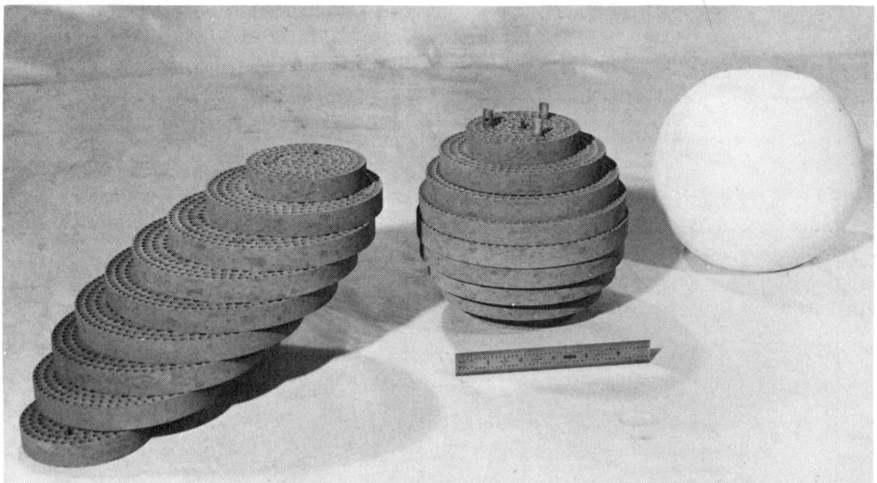

FIG. 7.4 Lightweight microwave lens constructed of artificial dielectric discs encapsulated in a low-loss, low-dielectric constant foam. (*Courtesy Emerson & Cuming, Inc.*)

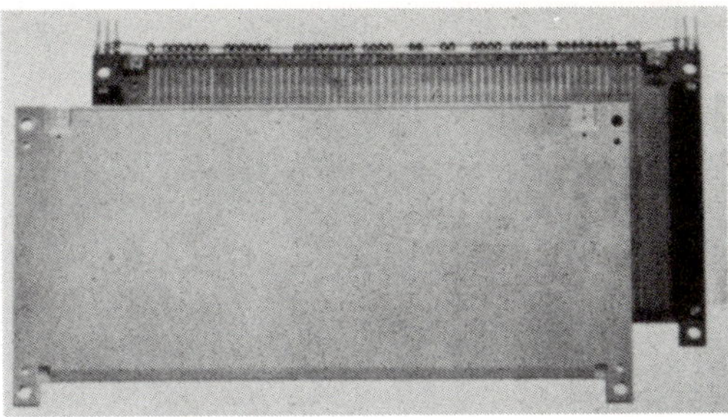

FIG. 7.5 Absolute flatness and 25-year dimensional stability is achieved by the use of glass bonded mica for the substrate in the Bell System's revolutionary electronic switching system. No point in its $6'' \times 12''$ surface is allowed to vary from any other measured point by more than $.006''$. (*Courtesy Mykroy Ceramics Co.*)

sional change are permissible. These materials include polyphenylene oxide, polysulfone, phenoxy, mineral-filled phenolic, diallyl phthalate, epoxy, rigid vinyl and styrene. Such products will gain from an after bake for dimensional stabilization. Glass fillers, may improve the dimensional stability of all plastics.

One of the most important characteristics in plastics applications requiring good dimensional stability is the absence of creep. Initially, when stressed, a plastics deforms or strains an amount which is predicted roughly by its modulus of elasticity. This is applicable only for a short period. With prolonged stress, there is a steady increase in strain (creep) until the rupture level is reached.

The rate of long term creep varies for a plastics material with its composition, load, temperature and other environmental factors. Processing variables can affect the creep. As a general rule, the rate of creep is inversely proportional to the elastic modulus of the material. The creep properties of the thermoplastics are obviously affected by the temperature. Table A.4A, in the Appendix shows the creep properties as measured by the apparent modulus after 1000 hours at various loads for some representative thermoplastics. (*Courtesy Materials Engineering*).

All materials using plasticizers must be avoided. Materials which exhibit substantial moisture absorption are not stable dimensionally. Many organic plastics show a high thermal expansion differential in comparison with mating metal products, and this can cause serious trouble if tight dimensional relations and tightly bonded inserts must be maintained. The thermal expansion of 400 series stainless steel matches glass bonded mica closely. A phenolic glass and diallyl phthalate glass material are available with very low shrinkage. Glass and other mineral fillers minimize the thermal expansion differential problem. Phenoxy and polyphenylene oxide are low in shrinkage and thermal expansion. The thermoplastics change dimensions rapidly as they approach the *cold flow* point. Great care must be taken in the selection of critical dimensional control parts for machines and instruments. Critical dimensions may be held best by included or assembled inserts in materials which have questionable stability.

Weathering. Many plastics have short life when exposed to outdoor conditions, as tabulated in Table 7.2. The better materials include acrylic, chlorotrifluoroethylene, vinylidene fluoride, chlorinated polyether, polyester, alkyd, black linear polyethylene. Black materials are best for outdoor service. Some of the styrene copolymers are suitable for certain outdoor uses. Rubber compounds vary greatly in their ability to resist weathering and must be selected with care. Polyether is one of the worst and silicone rubber one of the best for weathering.

Odor and Taste. Food packaging and refrigerating conditions will also eliminate many plastics. Melamine and urea compounds are suitable for this service, while polystyrene, styrene acrylonitrile, polyethylene, acrylic, ABS, polysulfone, EVA, polyphenylene oxide and many other thermoplastics are satisfactorily odor-free. FDA* *approvals* are available for many of these plastics, and these data will be most helpful when taste or odor is a problem.

Production Procedure. Many product designs are inherently limited by the economics of the process that must be used to make them. For example, thermosetting materials may not be blow-molded, and they have limited extrusion possibilities. Many hollow parts or large pieces may be produced more economically by the rotational process than by blow-molding. The quantity of parts required may eliminate the molding process and indicate a casting process. Heavy or thick cross section products are best produced from thermosets. Certain products are most economically produced by fabricating them with conventional machining from laminates or extruded sheets, rods or tubes. Inserts greatly increase the cost of thermoplastic parts, and it may be advantageous to design the product for post-molding assembly of inserts to gain the benefit of fully automatic molding and automatic insert installation. These fabrication studies will eliminate many plastics materials and further limit the choice. A conference with the molder or fabricator will provide the best guidance at this point.

Design Limits. The molder and fabricator place no limits on the design. There is a way to make the part—if the values justify the price. Any job can be done—at a price. The truly limiting factors are tool design considerations, material shrinkage, subsequent assembly or finishing operations, dimensional tolerance allowance, undercuts, insert inclusions, parting lines, fragile sections, production rate, and the essential selling price.

Material Cost. Some high priced materials mold at a very high speed, and are thus inexpensive to mold. Other material will run automatically and the adaptability of the material to fully automatic production methods often justifies a

* Food and Drug Administration.

TABLE 7.2 EFFECT OF WEATHER EXPOSURE ON APPEARANCE, THICKNESS, AND ELECTRICAL PROPERTIES*

Material	Surface Condition	Thickness Change, in.[a]	Electric Str Change, %[a]
THERMOSETTING PLASTICS			
Epoxy Casting (TETA Cured)	Frosted surface (600 hr)	—	+6.7 (3988 hr)
Melamine Laminate (G-5)	Loss of gloss (2500 hr)	0 (3188 hr)	−55.1 (3188 hr)
Melamine Molding (Cellulose Filled)	Erosion (2500 hr)	−0.001 (3130 hr)	+7.2 (3130 hr)
Phenolic Laminate (XXXP)	Fibers exposed (900 hr)	−0.001 (1611 hr)	+4.7 (1611 hr)
Phenolic Molding (Wood Flour Filled)	Filler exposed (1200 hr)	−0.002 (3130 hr)	+32.9 (3130 hr)
Polyester Laminate (FR)	Fibers exposed (600 hr)	0 (1597 hr)	+13.6 (1597 hr)
Polyester Laminate (GP)	Fibers exposed (600 hr)	0 (1597 hr)	+12.5 (1597 hr)
Silicone Laminate (G-7)	None (2500 hr)	0 (4002 hr)	−14.4 (4002 hr)
Urea Molding (Wood Flour Filled)	Filler exposed (1600 hr)	0 (3130 hr)	+7.5 (3130 hr)
THERMOPLASTICS			
ABS	Erosion (2500 hr)	−0.002 (3130 hr)	+4.7 (3130 hr)
Acetal	Severe erosion (900 hr)	−0.005 (1611 hr)	+6.1 (1611 hr)
Acetal (Stabilized)	Erosion (2500 hr)	−0.003 (3971 hr)	+18.7 (3971 hr)
Acetal (Carbon Black Filled)	Erosion 1200 (hr)	−0.003 (3188 hr)	+13.0 (3188 hr)
Acrylic (PMMA)	None (2500 hr)	−0.001 (3988 hr)	+3.6 (3988 hr)
Acrylic (Modified, Type A)	None (2500 hr)	—	—
Acrylic (Modified, Type M)	None (2500 hr)	—	—
Cellulose Acetate	Frosted surface (600 hr)	0 (1580 hr)	−13.2 (1580 hr)
Cellulose Acetate Butyrate	Erosion & frosting (2500 hr)	+0.002 (3893 hr)	−6.7 (3893 hr)
Fluorocarbon (PTFE)	None (2500 hr)	0 (4002 hr)	−3.6 (4002 hr)
Fluorocarbon (PCTFE, Type G)	None (2500 hr)	0 (4002 hr)	+3.4 (4002 hr)
Fluorocarbon (PCTFE, Type H)	None (2500 hr)	—	—
Nylon (Type 6/6)	None (2500 hr)	−0.001 (3988 hr)	−3.4 (3988 hr)
Nylon (6/6, Filled with MoS_2)	None (1600 hr)	0 (1611 hr)	−6.3 (1611 hr)
Polycarbonate	Frosted surface (900 hr)	0 (2224 hr)	+5.0 (2224 hr)
Polyethylene (Low Density)	None (1600 hr)	+0.001 (1611 hr)	−17.1 (1611 hr)
Polyethylene (High Density)	None (1600 hr)	0 (1611 hr)	+9.4 (1611 hr)
Polypropylene	None (1600 hr)	−0.001 (1580 hr)	+15.1 (1580 hr)
Polyvinyl Chloride (Rigid)	Erosion (2500 hr)	−0.001 (3893 hr)	−1.8 (3893 hr)
Polystyrene (General Purpose)	Frosted surface (1600 hr)	0 (1580 hr)	+10.0 (1580 hr)

[a] Values based on difference between shielded and exposed portions of panels. Figures in parentheses are length of exposure when observation or measurement was made.

*Courtesy Westinghouse Electric Corp.

premium material. The product engineer must evaluate the essential benefits demanded by the application and find a design and the material to deliver these benefits at minimum cost by this process of elimination. The ultimate material selection is a compromise choice, based on the most favorable balance of properties as shown in Fig. 7.3. In every case, the wrong material is by far the most costly, and the price per pound tells only part of the story.

TABLE 7.3 SOME CHOICE MATERIALS

Property	Thermoplastics	Thermosets
Low temperature	TFE, GBM*	DAP
Low cost	polystyrene	phenolic
Low gravity	polypropylene methylpentene	phenolic/nylon
Thermal expansion	phenoxy glass GBM	epoxy-glass
Volume resistivity	TFE	DAP
Dielectric strength	PVC	DAP
Elasticity	EVA	—
Moisture absorption	chlorotrifluoroethylene	alkyd-glass
Steam resistance	polysulfone	DAP
Flame resistance	TFE–GBM	melamine
Water immersion	chlorinated polyether	DAP
Stress craze resistance	polypropylene	all
High temperature	TFE–GBM	silicones
Gasoline resistance	acetal	phenolic
Impact	ultra high molecular weight PE	epoxy-glass
Cold flow	polysulfone	melamine-glass
Chemical resistance	TFE, FEP, PE, PP	epoxy
Scratch resistance	acrylic	allyl diglycol carbonate
Abrasive wear	polyurethane	phenolic-canvas
Colors	acetate, styrene	urea, melamine

*GBM—glass-bonded mica.

PLASTICS AND THE ENVIRONMENT*

In the past hundred years, and particularly in the last fifteen or twenty, plastics and plastics items have become an important element in our total environment. It is timely to consider the effect of plastics on the environment as well as the effect of the environment on plastics since the first will determine if plastics are to be used, and the second consideration is, of course, important to the design of useful plastics products.

Plastics can have a major effect on the ecological balance (the total life interaction system). For example, the widespread introduction of synthetic fibers and plastic furniture coverings has been an ecological disaster for moths who fed

* This section prepared by Sidney Levy, P.E., The Leal Company, Camden, N.J.

on the natural fibers. From an important economic waste, moth and larvae damage has been relegated to a curiosity. Another area where plastics has made a major change in the ecological balance is in its use in food packaging. It is obvious that many food spoiling microbes have been drastically checked and perhaps others that can cope more effectively with the "plastics" environment may be flourishing.

To convert one ton of raw material into usable plastics consumes many millions of Btu's of energy less than to convert one ton of ore into steel. Plastics materials are converted directly and their long life and salvage value contribute also to preserving our natural resources.

In the solid waste area, plastics are a minor part of our solid waste. Present methods of handling the waste are compatible with the properties of plastics. Plastics are useful in sanitary land fill. Although incineration in general is an undesirable method of disposal, most plastic materials burn cleanly, give up their total energy, and supply fuel value for the incineration process. The thermoplastic materials can be recycled into useful products where the economics of retrieval are rational. Most plastics waste materials are thermoplastics and can be recycled.

There has been work done on making disposable plastics products self-destructible. Biodegradable plastics are here. A concentrate (see Chapter 1, Biodegradable) can be added to some thermoplastics so that they can be digested by biological organisms in the presence of sunlight. This process converts the plastics into a fine dust in 90 to 150 days. This is an interesting approach, but not necessarily productive of positive effects. The degradation is bound to release gaseous and liquid components which will contribute to water pollution. A better trend might be to make plastics items as permanent as possible to aid repeated use and make them easily handled by the present solid waste management systems.

Plastics can be a positive aid in reducing air pollution and thermal pollution. A large portion of our energy consumption, made by consuming fossil fuels, goes toward heating our dwelling and work places in the winter, and cooling them in the summer. The use of the highly efficient plastics foam insulation such as expanded polystyrene and polyurethane foam greatly reduces this heat load. By reducing the heat load, the contribution to air pollution from power plants and home heating units is sharply reduced. The heat pollution to rivers from the power plants and the air conditioners would also be reduced.

In another area, plastics structures and elements are major factors in equipment used to reduce emission from industrial plants. They are used in cooling towers and in devices for removing particulate and gaseous emissions. The other major source of air pollution is the automobile. Plastics are under study and will be used in major functional roles in the electrical powered and external combustion vehicles which are necessary replacements for our present wasteful machines if air quality is to be maintained and improved.

Plastics are also making substantial contributions to improving the quality of water. They are major structural components in water treatment units for both industrial and municipal wastes. Membranes made of plastics materials are used in water purification systems to make fresh water from brackish water and seawater. Plastics films and pellets are used to hold ground waters from seepage and to prevent loss by evaporation, contributing substantially to the available fresh water supplies.

Design for the Environment

Assuming that a plastics product is desirable for economic value and utility, the designer is faced with designing it to resist the environment to which it is exposed, and to minimize its effects on the surrounding environment. The application of plastics to medical prosthesis is one dramatic area of use. A body implant, whether a heart valve, artery support, or pacemaker is exposed to a combination of body fluids whose function is to sustain tissue and reject foreign matter. This results in vigorous attack on implanted units. In addition, the implant can stimulate antibody reactions in the body which result in trauma, infection and even malignant cell reactions.

The medical application area is one in which the effects are important to such an extent that careful study, design, and test are essential. It is equally important in other areas of application. Unfortunately, the same cautious design procedures are not always followed. A critical part in an automobile, aircraft, or appliance can be just as potentially hazardous as a medical device. Frequently, only superficial attention is given to the in-use hazards and dangerous product failures result.

When the design engineer selects a plastics part for a specific application, he must justify it both on an economic and functional basis. In addition, it should outlast and outwear its non-plastics counterpart; plastics are generally considered the "new" material. Like the physician with a patient, the plastics engineer has not seen the particular material-application combination before and has little or no data on the combination of a specific material with a given set of environmental factors. Both the doctor and the engineer must rely on their background and their judgment based on past experience in estimating the probable success in applying a particular combination.

The engineer does have useful criteria to guide his judgment. One set of concepts is based on energy relationships between the material and the surroundings since the energy flows are a major source of the changes induced by the surroundings on the material. By analyzing the kinetics and the thermodynamics of these interactions, he gains insight to predict the probable performance of a polymer under a particular environmental stress. Energy relationships may also be used to forecast accurately plastics properties from creep endurance to weather resistance.

In order to intelligently apply these relationships, it is necessary to know what

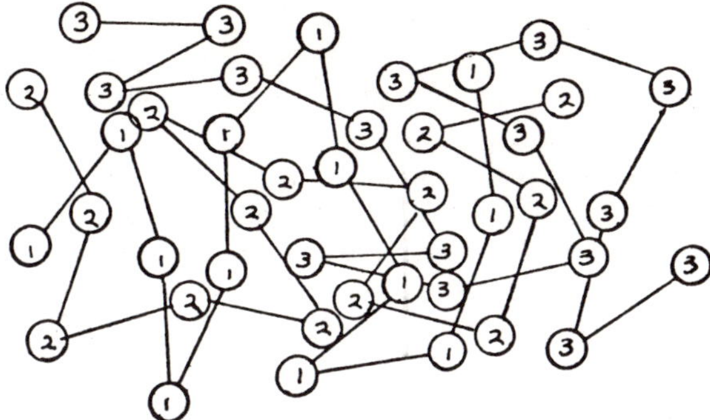

FIG. 7.6 Section of polymer structure with three chains illustrates the three different types of response to load.

a "plastic" is from a molecular structure standpoint. Most plastics are high polymer chains made up of repeating links of the monomer(s) from which they are formed. These chains exist in polymers which have been cast from monomers or they melt as a tangled network (see Fig. 7.6). In a thermoset material, the chains are crosslinked to each other which has the effect of partially immobilizing the relative movement of the chains. This can be useful in predicting the effects of energy flows on plastics material.

Performance Under Mechanical Stress

It is known that unlike metals and ceramics, plastics do not exhibit a linear stress-strain relationship. This characteristic is ascribed to the viscoelastic behavior of polymer materials. Structurally (Fig. 7.7) it is apparent that if the material is stretched, some of the stretching occurs by bond bending which is elastic, some of the stretching occurs by moving the molecules past each other

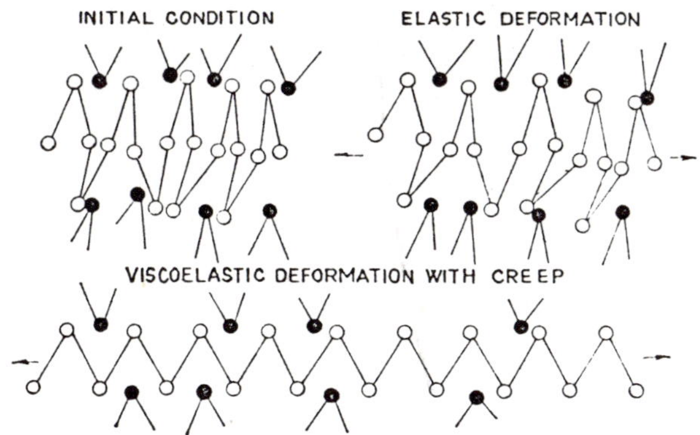

FIG. 7.7 Stages in polymer reaction to stress.

which, like flow in a fluid, is viscous flow, and some of the stretching occurs by major movement of the chains past each other. On relaxation of the stress, the elastic part is recovered rapidly. The short range viscous deformation is recovered slowly over a period of time by a diffusion process to a lower energy state and the long range molecular movement is not recovered at all. Hence, the typical plastics exhibits viscoelastic stress-strain characteristic with creep. The creep is reduced in crosslinked structures which prevent long range movement of the molecules and in crystalline polymers where the crystallite areas also inhibit long range movement. From this we deduce correctly that crosslinked materials like phenolics and polyesters or crystalline polymers such as Nylon and acetal are superior engineering materials with minimum creep and cold flow.

The fact that diffusion processes are a major effect in stress-strain relationships has led to equations predicting the stress-strain relationship of plastics and long term stress rupture performance and creep behavior. By assuming the process follows zero order reaction kinetics, they have been able to make accurate predictions based on easily acquired data such as melting points, softening point, crystalline melting points, and glass transition points. One major advance has been in making a master creep curve from creep data at one temperature and making the data applicable over a wide range of temperatures.

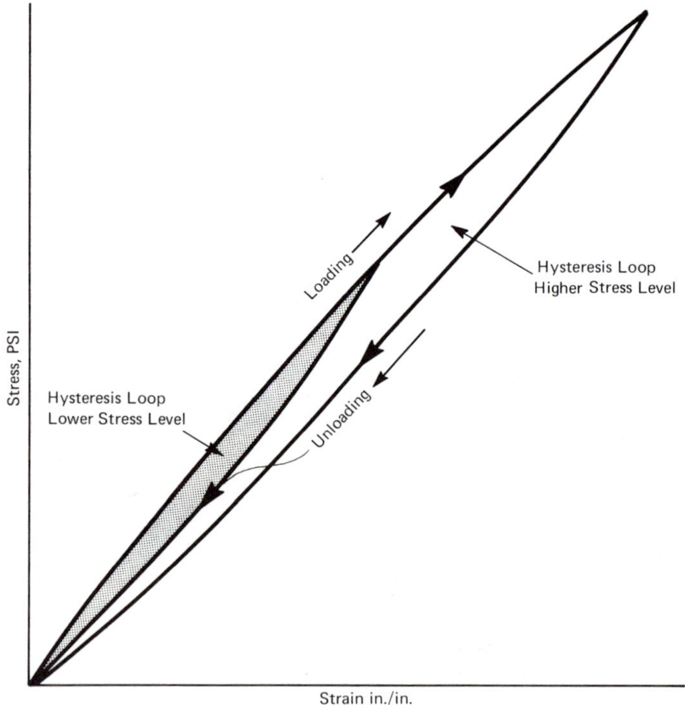

Fig. 7.8 Hysteresis curves.

The complete stress-strain loop curve which includes the stress relaxation as well as the stress application, Fig. 7.8, is useful in predicting the performance of plastics in a vibrational energy (varying stress) field. Energy stored in the mechanical hysteresis loop is converted to sensible heat in the plastic part. If it is not dissipated, it is a major cause of fatigue failure. (The hysteresis loop can also be used to predict the heating effects caused by high frequency electrical fields on polar plastics materials.) Using the hysteresis loop, the engineer can predict which materials will perform best under oscillating loads. Polypropylene with its low hysteresis will do much better than polyethylene with its high hysteresis. Knowing that the effect is one of heating, the engineer can also predict that the effect will lead to catastrophic failure since raising the temperature will increase the hysteresis energy per cycle, and increasing the heat generation rate in a chain reaction to failure. The engineer can apply his knowledge of heat transfer to design the part for best heat dissipation to reduce the possibility of failure.

Signposts in the Molecule

Going back to the polymer structure, we can see that the structure is held together by chemical bonds of various kinds. The bond energy relationship can be used to predict the effects of heat, infrared and actinic radiation on the polymer structure. Table 7.4 lists the bonding energy represented by typical structural configurations that make up the molecules of many polymers.

The action of heat at the normal temperature ranges is assumed to have little effect in breaking polymer bonds that have high bonding energy, and this holds true. Bond energy relationships can be used to predict the pyrolitic decomposition of polymers at elevated temperature. If the polymer bond formed by the polymerization is the weakest bond, the material will depolymerize to the monomer on heating. If the side group bonds are weaker, the material will carbonize rather than depolymerize. Both effects may be present. A classic case of the depolymerization is the heating of polymethylmethacrylate with a total conversion to the monomer. Polyvinyl chloride, on the other hand, converts to polyvinylene with the loss of HCL. There are other modes of polymer degradation under heat and, in each case, the relative bond energy is an accurate guide to the mechanism of breakdown.

Attack by Actinic Light

The effects of actinic light (UV) can be predicted from the table. When a quanta of the incident light has energy equal to the bond energy, there is a high probability of the bond rupturing. Materials with structures in this range must be protected from light by the use of masking pigments that absorb the light or UV stabilizers that either inhibit the bond rupture by chemical means or degrade the energy to lower levels that do not attack the bonds.

There may be specific structure in the polymer that responds to radiant energy

TABLE 7.4 BOND ENERGIES*

Bond	Bond Energy Kcal/mole	Wave Length for Corresponding Energy
1. $C\equiv N$ nitrile	209	
2. $C\equiv C$	200	
3. $C=O$	174	
4. $C=C$	145	
5. $C\equiv S$	129	
6. C—C aromatic	124	
7. C—H acetylene	121	
8. C—F	119	
9. O—H	110	2537 A° (Short wave length UV)
10. C—H ethylene	106	
11. C—H methane	98	
12. Si—O	89	
13. C—O	87	
14. S—H	87	
15. N—H	84	3650 A° (Long wave length UV)
16. C—C aliphatic	80	
17. C—O ether	79	
18. C—Cl	78	
19. $S=S$	76	
20. Si—H	75	
21. Si—C	70	4000 A° (Blue Light)
22. C—N nitromethane	68	
23. C—S	66	
24. O—O peroxide	64	5500 A° (Yellow Light)
25. N—N hydrazine	37	

*From R. Houwink, "Elastomers and Plastomers," New York, 1950, Elsevier Pub. Co., Vol. 1, p. 196 and Mark and Tobolsky, "Physical Chemistry of High Polymeric Systems," New York, 1950, Interscience, p. 15.

in a resonant phenomenon where lower energy quanta can have drastic effects. Certain epoxy resins are drastically affected by infrared energy in the 1 to 5 micron range because of the resonance absorption and must be protected from the IR exposure.

Ionizing radiation-alpha, beta, gamma and X-rays have a complex effect on plastics materials. They produce free radicals over a wide path in the material and these can rearrange the structure of the polymer. They cause crosslinking as well as chain scission and the predominant effect depends on the polymer as well as additives that may be present. Thermally stable polymers tend to be radiation stable but such classic examples as the excellent radiation stability of the polyethylene and the very poor radiation stability of TFE precludes any easy generalization. Extensive data exist on ionizing radiation effects and they are generally used empirically.

Past Environments—Flaws in Processing

A study of energy relationships gives the plastics engineer insight into how materials perform in their functioning environment. The product's formative environment—the processing to produce the product—represents drastic energy changes and may be responsible for flaws in its make-up. Formative stages in the making of a plastics are as important as they are in the life of a human being. Stresses produced in the processing, for instance, can shorten the useful life of plastics parts, from reinforced polyesters to high impact polystyrene. In reinforced plastics, poor resin formulation and inadequate control lead to premature stress and failure under the onslaught of bad weather. High impact polystyrene parts improperly cycled will distort at temperatures well below those the material would ordinarily withstand. Furthermore, molded-in strains will cause failure in polyethylene, a material which normally has good stress crack resistance. Many other premature failures can be traced to processing errors. Selection of easily processed materials for intricate parts exposed to severe stress is part of the designers job.

Behavior of plastics in fabrication, then, is as important to the engineer as selection of the material itself. He knows that some plastics materials will crack chemically at temperatures used for molding or forming. The lower molecular weight materials often lose many desirable properties in production. Susceptibility to this type of degradation is a weakness of some halocarbon polymers. In many thermosets, also, incomplete cure, overcure, poor mixing and other processing errors can result in weak parts that will fail prematurely when exposed to chemical or weathering environments.

New Materials, New Environments

Changes are brought into the whole picture of interactions between materials and environments as new materials—or improved formulations of standard ones—appear, and as new environments are found in the course of scientific advancements. For unusual resistance to corrosive environments, there are the more recent fluorocarbons which can be processed conventionally and easily. Chlorinated polyether compounds also show excellent long-term chemical stability.

Exotic environments of high vacuum space and radiation are becoming commonplace for plastics materials. It is well known that plastics are going undersea and underground. They are also now finding a place in man's own body where they are used inside living tissue for encapsulating delicate sensing devices for diagnosis and research. They are even being used to replace damaged and diseased parts of the human organism. Surgical implants will be a major field of application for plastics as techniques for repairing body tissue are more thoroughly developed. The implant mitral valve, as shown in Fig. 7.9, the artery reinforcements, the artificial kidney, the implant heart regulator, and the surgical prosthetic attachments are merely the beginning.

The new environment represented by body tissue is a complex chemical and biochemical one where both the effect of material on environment and environment on material are important. Silicones, epoxy resins, fluorocarbon resins and several polyesters have been employed with good results, and others have been tested. Some materials such as the olefins have had definitely adverse reactions.

Each new environment poses the same problems of material performance and points to new data requirements. What is needed is the development of broad theoretical concepts that will enable the engineer to predict performance and to pinpoint the type of information required to make the prediction. Such concepts are just beginning to evolve and, as they are applied and used widely, it can be expected that plastics will be recognized as true engineering materials and used with greater confidence to meet the demands of a variety of new applications.

The plastics engineer, faced with applications of new materials to new and untried environments, must use ingenuity in predicting success. The specific data required are usually unavailable and the practical considerations of time and

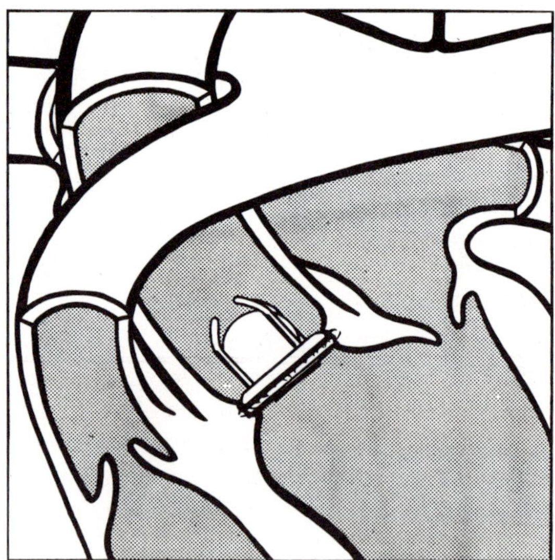

FIG. 7.9.A Sutureless aortic heart valve in position. Once the valve starts to function, the ball will seat firmly in the ring when the chamber contracts, thus allowing blood to flow in one direction only. Affixing needles can be seen at base of valve.

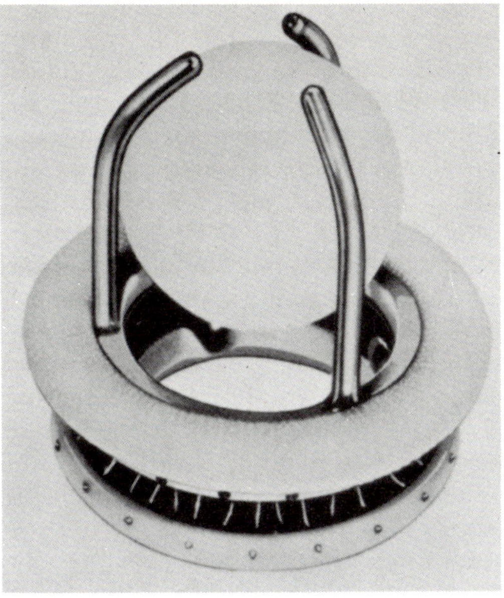

FIG. 7.9.B This replacement for a leaky heart valve consists of a TFE-rimmed titanium ring and a silicone plastics ball.

cost often preclude accurate testing to obtain the necessary information. The engineer must rely on a return to "First Principles," based on a knowledge of the materials and their response to various environmental stress, usually energy flows, to predict the performance of their designs. We have touched on some of these approaches, by way of example, so that the design engineer with good insight can produce useful products. Other areas such as biological, vacuum, and other environments are amenable to similar approaches by the skilled engineer.

MOLDABILITY

The molded piece must, of course, come out of the mold after it has hardened. For economical production, the product should be designed without undercuts or projections that lock it in the mold. It is possible to design molds for production of pieces that must contain undercuts, but this construction is uneconomical for many jobs and, therefore, it is best to avoid it whenever possible. All parts of the design must be checked to make sure that the product will "draw" out of the mold. When it is necessary to provide an undercut or side cored section, cam or hydraulically operated side cores, loose wedges or side-ram top-clamp molds are necessary. Volume jobs must be mechanized with automatically pulled side cores.

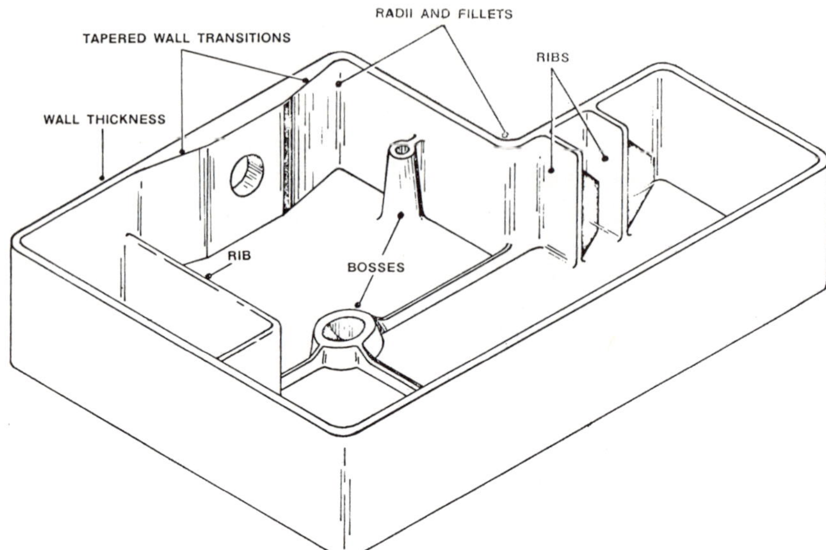

FIG. 7.10 Features of good plastics product design. (*Courtesy Marbon Div. Borg Warner Corp., Washington, West Va.*)

Radii and Fillets: Sharp internal corners are undesirable; they cause stress and should be avoided. A minimum radius recommended for internal corners is 0.0625″. External corners should be similarly rounded to maintain a uniform wall thickness. In general, the generous use of radii or fillets, whenever possible, can prevent surface blemishes and/or reduced performance capabilities of the molded part.

Wall Thickness: The wall thickness is generally dependent upon part size and shape and is further governed by its strength and rigidity requirements. Uniform wall thicknesses are ideal but variations are sometimes necessary. In this event, gradual transitions from one wall section to another (from thick to thin sections) via tapers, radii or fillets should be specified. Gradual changes in wall thickness improves material flow and minimizes flow turbulence which is commonly the cause of molded-in strains, voids or sink marks. Although these defects may not detract from the part's toughness, they are visually undesirable and should be prevented.

Sample molds or small volume jobs may use loose mold wedges for reduced tool cost. An example of a well designed part is shown in Fig. 7.10.

Loose wedges add considerably to mold and molding costs. Fins produced by these parts are costly to remove and present an unsightly appearance on many products. Loose mold wedges are to be avoided if possible, even at the expense of a major redesign of the product. Two piece assembly may solve this problem.

Internal undercuts often cannot be molded at reasonable cost, and must be eliminated in most cases. Many undercuts may be avoided by extending internal projections or bosses to the bottom and external projections or bosses to the

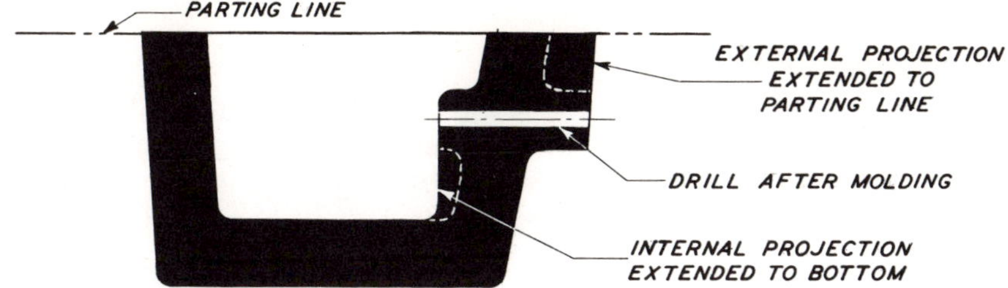

FIG. 7.11 Redesign of piece to facilitate molding. Dotted lines indicate original design.

parting line, as shown in Fig. 7.11. Simple undercuts may be machined at small cost. Products requiring internal undercuts should be redesigned as an assembly of two molded pieces or with the addition of a simple metal part or a machining operation. A side hole extended to the inner wall may serve to eliminate an internal undercut. Some undercut pieces, such as the one shown in Fig. 7.12, may be molded by sliding or turning the piece when removing it from the mold

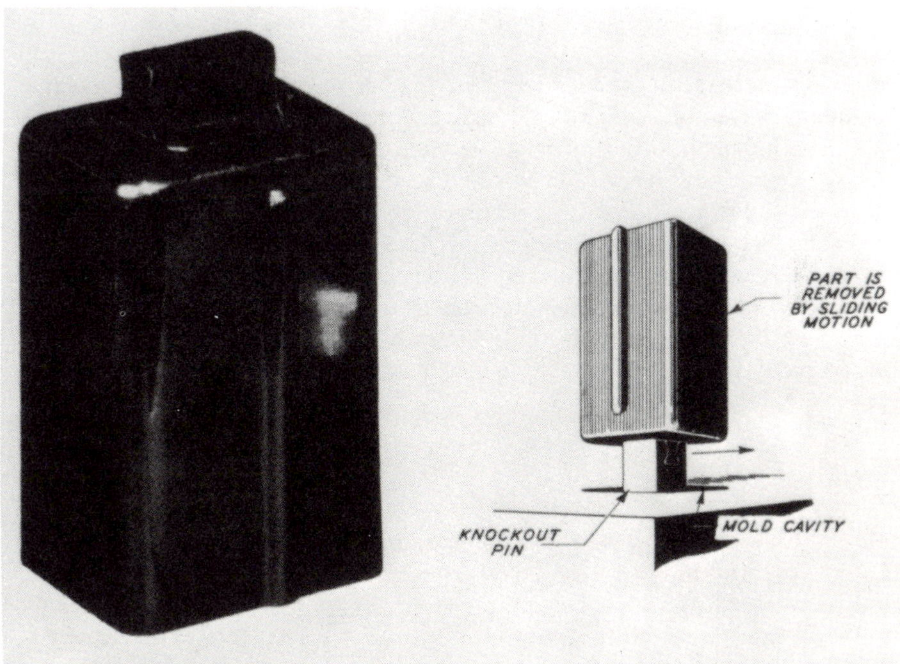

FIG. 7.12 Left: Cover unit with undercut handle molded without use of a split cavity. Right: Knockout bar pushes part out of cavity where it is released by a sideward sliding motion, as shown in sketch.

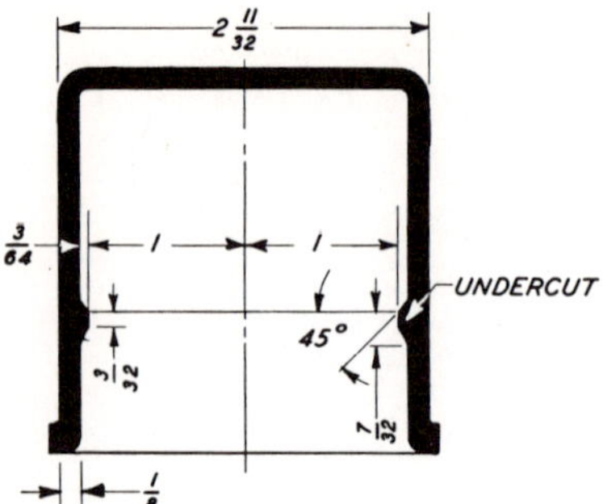

FIG. 7.13 Heavy undercuts may be stripped from mold if section is large enough to permit distortion without breaking part. Note tapered ends on undercut. These facilitate release from mold.

plunger. Side ram or angle presses are helpful for the production of pieces having side undercuts or projections, and they should be considered when such presses are available.

Parts with undercuts, Fig. 7.13, may sometimes be stripped from the mold without breaking off the projection which forms the undercut. This is especially true of small, shallow undercuts, or when using the tough, elastic materials, or when using the more flexible thermoplastics or the thermoplastic rubbers. Many plastics have sufficient elasticity to permit the use of snap fits as shown in Fig. 7.14.

It is less difficult to produce pieces having side cores by the injection molding method. Injection molds may be designed with cam-operated side cores for the automatic molding of many difficult jobs. Side cores for transfer and compression molds are generally operated manually, though in some cases they are operated by means of auxiliary pressure cylinders or cams on the side of the mold.

Location of gates, Fig. 7.15, must be determined for injection and transfer molded pieces, and some products will require redesign to facilitate gating and gate removal. Removal of the gate leaves a mark on the part which must be removed by grinding and buffing. Frequently, the character of the material or the design of the part will dictate the location of the gate, and slight changes may

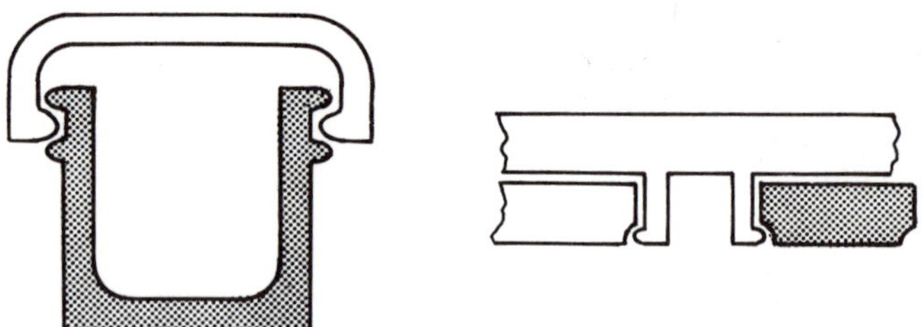

FIG. 7.14 Snap fits such as these are often used with nonrigid plastics.

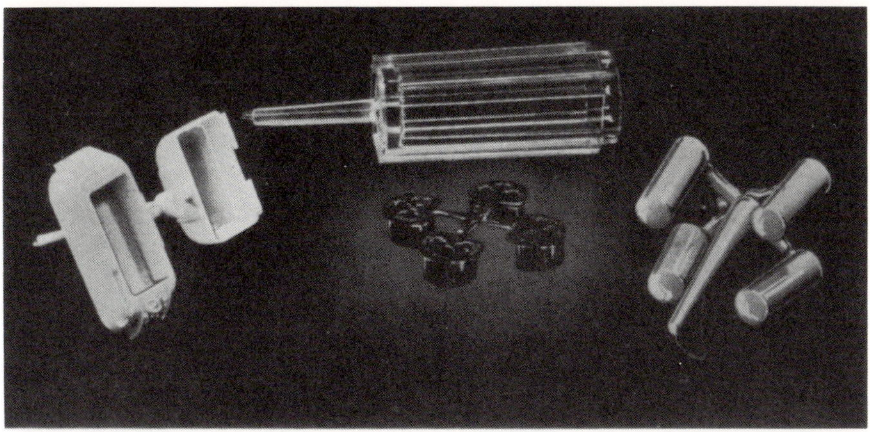

FIG. 7.15 The black pieces (center) were transfer molded and gated on the edge so gate mark would not show when parts were assembled. The other pieces were injection molded, their gates located so as to facilitate removal without disfiguring the piece.

be desirable to facilitate gate removal. In making this analysis, the mold designer should decide whether or not a slight design change facilitates removal of the gate or makes its appearance less objectionable. Parts should be gated on an accessible hidden surface when possible. If the part must be gated on an external surface, the gate should be removed from other projections and placed where the navel, or stub, may be ground and buffed when necessary. Gates for parts molded from thermoplastic materials should, when possible, be designed for removal by breaking, or by cutting in a die that will cut the gares from all cavities in one operation. When possible, self degating injection molds should be designed.

Venting is essential to good molded product quality and the gate location also is a function of good original product design. It is very important for the product engineer and the mold designer to give careful consideration to the location of vents, overflow needs, and the gate's location and design to get best appearance and cost. The removal of gate and vent marks from a surface can be very costly, and it is usually possible to modify a design to conceal such marks. These details are well covered in books on mold design and in the literature of the materials makers.

Parting Lines. The cost of finishing compression molded parts after molding often amounts to a large percentage of the direct labor cost of the part. The mold designer must make every effort to minimize the finishing costs through improved product design. Parting lines are a necessary evil in all molding. Flash must be removed at this line and, on many products, buffing will be necessary. Good designers minimize cleaning costs by using straight parting lines which

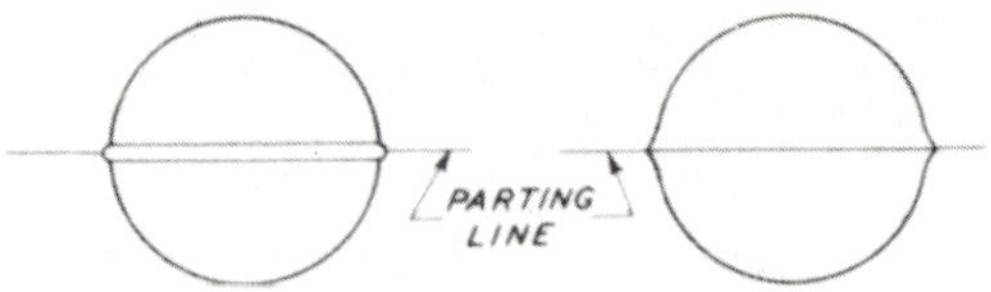

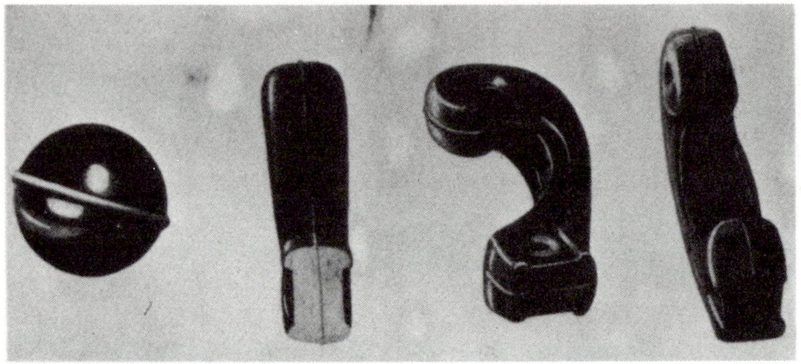

FIG. 7.16 Above: It is often desirable to "peak" or "bead" the parting line of thermosetting parts above surrounding surfaces to facilitate cleaning. When this is done, a knife or file will do a good job of cleaning without marring surrounding areas. Below, Finished parts show both peaked and beaded parting lines.

FIG. 7.17 The flush parting lines on this receiver-transmitter were undercut to give the piece a trim appearance and eliminate costly buffing. Runner shows at upper left.

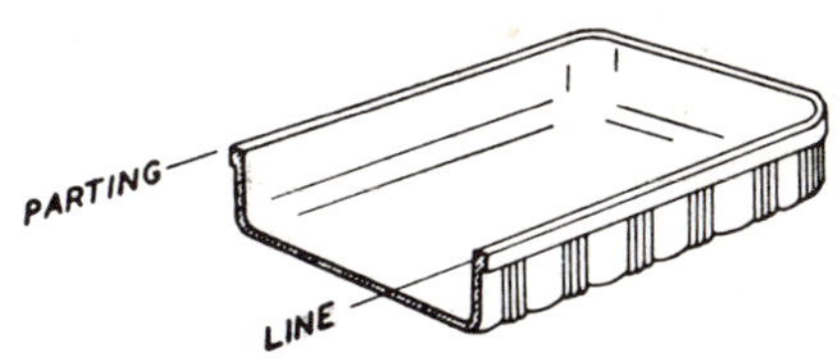

FIG. 7.18 Stopping decorative reeds below parting line makes it unnecessary to file thermosetting parts along each reed. The parting line shown above may be cleaned by a single straight stroke of the file. Molds are easy to maintain because thin reed sections are below parting line.

clean and buff easily. Molds which have straight parting lines cost much less than those which have curved or stepped parting lines. When the parting line does not come at the corner of the part, it should be elevated above surrounding surfaces, so these surfaces will not be marred or cut by the cleaning tools. Such parting lines are said to be "beaded" or "peaked," Fig. 7.16.

Flush parting lines are sometimes necessary in certain designs, and it may be desirable to fashion the product so that the parting line may be undercut in cleaning, as shown in Fig. 7.17. Many designers take advantage of this undercut parting line to add a stripe of color, painting it on after undercutting.

Reeds and flutes are frequently used as surface decorations for molded products. These flutes or reeds must stop immediately below the parting line, Fig. 7.18, so that cleaning may be done on a straight surface. A straight parting line eliminates the difficult job of following the contour of the decorations when filing the flash.

The use of a bead at the parting line is helpful in avoiding close alignment and contour matching of the cavity and plunger. This is an important factor in the cost of multiple-cavity molds. Stepped parting lines, Fig. 7.19, are extensively used to get clean, sharp lines that can be cleaned of flash by simple, inexpensive operations. The stepped parting line uses a step of 0.010 to 0.020 inch, and serves to eliminate extremely close and costly alignment of multiple-cavity molds.

FIG. 7.19 This molded phenolic valve handle makes use of a stepped parting line, thereby making it unnecessary to provide extremely close alignment between cavity and plunger and facilitating inexpensive cleaning. Note that the attachment insert is held in a molded recess.

The parting lines of most products normally will come at the edge or corner, where they will tumble clean or can be cleaned easily by filing or spindling. This corner will be radiused when the flash is removed, therefore drawings which show an extremely sharp corner at the parting line should not be approved without a check to make sure that the slight radius to be added in cleaning is not objectionable.

Well designed injection and transfer molds will produce flash-free parts. Compression molded pieces usually require a deflashing operation.

Loose wedges and side cores may produce fin lines which must be treated as parting lines. Loose wedge areas may be stepped back from the plane of surrounding surfaces to minimize cleaning costs (see Fig. 7.10). If the product design calls for irregular or otherwise difficult parting lines, it will be well to consider the use of injection or transfer molding processes which minimize flash lines. All the cloth-filled materials produce a heavy flash at parting lines when compression molded. Products which have irregular parting lines should be molded in transfer molds when cloth-filled materials are specified. The saving in finishing cost, in most cases, offsets the additional mold and material expense.

Finishing costs are an extremely important part of the total cost of products made from the thermosetting materials, and the mold designer should always design the mold in such a way as to minimize this expense. When pieces must be compression molded, lowest costs will be obtained from designs that will tumble free of flash.

Knockout Pins

Most molds use knockout or ejector pins or sleeves to push the part out of the cavity or off the mold force. It is possible to mold pieces without the aid of these pins but it is better to use them for all parts that are to be produced in volume at lowest cost. The mold cavity number and the molder's identification mark are often placed on the knockout pins. These knockout pins marks should appear on a hidden surface of the piece, and, as it is nearly always possible to design the mold so that the part will hold to either section, the knockout pins may be located accordingly. The point of location of the pins should be decided upon when the production drawings are made, as arbitrary placing of the pins may not be agreeable to the user, and some redesign may be necessary to effect a compromise. If the knockout pins are placed on an exposed surface, the problem should be referred to the stylist, who will decide upon the least objectionable spacing and decoration. Concentric rings and other simple designs are sometimes used to decorate or conceal the pin marks, as shown in Fig. 7.20. It is impossible to keep the knockout pin marks flush with the surface at reasonable cost. An understanding must be reached to govern the allowable variation above or below the surface. In general a total variation of ±0.015 inch is desirable for trouble-free production at no excess cost.

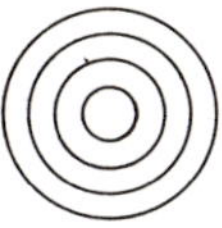  

FIG. 7.20 Knockout pins may be given some decorative treatment of the kind shown above, or they may carry molder's identification mark or cavity number.

Knockout pins should not press against thin areas. It is desirable to have these pins where they will push against ribs when possible. This is especially the case with the thermoplastic materials, which are often ejected while soft and easily distorted. Large area knockout pins are required for thin sections and soft materials. Cold-mold products often use a large portion of the cavity area as a knockout pad to minimize distortion during ejection.

Pickups

Shallow undercuts are often used to make the molded product hold to the mold force or remain in the mold cavity. These undercuts, Fig. 7.21, are usually from 0.005 to 0.010 inch high, and the length varies with the need to make the piece hold to the proper mold section. Before adding such pickups, the designer must make certain that these marks will not affect the performance or assembly of the product.

Shrinkage

The mold-maker can build molds to very close dimensional accuracy. Many variables in the art of molding and in molding materials, however, affect the dimensions of the finished part. Shrinkage variables introduce certain known dimensional changes, therefore it is desirable to check all dimensions immediately

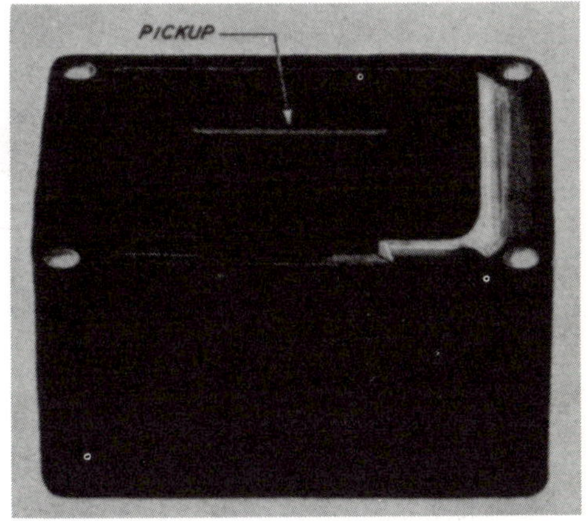

FIG. 7.21 Pickup marks serve to make molded piece adhere to plunger, thus assisting removal from mold cavity. The piece is then pushed off plunger by means of knockout pins.

and reject any designs which call for dimensions which cannot be met by conventional practice.

Molding materials shrink after they are taken out of the mold. Shrinkage depends on the character of the material and the final mold temperature. When the temperature and pressure are not closely controlled, other shrinkage variables are introduced. Minimum shrinkage occurs when the mold is chilled before removal of the part. In the cold-mold materials, a large dimensional change takes place after molding, as the curing or hardening is accomplished outside of the mold. Many of the conventional thermosetting and thermoplastic materials contineu to shrink for many months after molding. Figure 7.22 shows the shrinkage of urea molded products upon aging. Sometimes the ultimate shrinkage may be accelerated by a baking operation; in other cases it is necessary to allow for after-shrinkage in the fundamental design of the product.

Dimensions which cannot be met with sufficient accuracy by molding must be machined. Thermosetting parts which are to be machined to close dimensions should be baked before machining in order to achieve maximum shrinkage before the machining is done. If they are baked, the baking will also bring about dimensional changes which must be taken into account in the original design. The mold designer must make sure that when the baking shrinkage is fully accomplished, sufficient material remains for machining to the final close dimension. The baking shrinkage can be determined by test on a piece of similar shape, section, and type of insert. If large inserts are used, they may prevent

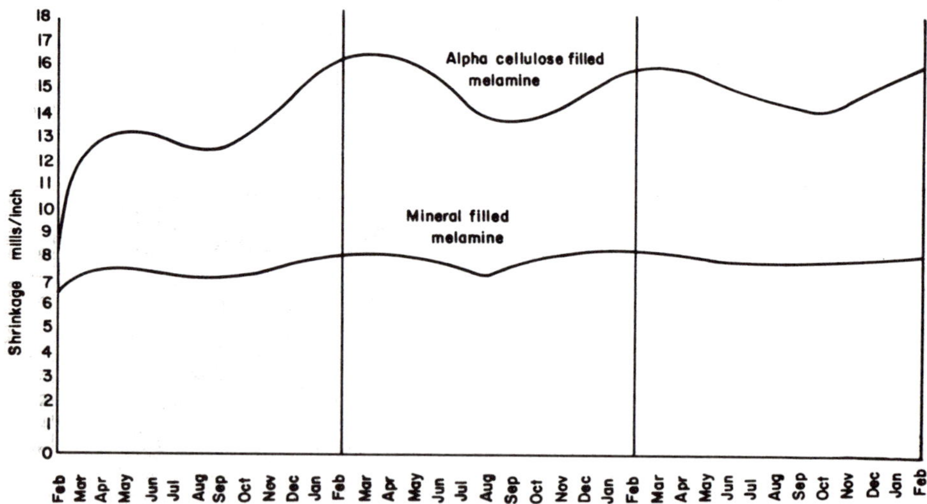

FIG. 7.22 Long term shrinkage at room temperature reflecting seasonal changes of temperature and humidity. (*Courtesy American Cyanamid Co.*)

normal shrinkage and reduce the amount of afterbake shrinkage. Material suppliers will furnish the data on normal material after shrinkage molding.

Standards for Tolerances on Molded Articles*

The purpose of this section is to indicate the magnitude of practical tolerances on the dimensions of articles molded from a variety of thermosetting and thermoplastic materials, other than laminates.

This information is given in the form of charts based upon data obtained from representative firms in the plastics molding industry by means of a questionnaire. It must be stressed that these charts are not to be construed as offering hard-and-fast rules applicable to all conditions. They can best be used as a basis for establishing standards for individual molded articles by agreement between the purchaser, or his design engineer, and the molder.

The questionnaire was based upon a hypothetical molded article, of which a crosssection is shown in the charts. The hypothetical article represents a variety of problems of tolerances on diameter, length, depth and thickness.

Charts of tolerances are given for only those materials which have been used widely enough, commercially, to provide significant data.

It should be recognized that extreme accuracy of dimensions in molded articles is expensive to achieve. The closer the tolerances demanded, the greater will be the cost of the molds, because of the precision required, and also the greater will be the operational costs of molding, because of the greater care required to maintain uniformity of conditions. In some cases a further expense arises from the need of using cooling fixtures after the molding.

Dimensional tolerances in a molded article are the allowable variations, plus and minus, from a nominal or mean dimension.

Fine tolerance represents the narrowest possible limits of variation obtainable under close supervision and control of production.

Standard tolerance is that which can be held under average conditions of manufacture.

Coarse tolerance is acceptable when accurate dimensions are not important.
(*Continued on page 278.*)

*From "Standards and Practices of Plastics Custom Molders, published by The Society of the Plastics Industry, Inc.

STANDARDS AND PRACTICES OF PLASTICS CUSTOM MOLDERS

Engineering and Technical Standards
ABS

NOTE: The Commercial values shown below represent common production tolerances at the most economical level. The Fine values represent closer tolerances that can be held but at a greater cost.

Drawing Code	Dimensions (Inches)	Plus or Minus in Thousands of an Inch 1 2 3 4 5 6 7 8 9 10 11 12 13 14 15 16 17 18 19 20 21 22 23 24 25 26 27 28	
A = Diameter (see Note #1) B = Depth (see Note #3) C = Height (see Note #3)	0.000 0.500 1.000 2.000 3.000 4.000 5.000 6.000		

Drawing Code	Dimensions (Inches)	Comm. ±	Fine ±
	6.000 to 12.000 for each additional inch add (inches)	.003	.002
D = Bottom Wall (see Note #3)		.004	.002
E = Side Wall (see Note #4)		.003	.002
F = Hole Size Diameter (see Note #1)	0.000 to 0.125	.002	.001
	0.125 to 0.250	.002	.001
	0.250 to 0.500	.003	.002
	0.500 & Over	.004	.002
G = Hole Size Depth (see Note #5)	0.000 to 0.250	.003	.002
	0.250 to 0.500	.004	.002
	0.500 to 1.000	.005	.003
Draft Allowance per side (see Note #5)		2°	1°
Flatness (see Note #4)	0.000 to 3.000	.015	.010
	3.000 to 6.000	.030	.020
Thread Size (class)	Internal	1	2
	External	1	2
Concentricity (see Note #4)	(T.I.R.)	.009	.005
Fillets, Ribs, Corners (see Note #6)		.025	.015
Surface Finish	(see Note #7)		
Color Stability	(see Note #7)		

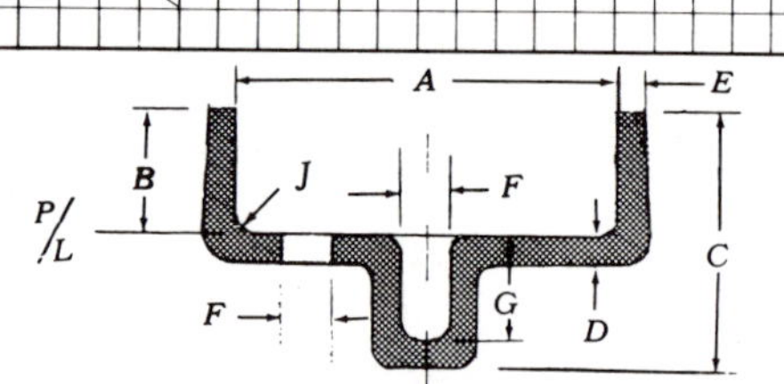

REFERENCE NOTES

1 — These tolerances do not include allowance for aging characteristics of material.

2 — Tolerances based on ⅛″ wall section.

3 — Parting line must be taken into consideration.

4 — Part design should maintain a wall thickness as nearly constant as possible. Complete uniformity in this dimension is impossible to achieve.

5 — Care must be taken that the ratio of the depth of a cored hole to its diameter does not reach a point that will result in excessive pin damage.

6 — These values should be increased whenever compatible with desired design and good molding technique.

7 — Customer-Molder understanding necessary prior to tooling.

STANDARDS AND PRACTICES OF PLASTICS CUSTOM MOLDERS

Engineering and Technical Standards
ACETAL

NOTE: The Commercial values shown below represent common production tolerances at the most economical level. The Fine values represent closer tolerances that can be held but at a greater cost.

Drawing Code	Dimensions (Inches)		Plus or Minus in Thousands of an Inch
A = Diameter (see Note #1) B = Depth (see Note #3) C = Height (see Note #3)	0.000 — 0.500 — 1.000 — 2.000 — 3.000 — 4.000 — 5.000 — 6.000		(graph: Fine / Commercial curves)

Drawing Code	Dimensions (Inches)	Comm. ±	Fine ±
	6.000 to 12.000 for each additional inch add (inches)	.004	.002
D = Bottom Wall (see Note #3)		.004	.002
E = Side Wall (see Note #4)		.004	.002
F = Hole Size Diameter (see Note #1)	0.000 to 0.125	.002	.001
	0.125 to 0.250	.003	.002
	0.250 to 0.500	.004	.002
	0.500 & Over	.006	.003
G = Hole Size Depth (see Note #5)	0.000 to 0.250	.004	.002
	0.250 to 0.500	.005	.003
	0.500 to 1.000	.006	.004
Draft Allowance per side (see Note #5)		1°	½°
Flatness (see Note #4)	0.000 to 3.000	.011	.006
	3.000 to 6.000	.020	.010
Thread Size (class)	Internal	1B	2B
	External	1A	2A
Concentricity (see Note #4)	(T.I.R.)	.010	.006
Fillets, Ribs, Corners (see Note #6)		.030	.015
Surface Finish	(see Note #7)		
Color Stability	(see Note #7)		

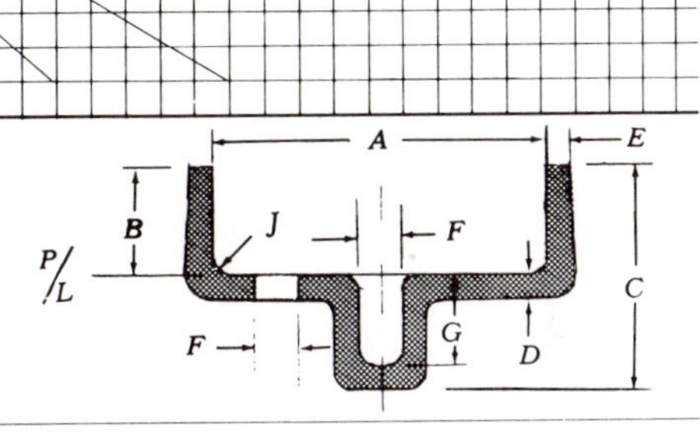

REFERENCE NOTES

1 — These tolerances do not include allowance for aging characteristics of material.

2 — Tolerances based on ⅛″ wall section.

3 — Parting line must be taken into consideration.

4 — Part design should maintain a wall thickness as nearly constant as possible. Complete uniformity in this dimension is impossible to achieve.

5 — Care must be taken that the ratio of the depth of a cored hole to its diameter does not reach a point that will result in excessive pin damage.

6 — These values should be increased whenever compatible with desired design and good molding technique.

7 — Customer-Molder understanding necessary prior to tooling.

STANDARDS AND PRACTICES OF PLASTICS CUSTOM MOLDERS

Engineering and Technical Standards
ACRYLIC

NOTE: The Commercial values shown below represent common production tolerances at the most economical level. The Fine values represent closer tolerances that can be held but at a greater cost.

Drawing Code	Dimensions (Inches)	Comm. ±	Fine ±
A = Diameter (see Note #1) B = Depth (see Note #3) C = Height (see Note #3)	0.000 — 0.500 — 1.000 — 2.000 — 3.000 — 4.000 — 5.000 — 6.000		
	6.000 to 12.000 for each additional inch add (inches)	.003	.001
D = Bottom Wall (see Note #3)		.005	.003
E = Side Wall (see Note #4)		.005	.003
F = Hole Size Diameter (see Note #1)	0.000 to 0.125	.003	.001
	0.125 to 0.250	.003	.002
	0.250 to 0.500	.004	.002
	0.500 & Over	.005	.003
G = Hole Size Depth (see Note #5)	0.000 to 0.250	.004	.002
	0.250 to 0.500	.004	.002
	0.500 to 1.000	.006	.003
Draft Allowance per side (see Note #5)		1°	½°
Flatness (see Note #4)	0.000 to 3.000	.010	.007
	3.000 to 6.000	.015	.010
Thread Size (class)	Internal	1	2
	External	1	2
Concentricity (see Note #4)	(T.I.R.)	.010	.006
Fillets, Ribs, Corners (see Note #6)		.030	.015
Surface Finish	(see Note #7)		
Color Stability	(see Note #7)		

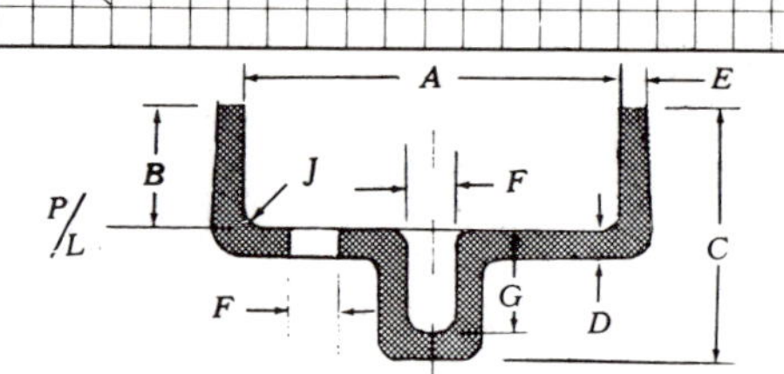

REFERENCE NOTES

1 — These tolerances do not include allowance for aging characteristics of material.

2 — Tolerances based on ⅛″ wall section.

3 — Parting line must be taken into consideration.

4 — Part design should maintain a wall thickness as nearly constant as possible. Complete uniformity in this dimension is impossible to achieve.

5 — Care must be taken that the ratio of the depth of a cored hole to its diameter does not reach a point that will result in excessive pin damage.

6 — These values should be increased whenever compatible with desired design and good molding technique.

7 — Customer-Molder understanding necessary prior to tooling.

STANDARDS AND PRACTICES OF PLASTICS CUSTOM MOLDERS	Engineering and Technical Standards ALKYD

NOTE: The Commercial values shown below represent common production tolerances at the most economical level. The Fine values represent closer tolerances that can be held but at a greater cost.

Drawing Code	Dimensions (Inches)	Plus or Minus in Thousands of an Inch 1 2 3 4 5 6 7 8 9 10 11 12 13 14 15 16 17 18 19 20 21 22 23 24 25 26 27 28
A = Diameter (see Note #1) B = Depth (see Note #3) C = Height (see Note #3)	0.000 0.500 1.000 2.000 3.000 4.000 5.000 6.000	

	Dimensions (Inches)	Comm. ±	Fine ±
	6.000 to 12.000 for each additional inch add (inches)	.0015	.001
D=Bottom Wall (see Note #3)		.002	.001
E = Side Wall (see Note #4)		.002	.001
F = Hole Size Diameter (see Note #1)	0.000 to 0.125	.002	.001
	0.125 to 0.250	.002	.002
	0.250 to 0.500	.002	.002
	0.500 & Over	.004	.003
G = Hole Size Depth (see Note #5)	0.000 to 0.250	.002	.002
	0.250 to 0.500	.002	.002
	0.500 to 1.000	.002	.002
Draft Allowance per side (see Note #5)		1°	½°
Flatness (see Note #4)	0.000 to 3.000	.010	.010
	3.000 to 6.000	.005	.005
Thread Size (class)	Internal	1	2
	External	1	2
Concentricity (see Note #4)	(T.I.R.)	.005	.005
Fillets, Ribs, Corners (see Note #6)		.062	.031
Surface Finish	(see Note #7)		
Color Stability	(see Note #7)		

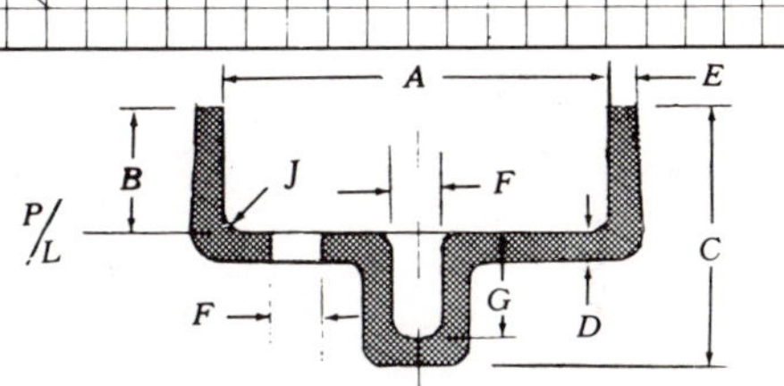

REFERENCE NOTES

1 — These tolerances do not include allowance for aging characteristics of material.

2 — Tolerances based on ⅛" wall section.

3 — Parting line must be taken into consideration.

4 — Part design should maintain a wall thickness as nearly constant as possible. Complete uniformity in this dimension is impossible to achieve.

5 — Care must be taken that the ratio of the depth of a cored hole to its diameter does not reach a point that will result in excessive pin damage.

6 — These values should be increased whenever compatible with desired design and good molding technique.

7 — Customer-Molder understanding necessary prior to tooling.

STANDARDS AND PRACTICES OF PLASTICS CUSTOM MOLDERS

Engineering and Technical Standards
CELLULOSICS

NOTE: The Commercial values shown below represent common production tolerances at the most economical level. The Fine values represent closer tolerances that can be held but at a greater cost.

Drawing Code	Dimensions (Inches)		Comm. ±	Fine ±
A = Diameter (see Note #1) B = Depth (see Note #3) C = Height (see Note #3)	0.000 / 0.500 / 1.000 / 2.000 / 3.000 / 4.000 / 5.000 / 6.000	Plus or Minus in Thousands of an Inch (Commercial / Fine)		
	6.000 to 12.000 for each additional inch add (inches)		.003	.002
D=Bottom Wall (see Note #3)			.003	.002
E = Side Wall (see Note #4)			.003	.002
F = Hole Size Diameter (see Note #1)	0.000 to 0.125		.002	.001
	0.125 to 0.250		.003	.002
	0.250 to 0.500		.004	.002
	0.500 & Over		.005	.003
G = Hole Size Depth (see Note #5)	0.000 to 0.250		.003	.002
	0.250 to 0.500		.004	.002
	0.500 to 1.000		.005	.003
Draft Allowance per side (see Note #5)			1°	½°
Flatness (see Note #4)	0.000 to 3.000		.007	.005
	3.000 to 6.000		.010	.007
Thread Size (class)	Internal		1B	2B
	External		1A	2A
Concentricity (see Note #4)	(T.I.R.)		.005	.004
Fillets, Ribs, Corners (see Note #6)			.005	.005
Surface Finish	(see Note #7)			
Color Stability	(see Note #7)			

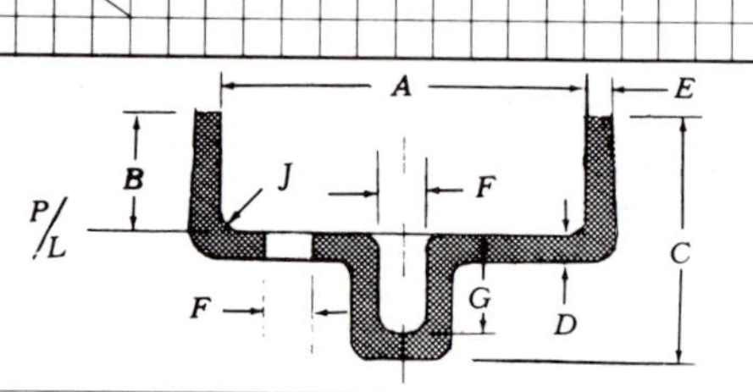

REFERENCE NOTES

1 — These tolerances do not include allowance for aging characteristics of material.

2 — Tolerances based on ⅛″ wall section.

3 — Parting line must be taken into consideration.

4 — Part design should maintain a wall thickness as nearly constant as possible. Complete uniformity in this dimension is impossible to achieve.

5 — Care must be taken that the ratio of the depth of a cored hole to its diameter does not reach a point that will result in excessive pin damage.

6 — These values should be increased whenever compatible with desired design and good molding technique.

7 — Customer-Molder understanding necessary prior to tooling.

STANDARDS AND PRACTICES OF PLASTICS CUSTOM MOLDERS

Engineering and Technical Standards
DIALLYL PHTHALATE

NOTE: The Commercial values shown below represent common production tolerances at the most economical level. The Fine values represent closer tolerances that can be held but at a greater cost.

Drawing Code	Dimensions (Inches)		Comm. ±	Fine ±
A = Diameter (see Note #1) B = Depth (see Note #3) C = Height (see Note #3)	0.000 / 0.500 / 1.000 / 2.000 / 3.000 / 4.000 / 5.000 / 6.000	Plus or Minus in Thousands of an Inch (Fine / Commercial graph)		
	6.000 to 12.000 for each additional inch add (inches)		.002	.001
D = Bottom Wall (see Note #3)			.005	.003
E = Side Wall (see Note #4)			.003	.002
F = Hole Size Diameter (see Note #1)	0.000 to 0.125		.002	.001
	0.125 to 0.250		.002	.001
	0.250 to 0.500		.002	.001
	0.500 & Over		.0025	.002
G = Hole Size Depth (see Note#5)	0.000 to 0.250		.002	.001
	0.250 to 0.500		.003	.002
	0.500 to 1.000		.005	.003
Draft Allowance per side (see Note #5)			1°	½°
Flatness (see Note #4)	0.000 to 3.000		.010	.005
	3.000 to 6.000		.012	.008
Thread Size (class)	Internal		1	2
	External		1	2
Concentricity (see Note #4)	(T.I.R.)		.005	.003
Fillets, Ribs, Corners (see Note #6)			.062	.031
Surface Finish	(see Note #7)			
Color Stability	(see Note #7)			

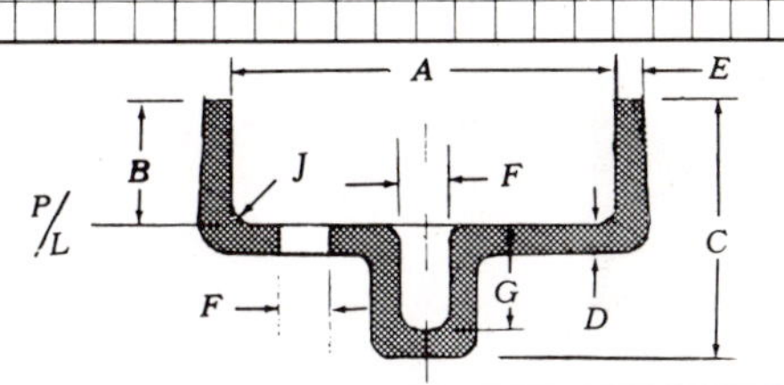

REFERENCE NOTES

1 — These tolerances do not include allowance for aging characteristics of material.

2 — Tolerances based on ⅛″ wall section.

3 — Parting line must be taken into consideration.

4 — Part design should maintain a wall thickness as nearly constant as possible. Complete uniformity in this dimension is impossible to achieve.

5 — Care must be taken that the ratio of the depth of a cored hole to its diameter does not reach a point that will result in excessive pin damage.

6 — These values should be increased whenever compatible with desired design and good molding technique.

7 — Customer-Molder understanding necessary prior to tooling.

<table>
<tr><td colspan="2">STANDARDS AND PRACTICES OF
PLASTICS CUSTOM MOLDERS</td><td>Engineering and
Technical Standards
EPOXY</td></tr>
</table>

NOTE: The Commercial values shown below represent common production tolerances at the most economical level. The Fine values represent closer tolerances that can be held but at a greater cost.

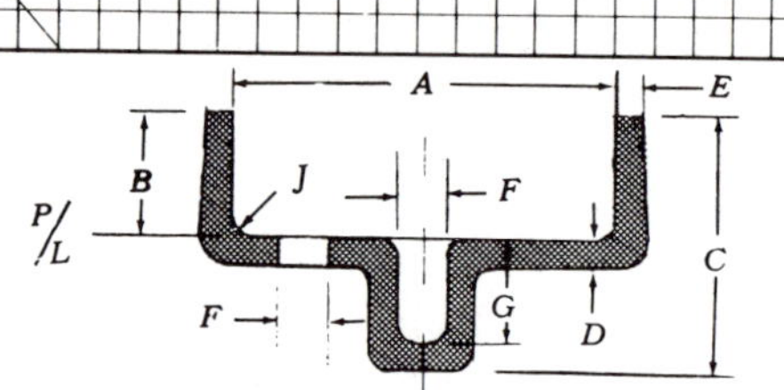

Drawing Code	Dimensions (Inches)		Comm. ±	Fine ±
A = Diameter (see Note #1) B = Depth (see Note #3) C = Height (see Note #3)	0.000 — 0.500 — 1.000 — 2.000 — 3.000 — 4.000 — 5.000 — 6.000			
	6.000 to 12.000 for each additional inch add (inches)		.0015	.001
D = Bottom Wall (see Note #3)			.002	.001
E = Side Wall (see Note #4)			.002	.001
F = Hole Size Diameter (see Note #1)	0.000 to 0.125		.002	.001
	0.125 to 0.250		.002	.002
	0.250 to 0.500		.002	.002
	0.500 & Over		.004	.003
G = Hole Size Depth (see Note #5)	0.000 to 0.250		.002	.002
	0.250 to 0.500		.002	.002
	0.500 to 1.000		.002	.002
Draft Allowance per side (see Note #5)			1°	½°
Flatness (see Note #4)	0.000 to 3.000		.010	.010
	3.000 to 6.000		.010	.015
Thread Size (class)	Internal		1	2
	External		1	2
Concentricity (see Note #4)	(T.I.R.)		.005	.005
Fillets, Ribs, Corners (see Note #6)			.062	.031
Surface Finish	(see Note #7)			
Color Stability	(see Note #7)			

REFERENCE NOTES

1 — These tolerances do not include allowance for aging characteristics of material.

2 — Tolerances based on ⅛″ wall section.

3 — Parting line must be taken into consideration.

4 — Part design should maintain a wall thickness as nearly constant as possible. Complete uniformity in this dimension is impossible to achieve.

5 — Care must be taken that the ratio of the depth of a cored hole to its diameter does not reach a point that will result in excessive pin damage.

6 — These values should be increased whenever compatible with desired design and good molding technique.

7 — Customer-Molder understanding necessary prior to tooling.

STANDARDS AND PRACTICES OF PLASTICS CUSTOM MOLDERS

Engineering and Technical Standards
MELAMINE-UREA

NOTE: The Commercial values shown below represent common production tolerances at the most economical level. The Fine values represent closer tolerances that can be held but at a greater cost.

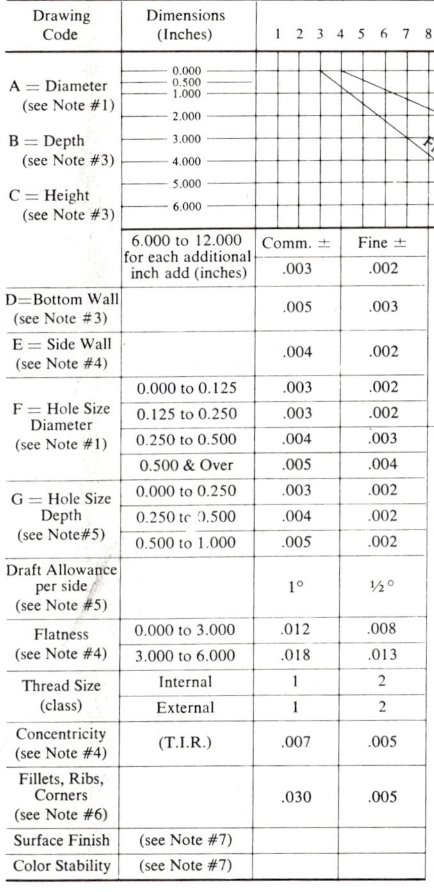

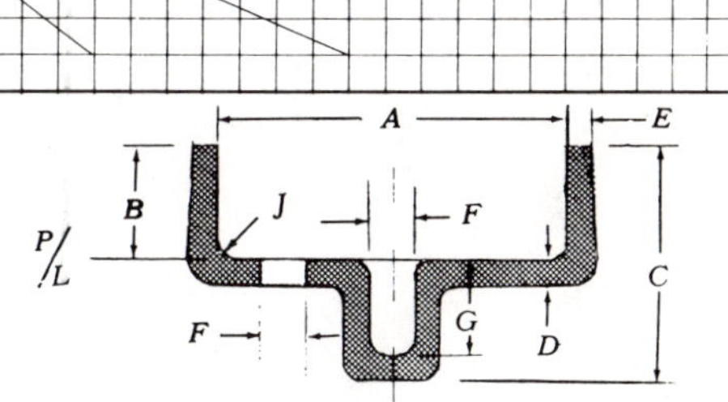

Drawing Code	Dimensions (Inches)	Comm. ±	Fine ±
A = Diameter (see Note #1) B = Depth (see Note #3) C = Height (see Note #3)	0.000 / 0.500 / 1.000 / 2.000 / 3.000 / 4.000 / 5.000 / 6.000		
	6.000 to 12.000 for each additional inch add (inches)	.003	.002
D = Bottom Wall (see Note #3)		.005	.003
E = Side Wall (see Note #4)		.004	.002
F = Hole Size Diameter (see Note #1)	0.000 to 0.125	.003	.002
	0.125 to 0.250	.003	.002
	0.250 to 0.500	.004	.003
	0.500 & Over	.005	.004
G = Hole Size Depth (see Note #5)	0.000 to 0.250	.003	.002
	0.250 to 0.500	.004	.002
	0.500 to 1.000	.005	.002
Draft Allowance per side (see Note #5)		1°	½°
Flatness (see Note #4)	0.000 to 3.000	.012	.008
	3.000 to 6.000	.018	.013
Thread Size (class)	Internal	1	2
	External	1	2
Concentricity (see Note #4)	(T.I.R.)	.007	.005
Fillets, Ribs, Corners (see Note #6)		.030	.005
Surface Finish	(see Note #7)		
Color Stability	(see Note #7)		

REFERENCE NOTES

1 — These tolerances do not include allowance for aging characteristics of material.

2 — Tolerances based on ⅛″ wall section.

3 — Parting line must be taken into consideration.

4 — Part design should maintain a wall thickness as nearly constant as possible. Complete uniformity in this dimension is impossible to achieve.

5 — Care must be taken that the ratio of the depth of a cored hole to its diameter does not reach a point that will result in excessive pin damage.

6 — These values should be increased whenever compatible with desired design and good molding technique.

7 — Customer-Molder understanding necessary prior to tooling.

STANDARDS AND PRACTICES OF PLASTICS CUSTOM MOLDERS

Engineering and Technical Standards

NYLON

NOTE: The Commercial values shown below represent common production tolerances at the most economical level. The Fine values represent closer tolerances that can be held but at a greater cost.

Drawing Code	Dimensions (Inches)		Plus or Minus in Thousands of an Inch
A = Diameter (see Note #1) B = Depth (see Note #3) C = Height (see Note #3)	0.000 / 0.500 / 1.000 / 2.000 / 3.000 / 4.000 / 5.000 / 6.000		*(graph: 1–28; Fine and Commercial lines)*
	6.000 to 12.000 for each additional inch add (inches)	Comm. ±	Fine ±
		.003	.002
D=Bottom Wall (see Note #3)		.004	.003
E = Side Wall (see Note #4)		.005	.003
F = Hole Size Diameter (see Note #1)	0.000 to 0.125	.002	.001
	0.125 to 0.250	.003	.002
	0.250 to 0.500	.003	.002
	0.500 & Over	.005	.003
G = Hole Size Depth (see Note #5)	0.000 to 0.250	.004	.002
	0.250 to 0.500	.004	.003
	0.500 to 1.000	.005	.004
Draft Allowance per side (see Note #5)		1½°	½°
Flatness (see Note #4)	0.000 to 3.000	.010	.004
	3.000 to 6.000	.015	.007
Thread Size (class)	Internal	1	2
	External	1	2
Concentricity (see Note #4)	(T.I.R.)	.010	.006
Fillets, Ribs, Corners (see Note #6)		.020	.012
Surface Finish	(see Note #7)		
Color Stability	(see Note #7)		

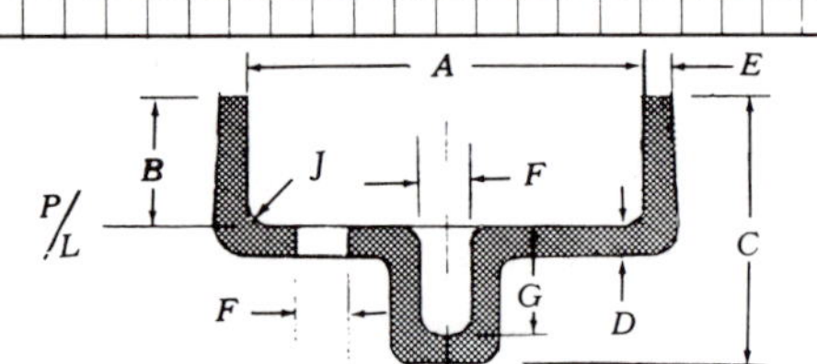

REFERENCE NOTES

1 — These tolerances do not include allowance for aging characteristics of material.

2 — Tolerances based on ⅛″ wall section.

3 — Parting line must be taken into consideration.

4 — Part design should maintain a wall thickness as nearly constant as possible. Complete uniformity in this dimension is impossible to achieve.

5 — Care must be taken that the ratio of the depth of a cored hole to its diameter does not reach a point that will result in excessive pin damage.

6 — These values should be increased whenever compatible with desired design and good molding technique.

7 — Customer-Molder understanding necessary prior to tooling.

STANDARDS AND PRACTICES OF PLASTICS CUSTOM MOLDERS

Engineering and Technical Standards
FIBRE-FILLED PHENOLIC

NOTE: The Commercial values shown below represent common production tolerances at the most economical level. The Fine values represent closer tolerances that can be held but at a greater cost.

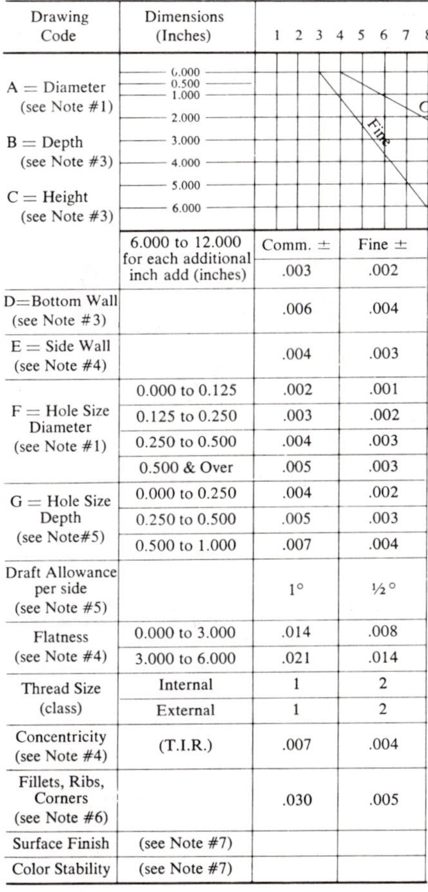

Drawing Code	Dimensions (Inches)	Comm. ±	Fine ±
A = Diameter (see Note #1) / B = Depth (see Note #3) / C = Height (see Note #3)	Plus or Minus in Thousands of an Inch 1 2 3 4 5 6 7 8 9 10 11 12 13 14 15 16 17 18 19 20 21 22 23 24 25 26 27 28		
	6.000 to 12.000 for each additional inch add (inches)	.003	.002
D = Bottom Wall (see Note #3)		.006	.004
E = Side Wall (see Note #4)		.004	.003
F = Hole Size Diameter (see Note #1)	0.000 to 0.125	.002	.001
	0.125 to 0.250	.003	.002
	0.250 to 0.500	.004	.003
	0.500 & Over	.005	.003
G = Hole Size Depth (see Note #5)	0.000 to 0.250	.004	.002
	0.250 to 0.500	.005	.003
	0.500 to 1.000	.007	.004
Draft Allowance per side (see Note #5)		1°	½°
Flatness (see Note #4)	0.000 to 3.000	.014	.008
	3.000 to 6.000	.021	.014
Thread Size (class)	Internal	1	2
	External	1	2
Concentricity (see Note #4)	(T.I.R.)	.007	.004
Fillets, Ribs, Corners (see Note #6)		.030	.005
Surface Finish	(see Note #7)		
Color Stability	(see Note #7)		

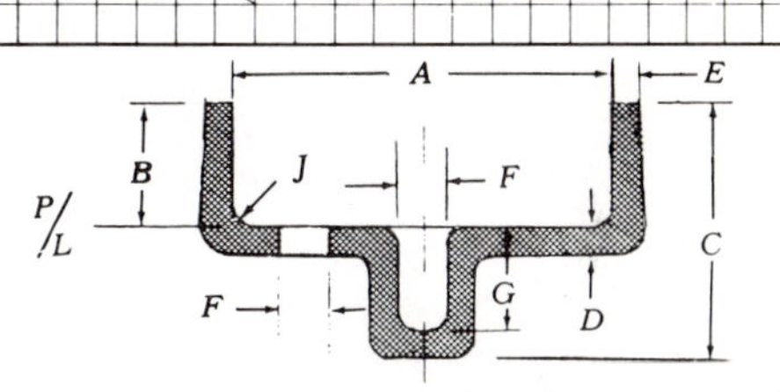

REFERENCE NOTES

1 — These tolerances do not include allowance for aging characteristics of material.

2 — Tolerances based on ⅛″ wall section.

3 — Parting line must be taken into consideration.

4 — Part design should maintain a wall thickness as nearly constant as possible. Complete uniformity in this dimension is impossible to achieve.

5 — Care must be taken that the ratio of the depth of a cored hole to its diameter does not reach a point that will result in excessive pin damage.

6 — These values should be increased whenever compatible with desired design and good molding technique.

7 — Customer-Molder understanding necessary prior to tooling.

STANDARDS AND PRACTICES OF PLASTICS CUSTOM MOLDERS

Engineering and Technical Standards
GENERAL PURPOSE PHENOLIC

NOTE: The Commercial values shown below represent common production tolerances at the most economical level. The Fine values represent closer tolerances that can be held but at a greater cost.

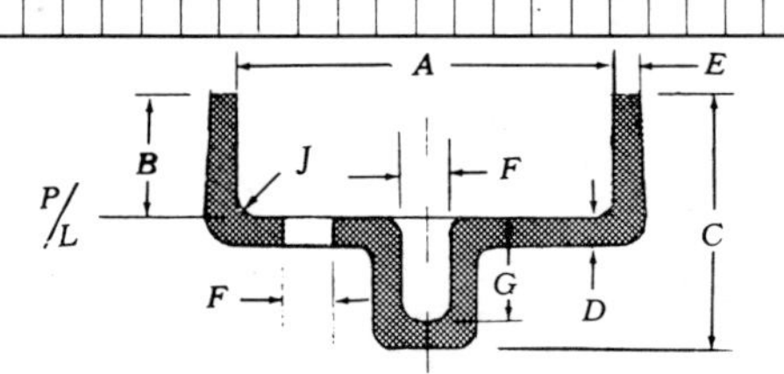

Drawing Code	Dimensions (Inches)	Comm. ±	Fine ±
A = Diameter (see Note #1)	0.000 / 0.500 / 1.000 / 2.000		
B = Depth (see Note #3)	3.000 / 4.000		
C = Height (see Note #3)	5.000 / 6.000		
	6.000 to 12.000 for each additional inch add (inches)	.002	.001
D = Bottom Wall (see Note #3)		.008	.005
E = Side Wall (see Note #4)		.005	.003
F = Hole Size Diameter (see Note #1)	0.000 to 0.125	.002	.001
	0.125 to 0.250	.002	.001
	0.250 to 0.500	.003	.002
	0.500 & Over	.003	.002
G = Hole Size Depth (see Note #5)	0.000 to 0.250	.004	.002
	0.250 to 0.500	.004	.002
	0.500 to 1.000	.005	.003
Draft Allowance per side (see Note #5)		1°	½°
Flatness (see Note #4)	0.000 to 3.000	.010	.005
	3.000 to 6.000	.012	.010
Thread Size (class)	Internal	1	2
	External	1	2
Concentricity (see Note #4)	(T.I.R.)	.005	.003
Fillets, Ribs, Corners (see Note #6)		.062	.031
Surface Finish	(see Note #7)		
Color Stability	(see Note #7)		

The chart columns read "Plus or Minus in Thousands of an Inch" graduated 1 through 28, showing the Fine and Commercial tolerance lines.

REFERENCE NOTES

1 — These tolerances do not include allowance for aging characteristics of material.

2 — Tolerances based on ⅛″ wall section.

3 — Parting line must be taken into consideration.

4 — Part design should maintain a wall thickness as nearly constant as possible. Complete uniformity in this dimension is impossible to achieve.

5 — Care must be taken that the ratio of the depth of a cored hole to its diameter does not reach a point that will result in excessive pin damage.

6 — These values should be increased whenever compatible with desired design and good molding technique.

7 — Customer-Molder understanding necessary prior to tooling.

STANDARDS AND PRACTICES OF PLASTICS CUSTOM MOLDERS

Engineering and Technical Standards
POLYCARBONATE

NOTE: The Commercial values shown below, represent common production tolerances at the most economical level. The Fine values represent closer tolerances that can be held but at a greater cost.

Drawing Code	Dimensions (Inches)		Plus or Minus in Thousands of an Inch (1–28)
A = Diameter (see Note #1)	0.000 / 0.500 / 1.000 / 2.000		
B = Depth (see Note #3)	3.000 / 4.000		
C = Height (see Note #3)	5.000 / 6.000		

Drawing Code	Dimensions (Inches)	Comm. ±	Fine ±
	6.000 to 12.000 for each additional inch add (inches)	.003	.0015
D = Bottom Wall (see Note #3)		.003	.002
E = Side Wall (see Note #4)		.003	.002
F = Hole Size Diameter (see Note #1)	0.000 to 0.125	.002	.001
	0.125 to 0.250	.002	.0015
	0.250 to 0.500	.003	.002
	0.500 & Over	.003	.002
G = Hole Size Depth (see Note #5)	0.000 to 0.250	.002	.002
	0.250 to 0.500	.003	.002
	0.500 to 1.000	.004	.003
Draft Allowance per side (see Note #5)		1°	½°
Flatness (see Note #4)	0.000 to 3.000	.005	.003
	3.000 to 6.000	.007	.004
Thread Size (class)	Internal	1B	2B
	External	1A	2A
Concentricity (see Note #4)	(T.I.R.)	.005	.003
Fillets, Ribs, Corners (see Note #6)		.015	.015
Surface Finish	(see Note #7)		
Color Stability	(see Note #7)		

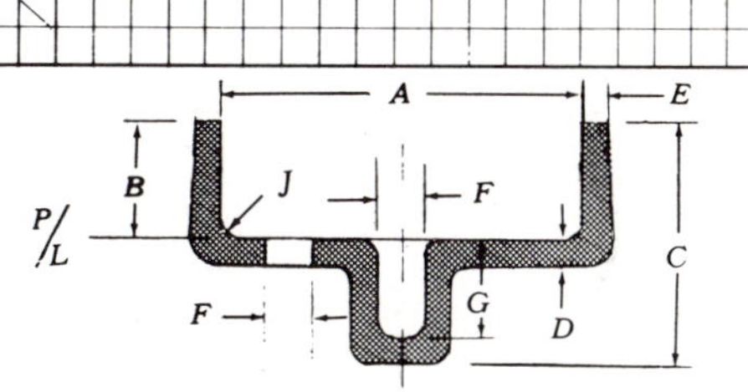

REFERENCE NOTES

1 — These tolerances do not include allowance for aging characteristics of material.

2 — Tolerances based on ⅛″ wall section.

3 — Parting line must be taken into consideration.

4 — Part design should maintain a wall thickness as nearly constant as possible. Complete uniformity in this dimension is impossible to achieve.

5 — Care must be taken that the ratio of the depth of a cored hole to its diameter does not reach a point that will result in excessive pin damage.

6 — These values should be increased whenever compatible with desired design and good molding technique.

7 — Customer-Molder understanding necessary prior to tooling.

STANDARDS AND PRACTICES OF PLASTICS CUSTOM MOLDERS

Engineering and Technical Standards
HIGH DENSITY POLYETHYLENE

NOTE: The Commercial values shown below represent common production tolerances at the most economical level. The Fine values represent closer tolerances that can be held but at a greater cost.

Drawing Code	Dimensions (Inches)	Comm. ±	Fine ±
A = Diameter (see Note #1) / B = Depth (see Note #3) / C = Height (see Note #3)	0.000 / 0.500 / 1.000 / 2.000 / 3.000 / 4.000 / 5.000 / 6.000 (Plus or Minus in Thousands of an Inch — graph 1–28)		
	6.000 to 12.000 for each additional inch add (inches)	.006	.003
D = Bottom Wall (see Note #3)		.006	.004
E = Side Wall (see Note #4)		.006	.004
F = Hole Size Diameter (see Note #1)	0.000 to 0.125	.003	.002
	0.125 to 0.250	.005	.003
	0.250 to 0.500	.006	.004
	0.500 & Over	.008	.005
G = Hole Size Depth (see Note #5)	0.000 to 0.250	.005	.003
	0.250 to 0.500	.007	.004
	0.500 to 1.000	.009	.006
Draft Allowance per side (see Note #5)		2°	¾°
Flatness (see Note #4)	0.000 to 3.000	.023	.015
	3.000 to 6.000	.037	.022
Thread Size (class)	Internal	1	2
	External	1	2
Concentricity (see Note #4)	(T.I.R.)	.027	.010
Fillets, Ribs, Corners (see Note #6)		.025	.010
Surface Finish	(see Note #7)		
Color Stability	(see Note #7)		

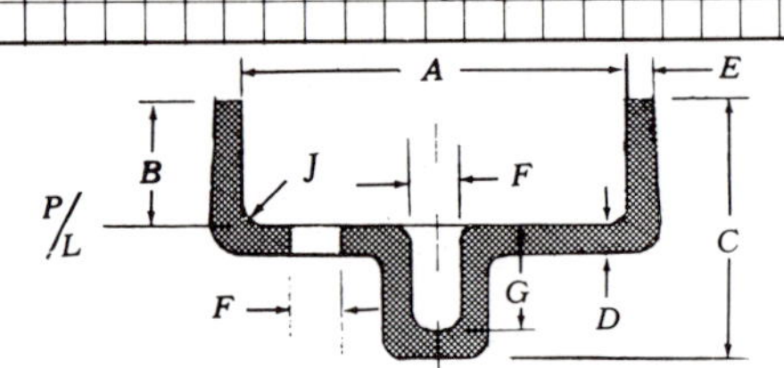

REFERENCE NOTES

1 — These tolerances do not include allowance for aging characteristics of material.

2 — Tolerances based on ⅛″ wall section.

3 — Parting line must be taken into consideration.

4 — Part design should maintain a wall thickness as nearly constant as possible. Complete uniformity in this dimension is impossible to achieve.

5 — Care must be taken that the ratio of the depth of a cored hole to its diameter does not reach a point that will result in excessive pin damage.

6 — These values should be increased whenever compatible with desired design and good molding technique.

7 — Customer-Molder understanding necessary prior to tooling.

STANDARDS AND PRACTICES OF PLASTICS CUSTOM MOLDERS

Engineering and Technical Standards
LOW DENSITY POLYETHYLENE

NOTE: The Commercial values shown below represent common production tolerances at the most economical level. The Fine values represent closer tolerances that can be held but at a greater cost.

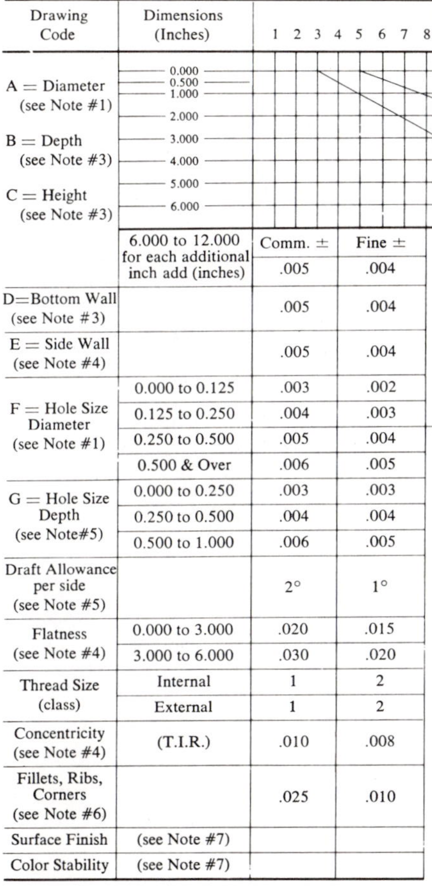

Drawing Code	Dimensions (Inches)	Comm. ±	Fine ±
A = Diameter (see Note #1)			
B = Depth (see Note #3)			
C = Height (see Note #3)			
	6.000 to 12.000 for each additional inch add (inches)	.005	.004
D=Bottom Wall (see Note #3)		.005	.004
E = Side Wall (see Note #4)		.005	.004
F = Hole Size Diameter (see Note #1)	0.000 to 0.125	.003	.002
	0.125 to 0.250	.004	.003
	0.250 to 0.500	.005	.004
	0.500 & Over	.006	.005
G = Hole Size Depth (see Note#5)	0.000 to 0.250	.003	.003
	0.250 to 0.500	.004	.004
	0.500 to 1.000	.006	.005
Draft Allowance per side (see Note #5)		2°	1°
Flatness (see Note #4)	0.000 to 3.000	.020	.015
	3.000 to 6.000	.030	.020
Thread Size (class)	Internal	1	2
	External	1	2
Concentricity (see Note #4)	(T.I.R.)	.010	.008
Fillets, Ribs, Corners (see Note #6)		.025	.010
Surface Finish	(see Note #7)		
Color Stability	(see Note #7)		

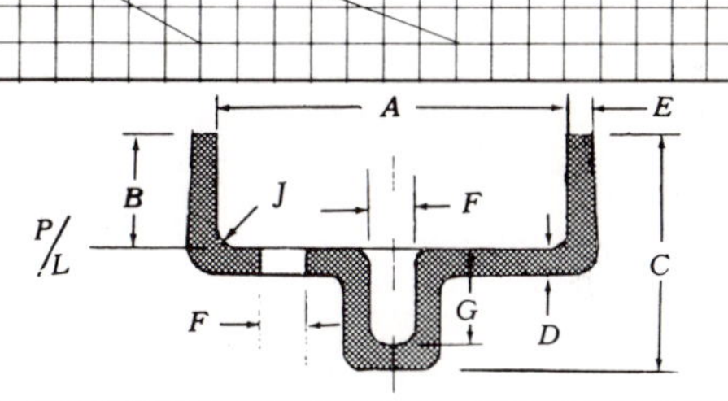

REFERENCE NOTES

1 — These tolerances do not include allowance for aging characteristics of material.

2 — Tolerances based on ⅛″ wall section.

3 — Parting line must be taken into consideration.

4 — Part design should maintain a wall thickness as nearly constant as possible. Complete uniformity in this dimension is impossible to achieve.

5 — Care must be taken that the ratio of the depth of a cored hole to its diameter does not reach a point that will result in excessive pin damage.

6 — These values should be increased whenever compatible with desired design and good molding technique.

7 — Customer-Molder understanding necessary prior to tooling.

STANDARDS AND PRACTICES OF PLASTICS CUSTOM MOLDERS

Engineering and Technical Standards
POLYPROPYLENE

NOTE: The Commercial values shown below represent common production tolerances at the most economical level. The Fine values represent closer tolerances that can be held but at a greater cost.

Drawing Code	Dimensions (Inches)		Plus or Minus in Thousands of an Inch
A = Diameter (see Note #1) B = Depth (see Note #3) C = Height (see Note #3)	0.000 — 0.500 — 1.000 — 2.000 — 3.000 — 4.000 — 5.000 — 6.000		(graph: Commercial / Fine)

Drawing Code	Dimensions (Inches)	Comm. ±	Fine ±
	6.000 to 12.000 for each additional inch add (inches)	.005	.003
D = Bottom Wall (see Note #3)		.006	.003
E = Side Wall (see Note #4)		.006	.003
F = Hole Size Diameter (see Note #1)	0.000 to 0.125	.003	.002
	0.125 to 0.250	.004	.003
	0.250 to 0.500	.005	.004
	0.500 & Over	.008	.006
G = Hole Size Depth (see Note#5)	0.000 to 0.250	.005	.003
	0.250 to 0.500	.006	.004
	0.500 to 1.000	.009	.006
Draft Allowance per side (see Note #5)		1½°	½°
Flatness (see Note #4)	0.000 to 3.000	.021	.014
	3.000 to 6.000	.035	.021
Thread Size (class)	Internal	1	2
	External	1	2
Concentricity (see Note #4)	(T.I.R.)	.016	.013
Fillets, Ribs, Corners (see Note #6)		.028	.015
Surface Finish	(see Note #7)		
Color Stability	(see Note #7)		

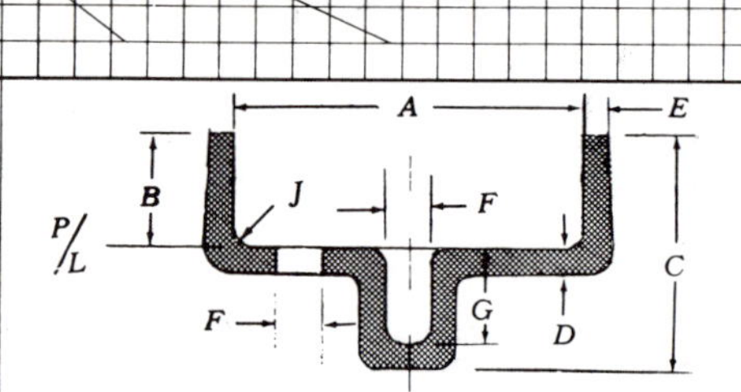

REFERENCE NOTES

1 — These tolerances do not include allowance for aging characteristics of material.

2 — Tolerances based on ⅛″ wall section.

3 — Parting line must be taken into consideration.

4 — Part design should maintain a wall thickness as nearly constant as possible. Complete uniformity in this dimension is impossible to achieve.

5 — Care must be taken that the ratio of the depth of a cored hole to its diameter does not reach a point that will result in excessive pin damage.

6 — These values should be increased whenever compatible with desired design and good molding technique.

7 — Customer-Molder understanding necessary prior to tooling.

STANDARDS AND PRACTICES OF PLASTICS CUSTOM MOLDERS

Engineering and Technical Standards
POLYSTYRENE

NOTE: The Commercial values shown below represent common production tolerances at the most economical level. The Fine values represent closer tolerances that can be held but at a greater cost.

Drawing Code	Dimensions (Inches)	Comm. ±	Fine ±
A = Diameter (see Note #1) B = Depth (see Note #3) C = Height (see Note #3)	Plus or Minus in Thousands of an Inch (see graph)		
	6.000 to 12.000 for each additional inch add (inches)	.004	.002
D=Bottom Wall (see Note #3)		.0055	.003
E = Side Wall (see Note #4)		.007	.0035
F = Hole Size Diameter (see Note #1)	0.000 to 0.125	.002	.001
	0.125 to 0.250	.002	.001
	0.250 to 0.500	.002	.0015
	0.500 & Over	.0035	.002
G = Hole Size Depth (see Note #5)	0.000 to 0.250	.0035	.002
	0.250 to 0.500	.004	.002
	0.500 to 1.000	.005	.003
Draft Allowance per side (see Note #5)		1½°	½°
Flatness (see Note #4)	0.000 to 3.000	.007	.004
	3.000 to 6.000	.013	.005
Thread Size (class)	Internal	1	2
	External	1	2
Concentricity (see Note #4)	(T.I.R.)	.010	.008
Fillets, Ribs, Corners (see Note #6)		.015	.010
Surface Finish	(see Note #7)		
Color Stability	(see Note #7)		

The dimension graph plots Plus or Minus in Thousands of an Inch (1–28) against Dimensions (Inches): 0.000, 0.500, 1.000, 2.000, 3.000, 4.000, 5.000, 6.000, with curves labeled "Fine" and "Commercial."

Diagram labels: A, B, C, D, E, F, G, J, P/L

REFERENCE NOTES

1 — These tolerances do not include allowance for aging characteristics of material.

2 — Tolerances based on ⅛″ wall section.

3 — Parting line must be taken into consideration.

4 — Part design should maintain a wall thickness as nearly constant as possible. Complete uniformity in this dimension is impossible to achieve.

5 — Care must be taken that the ratio of the depth of a cored hole to its diameter does not reach a point that will result in excessive pin damage.

6 — These values should be increased whenever compatible with desired design and good molding technique.

7 — Customer-Molder understanding necessary prior to tooling.

STANDARDS AND PRACTICES OF PLASTICS CUSTOM MOLDERS

Engineering and Technical Standards

FLEXIBLE VINYL

NOTE: The Commercial values shown below represent common production tolerances at the most economical level. The Fine values represent closer tolerances that can be held but at a greater cost.

Drawing Code	Dimensions (Inches)	Plus or Minus in Thousands of an Inch
A = Diameter (see Note #1)	0.000 / 0.500 / 1.000 / 2.000	
B = Depth (see Note #3)	3.000 / 4.000	
C = Height (see Note #3)	5.000 / 6.000	

Drawing Code	Dimensions (Inches)	Comm. ±	Fine ±
	6.000 to 12.000 for each additional inch add (inches)	.005	.003
D = Bottom Wall (see Note #3)		.007	.003
E = Side Wall (see Note #4)		.007	.003
F = Hole Size Diameter (see Note #1)	0.000 to 0.125	.004	.003
	0.125 to 0.250	.005	.004
	0.250 to 0.500	.006	.005
	0.500 & Over	.008	.006
G = Hole Size Depth (see Note #5)	0.000 to 0.250	.004	.003
	0.250 to 0.500	.005	.004
	0.500 to 1.000	.006	.005
Draft Allowance per side (see Note #5)		1½°	1°
Flatness (see Note #4)	0.000 to 3.000	.010	.007
	3.000 to 6.000	.020	.015
Thread Size (class)	Internal	(Not Recommended)	
	External	(Not Recommended)	
Concentricity (see Note #4)	(T.I.R.)	.015	.010
Fillets, Ribs, Corners (see Note #6)		.030	.010
Surface Finish	(see Note #7)		
Color Stability	(see Note #7)		

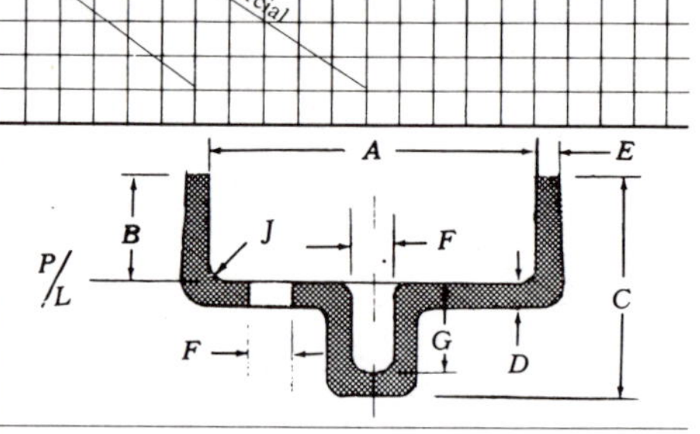

REFERENCE NOTES

1 — These tolerances do not include allowance for aging characteristics of material.

2 — Tolerances based on ⅛″ wall section.

3 — Parting line must be taken into consideration.

4 — Part design should maintain a wall thickness as nearly constant as possible. Complete uniformity in this dimension is impossible to achieve.

5 — Care must be taken that the ratio of the depth of a cored hole to its diameter does not reach a point that will result in excessive pin damage.

6 — These values should be increased whenever compatible with desired design and good molding technique.

7 — Customer-Molder understanding necessary prior to tooling.

STANDARDS AND PRACTICES OF PLASTICS CUSTOM MOLDERS

Engineering and Technical Standards
RIGID VINYL

NOTE: The Commercial values shown below represent common production tolerances at the most economical level. The Fine values represent closer tolerances that can be held but at a greater cost.

Drawing Code	Dimensions (Inches)		Plus or Minus in Thousands of an Inch
A = Diameter (see Note #1) B = Depth (see Note #3) C = Height (see Note #3)	0.000 / 0.500 / 1.000 / 2.000 / 3.000 / 4.000 / 5.000 / 6.000		*(chart: Fine and Commercial tolerance lines across scale 1–28)*

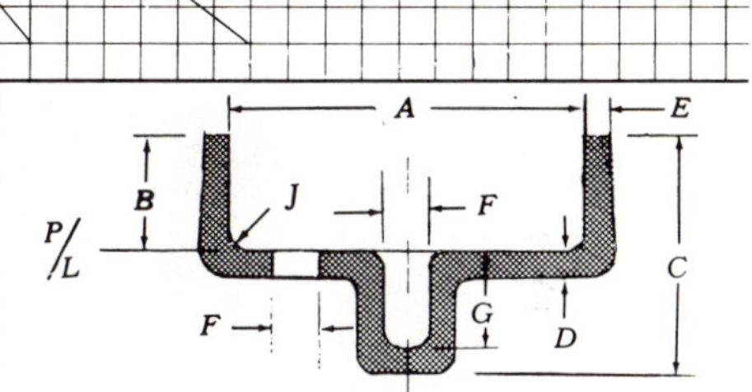

Drawing Code	Dimensions (Inches)	Comm. ±	Fine ±
6.000 to 12.000 for each additional inch add (inches)		.005	.003
D=Bottom Wall (see Note #3)		.007	.003
E = Side Wall (see Note #4)		.007	.003
F = Hole Size Diameter (see Note #1)	0.000 to 0.125	.004	.003
	0.125 to 0.250	.004	.003
	0.250 to 0.500	.005	.004
	0.500 & Over	.006	.005
G = Hole Size Depth (see Note#5)	0.000 to 0.250	.004	.003
	0.250 to 0.500	.005	.004
	0.500 to 1.000	.006	.005
Draft Allowance per side (see Note #5)		1°	½°
Flatness (see Note #4)	0.000 to 3.000	.015	.010
	3.000 to 6.000	.020	.015
Thread Size (class)	Internal	1	2
	External	1	2
Concentricity (see Note #4)	(T.I.R.)	.010	.005
Fillets, Ribs, Corners (see Note #6)		.035	.030
Surface Finish	(see Note #7)		
Color Stability	(see Note #7)		

REFERENCE NOTES

1 — These tolerances do not include allowance for aging characteristics of material.

2 — Tolerances based on ⅛" wall section.

3 — Parting line must be taken into consideration.

4 — Part design should maintain a wall thickness as nearly constant as possible. Complete uniformity in this dimension is impossible to achieve.

5 — Care must be taken that the ratio of the depth of a cored hole to its diameter does not reach a point that will result in excessive pin damage.

6 — These values should be increased whenever compatible with desired design and good molding technique.

7 — Customer-Molder understanding necessary prior to tooling.

Warpage

Large flat areas are to be avoided since they cannot be held flat and therefore will present a warped appearance. Such surfaces should be slightly convex or "crowned." A crowned surface constitutes a structural effect that reduces warpage and, because it offers better material flow, improves appearance. Large area surfaces should be crowned a minimum of 0.003 in./in. The warpage variable with the best of materials may run as much as plus or minus 0.0025 in./in. Large flat areas should be reinforced at the edges when possible, and ribs located on the underside will be helpful in reducing warpage. When an area of substantial size is to be mounted on a flat surface, a depressed center section, Fig. 7.23, should be

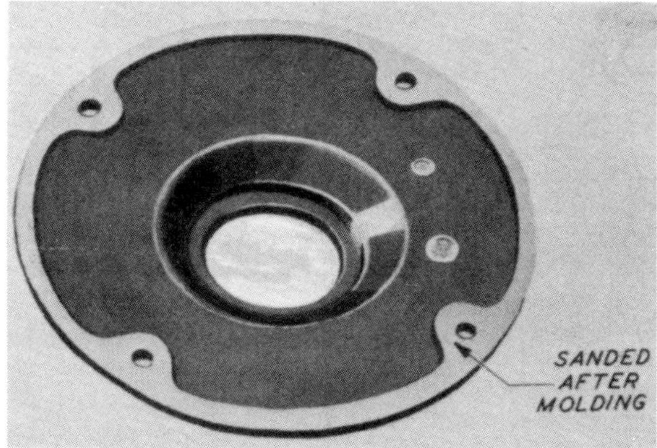

FIG. 7.23 Pieces that must offer a flat surface after molding should have a "bead" which can be cut in sanding to eliminate sanding of entire surface. All-over sanding increases tendency of piece to warp.

used to avoid all-over flatness. When necessary, it is then possible to grind the projecting edges on a sand belt to overcome the effects of warpage and insure a good fit.

Draft

Draft is provided for the removal of parts from the mold. It is possible, when absolutely necessary, to produce some surfaces without draft. In most jobs, however, failure to provide adequate draft will be the source of many difficult mold-

FIG. 7.24 Vertical walls should have a taper of 1° per side to facilitate removal from mold. The minimum practical draft is $1/2$°.

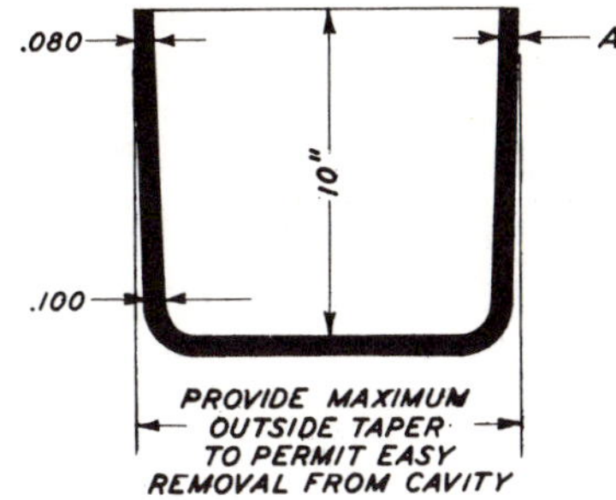

FIG. 7.25 As compression mold closes, materials flow up side wall. This requires maximum cross section at bottom to insure full density at top (A).

ing problems. A minimum of $1/2°$ taper per side is generally satisfactory, although $1°$ per side is most desirable for production jobs, Fig. 7.24. Side wall taper of large units must permit flow of material from the bottom of the mold up to the parting line when compression molds are used, Fig. 7.25.

Wall Thickness

The wall thickness of all new products must be checked to make sure that it is adequate and uniform. Especially thin sections that must be made from thermosetting materials may require the use of resin-treated paper or cloth (molded laminated). Good molding practice requires uniform section and minimum wall thickness to produce a fast and complete cure. Heavy sections attached to thin sections are troublesome and produce distortion and undercured and overcured parts. Thermosetting materials that are heavily filled with glass or other inorganic fillers offer a better answer for injection molded pieces with heavy and light sections. Thermoplastic products will often show concave depressions or "sinks" on heavy sections. Sinks are caused by internal shrinkage that takes place as the center hardens, thereby pulling in the outer surface. If thermosetting materials are specified, transfer molds are best for a combination of heavy and light sections. All heavy sections should be cored out if permissible. Frequently, this is accomplished by locating blind holes on the underside, Fig. 7.26, the round pins

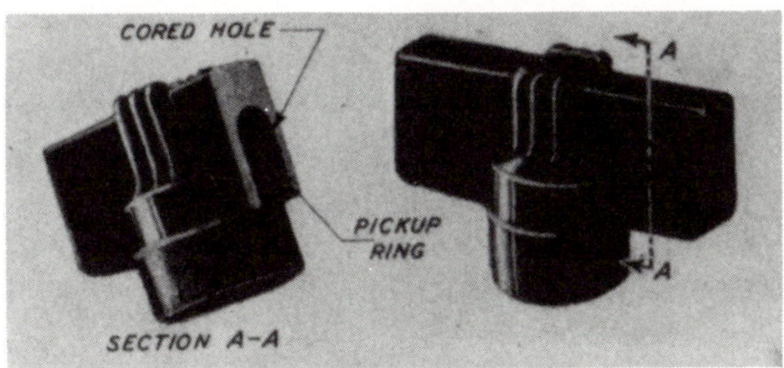

FIG. 7.26 The pins used to core out this piece also serve as pickups by providing a "hold" for lifting piece out of cavity.

forming these holes being easily inserted in the mold. Designers should always core out the underside of heavy sections to expedite the cure and save material.

A wall thickness of $^1/_{16}$ inch may be used on small pieces made from thermosetting materials, but $^3/_{32}$ inch is a much safer minimum. Large pieces such as business machines use thicknesses from $^1/_8$ to $^3/_8$ inch. Thermoplastic materials may be used in very thin sections, and many pieces have been produced successfully with a wall only 0.010 inch thick.

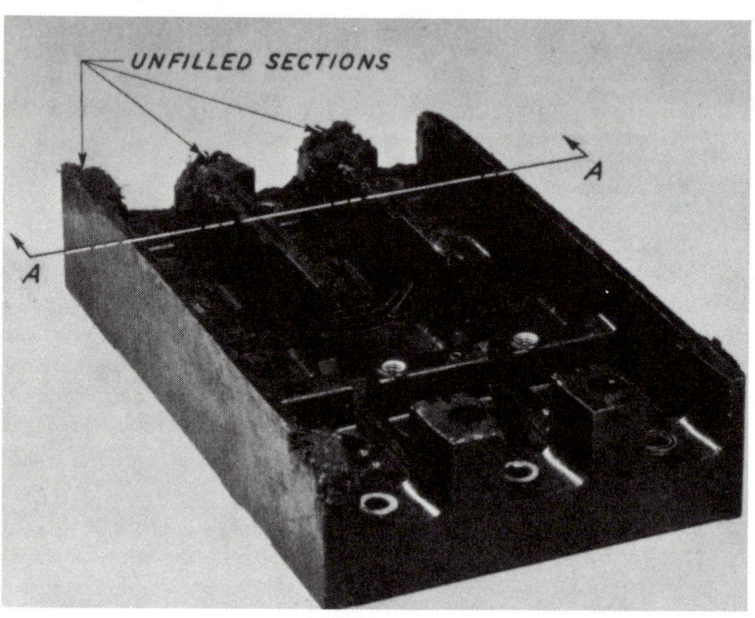

FIG. 7.27A Molded circuit breaker formed from fabric-filled material in a mold designed for wood flour-filled material. Lack of proper fillets and tapered cavities, and insufficient pressure, have caused many sections to remain unfilled.

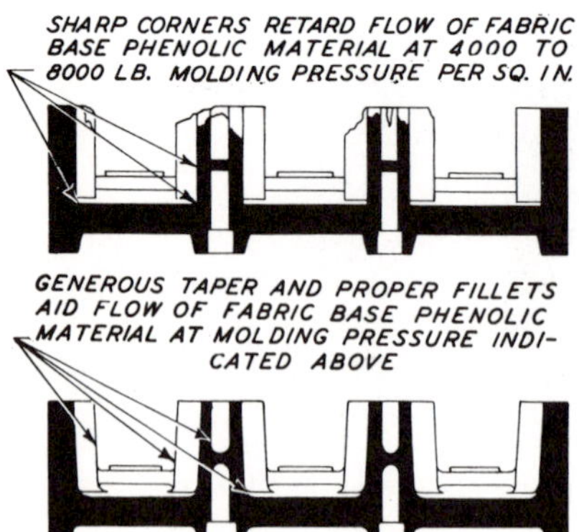

FIG. 7.27B Above: Thermosetting molded piece shown in Fig. 7.27A here shows sharp corners responsible for unfilled sections. Below: Redesign of same piece. Generous taper and proper fillets assure even flow of compound to all sections. Note holes are partially molded from each side for completion by drilling.

The cloth-filled phenolic materials require considerably thicker sections to facilitate material flow. Ribs should be added whenever possible to act as feeders. Transfer molds produce better results when thin sections and cloth-filled materials are specified.

Brittle materials, such as the cold-molded and mineral-filled phenolics, require a heavy wall thickness; not less than $1/8$ inch should be considered. In the absence of previous experience, it would be well to ask for recommendations from the raw material supplier. Considerable loss may be experienced by the use of over-thin wall sections.

Fillets

Radii or fillets must be used in all inside corners to assist material flow and strengthen the part (see Figs. 7.27A and 7.27B), as fractures start easily from sharp corners. Sharp corners on interior surfaces of all plastics pieces should be avoided. Molds which require sharp corners are frequently more costly and fragile than those made with generous radii in the corners. Mold designers should eliminate troublesome features of this kind when redesigning the product unless the mold construction or assembly requires sharp corners where the sections join.

Designers frequently specify radii which cannot be machined. These may be changed in most cases to radii which are machinable. In those few cases where there is no alternative, such radii must be cut in the mold by hand—a slow and extremely costly procedure. Figure 7.28 (A) shows a typical radius as designed by the product engineer, and, (B), the same detail is simplified by the tool designer. Designers must always look for, and change, when permissible, corners which cannot be machined. Product designs are guided by mold maker.

Note that the product called for a vertical projection with a $1/8$-inch corner radius. The dotted line at (A) shows the $1/4$-inch cutter needed to cut the $1/8$-inch radius in the mold. It can be seen that the cutter indicated will extend beyond the $3/16$-inch wall limit and cut away steel that should remain in the mold. The tool designer secured permission to change the design to a $3/32$-inch

FIG. 7.28 Typical redesign to simplify machining operations. Dotted line, in (A), shows area cut by $1/4$ in. cutter or end mill. Note that this radius extends beyond the desired area and this will necessitate hand chiseling. Revised design, shown at (B), permits radius to be machined easily with a $3/16$ in. cutter. Radii should be uniform on corners wherever practical.

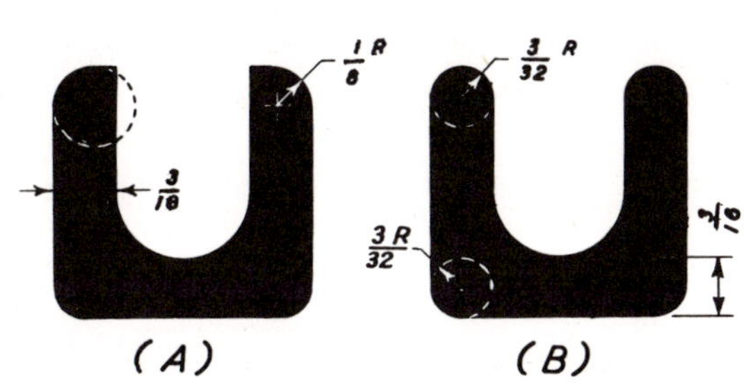

radius, as shown at (B), which can be cut with a ³/₁₆-inch cutter. If this change had not been made, the tool-maker would have been forced to chisel out the sharp corner by hand.

Sharp Outside Corners

Sharp corners are required at the parting lines, but are usually undesirable at all other points. All outside corners should be radiused as much as possible to help material flow, reduce mold cost, and avoid the sharp molded corners that chip and break easily during finishing operations. Where beveled or rounded edges are required, a flat area must be provided at the edge to eliminate feather-edge mold sections, which are easily broken, Fig. 7.29.

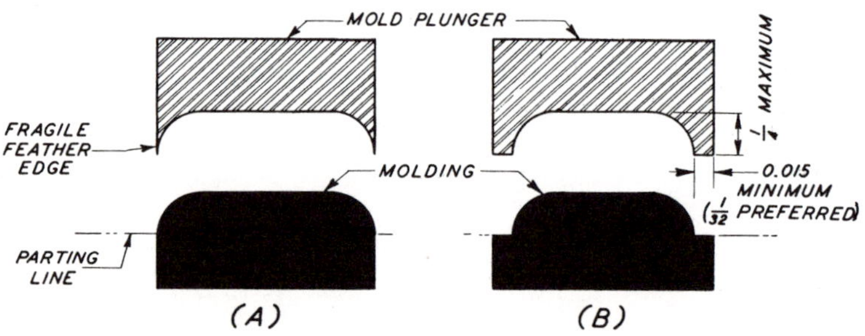

FIG. 7.29 Construction shown at (A) has produced a feather-edge that is very undesirable from standpoint of maintenance. Stepped corner, at (B), is preferable because its maintenance is negligible.

Mold designers must correct all corners and edges which would produce a knife- or feather-edge in the part.

Ribs and Bosses

Ribs and bosses are frequently used on molded products and they should be designed properly to get the best results. Parts which are not properly designed are weak at the base where they join the body of the piece, Fig. 7.30. When the height is too great in proportion to the width, the pieces do not fill out well, and this lack of density greatly reduces their strength. Table 7.5 shows the

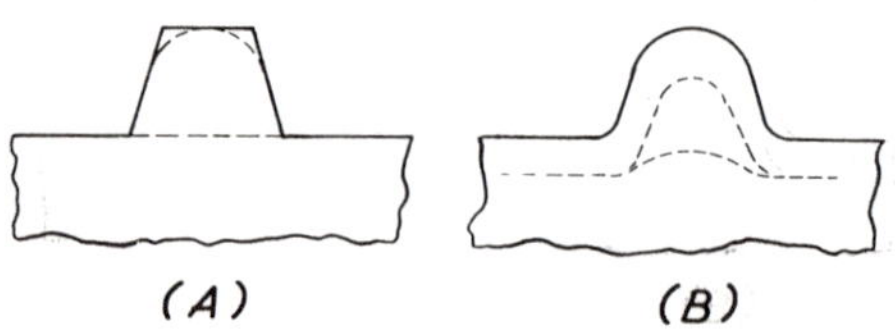

FIG. 7.30 Sharp corners must not be used on small ribs and bosses. At (A), flow of material tends to follow dotted line, leaving a weak structure. Correct design is shown at (B), with material flow greatly assisted.

TABLE 7.5 RECOMMENDED PROPORTIONS FOR RIBS

A	B	C	D	E	F	G
10°	5°	n	n/2	3n	n/4	n/4

proportions for ribs which have been found to be generally satisfactory. Table 7.6 shows the correct proportions for bosses.

If many ribs are to be inserted, they should be staggered to reduce the distortion caused by unequal shrinkage, Fig. 7.31. Ribs should not be located in the

TABLE 7.6 RECOMMENDED PROPORTIONS FOR BOSSES

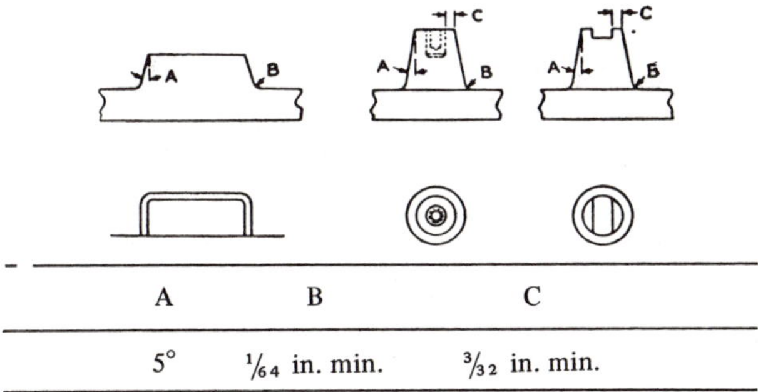

A	B	C
5°	¼₆₄ in. min.	³⁄₃₂ in. min.

center of large areas unless so required by the design. When central ribs must be inserted, reeds or flutes should be added to the outer surface of the piece to conceal resulting flow lines, as shown in Fig. 7.32.

The mold designer must make sure that the design of the part will provide

FIG. 7.31 Design at (A) is not as good as that at (B). Staggered rib construction minimizes distortion caused by shrinkage.

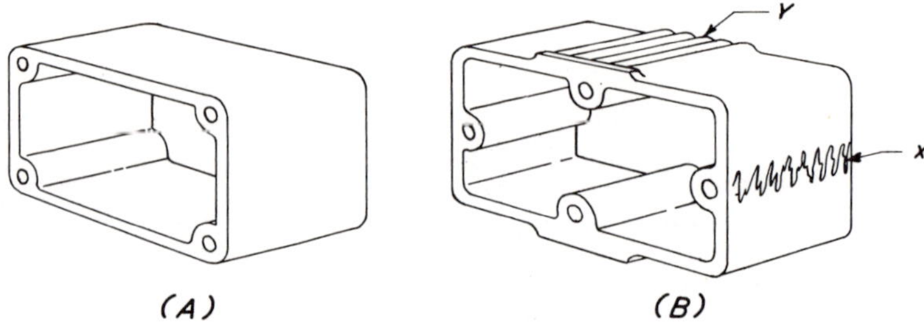

FIG. 7.32 Ribs on which lugs are to be mounted should be located in corners, as at (A). When ribs are located below plane surfaces as in (B), flow lines formed by uneven cross section will be unsightly, as indicated at x. When this construction is unavoidable, the exterior surface should be broken up by a reed or flute design, as at y, to conceal flow lines.

adequate tool strength. Products which are to contain ribs or bosses are frequently designed with holes located too near these sections. Figure 7.33 presents a typical example of this and indicates the correction. The distance A had to be increased from the original layout. In this case, the projections which form the holes could be machined as a solid part of the mold cavity, but this is undesirable because of the danger of breaking the projections, thus making it necessary to rebuild the mold cavity. The correct method is illustrated in Fig. 7.33, wherein the design is seen to make use of mold pins. The holes for the pins must not be too close to the depression for the rib, because this will produce a thin mold section which may crack in hardening or in service. For best tool design, the minimum dimension A will be determined by the diameter of the pin and the height of the rib. If the rib is very high, the distance A should be at least one half the diameter of the pin. A good general rule to follow is to make the dimension A from $\frac{1}{3}$ to $\frac{1}{2}$ the diameter of the molded hole.

Holes

Flow of compound around mold pins causes considerable wear and some distortion. Mold pins should be built to the maximum diameter and replaced when they wear down to the minimum. Pins are frequently broken when flow condi-

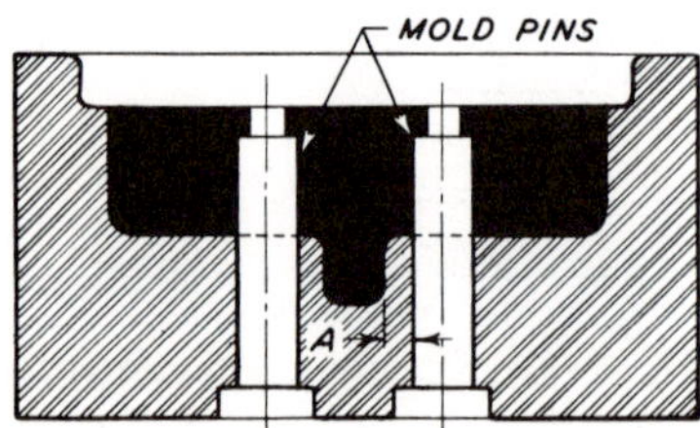

FIG. 7.33 Adequate mold steel between holes and ribs is necessary to good design. The minimum dimension A will be determined by diameter of pin and height of molded rib. A workable rule is to make the dimension A from $\frac{1}{3}$ to $\frac{1}{2}$ the diameter of the molded hole.

TABLE 7.7 HOLE DIMENSIONS AND WALL THICKNESSES

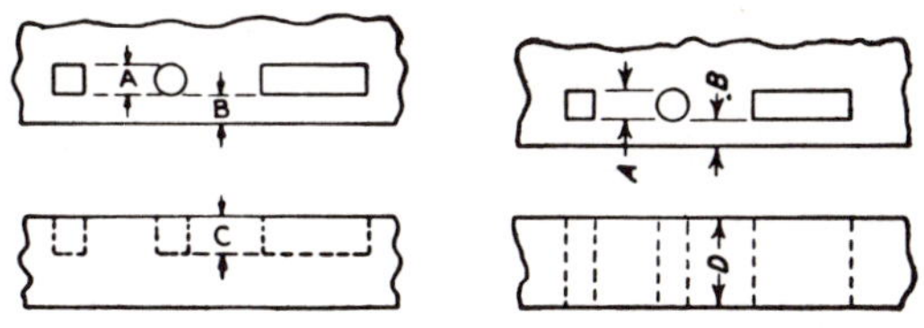

Minimum Hole Diameter A in Inches	Maximum Hole Depth		Minimum Side Wall B in Inches	
	C in Inches	D in Inches	Hot-molded	Cold-molded
1/16	1/16	1/8	1/16	3/32
5/64	3/32	3/16	1/16	3/32
3/32	1/8	1/4	1/16	3/32
7/64	5/32	5/16	3/32	3/32
1/8	3/16	3/8	3/32	1/8
5/32	1/4	1/2	3/32	1/8
3/16	5/16	5/8	1/8	1/8
7/32	3/8	3/4	1/8	3/16
1/4	7/16	7/8	1/8	3/16
5/16	9/16	1 1/8	5/32	1/4
3/8	11/16	1 3/8	5/32	1/4
7/16	13/16	1 5/8	3/16	5/16
1/2	15/16	1 7/8	3/16	3/8

tions are particularly bad. Pin breakage is more prevalent in compression molding than in transfer and injection molding. Table 7.7 shows the proportions of molded holes for most satisfactory and trouble-free production. It is desirable to stay within these limits for molds which must stand up under heavy production schedules. Since there is no means of predicting the flow conditions which may exist in a new mold, holes which come outside these limits must be tried and the pins shortened to the maximum length that may be expected to stand up. It is common practice to mold deep holes to their maximum molding depth, as shown in Table 7.7 and drill them to the full depth after molding.

Note that through holes may be molded to twice the depth of blind holes. The through hole pins may butt in the center or be supported at each end by entering the matching section of the mold.

Mold builders frequently use two pins which butt in the center to produce deep holes. It is desirable to make one pin 0.030 inch larger than the other to compensate for mold wear, misalignment and deflection of the pins, Fig. 7.34.

When holes are to be drilled after molding they should generally be spotted by the mold. This facilitates drilling without the use of drill jigs. A tapered

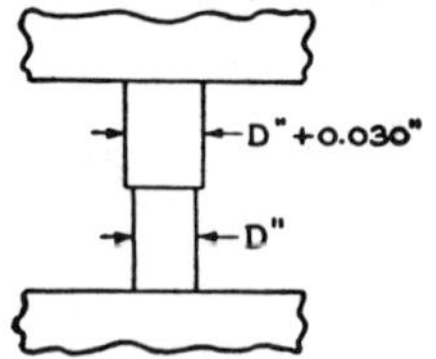

FIG. 7.34 When butt pins are used for deep holes, one pin should be 0.030 in. larger than the other to compensate for misalignment, mold wear and deflection of pins.

section, as shown in Fig. 7.35, will minimize chipping when the drill comes through.

Use of flat-head screws requiring countersunk holes should be avoided whenever possible if brittle plastics materials are specified. When these materials are used, the angle must be right or the wedging action which takes place when the screw is tightened may break the piece. When countersunk holes are required, and the screw head must seat below the surface, $1/64$ in. minimum clearance, Fig. 7.36, should be provided to compensate for variation in the screws.

Table 7.8 shows the proper size of clearance holes for counterbores, screws and nuts.

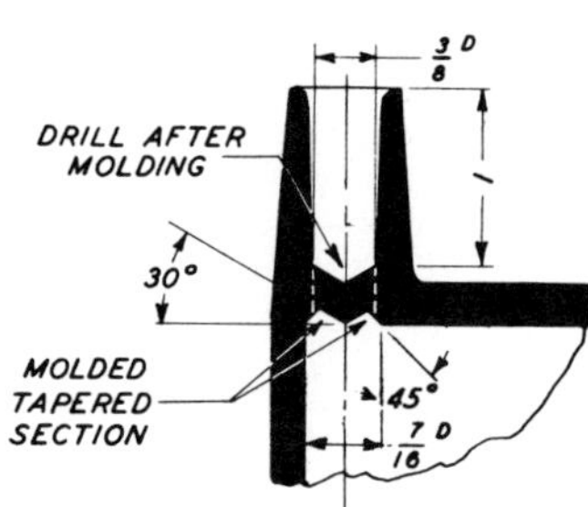

FIG. 7.35 Section designed with taper to minimize chipping in drilling operations.

Bosses should be added around screw holes if the product is to be mounted on an uneven surface, Fig. 7.37. Three holes or mounting points are best, as three points will always make contact on an uneven surface. If more than three mounting points are required, the bosses may be cut by sanding on a flat surface to insure a fit without distortion of the piece.

Blind holes are frequently molded in plastics materials and are tapped after molding. Experience has shown that complaints often result from insufficient depth, which condition fails to provide adequate space for chips that accumulate in tapping. Since the depth of hole required is usually beyond the limits of economical production, holes are generally molded to the maximum molding

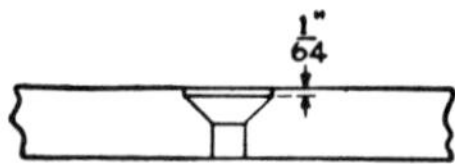

FIG. 7.36 Minimum clearance of $1/64$ in. for countersunk holes will compensate for variation in screws.

TABLE 7.8 RECOMMENDED CLEARANCE HOLES

| Screw | | Hole | | Counterbores in In. | | |
Size (in.)	Dia. (in.)	Std. Dia. (in.)	Min. Dia. (in.)	Standard	Socket Wrench Hex Head	Socket Wrench Hex Nut
No. 4	.112	+.000 .128−.003	+.003 .115−.000	$11\!/\!32$	—	—
No. 6	.138	+.000 .157−.007	+.003 .141−.000	$13\!/\!32$	—	—
No. 8	.164	+.000 .182−.007	+.003 .167−.000	$7\!/\!16$	—	$21\!/\!32$
No. 10	.190	+.000 .209−.007	+.003 .193−.000	$1\!/\!2$	—	$11\!/\!16$
No. 12	.216	+.000 .234−.007	+.003 .219−.000	$9\!/\!16$	—	—
$1\!/\!4$	$1\!/\!4$	+.000 .265−.007	+.004 .253−.000	$5\!/\!8$	$3\!/\!4$	$15\!/\!16$
$5\!/\!16$	$5\!/\!16$	$11\!/\!32$	+.004 .317−.000	$3\!/\!4$	$15\!/\!16$	$11\!/\!16$
$3\!/\!8$	$3\!/\!8$	$13\!/\!32$	+.004 .380−.000	$13\!/\!16$	$11\!/\!16$	$13\!/\!16$
$1\!/\!2$	$1\!/\!2$	$17\!/\!32$	+.004 .506−.000	$11\!/\!16$	$11\!/\!4$	$13\!/\!8$

FIG. 7.37 Parts which are to be attached to or mounted on a base or other unit should use three mounting holes, as by this means it is always possible to obtain three-point contact on an irregular surface. Bosses placed around mounting holes serve to reduce breakage caused by distortion.

TABLE 7.9 PREFERRED DEPTH OF HOLES FOR THREADED PARTS

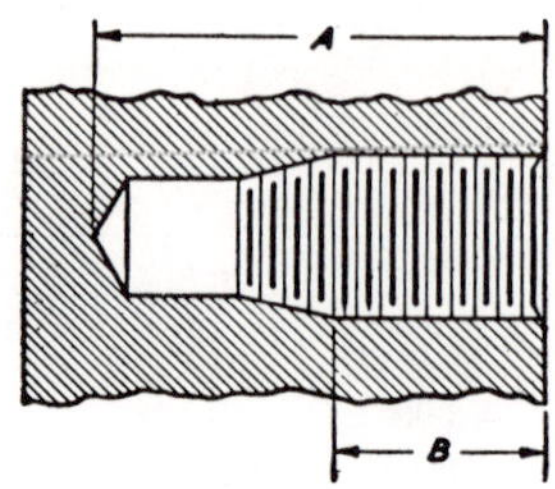

* Size Tap No.	Min. Molding or Drilling Depth A	Min. Depth of Full Threads B	* Size Tap No.	Min. Molding or Drilling Depth A	Min Depth of Full Threads B
10	$^{21}\!/_{32}$	$\frac{1}{4}$	$\frac{1}{2}$	$1^{11}\!/_{16}$	$\frac{5}{8}$
12	$^{25}\!/_{32}$	$^{9}\!/_{32}$	$^{9}\!/_{16}$	$1^{11}\!/_{16}$	$^{23}\!/_{32}$
$\frac{1}{4}$	$^{29}\!/_{32}$	$^{5}\!/_{16}$	$\frac{5}{8}$	$1^{27}\!/_{32}$	$^{25}\!/_{32}$
$^{5}\!/_{16}$	$1^{1}\!/_{32}$	$^{13}\!/_{32}$	$\frac{3}{4}$	$2^{9}\!/_{32}$	$^{15}\!/_{16}$
$\frac{3}{8}$	$1\frac{1}{4}$	$^{15}\!/_{32}$	$\frac{7}{8}$	$2^{5}\!/_{16}$	$1^{3}\!/_{32}$
$^{7}\!/_{16}$	$1^{15}\!/_{32}$	$^{9}\!/_{16}$	1	$2^{5}\!/_{8}$	$1\frac{1}{4}$

* American Standard coarse and fine threads.
Courtesy of General Electric Co., Schenectady, N.Y.

depth and drilled to full depth before tapping. Table 7.9 shows the preferred practice for trouble-free production. Holes which are to be tapped should be provided with a molded countersink to minimize edge chipping during the tapping operation.

Many products now use molded holes for self-tapping screws. It is frequently desirable to drill these molded holes for additional depth. Table 7.10 shows the correct molded hole size, drilled hole size, driving torque and holding power of such screws. It will be well to check the hole sizes on products that are to use self-tapping screws.

Holes made to receive rectangular shafts should be designed to allow for mold pin wear. Figure 7.38, at (A), shows a common type of molded hole which provides for wear caused by the sharp corners of square mold pins. If strict accuracy of fit is required, the hole should be designed as shown at (B). This will permit broaching to close dimensions after molding.

Rectangular holes should, when possible, have rounded ends for increased life and to prevent cracks from starting at the sharp corners. In making rectangular mold pins, round rod is used because of the ease of mounting the fixed end in a drilled hole. Rectangular holes should be designed so that the short side is formed by the outside diameter of the pin, as shown in Fig. 7.39

TABLE 7.10 SELF-TAPPING SCREWS FOR PLASTICS

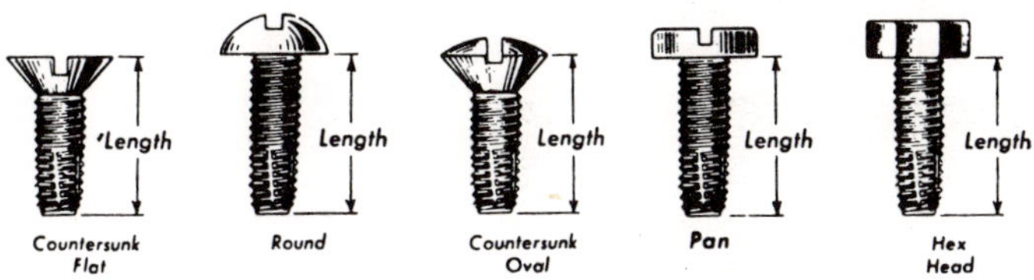

The holes for the screws may be molded or drilled, whichever is most practical. If the material is brittle or friable, molded holes should be formed with a rounded chamfer and drilled holes should be machine chamfered as shown.

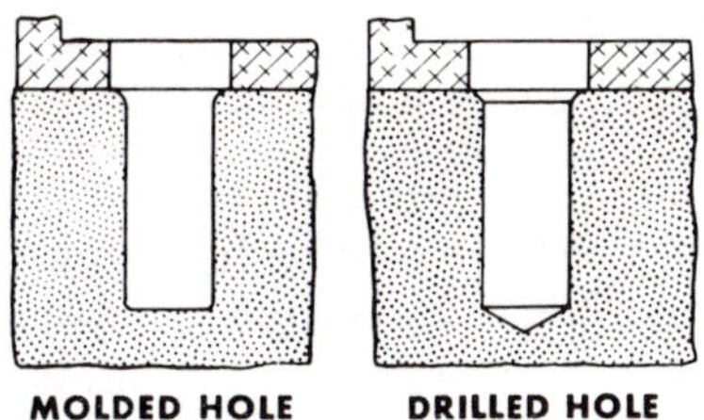

FIG. A This minimizes spalling. In all cases a clearance hole should be provided in the part to be fastened. Depth of penetration should be held within the "Minimum and Maximum" limits recommended. The depth of the hole should be deeper than the screw penetration. In some cases it is desirable to apply a lubricant* to the screws to facilitate driving and to prevent cracking the material.

Screw Diameter	*Thermosetting Materials*				*Thermoplastic Materials*			
	Hole Required (in.)	Drill Size No.	Depth of Penetration * Minimum (in.)	Maximum (in.)	Hole Required (in.)	Drill Size No.	Depth of Penetration * Minimum (in.)	Maximum (in.)
No. 2-56	.073	49	$7/32$	$3/8$	NOT RECOMMENDED (See Type Z)			
No. 4-40	.098	40	$1/4$	$5/16$	.093	42	$1/4$	$5/16$
No. 6-32	.116	32	$1/4$	$5/16$	.116	32	$1/4$	$5/16$
No. 8-32	.144	27	$5/16$	$1/2$	.144	27	$5/16$	$1/2$
No. 10-32	.166	19	$3/8$	$1/2$	.166	19	$3/8$	$1/2$
No. 10-24	.161	20	$3/8$	$1/2$	.161	20	$3/8$	$1/2$
$1/4''$-20	.228	1	$3/8$	$5/8$	.228	1	$3/8$	1

* Includes taper.

Note: For satisfactory results, holes must be neither too large nor too small. Size of hole depends upon kind of material, its hardness, uniformity, etc. In most cases hole sizes shown are suitable. If material is very hard a size larger drill might be necessary; if material is very soft a size smaller drill should be used.

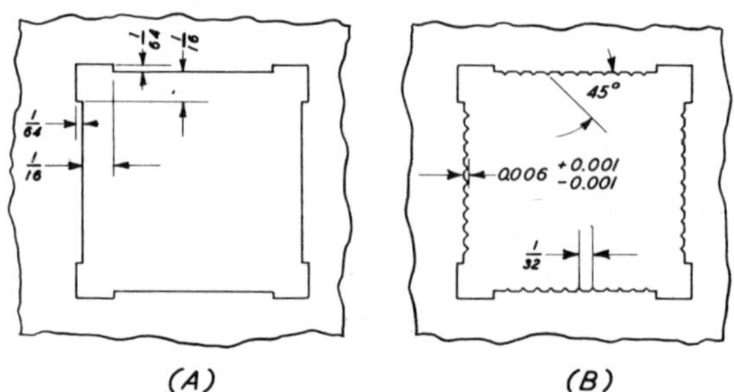

(A) (B)

FIG. 7.38 A molded hole design for close fit to square pins is shown at (A). Enlarged corners permit insertion of square shaft even though corners of mold pins wear considerably. (B) illustrates design which permits projections to be cut to the close dimensions needed. This is done in a broaching operation.

at (A). This is better than the pin construction shown at (B), since the latter requires a milled pin shape, while, in (A), the pin is formed by cutting the two flat sides, and filing the sharp corners which remain. If many pins are required, this small design change will effect a considerable saving in mold cost.

Very thin mold pins for making rectangular slots should be strengthened by an increased section, as shown in Fig. 7.40. Since such holes are usually intended to serve as a guide for metal strips, designers will usually grant this change.

Holes which must enter the molded piece at right angles to the direction of mold closing (cross or side holes) introduce many problems. Injection molds can be built that will core these side holes and pull out the mold pins by a cam action before ejecting the part. Some automatically operated side core pins are used to transfer and compression molding, but manual or hydraulic operation of such side pins has been found to be more satisfactory. It is usually possible to follow the depth to diameter ratios given in Table 7.4 for molded side holes. Again, flow conditions are unpredictable and it may be necessary to determine the maximum moldable side hole depth by trial.

In compression molding, oblique or side holes create a real problem. It is often much less costly to drill such side holes after molding, and this practice is followed for all small holes. When long side holes are specified, additional support must be given to the mold pin. Pins supported in the mold at both ends are molded satisfactorily when their length does not exceed $2\frac{1}{2}$ times their diameter. If longer holes are necessary, transfer or injection molding may be

(A)

(B)

FIG. 7.39 The simplest and least costly design for rectangular pins is shown at (A). The design at (B) calls for a milled pin shape that will add considerable expense if many pins are required.

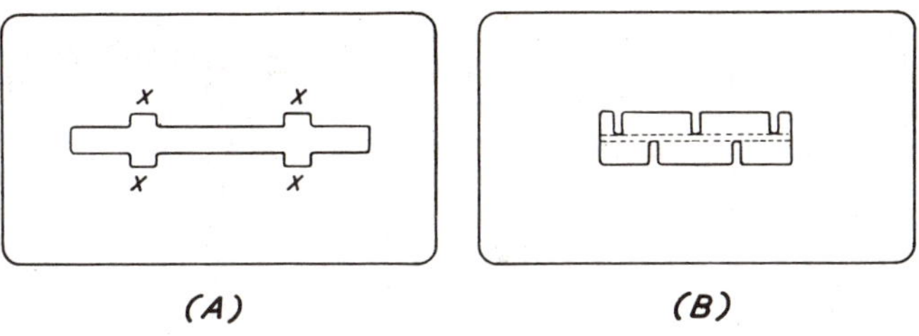

FIG. 7.40 (A) Long, narrow slots require thin mold pins, which are fragile. Addition of material at x will strengthen the pin forming the slot and still permit use of slot in most applications. Note that in each case the hole outlines the shape of mold pin to be used. (B) When extremely narrow slots are required, it is better to use a large opening with projections to form slot edges. The dotted line shows that a metal stamping is to pass through slot in plastics molding. Because strip is too thin to permit a molded slot to be used, the hole will be made ample in size and will contain internal projections for support.

used with success. Large side or cross holes may be molded in the part by the use of the set-back construction shown in Fig. 7.41. Half of this hole is produced by the mold cavity and half by the mold force. This is a convenient production method that can be used to great advantage when design changes or restyling are permissible.

When plastic materials are forced to flow around a pin in a constricted area, they may not knit well on the far side of the hole, Fig. 7.42. The molded pieces may look all right at first, but cracks are liable to develop in these knit lines. Products of this type must have openings uniformly spaced and as far

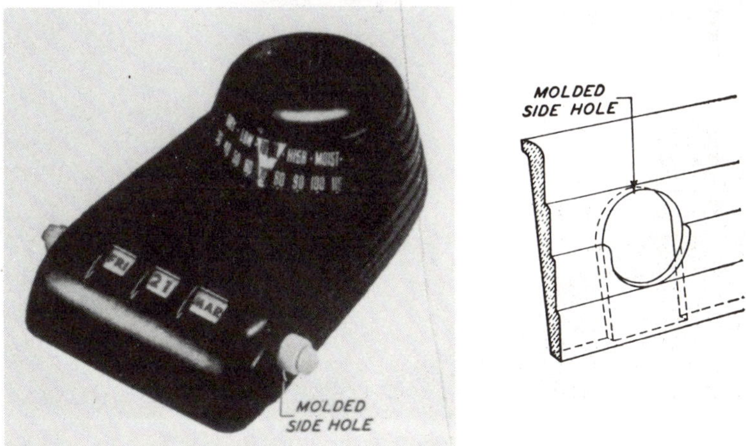

FIG. 7.41 Side hole molded without side pins. Half of hole is produced by mold plunger and half by mold force. (*Courtesy Autopoint Co., Chicago, Ill.*)

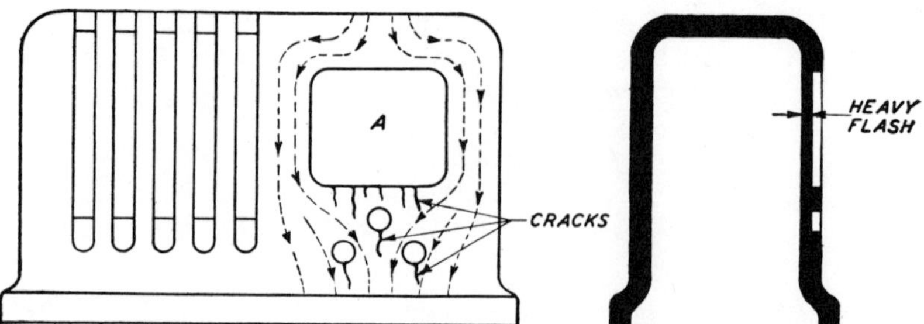

FIG. 7.42 Radio cabinet molded with side cores to produce opening A and holes below. Material flow is indicated by broken lines. A weak section will form below these openings where divided flow meets on far side of pin. This often causes cracking to occur during aging, because compound fails to "knit" compactly. To avoid this, a heavy flash may be used in openings so compound will flow over pin as well as around it.

apart as possible to permit adequate material flow. It may be necessary to leave a heavy flash in large holes to remove the knit line and minimize subsequent aftercracking.

Many of the thermoplastic and laminated phenolic materials will withstand punching operations. Designers frequently take advantage of this and lay out the side holes for punching after molding, as shown in Fig. 7.43. Louvres and similar side openings are produced by mold construction such as that shown in Figs. 7.43 and 7.44.

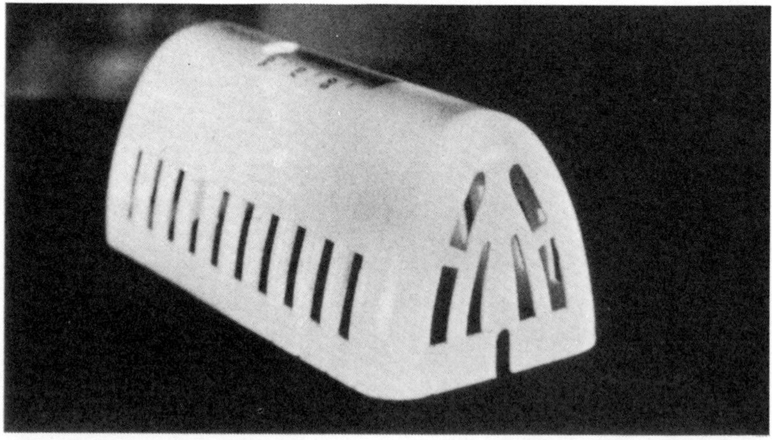

FIG. 7.43 Thermostat housing molded from thermoplastics. The end slots were punched after molding to simplify mold construction and eliminate necessity for "side pull" in mold.

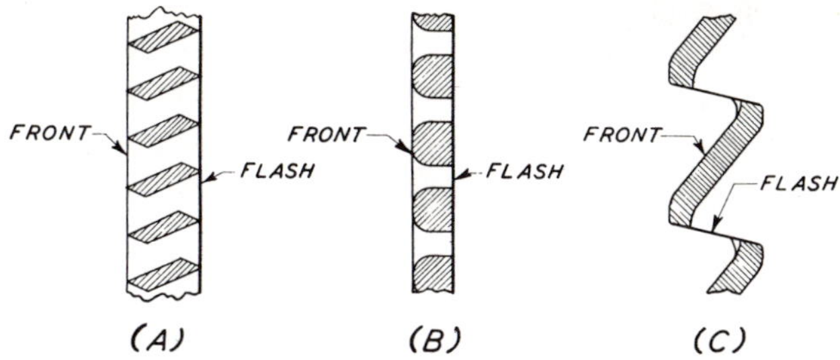

FIG. 7.44 Simple louvres are produced as shown at (A). The angle should be less than 10°. The louvres shown at (B) are satisfactory since narrow openings prevent flash line from showing. In design (C), flash lines are hidden from external view. (*Courtesy Chicago Molded Products Co., Chicago, Ill.*)

Molded Threads

Internal and external threads may be molded on plastics products. The molded piece must be unscrewed from the mold member or vice versa. It is frequently less costly to mold in place an insert having the required thread, or to tap a molded hole. Shallow threads, Fig. 7.45, may be molded and stripped from the mold without unscrewing. Such threads are frequently used on cosmetic packages. The Glass Container Association of America has adopted a special round profile thread for closures, Fig. 7.46. This design is rugged and compensates for dimensional variables in the molded glass threads. The conventional V-type thread is generally used for mechanical assemblies. Thirty-two threads per in. is the maximum for production jobs. Close-fitting threaded parts are to be avoided on products molded from materials that are poor in dimensional stability. A Class 2 fit is the best that can be expected in routine production with the better materials. Threads should be gauged according to specifications given in the *National Bureau of Standards Handbook H-28*. Mold designers must also make sure that the design provides a grip or slot by means of which the part may be unscrewed from the mold section, Fig. 7.47. Flats or straight knurls are often added to round surfaces for this purpose, as may be seen in Fig. 7.48. For products requiring a flat surface, holes may be inserted to permit unscrewing with a spanner wrench.

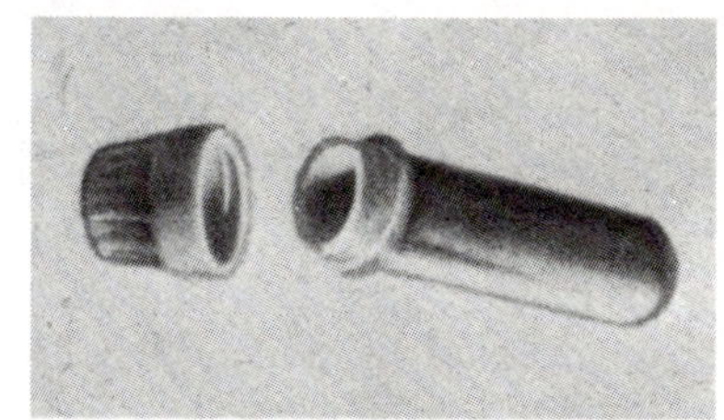

FIG. 7.45 Very shallow threads may be stripped from mold without unscrewing. The pieces shown were molded from urea formaldehyde compound.

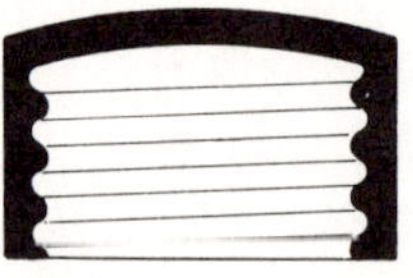

FIG. 7.46 Round profile thread used on closures for fitting molded glass threads.

Long threads are undesirable for products that are to be molded from material having a high shrinkage. A result of the shrinkage of such material is that the thread will have less than standard pitch and therefore will not fit a long piece on which the thread is standard. All long threads should be eliminated in favor of tapped holes, cut threads or inserts.

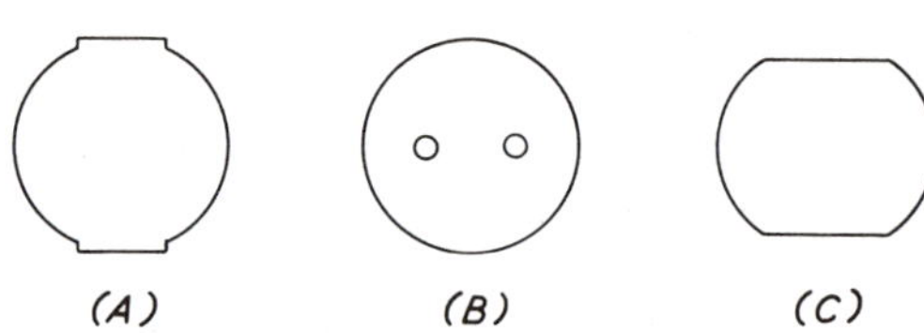

FIG. 7.47 Preferred types of grips or "holds" for unscrewing a round part from a mold. The shape shown at (A) can be machined easily as the diameter may be turned and the flats milled on the sides of hole. The holes shown in (B) provide slots for unscrewing with a spanner wrench. Irregular shapes, such as that shown at (C), are to be avoided as they increase mold construction cost.

External and internal threads are often produced in the mold by means of a thread ring mold section which may or may not come out of the mold with the piece, Fig. 7.49. The part can be unscrewed from the ring. Frequently, two sets of rings are used so that the mold may be reloaded while the previously molded part is unscrewed from the ring. A slight delay in unscrewing an external

FIG. 7.48 Typical unscrewing grips used to remove pieces from mold sections.

FIG. 7.49 The mold section shown at the left forms thread in molded piece. Piece is unscrewed from section to remove it from mold.

thread is advantageous, since the material shrinkage frees the part from the mold. Many external threads are unscrewed from the mold section while it is in the press.

Threads are often cut in each half of the mold when they are perpendicular to the parting line. The thread may be cleaned of flash and some eccentricity corrected by running the pieces through a sizing die. For many jobs this type of thread is not altogether satisfactory because the mold wear will increase the thickness of the parting line and the thread diameter will become eccentric.

Molded threads must not run out to a sharp edge because the resulting "feather-edge" thread end would be weak and the sharp, thin steel mold section at this point would be fragile and impossible to maintain. Molded threads must be started and ended as shown in Fig. 7.50. When the molded external thread must enter a tapped piece and meet at a shoulder, as shown in Fig. 7.51, a depression is provided at the end of the molded thread to permit a proper thread ending. This construction provides a strong mold section.

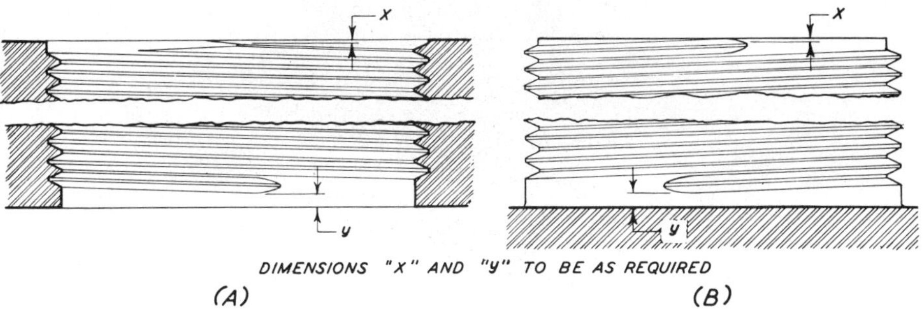

FIG. 7.50 Method of starting and ending threads on a mold section to produce a right-hand thread. (A) Molded external thread; (B) molded internal thread.

FIG. 7.51 When a molded external thread must enter a tapped or cut inside thread and engage at shoulder, a depression is provided at base of thread to permit a proper thread ending, thus eliminating the "feather edge" that would result if thread were permitted to run out at shoulder.

Internal threads are often produced by loose mold pins which come out of the mold with the part and are then unscrewed. Round pieces which have no locking projections may be unscrewed directly from a center mold pin. Internal threaded parts must be unscrewed immediately after molding, otherwise shrinkage tightens the part on the mold section, making release difficult. Mold designs involving internal threads must permit quick removal of the piece, and the number of cavities must be kept small. Multiple cavity molds are operated effectively with unscrewing fixtures designed to unscrew all pins at once while the pieces are still held by one mold member.

Inserts and Fasteners

Inserts are used extensively in plastics products to provide anchorage, bearings and shafts, internal or external threaded sections, thermal or electrical conductivity, terminals, arc resistance, nameplates, positioning studs, reinforcement, hinging, magnetic laminations, and for many functional and decorative purposes. (Typical inserts are shown in Figs. 7.52, 7.53, 7.54 and 7.55.)

FIG. 7.52 The capacitor cover shown above was produced by molding plastics section with central contact inserts in opening provided in stamped metal cover. Special treatment is needed to get a gas-tight seal.

FIG. 7.53 Inserts which use a knurl to improve bond to plastics.

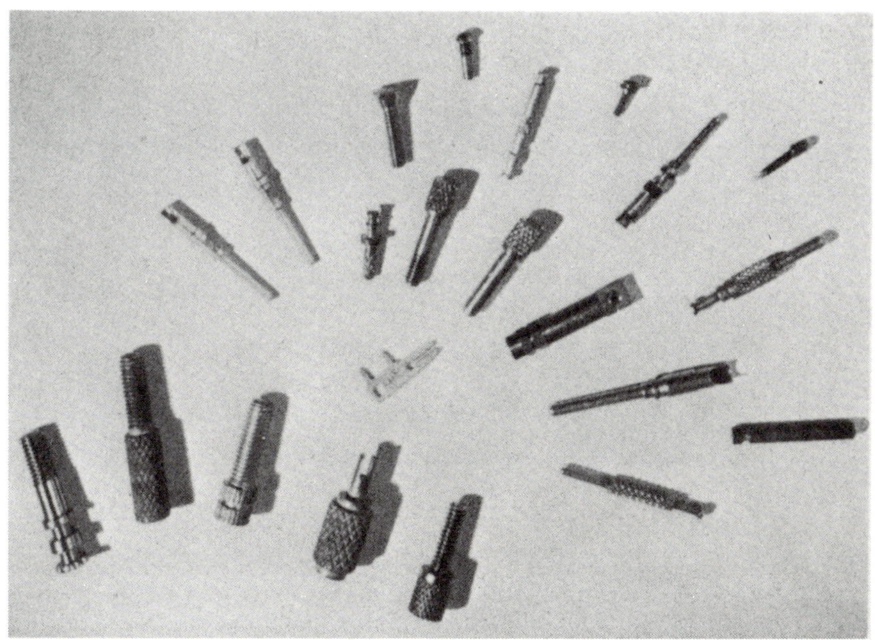

FIG. 7.54 Stud-type inserts commonly used in plastics products.

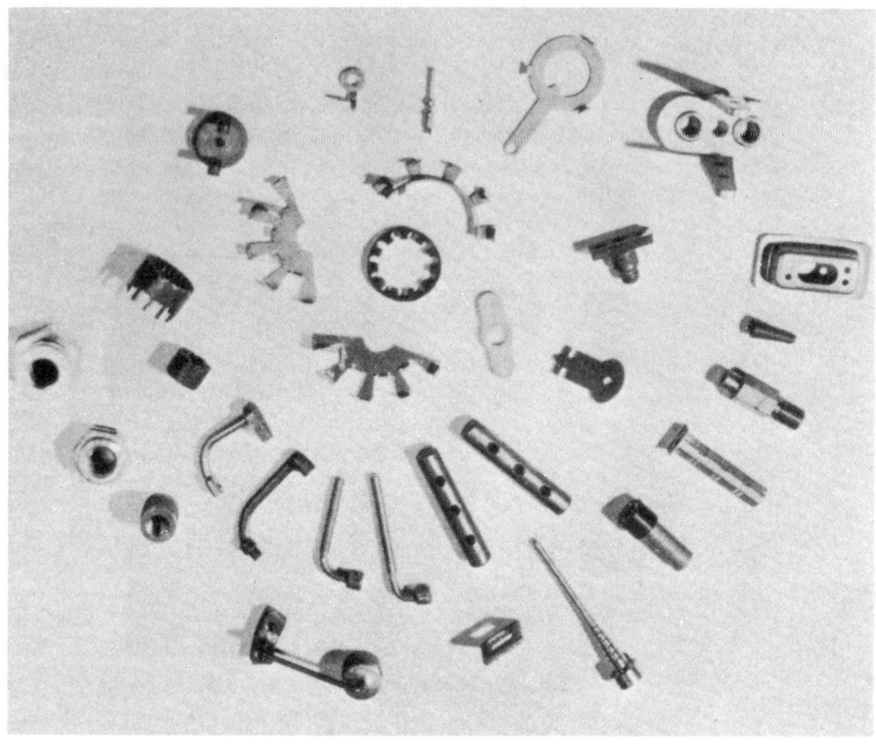

FIG. 7.55 Special-purpose molded-in type inserts for plastics.

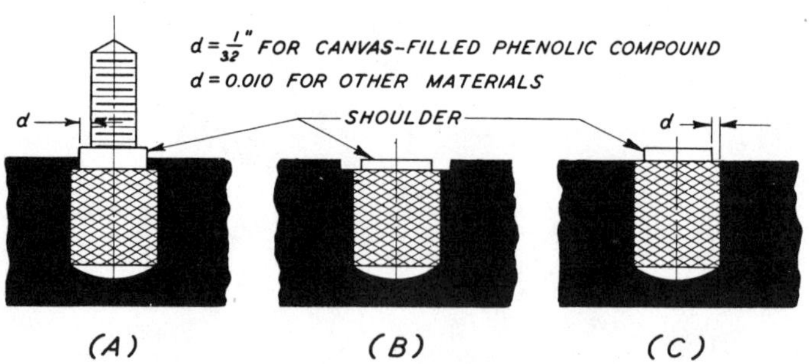

FIG. 7.56 Projecting stud-type insert (A) uses a shoulder to keep compound from flowing into threaded portion. Female inserts (B, C) frequently use a shoulder which enters mold pin, thus preventing compound from flowing into thread. When insert must be flush with the surface, design (B) is used.

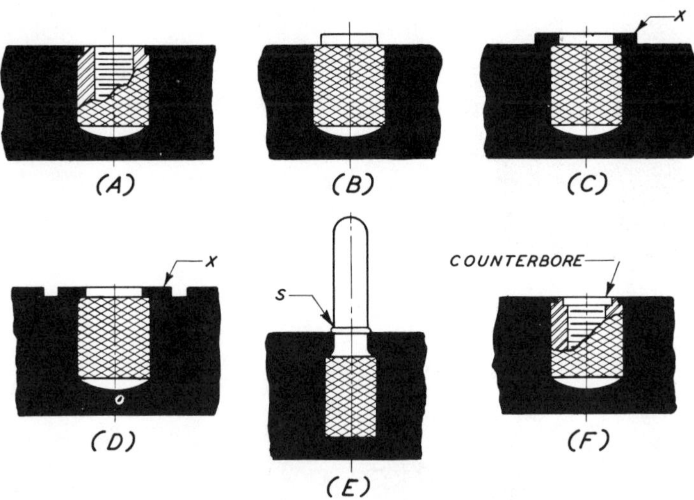

FIG. 7.57 Several methods which prevent material from filling thread. Compound flows easily over end of insert (A) into threaded portion and insert must be retapped after molding. Insert shoulder enters mold pin (B) to prevent compound from flowing into insert. (This is the best construction.) Compound hardens quickly at thin sections "x" (C, D) stopping flow of compound over end of inserts and into thread. Squeeze ring "s" is crushed (E) when rod type insert is driven into mold insert pin thus sealing compound from exterior surface of insert. Internal counterbore (F) also prevents compound from entering thread. The mold pin fits the counterbore and inside diameter of the threaded section.

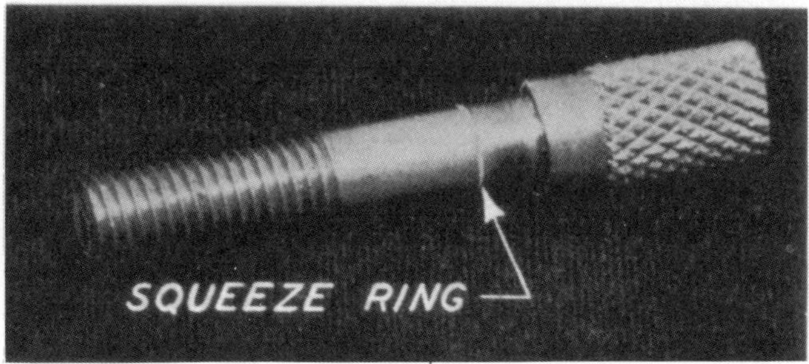

FIG. 7.58 Squeeze ring near center of insert is crushed to form a tight seal when the insert is driven into the mold pin. This prevents flash from flowing past the ring and into the thread. Ring also holds insert in mold member when loaded in upper half of mold.

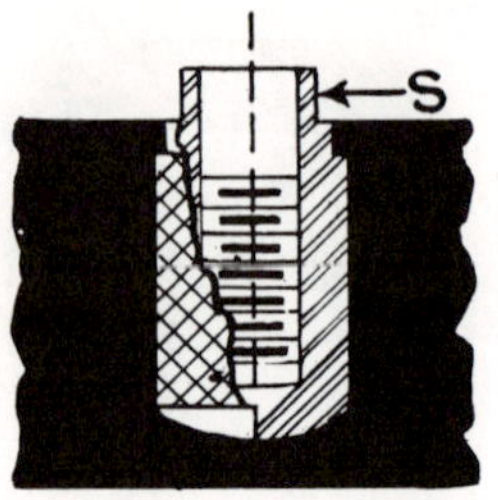

FIG. 7.59 Rivet-type insert for fastening contact strips and other assembly parts to molded inserts. Shoulder "s" is spun down on strip leaving tapped insert available for wire attachments.

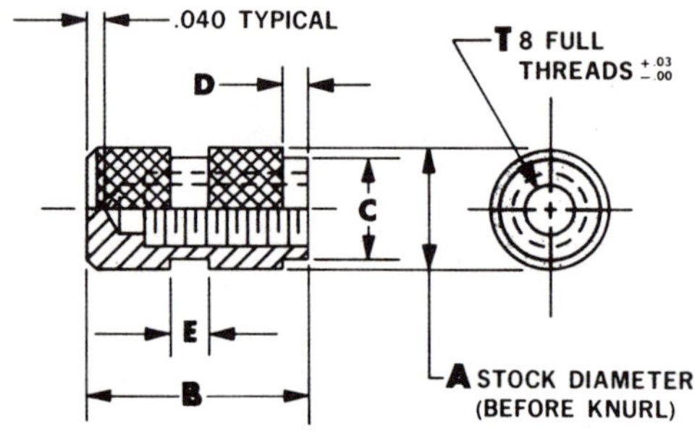

TAP DRILL	KNURL PITCH (TPI)	A	B	C	D	E
50 (.070) 49 (.073)	50	.156	.250	.142	.031	.031
47 (.078) 45 (.082)	50	.188	.281	.171	.031	.031
43 (.089) 42 (.093)	40	.188	.344	.171	.031	.031
38 (.101) 37 (.104)	40	.219	.344	.202	.031	.031
35 (.110) 33 (.113)	30	.219	.406	.202	.062	.031
29 (.136) 28 (.140)	30	.250	.406	.226	.062	.046
25 (.149) 21 (.159)	30	.281	.531	.259	.062	.046
16 (.177) 14 (.182)	30	.312	.531	.288	.062	.046
7 (.201) 3 (.213)	20	.344	.625	.321	.094	.062
F (.257) I (.272)	20	.438	.719	.404	.094	.078
5/16 (.312) Q (.332)	20	.500	.812	.466	.094	.094

FIG. 7.60 Standard general-purpose female insert design for plastics materials. Because it is designed for flush molding no sealing ring is provided. (*Courtesy Standard Insert Co., Pacoima, Calif.*)

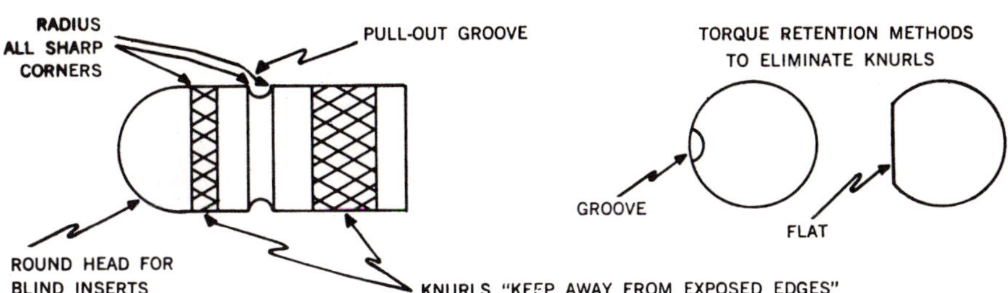

FIG. 7.61 Recommended form of a blind insert for thermoplastics. All sharp corners must be removed for high voltage application inserts.

Steel, brass, copper, gold, silver, aluminum, laminated plastics, fiber, glass, ceramics, glass-bonded mica, and many other materials are used to make inserts. Well designed inserts are easy to load and readily come free when a part is removed from the mold. Tolerances must be tight to keep molding compound out of threads yet there must be sufficient compound around them to prevent cracking, since the insert load is carried by adjacent compound. Anchorage is provided by roughened surfaces such as the diamond knurl, annular grooves in a hexagonal body, holes, slots, swaged sections, etc. (Figs. 7.56 to 7.61).

Radio tube pins, jacks, terminals, and contacts are often produced by a multiple swage process in ¼ in. to 1½ inch lengths. These drawn pins (Fig. 7.62) are low in cost and can be molded or assembled automatically in finished parts. When molded in, the pin may be flared out to provide anchorage and sealing by a suitable mold design. These pins are used extensively for assembly in fabricated parts and printed circuits.

The fully automatic molding methods used for high volume molded parts has greatly decreased the use of molded-in inserts. Several types of inserts have been developed for post-molding assembly in all plastics materials. Many conventional and special inserts are pressed in thermosetting products after molding; these

FIG. 7.62 Multi-swage terminals which may be assembled by any of the usual methods. (*The Bead Chain Mfg. Co., Bridgeport, Conn.*)

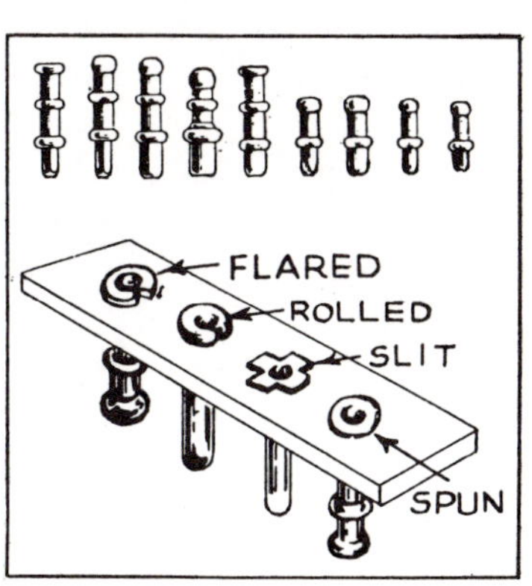

CONE-SPREAD INSERT

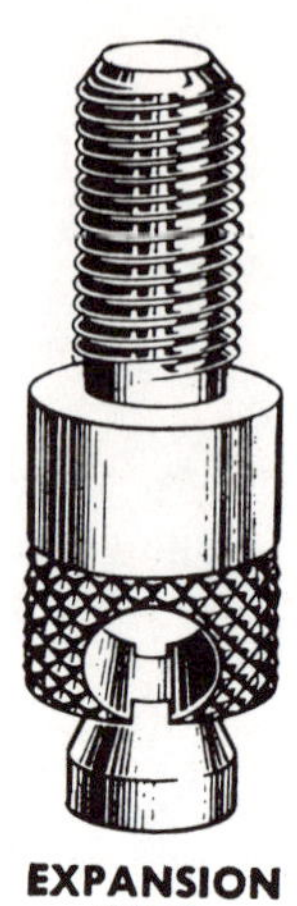

**EXPANSION
STUD**

**STANDARD
EXPANSION INSERT**

FIG. 7.63

materials are often slightly pliable when first ejected from the mold, and inserts with good holding power may be pressed in place with fixtures at minimum cost. Vertical external splines or a reverse external thread are used in such inserts to gain torque resistance.

A variety of special inserts has been developed for post-molding insertion in the thermoplastics as shown in Figs. 7.63, 7.64, 7.65, 7.66, 7.67, 7.68, 7.69 and 7.70.

Ultrasonic tools are extensively used for assembling inserts such as shown in Fig. 7.70 in the thermoplastics because this system minimizes boss crazing resulting from residual stresses and is readily adaptable to high volume automatic production.

Since molded-in inserts greatly increase costs, volume products should be designed without them. Usually it is possible to work out a design that will facilitate use of inserts that are automatically fastened in place after molding.

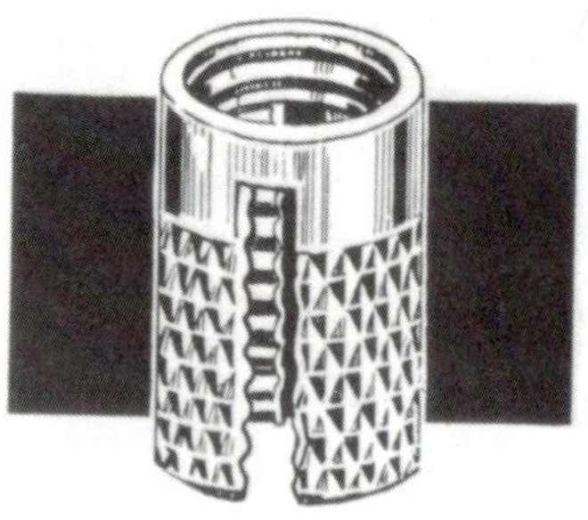

FIG. 7.64 Diamond knurled insert is designed to be pressed into molded parts after molding. (*Courtesy Boots Aircraft Nut Div., Norwalk, Conn.*)

FIG. 7.65 These inserts are pressed into molded or drilled holes; the collapsing end locks the piece in place. (*Courtesy Fastener Products, Inc., Southport, Conn.*)

Epoxy adhesives provide an excellent means of fastening to thermosetting materials and there are solvent-type adhesives that will bond to most of the thermoplastics.

When multiple contact inserts are used it is advantageous to stamp them from a single sheet with "ties" that maintain their relation to each other during molding operations (Fig. 7.71).

Use of surface-mounted inserts that are loaded in place during product assembly is economical. For example, a square nut can be placed in a square recess in the molded piece to provide an internal thread. Dovetail grooves will hold thin metal stampings.

Electroformed metal inserts can be produced with infinite detail and with tolerances of plus or minus two microns for precision screens and circuit.

Stamped spring steel clips (Fig. 7.72) that bite into plastic lugs are inexpensive to install, and useful in fastening components to molded products.

FIG. 7.66 Slotted insert is self-tapping and self-locking. (*Courtesy Groove-Pin Corp., Ridgefield, N.J.*)

FIG. 7.67 Very hard materials such as the laminates, thermosets and ceramo-plastics gain maximum thread strength by the installation of precision formed screw thread coils. The coils have a tang that is used for installation and then broken off. Installation is achieved by threading the coil into a tapped hole in the plastics product. (*Courtesy Heli-Coil Products Div., Danbury, Conn.*)

Self-tapping screws (Fig. 7.73) (see also Table 7.8) have excellent holding power, and are considered to be better than standard machine screws in tapped holes for single insertion applications. If the screw must be removed and re-inserted from time to time, a tapped hole or a metal insert is indicated.

Hinging plastics products is done by several methods. Where flexible plastics are usable, a self-hinged (integral hinge) product can be made. Figure 7.74 shows a variety of moldable hinges for inexpensive packages. An economical hinge which anchors itself when pressed into place is shown in Fig. 7.75.

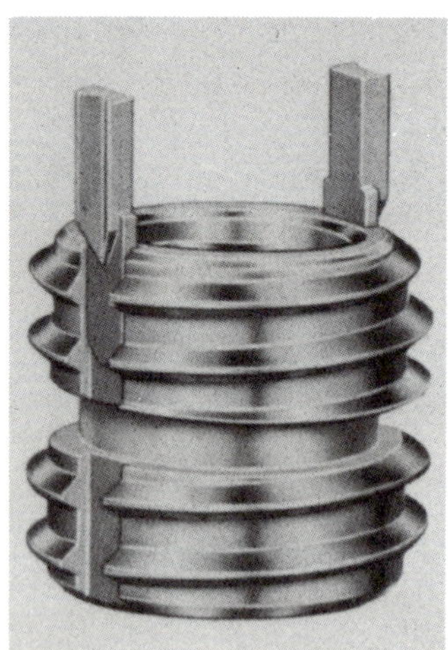

FIG. 7.68 Inserts of this type are available in several styles for thermoplastics and other soft materials. Rotational resistance is achieved by driving the keys down in to the threads of plastics after insertion. (*Courtesy Tridair Industries, Torrance, Calif.*)

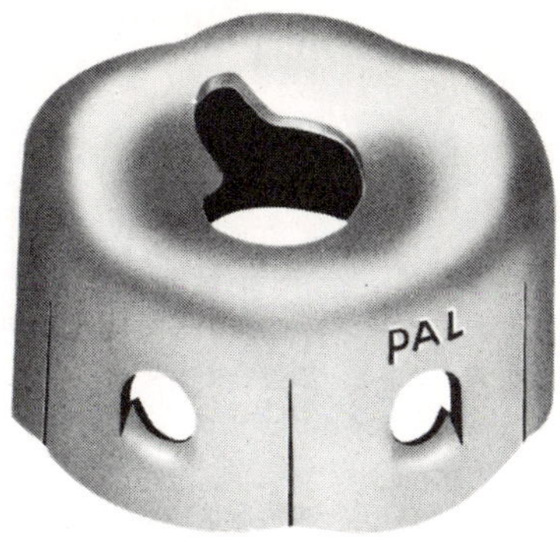

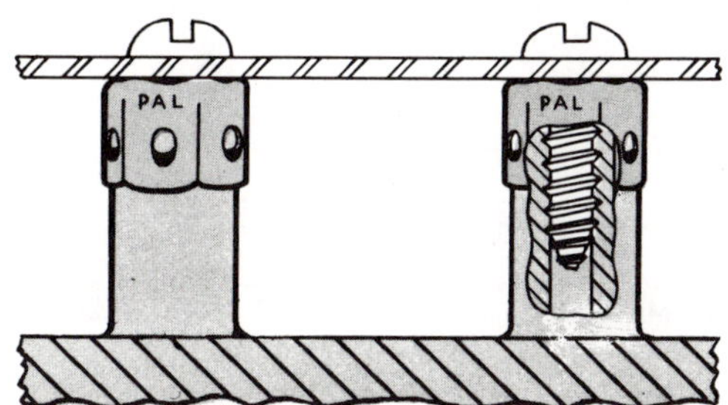

FIG. 7.69 Metal attachment devices that can be fastened to bosses are easily assembled on the plastics product. (*Courtesy The Palnut Co.*, Mountainside, *N.J.*)

The Integral Hinge

The unique flexibility and almost unlimited flex life of molded polypropylene, combined with the basic rigidity of the material, permits the molding of products with a new and highly functional feature—the molded-in hinge. Integral hinges are already finding wide commercial application in container lids, accelerator pedals, clamps, floor mops and many other products. They are now readily fabricated, without need for assembly, and at minimum cost of several thermoplastics.

With the few basic design principles outlined below, built-in hinges can be made to withstand well over a million flexes in service.

Land Length to Thickness Ratio at Least 3 to 1. Land length of the molded hinge should be at least three times its cross-sectional thickness. Figure 7.76 shows a container with its lid joined to the base by an integral hinge and the gate considerations are illustrated by Fig. 7.77. All corners of the hinge should be

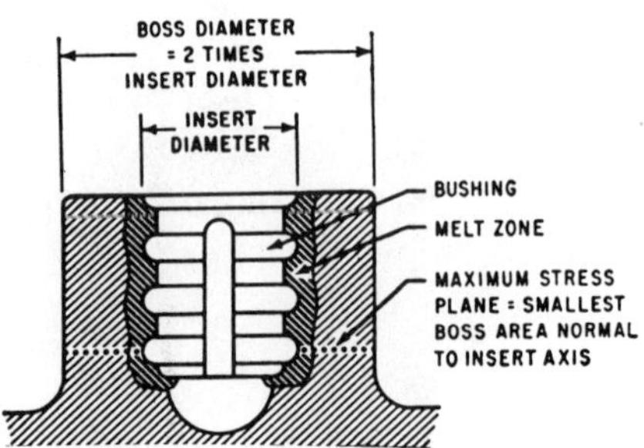

Design and performance

The diagram and table provide design and performance data on a type of bushing,* made by Cavitron Ultrasonics Inc., which is designed for ultrasonic insertion.

*Patented

Insert thread size	4-40	6-32	8-32	10-32
Hole diameter (in.)	.147	.180	.199	.234
Min. depth (in.)	.500	.500	.500	.500
Stripping torque (in. lbs.)	35	45	60	75
Pull-out force (lbs.)	340	400	420	500
Max. torque for boss stress plane (in. lbs.)	5	8	14	24
Insertion time (Sec.)	1	1.5	2.0	2.5

FIG. 7.70 Inserts of this type are especially designed for ultrasonic fastening. Pressure from the ultrasonic horn pushes the insert into position and the ultrasonic vibrations melt a thin film of resin at the interface. This action does not heat the insert so quick cooling leaves the insert firmly bonded.

The diagram and table provide design and performance data on a type of bushing,* made by Cavitron Ultrasonics Inc., which is designed for ultrasonic insertion.

smooth and rounded to gain maximum strength with a minimum radius of 0.010 inch. If the ratio is less than 3:1, knit or weld lines are likely to occur in the hinge area of the product, flexing will create considerable stress and flex life of the hinge will be drastically reduced.

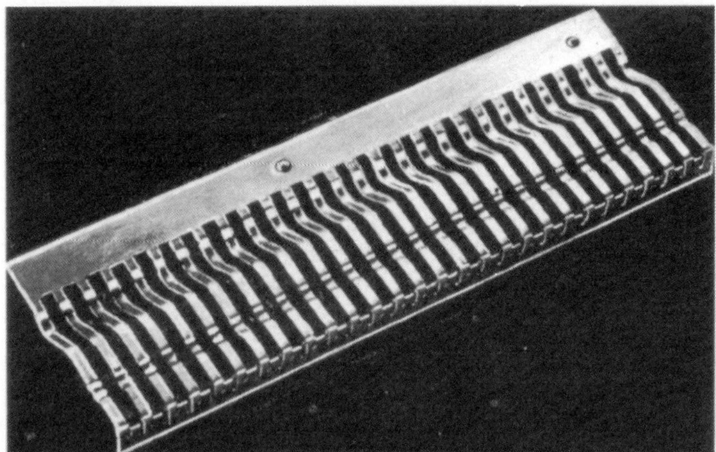

FIG. 7.71 Multiple contact insert. Note premolding ties that hold contacts accurately in position before molding. Ties are scored so as to break clean after the molded parts are cleaned. (See Fig. 7.4.)

When the land length to thickness ratio is less than the recommended 3:1, back pressure generated during mold filling will not be great enough to insure even flow of the polymer through the hinge area. The polymer flow pattern will be uneven across the hinge. (Similar irregular flow patterns are also caused by minute imperfections in the mold surface in the hinge area.)

When land length is too small (or there are imperfections in the mold surface in the hinge area), the molten polymer will tend to enter the hinge area at several points instead of flowing evenly across. This causes knit or weld lines

FIG. 7.72 This fastener is used extensively for making attachments to molded lugs on plastics products. (*Courtesy Tinnerman Products, Inc.*)

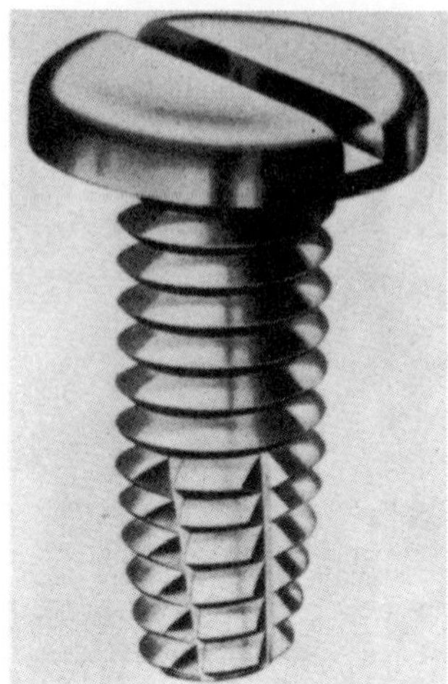

FIG. 7.73 Self-tapping screws have excellent holding power and are considered to be superior for single insertion applications in plastics products.

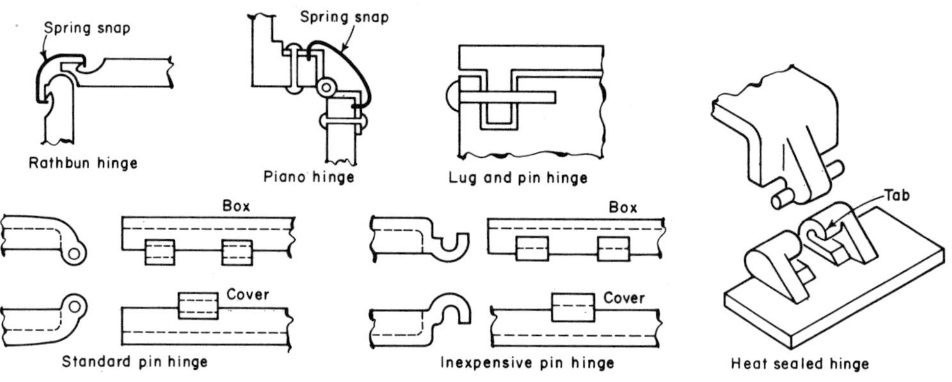

FIG. 7.74 Some of the better hinging procedures are shown here for the rigid materials that cannot use the integral hinge.

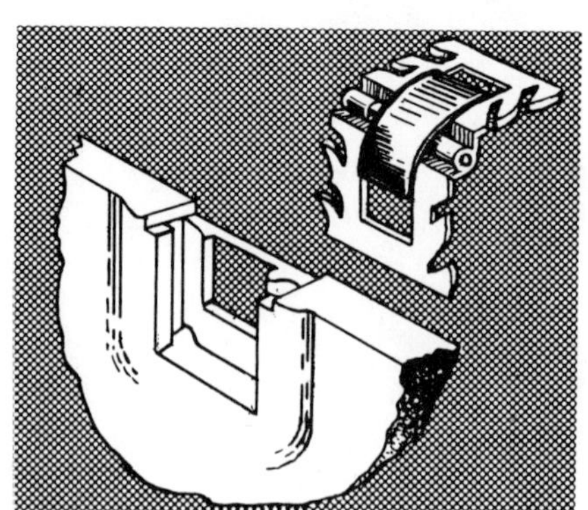

FIG. 7.75 Serrated-edge hinge bites into molded slots.

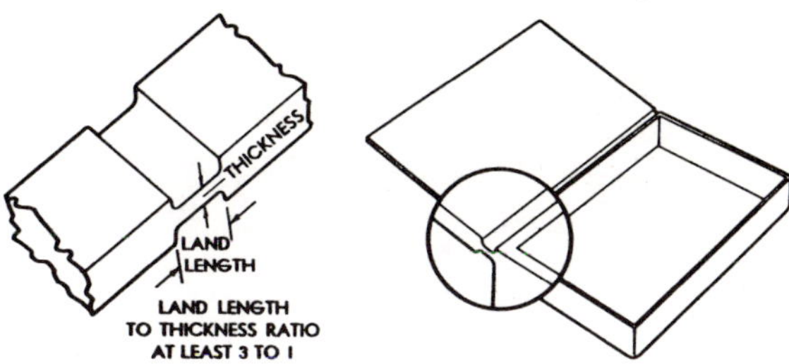

FIG. 7.76

along the downstream edge of the hinge and results in rapid breakdown of the hinge in service. Proper proportions are essential.

As a general rule, the hinge land length should always be at least three times its cross-sectional thickness, and preferably greater. If land length is increased at a constant cross-sectional thickness, back pressure generated during polymer flow will naturally increase. A ratio of less than 3:1 will almost always create a problem in obtaining the proper flow pattern, but there is also a limit to which the land length to thickness ratio can be increased beyond 3:1. Beyond a certain limit, depending on product design, it becomes difficult to fill the section of the mold across the hinge on the opposite side from the gate (in a one-gate mold). It may even be difficult to fill the hinge area itself.

Lettering on Molded Products

Molded letters, figures, and decorative designs are often used on molded parts. Lettering which must be applied perpendicular to the parting line or be put on side walls must provide sufficient draft for the letters to "draw" out of the mold. Letters may be raised or depressed, or they may face up from a depressed panel, as in Fig. 7.78. If the mold cavities are to be hobbed, the letters should be depressed. When cavities are to be machined, raised letters may be had at lowest cost. Depressed lettering requires raised mold letters, which are easily damaged and may be costly to maintain. These depressed letters are produced by forming raised letters in the mold. After the letters are formed the surrounding steel is

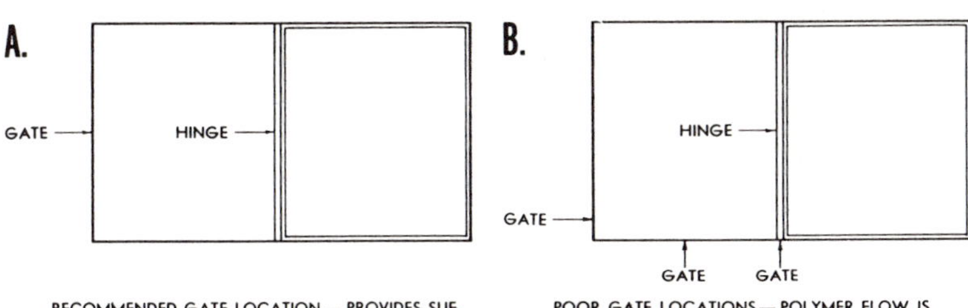

FIG. 7.77

FIG. 7.78   Showing how raised lettering may be used in a depressed panel. This construction is used when lettering is put on a separate block of steel to be inserted in the mold.

cut away. Raised letters are produced by stamping or engraving the letters on the mold surface. When mold cavities are hobbed, the letters are stamped or engraved in the hob. The hob forms raised letters in the mold cavity and, in this way, the molded part becomes a replica of the hob and contains depressed letters. These effects are easily achieved also by electrical discharge machining.

Designers usually call for depressed letters when paint is to be applied for better visibility. It is often possible to form depressed letters in machined cavities without cutting away the entire mold surface. This is accomplished by using a lettered insert panel which is set in the mold at the desired point. It is desirable to elevate the panel from surrounding surfaces when possible. Frequently, paint is wiped into the fin line during the painting operation, but by using a raised panel the paint is kept away from the fin lines. Lettering which may require change should be placed on removable plugs in the original mold construction to facilitate removal.

Small raised letters are generally 0.008 to 0.015 inch high. Larger letters may be $1/32$ to $1/64$ inch high. In all cases it is desirable to use considerable draft on the sides of letters, and to have a good fillet where they join the molded piece. Failure to observe this practice will produce letters that will be fragile and consequently liable to break off in the mold. Letters that are to be painted should have no greater width than $1/32$ inch. Wide depressed letters and lines that are to be painted should be avoided if possible, as paint will wipe out when the excess is removed from the surface. Wide letters are frequently simulated by forming them with a series of parallel lines.

It is often necessary to add letters to hardened molds. This can be done by etching the surface with acid, although this work is crude in appearance. Some moldmakers make use of a photoetching process. This process produces attractive work, and is convenient when special type is to be used, or when a considerable amount of lettering is required. Such letters are not deep, but are readily visible when contrasted with a highly polished mold background. Highly polished or

FIG. 7.79 This acrylic cardbox has three dimensional lettering molded in the underside of the lid and filled with paint.

dull backgrounds are commonly used to make molded letters stand out clearly without being painted.

The transparent thermoplastic materials offer unusual decorative possibilities. Lettering may be cut in on the underside of such products. Contrasting colors are used for the background and lettering to produce an attractive "third dimension" effect, as shown by Fig. 7.79.

Several types of metal inlays are available if distinctive lettering is desired. When such lettering is specified, the mold designer should get in touch with the vendor to get the details necessary for correct mold design. Rubber stamps are sometimes used for branding serial numbers or other special designation on molded products. Special inks which provide good adhesion are available for this work. The marking may be made more permanent if the details are given so the moldmaker can provide a grit-blasted mold surface for the marking area. Blasting should be on chrome plate.

Mold designers must consider carefully all designs which specify lettering. A considerable expenditure may be wasted in excess part cleaning or painting cost. Oftentimes the quality of the lettering makes the difference between an attractive or a poorly finished product.

Surface Finishes

The best finish for any mold is a highly polished surface protected by chromium plating. A highly polished mold surface minimizes sticking, improves flow condition in the mold and produces nicely finished pieces. Many special surface textures are demanded for decorative effect. The effectiveness of the finish depends on the ingenuity of the mold-maker. Dull or "sandblasted" finishes are often specified for the purpose of minimizing light reflection. These finishes are produced by grit blasting or acid etching. Surfaces treated in this manner will "polish up" in time, however, and because of this, better results may be obtained by grit blasting the chrome plate after the mold is finished and plated.

All special surface finishes should be limited to flat areas, or to sides which provide considerable draft.

A series of standard mold finish specimens has been prepared by SPE.

GLOSS

A highly polished mold will produce a high gloss finish when molding is done under the proper conditions. Where a very high gloss is a critical requirement, the mold should be polished by a specialist. Few moldmakers are prepared to provide the mold finish required for lenses, organ keys, scientific apparatuses, etc. High gloss is often achieved as a salvage operation by buffing, but it is very difficult to achieve product uniformity by buffing. Raw materials makers sometimes supply special high gloss compounds for critical applications.

Failure to produce a high gloss with a properly finished mold is the result of inadequate pressure, release of volatiles or moisture and differential thermal contraction of the elements of the compound. Proper filling of the cavity is essential to insure uniformity of gloss.

METAL PLATED PLASTICS

A very few plastics plating applications are predicated on electrical, chemical or permeability requirements—most are for decorative purposes. Plating amplifies appearance defects because of the higher light reflectivity, and this requires extra care. Flat areas should be avoided and a surface crowning with a curvature of 0.015 in./in. is very desirable.

Parting lines, gate and vent areas, etc., must be located on non-appearance areas. Sharp surface corners will build up plating problems as shown in Fig. 7.80, and are to be avoided. Uniform wall section is high desirable with a gradual change from thin to heavier areas.

Many plastics formulations are platable and these processes are described in Chapter 6. The same design considerations and precautions must be observed for either process. Chrome plated plastics are used in applications subject to handling. The vacuum deposition process may be used only for ornamental and decorative areas that will be protected or that have short life expectancy.

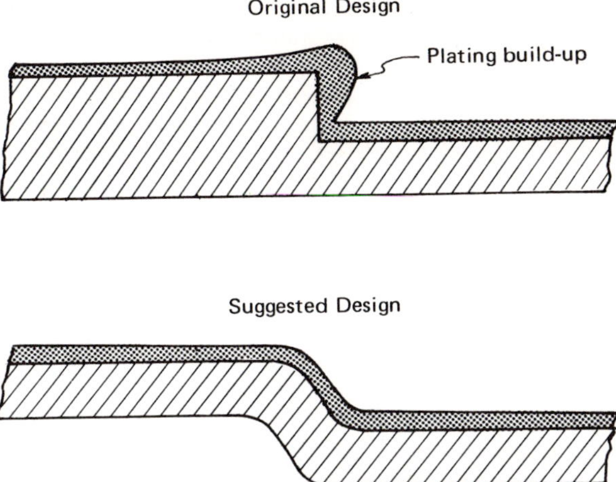

FIG. 7.80 A uniform wall thickness with gradual transitions from one wall section to another will enhance the plated part performance.

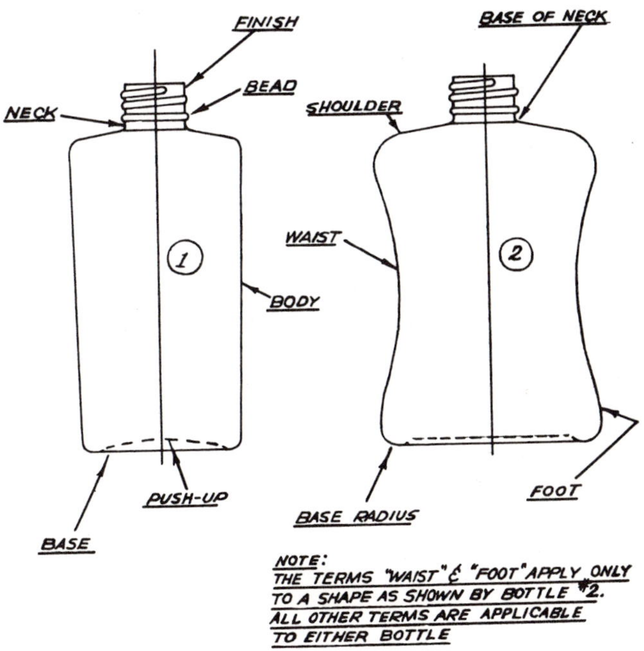

FIG. 7.81 Standard bottle-finish terminology.

TABLE 7.11 INSERT PERFORMANCE COMPARISON. (*COURTESY HELICOIL PRODUCTS DIV., DANBURY, CONN.*)

Fastening Method	Advantages	Disadvantages
Self-tapping screws	1. Least expensive 2. Fastest assembly	1. Doubtful holding power 2. Holding power loss after reassembly 3. Chipping and cracking unless tolerances are close 4. Threads easily damaged by improper reassembly
Molded or tapped threads with machine screws	1. Relatively inexpensive screws 2. Better holding power than self-tapping screws 3. Minimum stress in thread forming 4. Easy assembly	1. Drilling and tapping increase cost 2. Not as strong as molded-in or post-mold insert 3. Possible damage from improper screw entry 4. Molded-in threads adds to molding cost
Molded-in inserts	1. Very good holding power 2. Reassembly without holding power loss	1. Adds to molding time and cost 2. Insert missing or misplaced causes part to be rejected 3. Cleaning insert of flash increases cost 4. Post molding changes can affect holding power 5. Possibility of mold damage 6. Large inserts need preheating
Post-molding press-in or cement-in inserts	1. No interference with molding operation 2. No cleaning of flash 3. Simple installation without expensive tooling	1. Holding power less than molded-in inserts 2. Doubtful retention of press-in due to close tolerances needed and of cemented-in due to unclean surface 3. Press-in can crack material if tolerances are not held
Post-molding self-tapping inserts	1. Strong holding power 2. Good quality internal thread	1. Higher cost inserts 2. Slowest installation 3. Requires close tolerance control to avoid high driving torque or poor retention 4. Usually restricted to sizes No. 8 through $\frac{1}{2}''$

TABLE 7.11 (*Continued*)

Fastening Method	Advantages	Disadvantages
Post-molding inserts into tapped holes	1. Highest quality internal thread 2. Strongest holding power 3. Can use screw locking feature	1. Highest cost insert 2. Slow installation, with relatively expensive equipment needed for automatic installation
Post-molding expansion type inserts	1. Avoid all extra operation costs associated with molded-in inserts 2. Holding power adequate for most operations 3. Automatic installation at speeds to 60/min.	1. Automatic equipment required for fast installation 2. Strength exceeded by self-tapping post molding inserts
Post-molding inserts for thermoplastics	1. Avoids costs and problems of molded-in or tapped thread types 2. High strength, stress free 3. Useful with smaller bosses and thinner walls, and stress sensitive materials 4. Liberal tolerances 5. Fast installation in a variety of holes	1. Slightly higher initial cost than molded-in or expansion types 2. Requires special (but relatively low cost) equipment 3. Use restricted to thermoplastics

BOTTLE FINISH

The effectiveness of a plastics bottle as an ideal container depends on a seal so positive that it keeps the contents from escaping, yet permits easy opening and resealing. This seal is achieved in the great majority of bottles when the bottle cap is compatible with the bottle opening; that is, when the cap threads engage with the corresponding threads molded in the neck of the bottle.

Most advances in construction of bottle closures have come from standardization of these neck threads, or to use the technical glass term, bottle finish (so called because before automatic blowing, the mouth of the bottle was the last part finished). Through standardization of bottle finish, today's packagers, by specifying the finish number they need, are sure of effective seals, no matter how many manufacturing sources they draw upon.

Because the bottle closure field is one where rule-of-the-thumb methods and deviations will not produce effective seals, the following short discussion of bottle finish and terms is presented. The plastics industry has adopted the glass terminology for blown plastics bottles and the following definitions are commonly used: (Fig. 7.81)

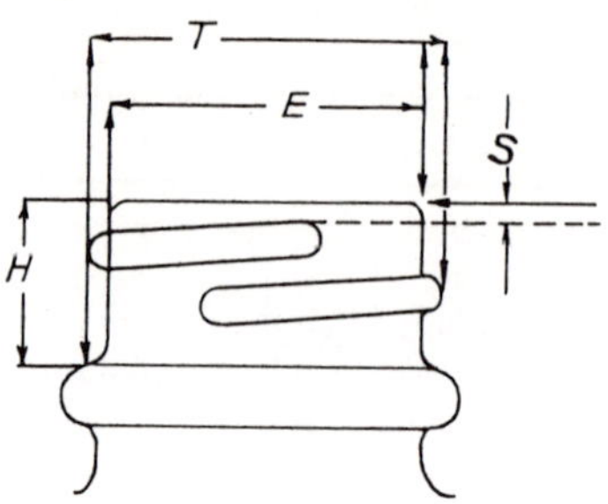

Size	T	E	H	S	Threads /in.	No. of Turns
18	.694±.010	.610±.010	.356+.008 −.007	.034+.013 −.012	8	1
20	.773±.010	.689±.010	.356+.008 −.007	.034+.013 −.012	8	1
22	.852±.010	.768±.010	.356+.008 −.007	.034+.013 −.012	8	1
24	.930±.010	.846±.010	.385+.008 −.007	.046+.016 −.015	8	1
28	1.075+.013 −.012	.981+.013 −.012	.385+.008 −.007	.046+.016 −.015	6	1
30	1.114+.013 −.012	1.020+.013 −.012	.388±.010	.046+.016 −.015	6	1
33	1.252+.013 −.012	1.158+.013 −.012	.388±.010	.046+.016 −.015	6	1
38	1.456±.020	1.360±.020	.388±.010	.046+.016 −.015	6	1
43	1.634±.020	1.540±.020	.388±.010	.046+.016 −.015	6	1
48	1.850±.020	1.756±.020	.388±.010	.046+.016 −.015	6	1
53	2.047±.020	1.953±.020	.393±.010	.046+.016 −.015	6	1
58	2.204±.020	2.110±.020	.393±.010	.046+.016 −.015	6	1
63	2.441±.020	2.347±.020	.393±.010	.046+.016 −.015	6	1
70	2.716±.020	2.622±.020	.408±.015	.046+.016 −.015	6	1
83	3.248±.020	3.128±.020	.472±.015	.063+.015 −.016	5	1
89	3.491±.020	3.371±.020	.520±.015	.063+.015 −.016	5	1
100	3.917±.020	3.797±.020	.570±.015	.063+.015 −.016	5	1
120	4.704±.020	4.584±.020	.670±.015	.063+.015 −.016	5	1

FIG. 7.82 Finish No. 400 (1 thread) (standard glass type thread).

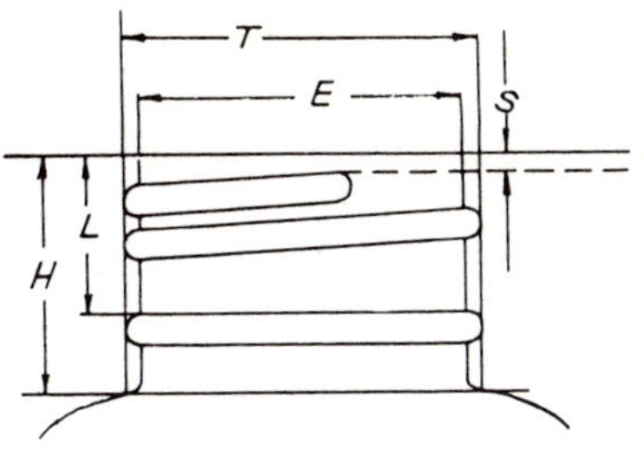

Size	T	E	H	S	L (min.)	Thread Turns
18	.694±.010	.610±.010	.508+.008 −.007	.034+.013 −.012	.345	1½
20	.773±.010	.689±.010	.539+.008 −.007	.034+.013 −.012	.345	1½
22	.852±.010	.768±.010	.570+.008 −.007	.034+.013 −.012	.360	1½
24	.930±.010	.846±.010	.631+.008 −.007	.046+.016 −.015	.421	2
28	1.075+.013 −.012	.981+.013 −.012	.693+.008 −.007	.046+.016 −.015	.405	1¼

FIG. 7.83 Finish No. 410 (1½ thread) (standard glass type thread).

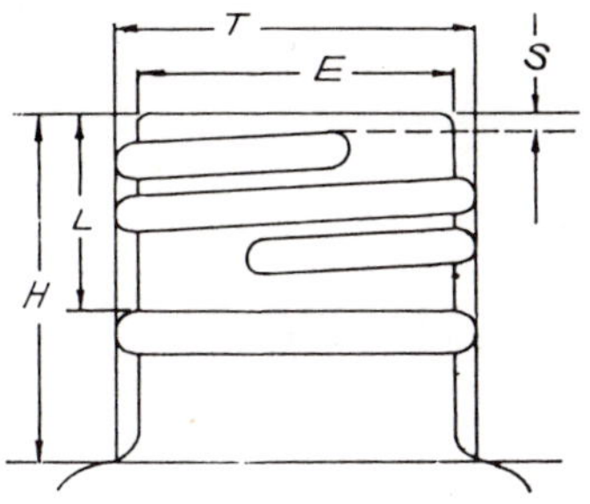

Size	T	E	H	S	L (min.)
13	.506+.008 −.007	.446+.008 −.007	.437+.008 −.007	.034+.013 −.012	.290
15	.573+.008 −.007	.513+.008 −.007	.542+.008 −.007	.034+.013 −.012	.332
18	.694±.010	.610±.010	.602+.008 −.007	.034+.013 −.012	.413
20	.773±.010	.689±.010	.727+.008 −.007	.034+.013 −.012	.440
22	.852±.010	.768±.010	.822+.008 −.007	.034+.013 −.012	.530
24	.930±.010	.846±.010	.942+.008 −.007	.046+.016 −.015	.545
28	1.075+.013 −.012	.981+.013 −.012	1.067+.008 −.007	.046+.016 −.015	.639

FIG. 7.84 Finish No. 415 (2 threads) (standard glass type thread).

Finish: The part of a bottle for holding the cap or closure. (Thus thread is part of finish)

Neck: The part of bottle between finish and shoulder.

Neck Ring: A metal mold part used to form the finish of a hollow glass or plastics container.

Shoulder: Not defined.

Base: The bottom of a bottle.

Heel Tap: An imperfection in which the base or bottom of a bottle is very thick in one area and very thin in another.

Bead: Any raised section extending around the article.

Cap Seat: The ledge inside the mouth of a milk-type bottle.

Most of the stock closures have been designed to take advantage of the standard glass bottle threads (Figs. 7.82–7.85). Standard glass finishes are numbered, such as Finish No. 400, which is a shallow continuous thread. Finish sizes are designated by millimeters of the outer diameter of the bottle or interior diameter of the closure.

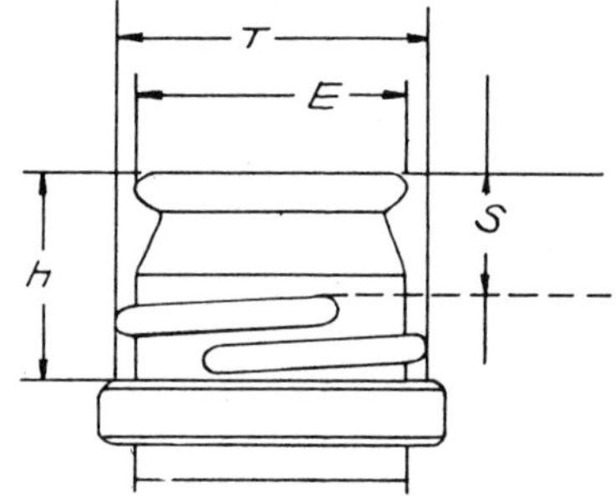

Size	T	E	H	S	L
18	.694±.010	.610±.010	.604±.010	.288+.008 −.007	9/32
20	.773±.010	.689±.010	.604±.010	.288+.008 −.007	9/32
22	.852±.010	.768±.010	.604±.010	.288+.008 −.007	9/32
24	.930±.010	.846±.010	.647±.010	.319+.008 −.007	5/16
28	1.075+.013 −.012	.981+.013 −.012	.724+.013 −.012	.351+.008 −.007	11/32
30	1.114+.013 −.012	1.020+.013 −.012	.760+.013 −.012	.377+.008 −.007	23/64
33	1.252+.013 −.012	1.158+.013 −.012	.775+.013 −.012	.377+.008 −.007	23/64
38	1.461±.015	1.367±.015	.946±.015	.552±.010	7/16

FIG. 7.85 Finish No. 430 (1 thread) (standard glass type thread).

TABLE 7.12   REPRESENTATIVE CLOSURE AND BOTTLE THREAD DIMENSIONS

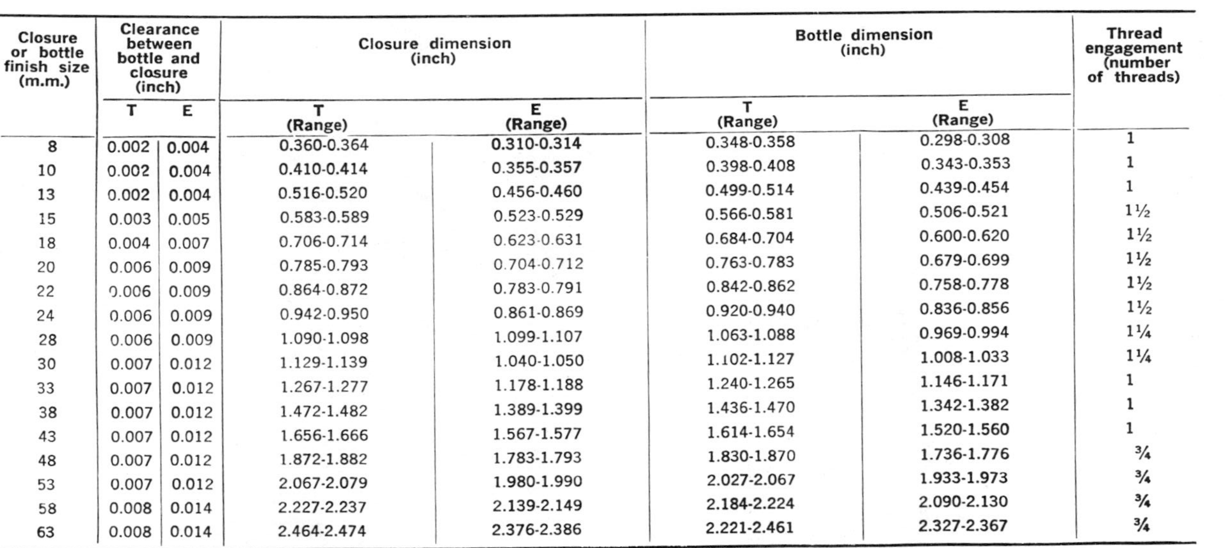

Closure or bottle finish size (m.m.)	Clearance between bottle and closure (inch)		Closure dimension (inch)		Bottle dimension (inch)		Thread engagement (number of threads)
	T	E	T (Range)	E (Range)	T (Range)	E (Range)	
8	0.002	0.004	0.360-0.364	0.310-0.314	0.348-0.358	0.298-0.308	1
10	0.002	0.004	0.410-0.414	0.355-0.357	0.398-0.408	0.343-0.353	1
13	0.002	0.004	0.516-0.520	0.456-0.460	0.499-0.514	0.439-0.454	1
15	0.003	0.005	0.583-0.589	0.523-0.529	0.566-0.581	0.506-0.521	1½
18	0.004	0.007	0.706-0.714	0.623-0.631	0.684-0.704	0.600-0.620	1½
20	0.006	0.009	0.785-0.793	0.704-0.712	0.763-0.783	0.679-0.699	1½
22	0.006	0.009	0.864-0.872	0.783-0.791	0.842-0.862	0.758-0.778	1½
24	0.006	0.009	0.942-0.950	0.861-0.869	0.920-0.940	0.836-0.856	1½
28	0.006	0.009	1.090-1.098	1.099-1.107	1.063-1.088	0.969-0.994	1¼
30	0.007	0.012	1.129-1.139	1.040-1.050	1.102-1.127	1.008-1.033	1¼
33	0.007	0.012	1.267-1.277	1.178-1.188	1.240-1.265	1.146-1.171	1
38	0.007	0.012	1.472-1.482	1.389-1.399	1.436-1.470	1.342-1.382	1
43	0.007	0.012	1.656-1.666	1.567-1.577	1.614-1.654	1.520-1.560	1
48	0.007	0.012	1.872-1.882	1.783-1.793	1.830-1.870	1.736-1.776	¾
53	0.007	0.012	2.067-2.079	1.980-1.990	2.027-2.067	1.933-1.973	¾
58	0.008	0.014	2.227-2.237	2.139-2.149	2.184-2.224	2.090-2.130	¾
63	0.008	0.014	2.464-2.474	2.376-2.386	2.221-2.461	2.327-2.367	¾

(Courtesy M. A. Delia Associates, Berkeley Heights, N.J.)

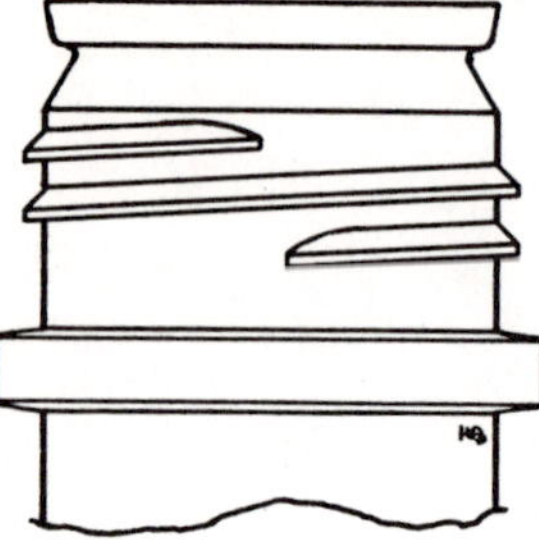

FIG. 7.86 Finish No. 415 with buttress-type thread for flexible bottles. This thread was designed to give improved holding of caps on plastics bottles.

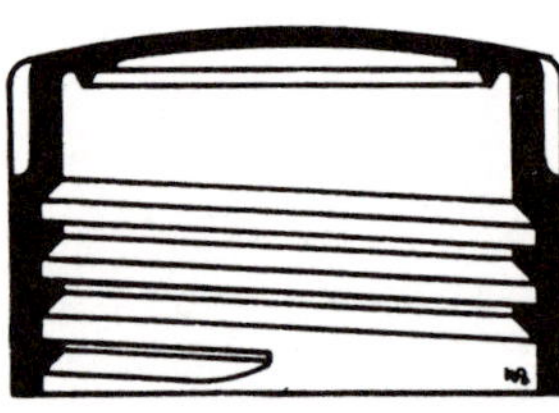

FIG. 7.87 Closure for buttress thread.

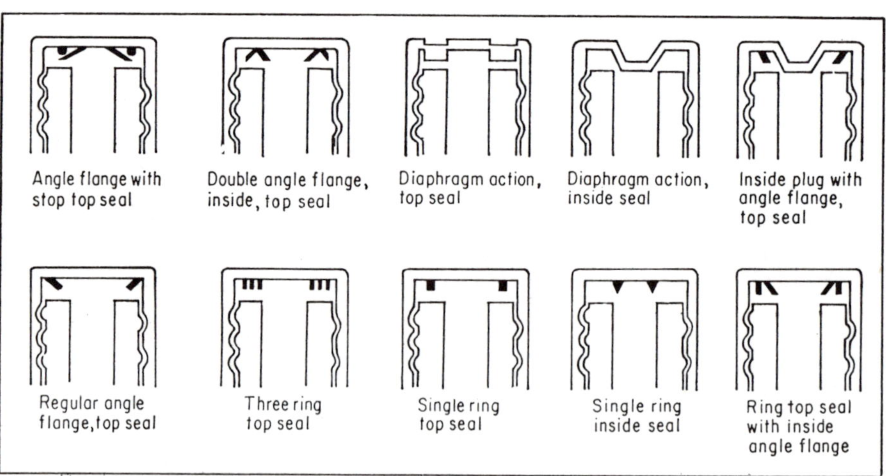

FIG. 7.88 A variety of sealing procedures are used to eliminate the liner in closures. The designs shown here will strip out of the mold when made from flexible compounds. (*Courtesy M. A. Delia Associates, Berkeley Heights, N.J.*)

TABLE 7.13 PRODUCT SPECIFICATION AND PROCEDURE CHECK LIST

1. Material Selection
   Environmental requirements
   Electrical
   Thermal
   Mechanical
   Chemical
   Machinability
   Dimensional Stability
   Thermal Expansion
   Radiation stability
   Arc Resistance
   Outgassing
   Moisture effects
   Competitive materials
   Material availability
   Product function
   Annual usage
2. Production Procedure
   Fabrication/forming
   Injection/transfer molding
   Compression molding
   Blow molding
   Casting
   Rotational Molding
   Other methods
   Allowable molding lubricants
   Platability
3. Colors and color latitude
   Percentage of regrind allowable
   Transparency/scratch resistance
4. Inspection requirements
   Certificate of Compliance
   Mil. Spec. Q.A. Level %
   Areas demanding 100% inspection
   Surface perfection
   Allowable sinks
   Clarity of transparency
   Knit line allowance
   Reasonableness of tolerances
   Inspection fixture responsibility
5. Dimensioning the product
   Tolerances and measurement method
   Flatness
   Concentricity
   Draft Allowance

   Wall Section
   Fillets
   Corner sharpness
   Location and allowance for KO marks
   Location and treatment of vent/gate areas
   Special inspection instructions
   Material designation and approved alternate
   SPI/SPE finish designation
   Gloss or buffing requirement
   Location of molders mark and cavity number
   Lettering details, raised or depressed—in panel
   Texture specification
6. Inserts
   Molded in place
   Responsibility for inserts
   Spoilage allowance
   Adequacy of standard inserts
   Post molding insert assembly
7. Post molding finishing
   Gate broken only
   Other acceptable gate removal plan
   Special polishing
   Drilling/tapping
   Post molding assembly
   Adhesives/solvents
   Printing/decorating
   Special handling for plating
8. Packing/shipping
   Fragility of parts
   Essential protective wrapping
   Protection to plater
   Shipping weight limit
   Carton size, separators and source
   Bulk packaging
   Carton identification
   Testing of shipping package
9. Review of all cost estimates
   Equipment availability
   Tool delivery review
   Reliability of vendor deliveries
   F.O.B. point

Rigid plastics bottles may use the conventional glass bottle thread in many applications without cold flow loss. When flexible plastics are used, a more rigid thread, known as the buttress thread (Figs. 7.86 and 7.87) is desirable since it reduces cold flow and provides more contact area to distribute cap pressure.

The Plastics Bottle Division of the Society of the Plastics Industry Inc. have established Bottle Finish Standards for the industry and their latest revision should be used at all times.

CLOSURE DESIGN FOR PLASTICS

Closures may incorporate a sealing feature in the molding operation eliminating the necessity for lining the caps. Typical designs are shown in Fig. 7.88.

Allowances have to be made between the closure thread and the bottle thread in order to gain proper relationship between these two parts. Table 7.12 presents a workable dimensional range based on information from closure suppliers and users.

8 HYDRAULICS FOR PLASTICS ENGINEERS

Most plastics molding machines are hydraulically driven mechanical devices. An understanding of the hydraulic elements and circuitry is an essential background for every plastics production and plant engineer.

PUMPS

The prime mover of any hydraulic system for the oil source is the pump. Rotary pumps are most commonly used, and among these are the gear, vane, and piston pumps, both axial and radial. Each of these will be described in turn.

Gear pumps are used to generate pressures to 1000 psi as a general maximum, although for special purposes, gear pumps are available to 2500 psi. The gear pump in its simplest form is shown in Fig. 8.1. As can be seen, the pump consists of a housing which is fitted with two close-fitting spur gears, one of which is driven by a motor. As the gears rotate, hydraulic fluid is carried from the suction side "A" in the open spaces between the teeth, around the outer periphery of each gear to the discharge side "B." Fine clearance permits pressure to be generated without appreciable leakage to the suction side. To enhance the pressure rating and output efficiency, helical, herringbone, and internal gear pumps have been developed. Their operating principle is the same as the spur gear pump just described.

The gear pump is a somewhat noisy, constant displacement machine, and finds relatively limited use in modern high-performance molding machines.

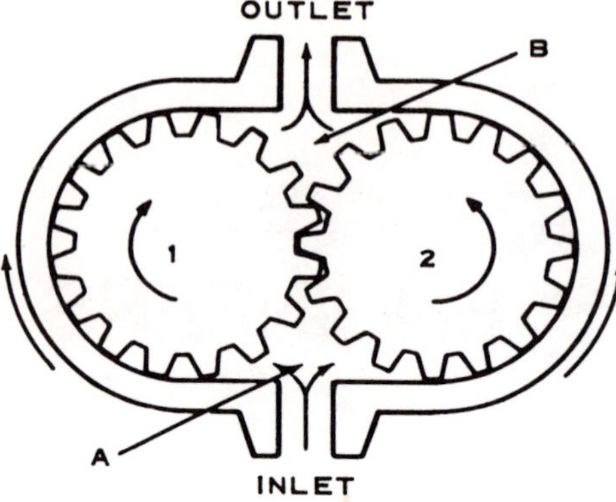

FIG. 8.1 The gear pump. (*Courtesy Vickers, Inc.*)

Vane pumps are available in two types. Figure 8.2 illustrates the unbalanced simple vane pump, and Fig. 8.3 depicts the balanced vane pump. Both types are in general use at pressures to 2000 psi and higher for intermittent use. The simple vane pump operates in basically the same way as a gear pump. The fluid is picked up on the entry side of the suction created by the continuously increasing volume of the vane pockets. The fluid is then conveyed with increasing pressure to the discharge side because of the constantly decreasing volume of the chambers between the vanes. The imbalance of forces on the vane rotor, and resulting wear on the shaft bearings, which causes reduced life lead to the development of the balanced vane pump is shown in Fig. 8.3. In the balanced vane pump, the oil pumping action is again created by the eccentricity of the rotor relative to the ring. The pumping action is dependent upon the elliptical shape of the ring. In operation, the fluid is drawn in at ports *a* and *b* (the light areas), and discharged at ports *c* and *d* (the dark areas), at diametrically opposite sides of the rotor, thus balancing the bearing loads. As can be seen, the operation is otherwise identical with the unbalanced vane pump.

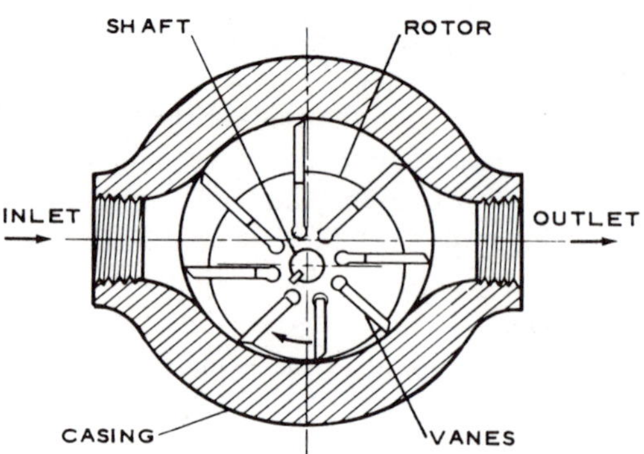

FIG. 8.2 Unbalanced simple vane pump. (*Courtesy Vickers, Inc.*)

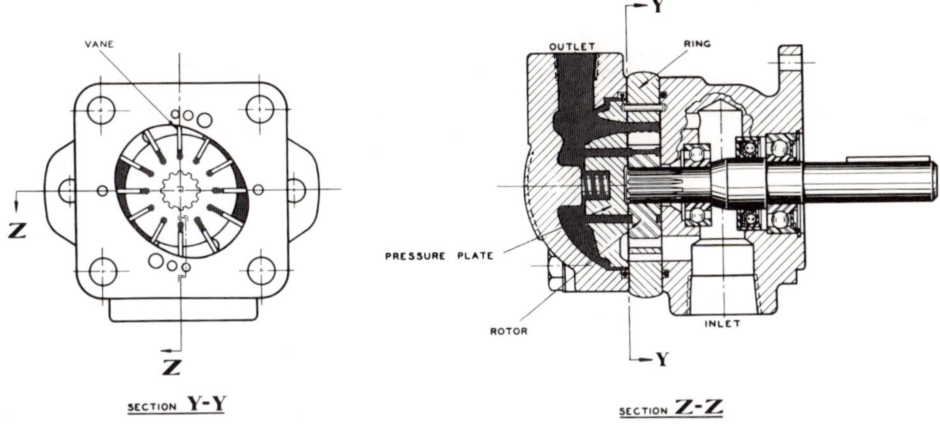

FIG. 8.3 Balanced vane pump. (*Courtesy Vickers, Inc.*)

Balanced vane pumps on the same shaft or housing are often used, such as one unit providing oil at a rate of 20 gpm for fast ram advance, then unloading through an integral unloading valve, at for example 500 psi, supplemented by the tandem unit which then provides the final squeeze at a rate of 2 gpm, with the built-in system relief valve set at 2000 psi. The advantage lies in the lower horsepower required to do the overall job. A Vickers pump of this type is shown in Fig. 8.4.

FIG. 8.4 Tandem vane pumps on a single shaft. (*Courtesy Vickers, Inc.*)

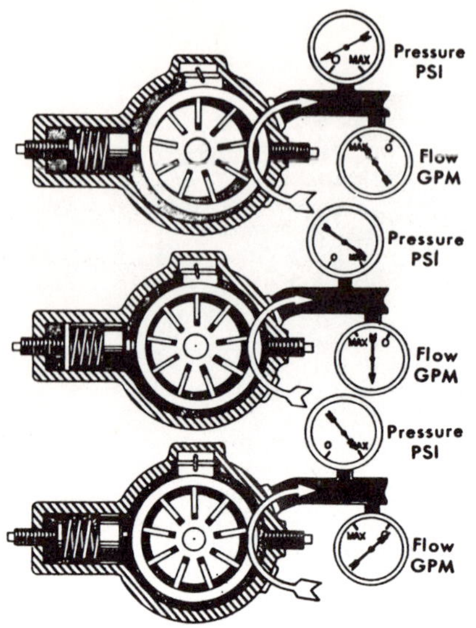

FIG. 8.5 Variable volume pumps in which the vane rotor turns within the moveable pressure ring. (*Courtesy Racine Hydraulics & Machinery, Inc.*)

Similar results can also be obtained from the variable volume pump made by Racine, as shown schematically in Fig. 8.5. Examination of this illustration will show that the variable volume vane pump is fitted with a moveable pressure ring in which the vane rotor turns. This pressure ring is moved relative to the rotor so that the pump delivers maximum volume at lowest pressure in the full eccentric position, or minimum volume at highest pressure when in the unloaded near-concentric position. In other words, the pumping action is a function of the magnitude of the ring eccentricity relative to the rotor.

Pressure ring actuation is accomplished by means of a spring which can be adjusted to develop a particular maximum pressure as volume requirements vary. The primary advantage of the pump just described is that the pump delivers the hydraulic fluid at the most efficient level, volume and pressure-wise throughout its operating range. This results in input horsepower economy. A secondary advantage lies in the fact that oil heating is at a minimum since the oil is not being dumped through a relief valve as it is when using a fixed volume pump.

Piston pumps used in hydraulic machinery are usually of the rotary type, and are available as axial piston and radial piston pumps. The radial is the simpler (mechanically) of the two. The axial is, however, smaller in diameter. The advantages gained in using piston pumps are high pressure ratings (to 10,000 psi), high volumetric efficiency (95 to 97 per cent), and high speed. They can also be used for directional control, as will be discussed later. Either design may be obtained as a fixed or variable volume device. In fact, the forms in which piston pumps are built are so numerous that a separate chapter could be written describing them. To illustrate their principle, only a few designs will be discussed.

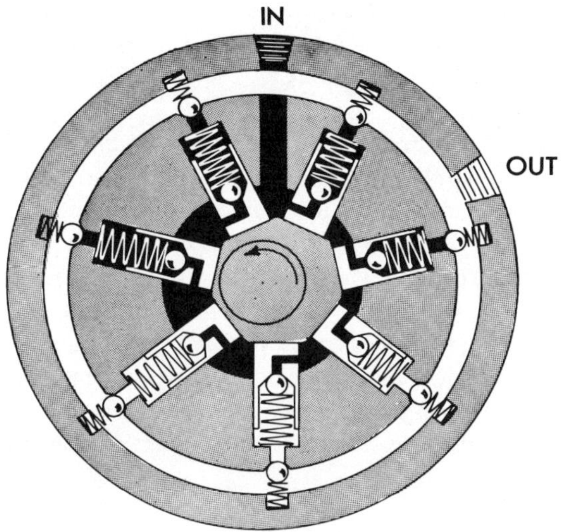

FIG. 8.6 Radial piston pumps. Schematic view. (*Courtesy Racine Hydraulics & Machinery, Inc.*)

The principle of the radial piston pump is shown schematically in Fig. 8.6. Figure 8.7 depicts the typical construction of such a pump by Racine-Seco. (Racine Hydraulics & Machinery, Inc.) As the eccentric on the drive shaft rotates, it causes the pistons to move in and out of the cylinders. When the piston is at the bottom of its stroke, the fluid enters the hollow interior through the suction port. Upon forward movement, the check ball seats and the fluid is compressed, building pressure until the exit check is unseated. The fluid is then discharging through the exit port.

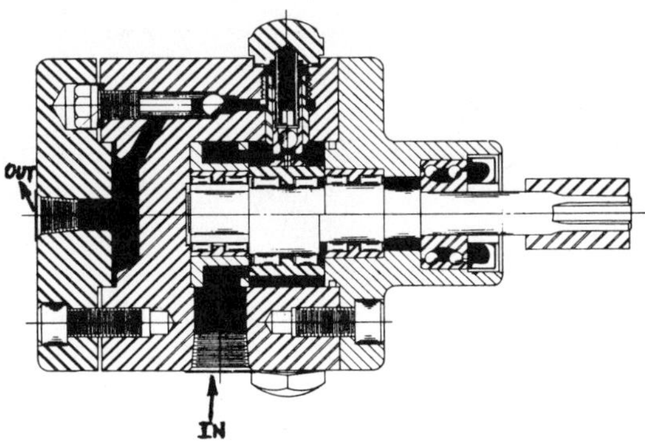

FIG. 8.7 Radial piston pump assembly view. (*Courtesy Racine Hydraulics & Machinery, Inc.*)

Since each piston discharges through its own port, it is possible to direct these individual quantities to perform individual tasks. The operation of a tandem vane pump to move a piston rapidly at low pressure-high volume and then hold at high pressure-low volume was described earlier. Exactly the same thing can be done by manifolding the output of the several pistons in the radial piston pump.

Figure 8.8 diagramatically shows how this is done with a Racine-Seco radial piston pump. The hydraulic fluid is conducted, for example, through the manifold plate "M" in such a fashion that a high-volume source is delivered at port "H.V." from six pump pistons. The low-volume fluid is delivered at port "L.V." from one pump piston. Other arrangements are of course possible, such as two pistons feeding "H.V.," and three supplying "L.V.," etc. The oil from the "H.V." port is generally passed through an unloading valve so that the work ceases at a preset low pressure. The high pressure low-volume fluid is then supplied from the "L.V." port. The advantages of such an arrangement are low pump input horse-

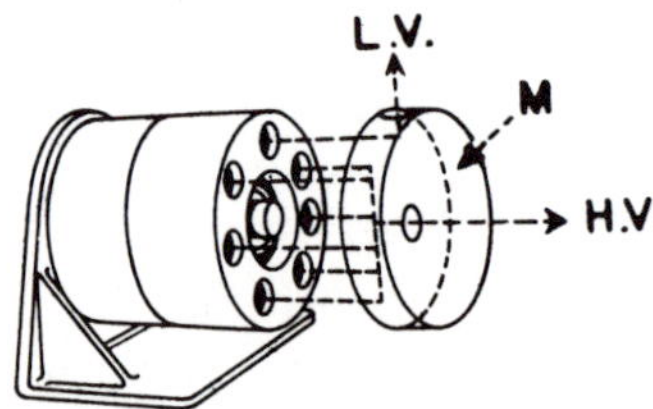

FIG. 8.8 The hydraulic fluid is conducted through the manifold plate M from 6 pistons to H.V. (high volume) port and from one piston to L.V. (low volume) port. (*Courtesy Racine Hydraulics & Machinery, Inc.*)

power and reduced fluid heating, resulting in enhanced pump life. Thus far the discussion of radial piston pumps has dealt with fixed volume units. The variable volume radial pump made by the Oilgear Company is illustrated in cross section in Fig. 8.9A, and planwise in Figs. 8.9B, C, and D. When rotary power is applied to the input shaft, the cylinder block and pistons rotate within the slideblock ring. The pistons are held in contact with the ring by centrifugal force and system return or supercharge pressure from the built-in gear pump. Oil is delivered to and from the pistons through drilled cross-holes in the cylinder and to and from the crescent ports in the flat valve.

The output and flow direction of the pump is dictated by the position of the slideblock relative to the rotating cylinder. When the slideblock is in the center position, shown in Fig. 8.9B, no oil is delivered.

As the slideblock is shifted to the left as in Fig. 8.9C, low-pressure oil enters the piston cylinders through the lower crescent, and high-pressure oil leaves through the upper crescent. The position shown illustrates the position for highest volume delivery. As the slideblock moves to the right toward neutral, the piston stroke is reduced, and consequently the volume is reduced.

Figure 8.9D shows the slideblock in the full right position. Here the low pressure oil enters the upper crescent and leaves through the lower crescent.

Thus the flow of oil through the pump is reversed. The reversal of direction is shockless. The shifting of the slideblock is accomplished hydraulically, mechanically, pneumatically, and by other means required in a particular circuit design. The flow reversing feature is extremely important. It permits design of very efficient and sophisticated hydraulic circuits, often allowing the elimination of directional control, flow control, and other valves from the circuit. (See Chapter 4 for further discussion of this feature.)

The other form of hydraulic piston pump is shown in Fig. 8.10. This is a fixed

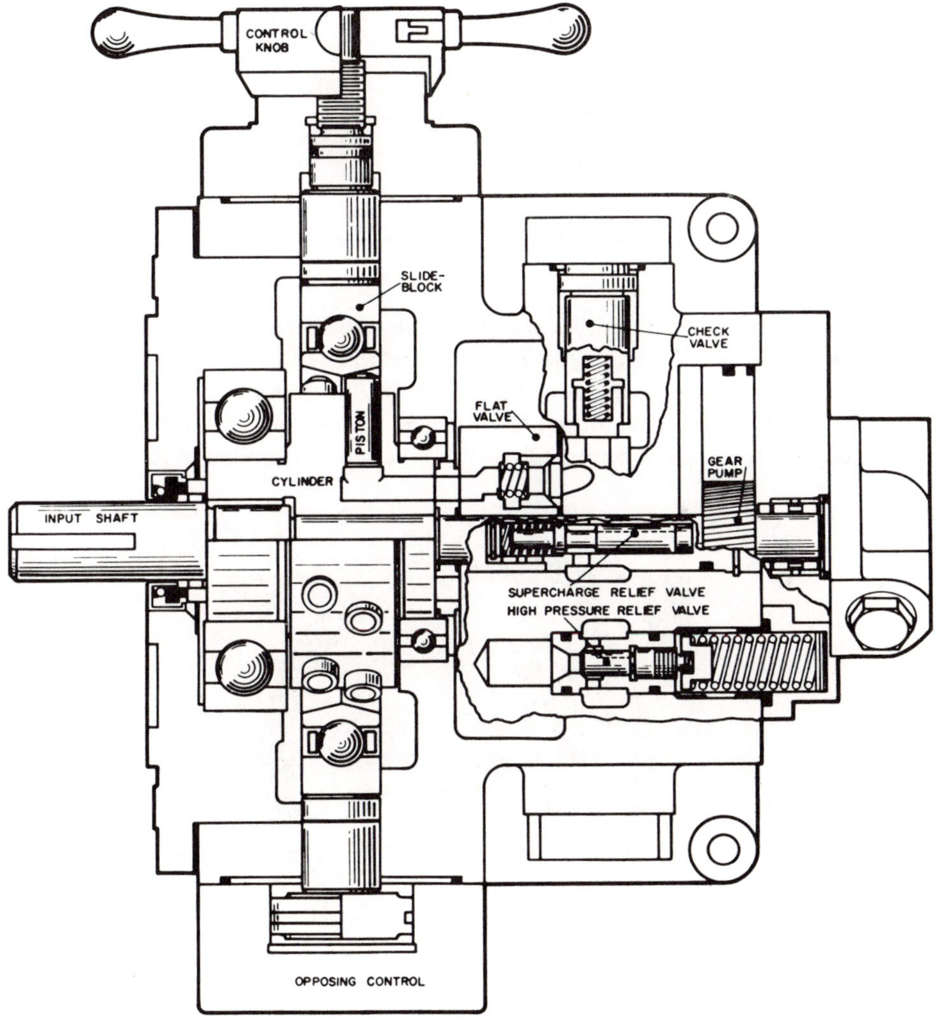

FIG. 8.9A Variable volume radial pump. (*Courtesy, The Oilgear Co.*)

FIG. 8.9B Variable volume pump in centered position. (*Courtesy The Oilgear Co.*)

volume axial piston pump. The reciprocating movement of the pistons is ob-tained from the rotation of the wobble or cam plate. The internal structure of the piston is shown in the lower portion of the illustration. As the wobble plate rotates, the pistons reciprocate. On the back stroke, the inlet ball check is un-seated and the hollow piston fills with oil because of the lower pressure in its

FIG. 8.9C Variable volume pump with high pressure leaving through upper crescent. (*Courtesy The Oilgear Co.*)

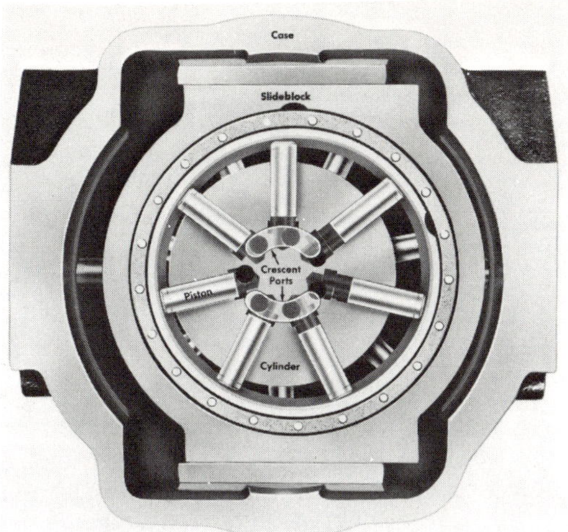

FIG. 8.9D Variable volume pump with high pressure leaving through lower crescent. (*Courtesy The Oilgear Co.*)

interior. On the forward stroke, the fluid is compressed and finally unseats the outlet check valve, discharging to the outlet port.

In the foregoing discussion of the radial piston pump, it was pointed out that the output of individual cylinders could be obtained for individual tasks. The same is true in the axial piston pump. One method is to replace the outlet check

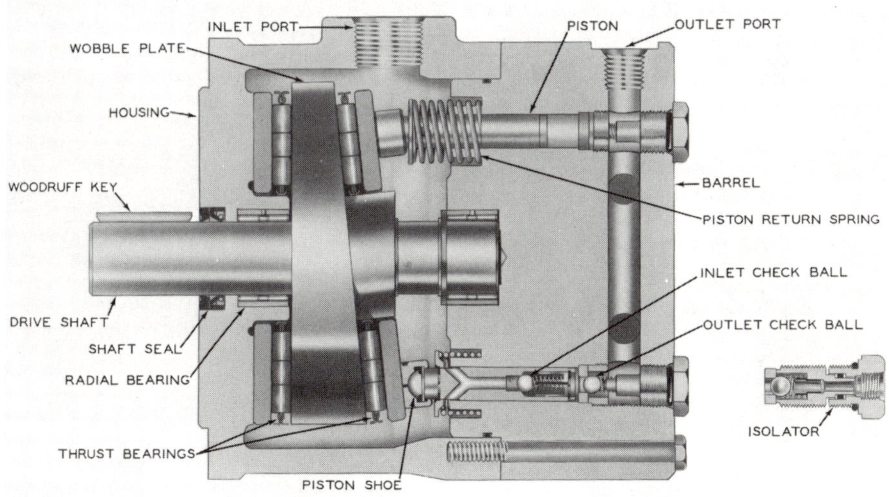

FIG. 8.10 Fixed volume axial piston pump. (*Courtesy Dynex Co.*)

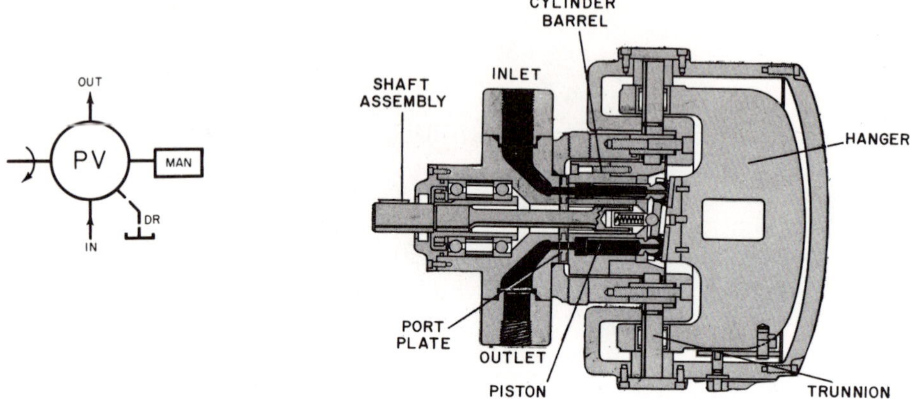

FIG. 8.11 Variable volume pump with axial pistons with variable wobble plate. (*Courtesy Denison Engineering Division, American Brake Shoe Co.*)

valve with a device called an "isolator," which is simply a check valve with a direct port to the outside of the pump. Axial piston pumps are also procurable as variable volume units. This characteristic is accomplished by varying the angle of the cam or wobble plate. One method for doing this is shown in Fig. 8.11, where a cam plate hangar can be swung through a 30 degree arc. This changes the angular relationship of the cam plate and the pistons. The resulting change in the piston strokes provides variable volume output. It should also be noted that this pump is valveless. The inlet and outlet fluid is conducted to and from the pump cylinders by a port plate, obviating the need for internal pump valves. The port plate is shown in Fig. 8.12. A cutaway of the actual pump is illustrated in Fig. 8.13. Movement of the hangar plate (i.e., cam angle) can be accomplished manually by handwheel, as shown, or in several other ways, such as hydraulically, by electric motor, pressure compensation, or by servo control.

PORT PLATE

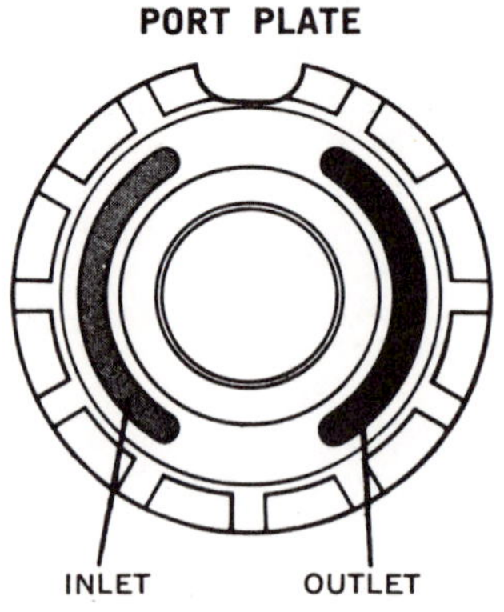

FIG. 8.12 Port plate for variable volume pump shown in Fig. 8.11. (*Courtesy Denison Engineering Division, American Brake Shoe Co.*)

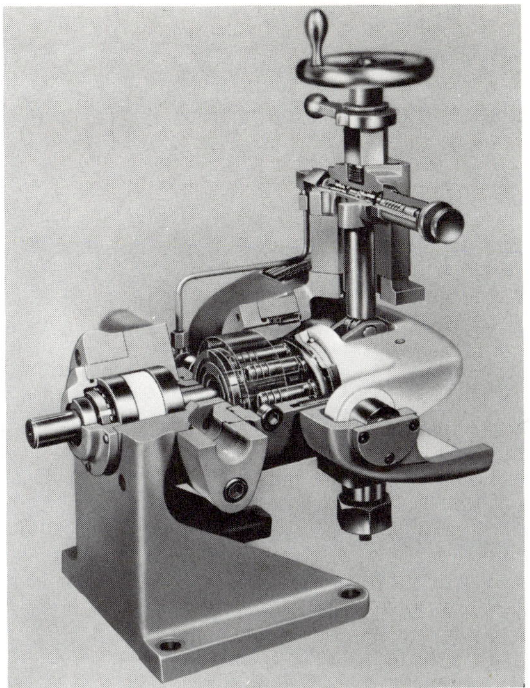

FIG. 8.13 Variable volume pump with axial pistons and hangar plate. (*Courtesy Denison Engineering Division, American Brake Shoe Co.*)

With the advent of the screw injection machine, numerous machines have appeared with hydraulic drives. The advantage lies in the infinite screw speeds that can be obtained within a particular speed range. With some resins and molding cycles, this has been found to be particularly important. Hydraulic motors are an inversion of the hydraulic pump in that the pressure fluid flows through the device in reverse to the pump. Motors are available in the same modes as pumps (i.e., gear, vane, radial, and axial piston types). The most commonly used are the axial piston type, because of their high efficiency, pressure handling characteristics, and long life.

THE NUTATING HYDRAULIC MOTOR

The newest development in hydraulic motors is characterized by offering high torque and low speed output without resorting to extensive gear trains. The speed of the nutating motor is dependent on the volume of oil passing through the motor with the torque being a function of the system pressure.

Figure 8.14 depicts a schematic of the nutating motor. There are eight pistons in this design, it is in fact essential that even multiples of two be used in the

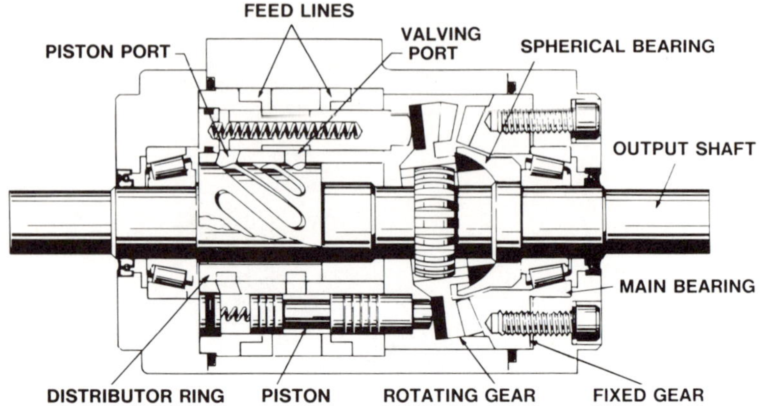

FIG. 8.14 Nutating motor schematic. (*Courtesy Parker Hannifin Corp.*)

nutating motor so that the integral valving distributor ring can function to drive the pistons in a sequential fashion.

The speed reduction is accomplished through a fixed and a rotating gear. The fixed gear is always cut with a lesser number of teeth than the rotating gear. For example if the fixed gear had 30 teeth and the rotating gear had 31 teeth, the ratio would be 30:1.

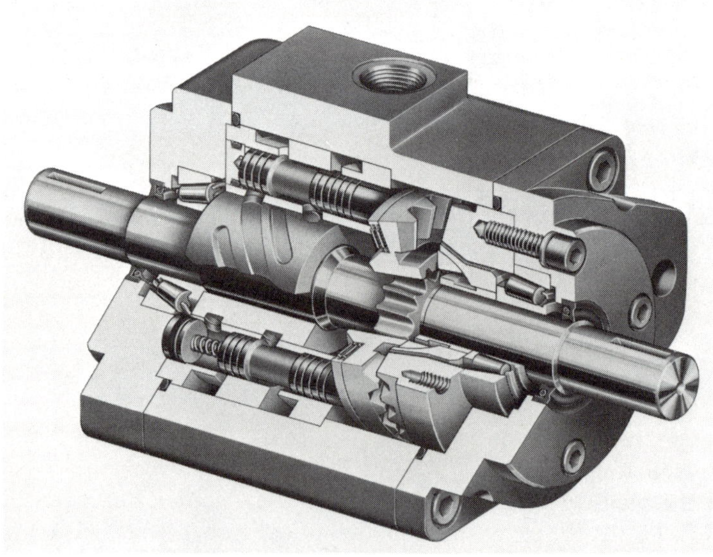

FIG. 8.15 Nutating motor section. (*Courtesy Parker Hannifin Corp.*)

$$\frac{30\,T\,\text{(fixed)}}{31\,T\,\text{(rotate)} - 30\,T\,\text{(fixed)}} = \frac{30}{1} = 30:1$$

The effect of other gear tooth combinations can be easily visualized.

As each piston is energized in sequence the rotating gear is driven to turn the shaft in the ratio of the gear system at a speed dependent on the oil flow rate. A cross-section of the actual motor is shown in Fig. 8.15.

PRESSURE CONTROLS

Now that an acquaintance with the source of power for the hydraulically operated machine has been developed, consideration must be given to the control of this force. Pressure, direction, and rate of flow to the cylinders that do the work in the machine must be regulated. Every hydraulic system must include some means for controlling the maximum pressure allowed by the construction of the machine, the work to be performed, and protection from damage for the components. This function is performed by a relief valve usually piped to the pump outlet. This valve can be direct acting, as shown schematically in Fig. 8.16 or as more commonly encountered in Fig. 8.17.

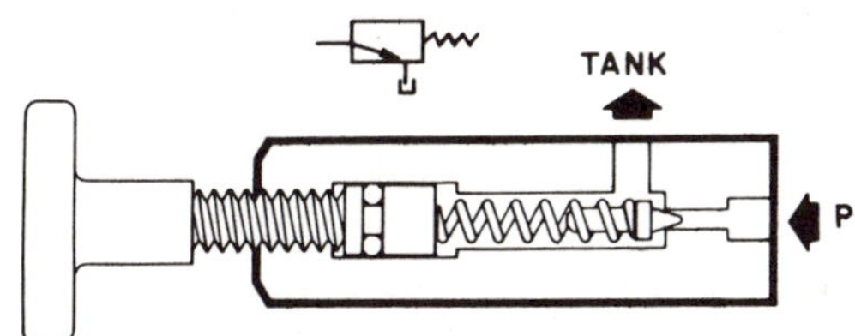

FIG. 8.16 Direct operated relief valve. (*Courtesy Machine Design*)

Referring to Fig. 8.16, it can be seen that the cone will be unseated if the pressure on it becomes greater than the force exerted by the spring. The fluid, of course, will then flow through the valve to a down stream point such as the tank. This type of valve is not used to handle substantial volumes of fluid, but is frequently utilized as a means for remote control on the piloted relief valve, which will now be discussed.

Figure 8.17 shows the relief valve in the closed position. As pressure increases, the fluid presses against the bottom of the spool, tending to shift it. Simultaneously, oil flows through the passageway in the spool and presses on the top of the spool and the cone of the pilot relief valve. The fluid force on top of the spool supplements the main spool spring force. Depending on the setting of the pilot spring, fluid will begin to pass the pilot cone and reduce the pressure on top of the piston. When this happens, the fluid pressure on the bottom of the

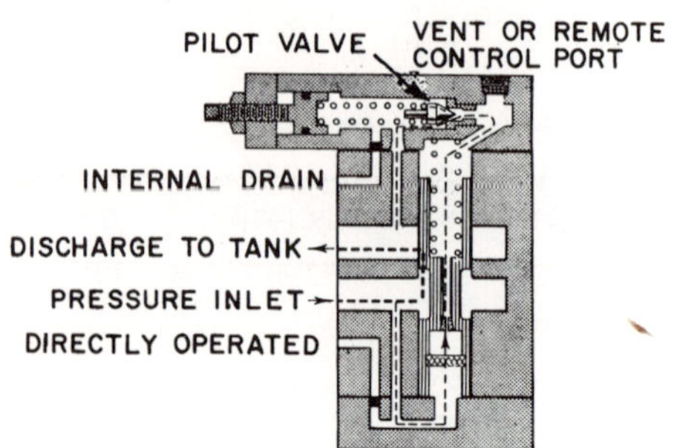

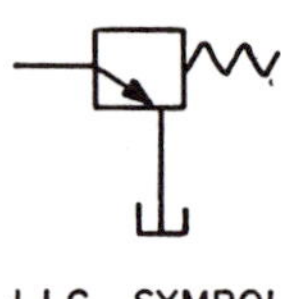

FIG. 8.17 Pilot-operated relief valve. (*Courtesy Racine Hydraulics & Machinery, Inc.*)

spool will force the spool upward, and fluid will flow to the discharge side of the valve. This will continue so long as the inlet pressure remains high enough to keep the pilot cone unseated. Thus the pressure is relieved to insure the safety or proper operation of the system. It was mentioned that the valve shown in Fig. 8.16 could be used as a remote control device for establishing lower pressure settings than could the integral pilot valve in the relief valve. This is done by connecting the inlet port of the relief valve shown in Fig. 8.16 to the remote control port of the pilot operated relief valve shown in Fig. 8.17. In this case, the integral pilot valve is set for the maximum system pressure desired. It is apparent that the remote control valve will now establish the desired pressure at any level lower than the maximum set in the integral pilot valve.

The next important member of the pressure control family is the unloading valve. This valve was briefly mentioned in the pump section in reference to unloading the high-volume low-pressure stage of the two stage vane pump. Figure 8.18 schematically shows how this functions. As the pressure builds up in the low-volume high-pressure section of the two-stage pump, pressure is applied to the spool of the unloading valve through the remote-pressure control-port, and finally unseats the spool to direct the flow (pressure inlet) to tank (discharge outlet), thus unloading the low pressure pump. The pressure at which this unloading occurs is dependent upon the spring setting. The illustration shows the spool in the unloading position.

The same valve with internal pilot pressure, Fig. 8.19, can be used as a sequence valve. Assume, for example, two circuits, the first requiring full pressure before the other. Pressure fluid, in this case, could flow through the valve. When pres-

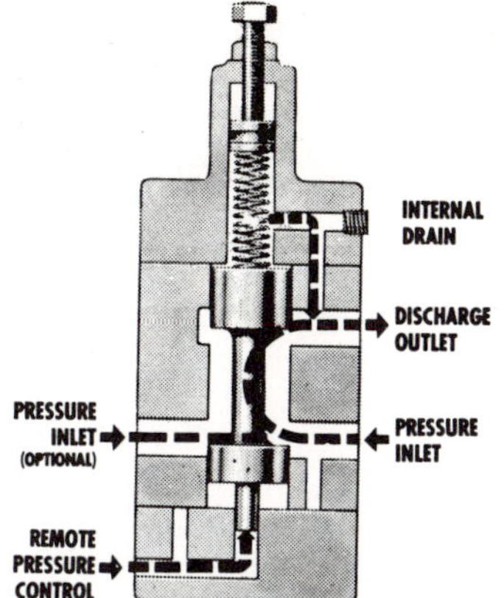

FIG. 8.18 The unloading valve. (*Courtesy Vickers, Inc.*)

sure is built up in the first circuit, the spool would be unseated by this pressure, thus causing the fluid to flow through the secondary outlet to the second cylinder at the same pressure as that in the primary circuit. Again, the pressure at which this would occur is dependent upon the spool spring setting. Unfortunately, with the arrangement just described, the return flow from the second circuit would be blocked unless other means for return flow are provided. One such arrangement to remove this problem is shown in Fig. 8.20. Note the check valve on the left hand side of the valve body. When pressure drops on the inlet side, the main spool closes. It can be seen that return flow can then pass the

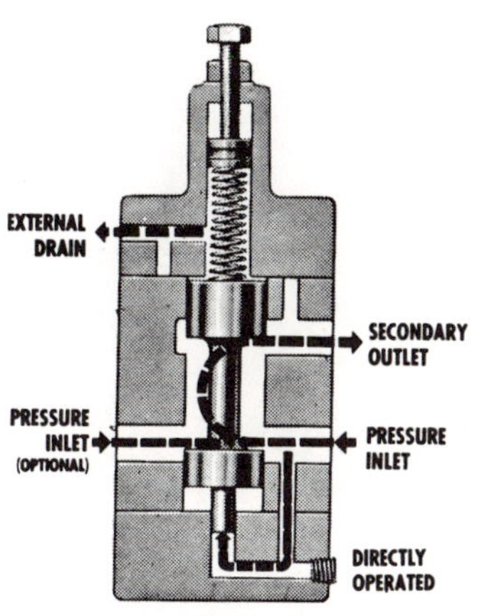

FIG. 8.19 Unloading type valve with internal pilot pressure for use as a sequence valve. (*Courtesy Vickers, Inc.*)

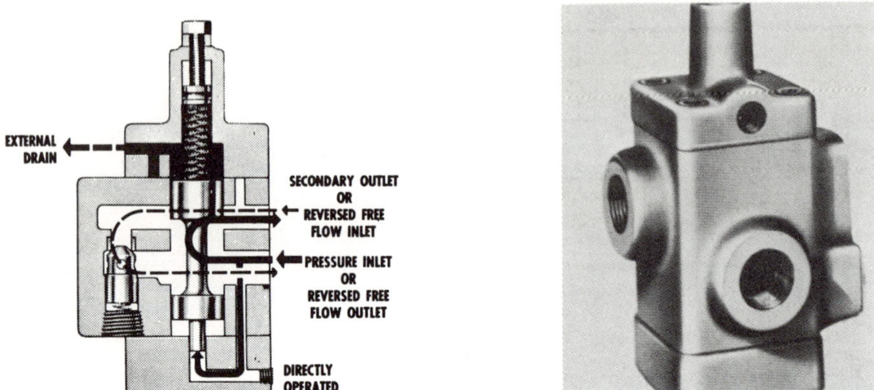

FIG. 8.20 Sequence and check valve.

check valve, since only light spring pressure need be overcome. Examination of these several valves should make it apparent that the same valve body, spool, etc., can be used to perform several functions by changing the pilot from external to internal. In fact, the sequence and check valve with an *external pilot*, as shown in Fig. 8.21, is used as a counterbalance valve. Other functional versions are also possible.

When it is necessary to operate a secondary circuit at a lower pressure than in the primary circuit, a pressure reducing valve is used. This is shown in Fig. 8.22. Fluid flows through this valve from the high pressure inlet to the low pressure outlet. The throttling action of the spool causes a pressure drop across the valve. As pressure increases in the low pressure side of the valve and on the bottom face

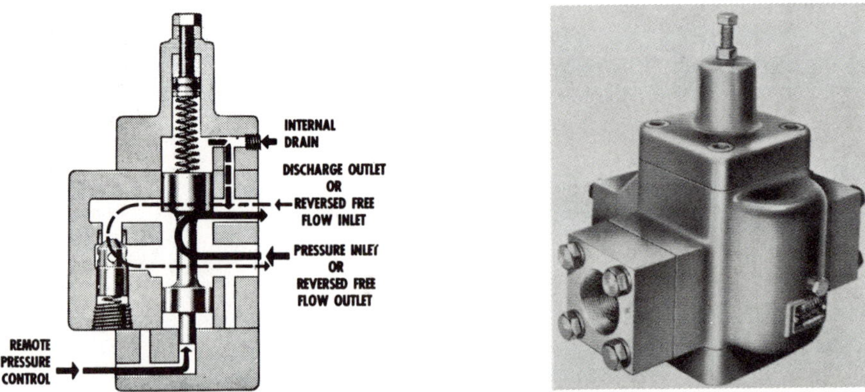

FIG. 8.21 Sequence and check valve with external pilot for use as a counterbalance valve.

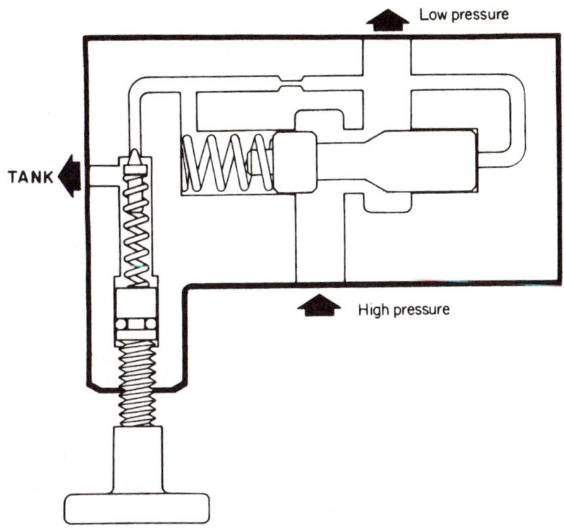

FIG. 8.22 Pressure-reducing valve. (*Courtesy Machine Design*)

of the spool, the spool moves towards a closed position. The orifice between spool and seat is thus reduced and pressure drop increases. Simultaneously the downstream pressure is acting on the top of the spool (additive to the fixed spring pressure) and on the pilot cone. If the pressure rises to a level greater than the fixed spring force, the valve closes and further reduces the downstream pressure. If the pressure rises to the level set by the pilot spring, the cone unseats and relieves the excess pressure to the tank. As a result of all of these actions, a constant downstream-reduced pressure is obtained.

Electrohydraulic pressure control valves were developed to be used in conjunction with the solid-state control systems for hydraulic machinery.

These valves are available to satisfy all modes of pressure control; relief, reducing, unloading and sequence. In the discussion just completed it has been shown that the required pressure is obtained by making suitable adjustment to the spool spring setting. This mechanical function is accomplished in the electrohydraulic valve by pilot oil pressure, whose pressure value in turn is determined by the magnitude of the pressure drop across the poppet controlled by the D.C. coil.

The schematic Fig. 8.23 shows the flow of the pilot pressure fluid. Note that the oil passes through the main spool as well as through the compensator and across the orifice created by the position of the D.C. coil poppet. The pressure drop across this orifice determines the balance of pressure on the piston which in turn establishes the position of the main spool. The position of the main spool of course creates the orifice between the inlet and outlet to control the final

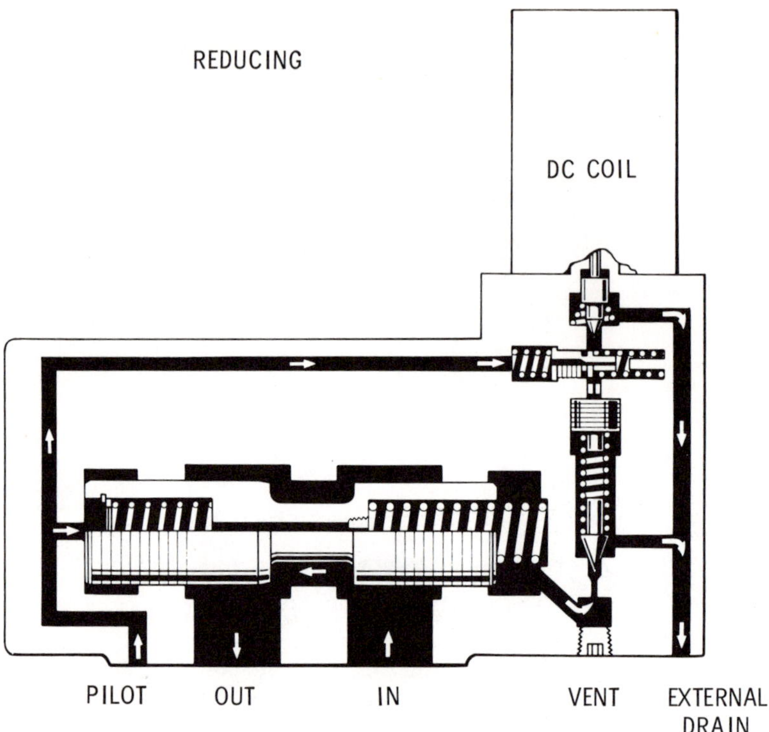

FIG. 8.23 Pressure control valve schematic. (*Courtesy Racine Hydraulics Inc.*)

system pressure. The magnitude of the coil force is thus directly related to system pressure.

The signal voltage and current to control the D.C. coil poppet position is established by the potentiometer setting in the solid-state control system discussed in Chapter 9, or on machines not so equipped, from an adjustable current control module available from the valve manufacturer. Figure 8.24 illustrates a cross-section of the actual valve.

DIRECTIONAL CONTROL

Directional valves in hydraulic systems control the fluid flow in the system. Essentially, this group consists of check valves, two-way, three-way, and four-way valves actuated manually, electrically, or by other means.

The check valve is a device which stops flow in one direction and permits free flow in the opposite direction. The most basic of these are the horizontal and angle check valves, shown in Fig. 8.25. For proper operation, these valves must be installed with the bonnet up, as indicated by the figure. Operation of these valves is identical. Fluid flows freely from the "in" to the "out" ports. If, how-

FIG. 8.24 Cross section of electro-hydraulic pressure control valve. (*Courtesy Racine Hydraulics Inc.*)

ever, the pressure on the "out" side rises above that of the "in" side, the valve poppet seats and reverse flow is prevented. To allow unrestricted installation position, a spring-loaded ball check valve (Fig. 8.26) is used. Its operation is the same as that of the valves shown in Fig. 8.25.

The check valves just described are strictly one-way flow valves. When free return flow is desired, a pilot-operated check valve is used. This is illustrated in Fig. 8.27. This valve operates on the same principle as the other valves described; however, when pilot pressure is applied and the check is thereby unseated, the fluid can flow in the opposite direction without restraint.

An important feature is the bleeder valve which is built into the main deck poppet. When the pilot piston rod moves forward it first unseats the bleeder poppet, causing a slow reduction in pressure before the main poppet unseats. The purpose is to reduce hydraulic shock, which can be quite severe in valves without this feature.

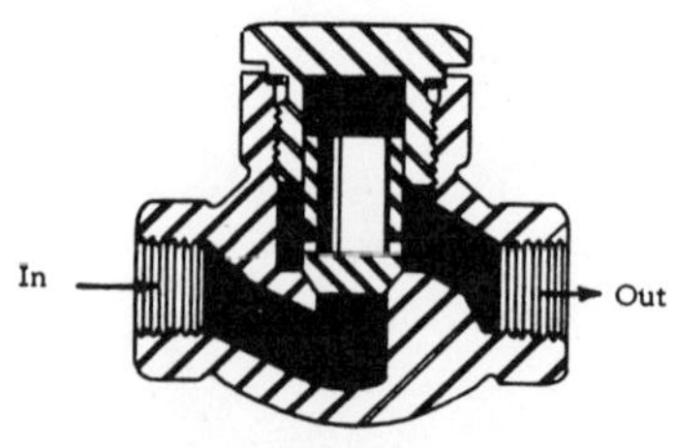

HORIZONTAL TYPE

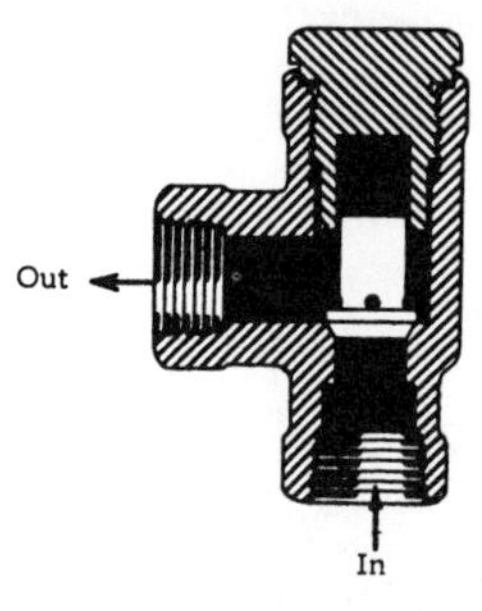

ANGLE TYPE

J. I. C. SYMBOL

FIG. 8.25 Check valves. (*Courtesy Racine Hydraulics & Machinery, Inc.*)

Two-way valves are used to direct the fluid from one circuit to another. Figure 8.28 shows a solenoid-pilot operated spring return a normally open two-way valve. They are also obtainable as normally closed valves. The operation is simple. When the solenoid is energized, the pilot valve is shifted, thereby allowing fluid pressure to act on the pilot piston. The pilot piston moves forward and shifts the

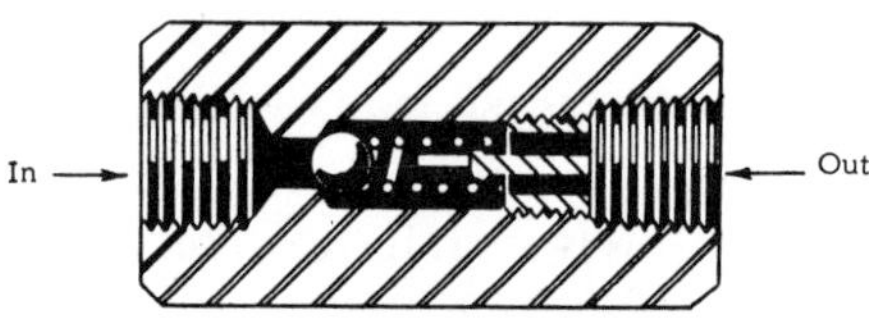

10,000 PSI TYPE

FIG. 8.26 Spring-loaded check valve. (*Courtesy Racine Hydraulics & Machinery, Inc.*)

J. I. C. SYMBOL

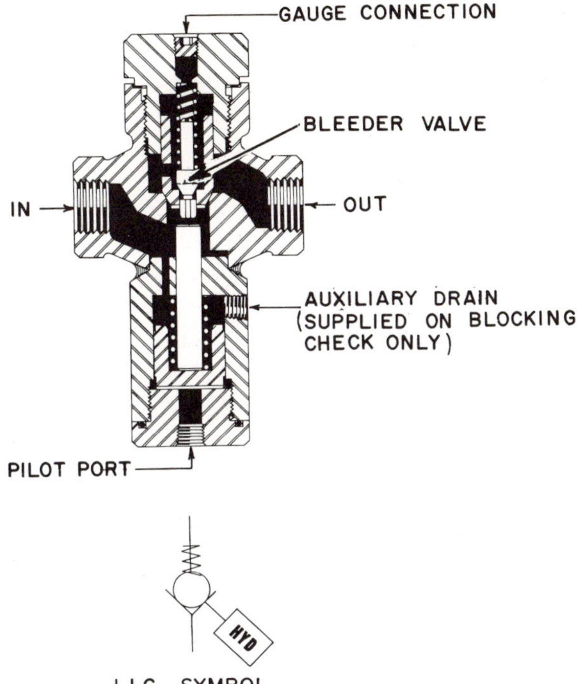

FIG. 8.27 Pilot-operated check valve. (*Courtesy Racine Hydraulics & Machinery, Inc.*)

spool to the closed position. This position is held so long as the solenoid is energized. When the solenoid is de-energized, the spring on the end of the spool returns it to the normally open position. It should also be noted that this particular valve has a built-in check valve which permits return flow even though the spool may be in a closed position.

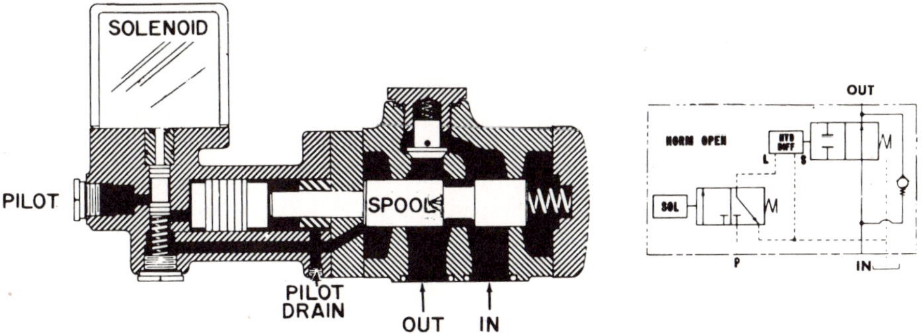

FIG. 8.28 Two-way directional valve. (*Courtesy Racine Hydraulics & Machinery, Inc.*)

Three- and four-way valves will be discussed as one, since by plugging one outlet on a four-way valve body, three-way valve operation can be obtained. It should also be noted that a four-way valve can be used as a two-way, merely by plugging two ports.

Four-way valves can be furnished with numerous spool configurations, able to perform many functions. It should be noted that the same body can be used in all cases. Through the courtesy of the Denison Engineering Division of the American Brake Shoe Company, Fig. 8.29 schematically illustrates some of these functions. Careful study of these is recommended to enable understanding of the control of hydraulic circuits.

Four-way valves are available either with threaded body, or as sub-plate types. Operation of the valves can be accomplished in many ways. Among these are: direct, manual, stem, cam, direct pilot, single and double solenoid direct, single and double solenoid pilot operated, etc. Perhaps an explanation of the J.I.C.* symbols shown above each of the valve bodies would be helpful at this point. A discussion of symbol Fig. 8.29.3, used as an example, follows.

Each of the squares in the symbol represents a position of the spool. As seen, the spool is shown in its normal centered position. All ports, P-pressure, T-tank, A-circuit A, and B-circuit B are blocked. If the left square is now moved (referred to as an "envelope") to the center, the flow now is indicated to be from "P" to "A," and from "B," to "T." Similarly, if the right envelope is moved to the center position, the flow is from "P" to "B" and from "A" to "T." All other J.I.C. symbols are evaluated in the same fashion.

FLOW OR VOLUME CONTROLS

Most hydraulic circuits are designed to provide means for achieving maximum conditions of pressure and flow. Oftentimes however, it is found that less than these maximums are necessary to obtain the desired results. Means for controlling pressure have already been examined; now consideration of flow control will be discussed.

Flow is regulated by throttling or diversion of the flow. Pressure and fluid temperature variations are also an important consideration. Valves compensating for these variables are in common use. The simplest form of flow control is achieved by the needle valve. A schematic of this is shown in Fig. 8.30. By raising or lowering the needle relative to the seat, the flow through the valve can be varied. This is known as a non-compensating (temperature or pressure) valve. The disadvantage of the needle valve is that the "through" flow varies as the oil pressure changes. Its advantage is low cost. Other versions of this basic type

*Graphical symbols were developed by Joint Industry Conference (J.I.C.) for hydraulic and pneumatic equipment, and first published in September, 1948.

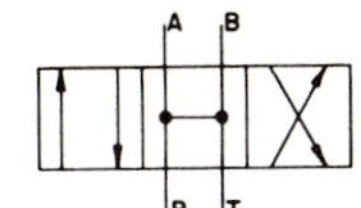

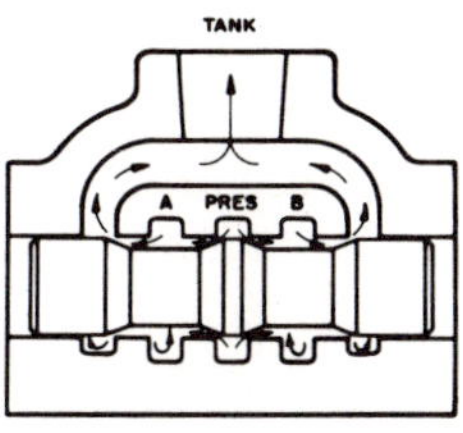

FIG. 8.29.1 This four-way directional control valve is often used in fluid motor circuits where no braking action is required, or with hydraulic cylinders when the ram need not be held in a fixed position. (*Courtesy Denison Engineering Division, American Brake Shoe Co.*)

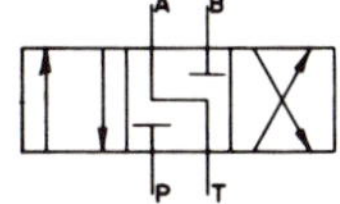

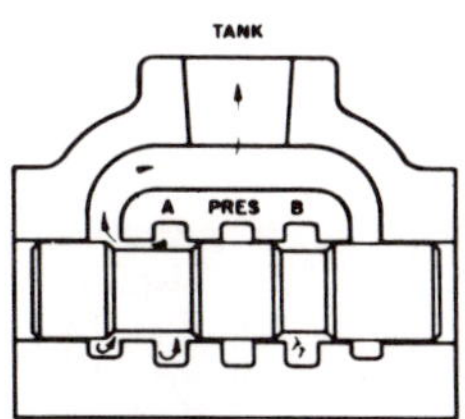

FIG. 8.29.2 Valves employing this spool permit locking movement of a hydraulic ram or fluid motor in one direction, while pressure is available to a second circuit.

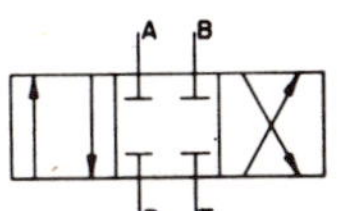

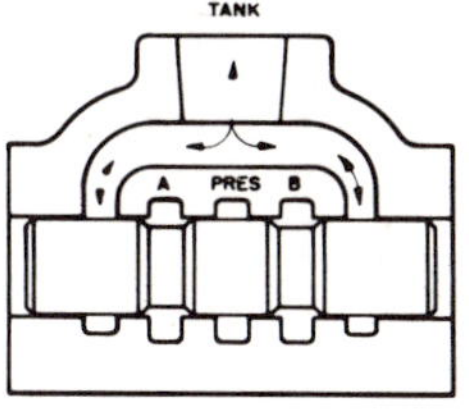

FIG. 8.29.3 Used in applications such as circuits involving a fluid motor or hydraulic cylinder ram which must be braked in either direction while, at the same time, flow to another portion of the circuit is permitted.

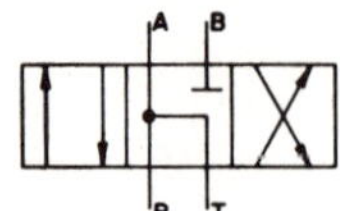

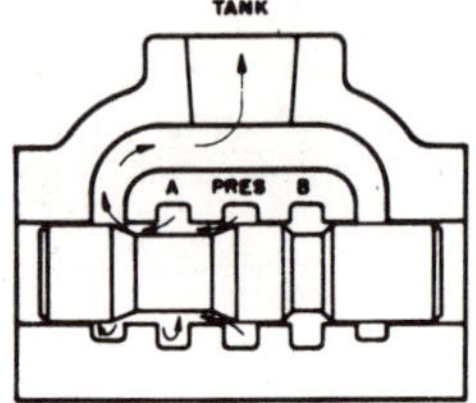

FIG. 8.29.4 This spool is used holding circuit applications, as when movement of a hydraulic press ram must be locked in one direction while pressure on the pump is relieved.

FIG. 8.29.5 This is used in holding circuit applications, as when movement of a hydraulic press ram must be locked in one direction while pressure on the pump is relieved.

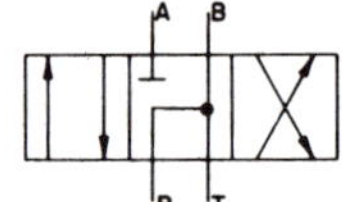

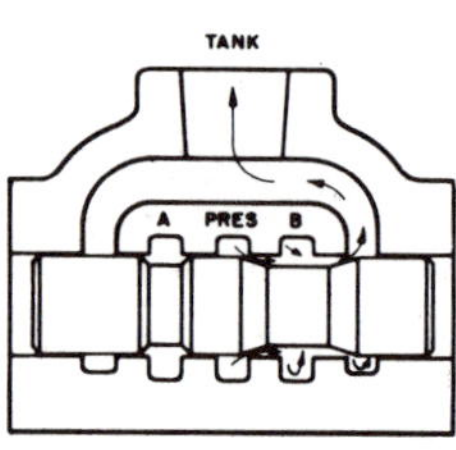

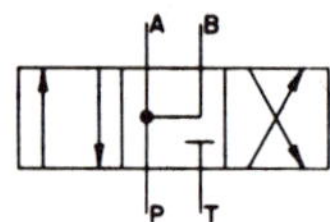

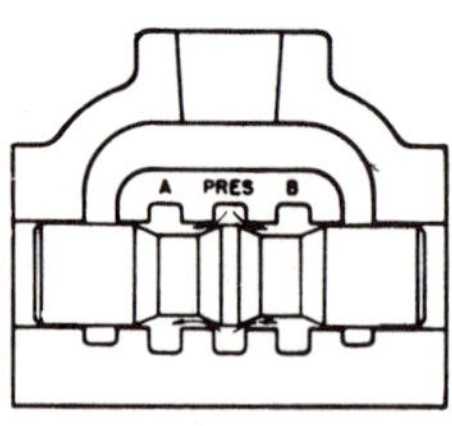

FIG. 8.29.6 Pilot valves of solenoid-controlled, pilot-operated, directional control valves employ this spool when the primary valve is pressure-centered type.

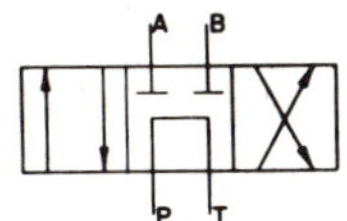

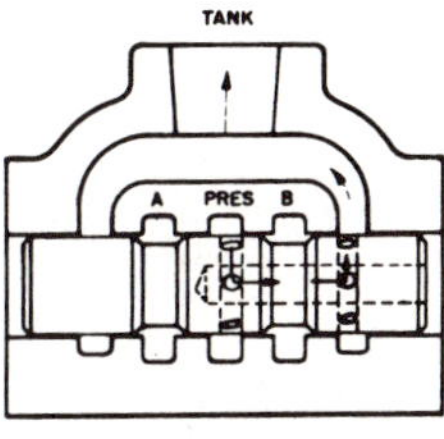

FIG. 8.29.7 Used to lock movement of a fluid motor or hydraulic ram in both directions and, at the same time, relieve pressure on the pump.

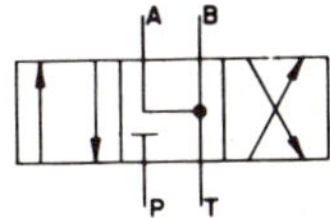

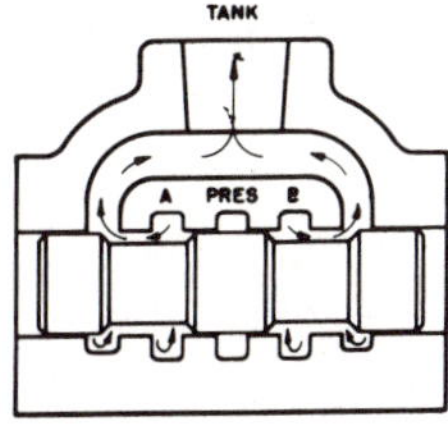

FIG. 8.29.8 This spool is often used where it is not necessary to lock movement of a fluid motor or hydraulic ram while flow is directed to a second circuit in the system.

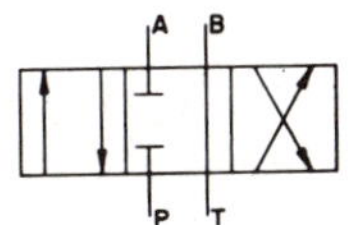

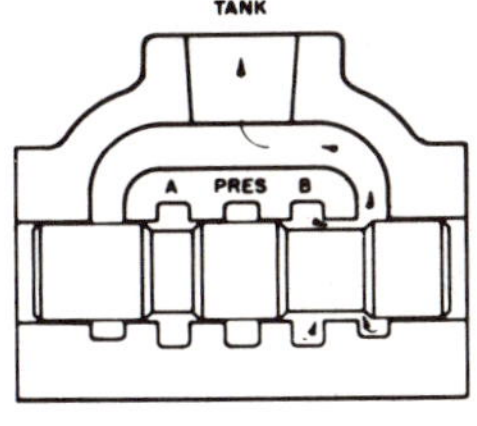

FIG. 8.29.9 Valves employing this spool permit locking movement of a hydraulic ram or fluid motor in one direction, while pressure is available to a second circuit.

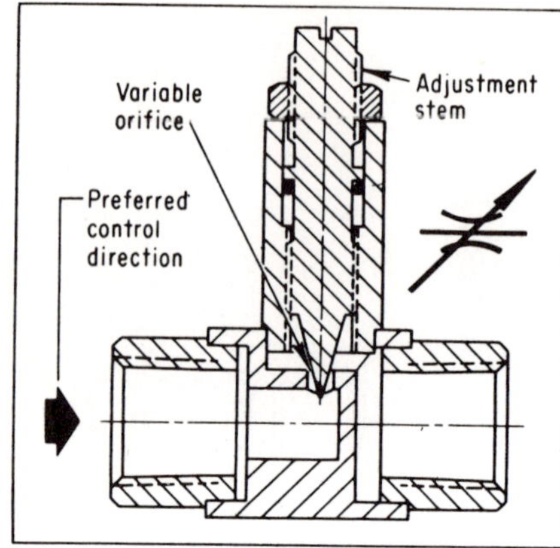

FIG. 8.30 Simple needle valve. (*Courtesy Machine Design*)

include a needle valve with built-in check valve which permits free flow return (Fig. 8.31).

Figure 8.32 schematically depicts a pressure compensated flow control valve and an illustration of its actual counterpart. The flow of hydraulic fluid through the valve is controlled by the size of the orifice "O." As pressure variations occur, the hydrostatic valve "H" moves to vary the orifice "P." This results in an unbalance of pressures on the several areas of the hydrostatic valve opposed

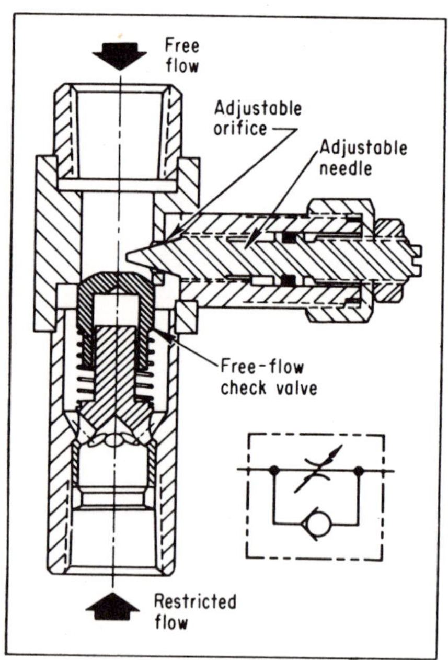

FIG. 8.31 Needle valve with integral check valve. (*Courtesy Machine Design*)

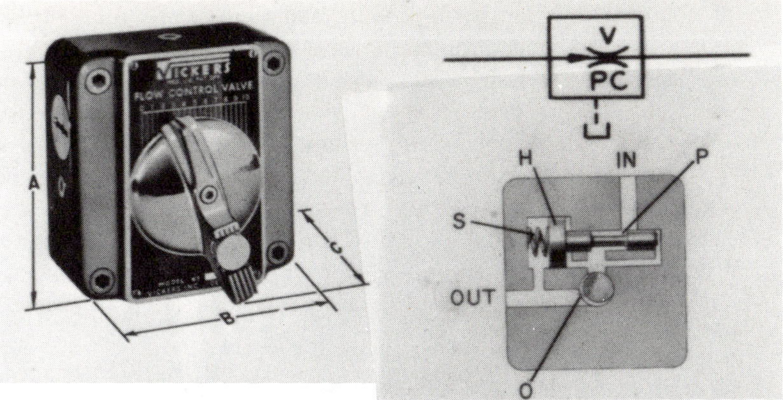

FIG. 8.32 Pressure compensated flow control valve. (*Courtesy Vickers, Inc.*)

by the spring pressure "S." The total effect is that as pressure varies, the hydrostatic valve orifice position varies, absorbing the increased pressure. With falling downstream pressure, the hydrostat opens, thus decreasing the pressure drop across it. Both movements have the effect of maintaining the pressure drop across the throttle "O." Figure 8.33 illustrates a flow control valve which is

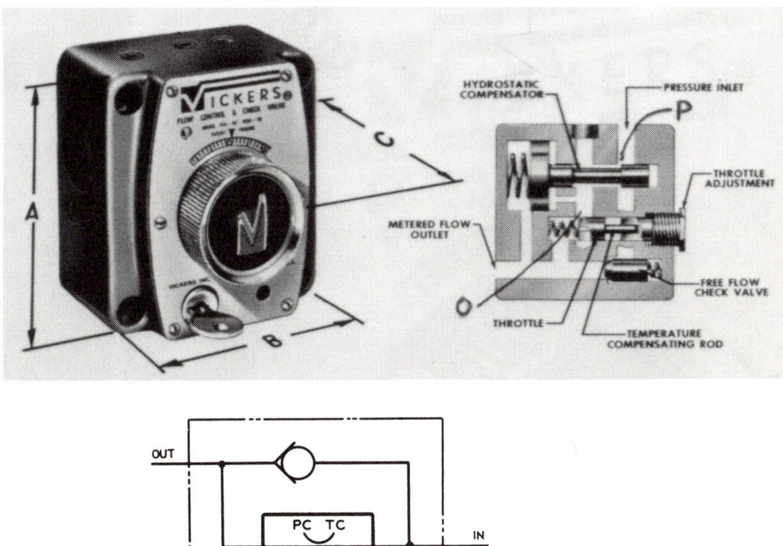

FIG. 8.33 Flow control valve which can compensate for pressure and fluid temperature variations. (*Courtesy Vickers, Inc.*)

capable of compensating for both pressure and fluid temperature variations. Again the basic flow of fluid through the valve is controlled by orifice "O" and the pressure variations are compensated by changes in orifice "P" as before. If fluid temperature changes occur, the temperature (hydrostatic compensator) compensating rod length changes, thus increasing or decreasing the orifice opening "O" to compensate for the change in fluid viscosity due to temperature change. An additional feature of this particular valve includes a check valve which permits free flow return of fluid from the downstream side. Another version of the flow control valve includes an overload relief valve, shown in Fig. 8.34, which is self-explanatory.

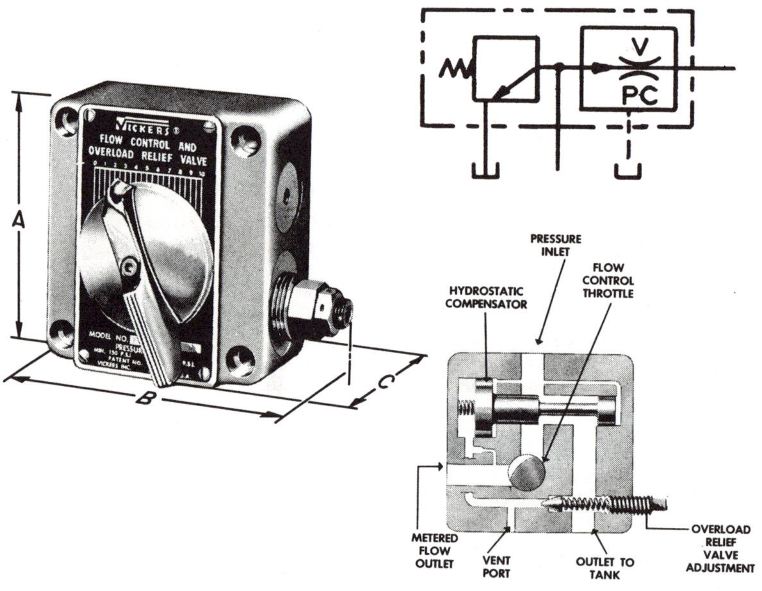

FIG. 8.34 Flow control with overload relief valve. (*Courtesy Vickers, Inc.*)

Electrohydraulic Flow Control Valve

The rate at which an injection plunger or a clamping cylinder piston operates is important to the success of the operation. The electrohydraulic flow control valve is a device which provides infinitely variable flow rates. These flow rates are obtained by regulating the magnitude of the direct current applied to the coil in the electrohydraulic control head.

Figure 8.35 schematically shows how this is accomplished. A constant, pressure compensated flow of oil is passed through the orifice created by the poppet in

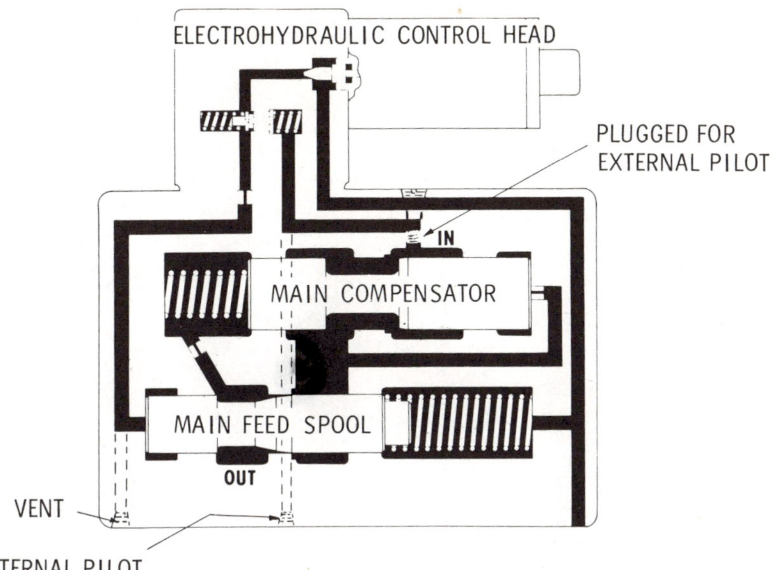

FIG. 8.35 Electro-hydraulic flow control valve schematic. (*Courtesy Racine Hydraulics Inc.*)

the electrohydraulic control head. The size of this orifice is inversely proportional to the current magnitude in the D.C. coil.

The resultant pressure which is remotely adjustable by varying this D.C. signal, serves to position the main feed spool against its bias spring. The spool, is directly proportional to the magnitude of the pressure established by the control head orifice, which of course is dependent on the value of the electrical signal.

The main stream flow, which is established by the position of the main feed spool, is automatically regulated, regardless of inlet pressure variations by the action of the main compensator.

The main compensator spool is spring offset to a normally open position, and by piloting from the differential pressure across the main feed spool, the compensator automatically modulates the size of its opening to the down stream side in series with the main feed spool and thus maintains a constant pressure differential across the opening established by the main feed spool.

Since the pressure differential will thus be maintained constant across the feed orifice, pressure compensation is accomplished and inlet pressure variations will have minimal effect on set flow rates.

Figure 8.36 depicts a valve of this type. The flow control valve is one of the most important of the electrohydraulic control valves since so many functions of hydraulically operated machines is flow rate dependent.

FIG. 8.36 The electrohydraulic flow control valve. (*Courtesy Racine Hydraulics Inc.*)

HYDRAULIC CIRCUITS

To illustrate circuit principles, several hypothetical hydraulic circuits will now be developed. A simple hydraulic circuit is shown in Fig. 8.37. This consists of a pump "A," whose maximum pressure is governed by relief valve "B." The oil to the pump passes through a strainer in the suction line. This is an important component in any hydraulic system, since pump life, valve and other system functions depend upon clean oil. The manually operated four-way valve is fitted with a spool which passes the fluid from the pump directly back to the tank when no work is required. This results in minimum oil heating and load on the driving motor.

When the spool is shifted to the right, fluid passes from "P" to "A," and the cylinder ram moves forward. Oil in the rod end of the cylinder flows from "B" to "T." When the ram bottoms or full work force builds up, the excess pump pressure discharges through the relief valve "B." Upon completion of the work cycle, the four-way valve is shifted to the left and the fluid flows from "P" to

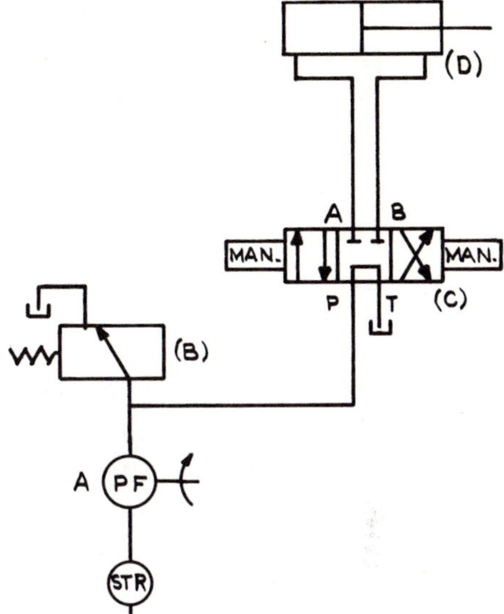

FIG. 8.37 Simple hydraulic circuit.

"B"; the return oil flows from "A" to "T." The ram is thus returned to its retracted position.

In Fig. 8.38 is shown a somewhat more complicated circuit. This is designed to control the movement of two cylinders. Cylinder (1) is required to complete its work stroke before Cylinder (2) moves forward. Examination of the schematic shows that the system is protected by a relief valve to monitor the maximum system pressure. Also shown is an unloading valve, whose function will be explained later.

The four-way valve is direct solenoid operated. When voltage is applied to the solenoid on the left side, the spool moves to the right, and oil flows from "P" to "A." Cylinder (1) thereupon moves forward at a rate dictated by the variable flow control valve (1) setting. When full pressure or work force is built up in Cylinder (1), the sequence valve spool is unseated, as controlled by its spring setting, and the oil flows through it to Cylinder (2), causing it to move forward to do its work. The rate of movement is controlled by the setting of the Flow Control Valve (2). While the rams are forward performing their work, the full volume output of the pump is unnecessary. The excess oil would be dumped through the relief valve resulting in undue oil heating. To circumvent this, an unloading valve is inlcuded in the system. Operation of this valve is initiated by the pilot pressure on the valve spool, and the pump output is returned with a minimum of restriction directly on the tank. This results in minimum oil heating. System pressure is held by check valve "V."

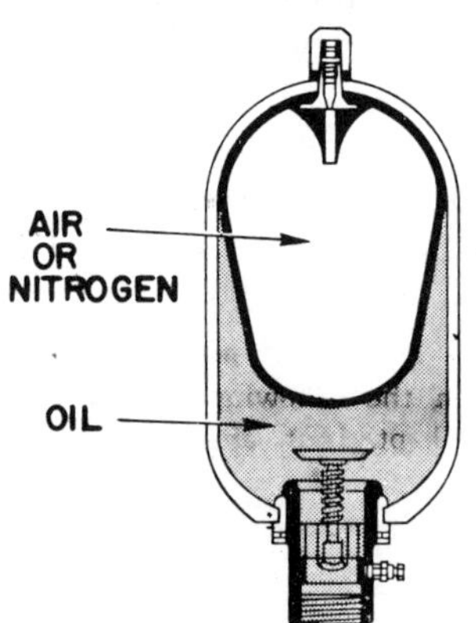

FIG. 8.38 Complex hydraulic circuit showing use of sequence and flow control valves.

The components of any hydraulic system have an internal leakage which will reduce the system pressure. In this circuit we have included an hydraulic accumulator to make up this leakage and pressure loss. A bag-type version of this device is illustrated in Fig. 8.39.

FIG. 8.39 Bag-type hydraulic accumulator. (*Courtesy Machine Design*)

An inert gas, such as nitrogen, is charged into the rubber bag to maintain a basic pressure. During the off-duty cycle of the system, rams retracted and not working and with the four-way valve spool centered, oil flows into the accumulator casing, causing the bag to collapse and compressing the nitrogen charge. The pressure builds to the system maximum and the accumulator is considered charged. When the four-way valve shifts, the accumulator discharges as the nitrogen expands, initiating the cylinder motion. As the pressure drops, the check valve is unseated and the pump supplies the oil for the completion of the cylinder work. When the rams bottom, the accumulator is again recharged, whereupon the check valve seats and the pump unloads as previously described. At this time the oil in the accumulator makes up the leakage losses, and maintains system pressure. Return of the rams is initiated by applying voltage to the right hand solenoid, which shifts the valve to the left, causing oil to flow from "P" to "B," returning from "A" to "T." The return flow from cylinder (2) is unimpeded by the unseated check valves in the flow control and sequence valves.

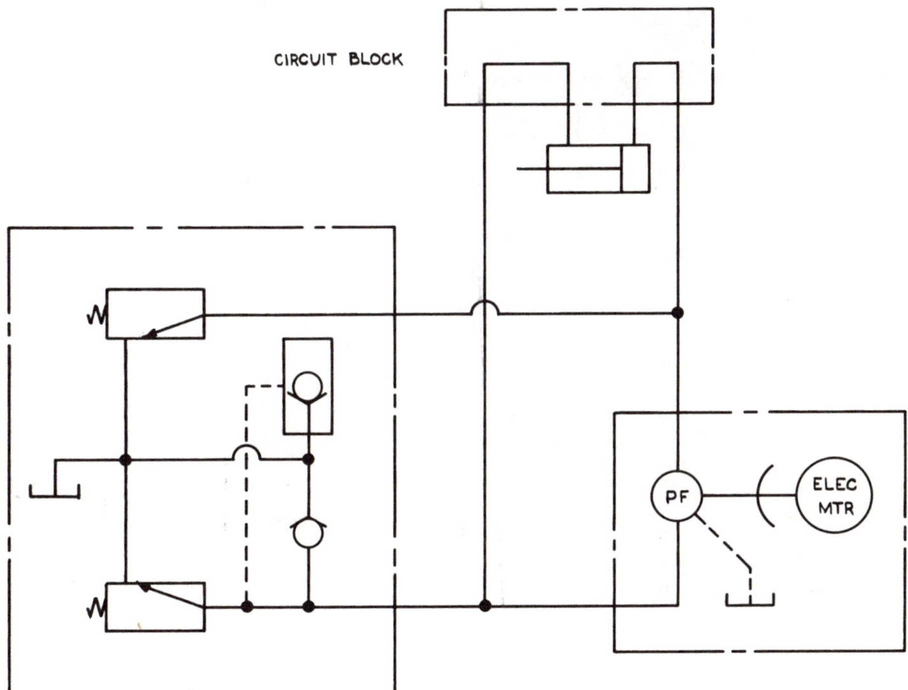

FIG. 8.40 On the advance stroke, fluid drawn from the reservoir is pumped to the blind end of the cylinder. Discharge fluid from the rod end adds to that flowing to the pump. On the retract stroke, fluid from the blind end of the cylinder is pumped to the rod end. As pressure builds up, the pilot-operated check valve opens, returning excess fluid to tank. The relief valves set different pressures on the advance and retract strokes. (*Courtesy The Sheffer Corp.*)

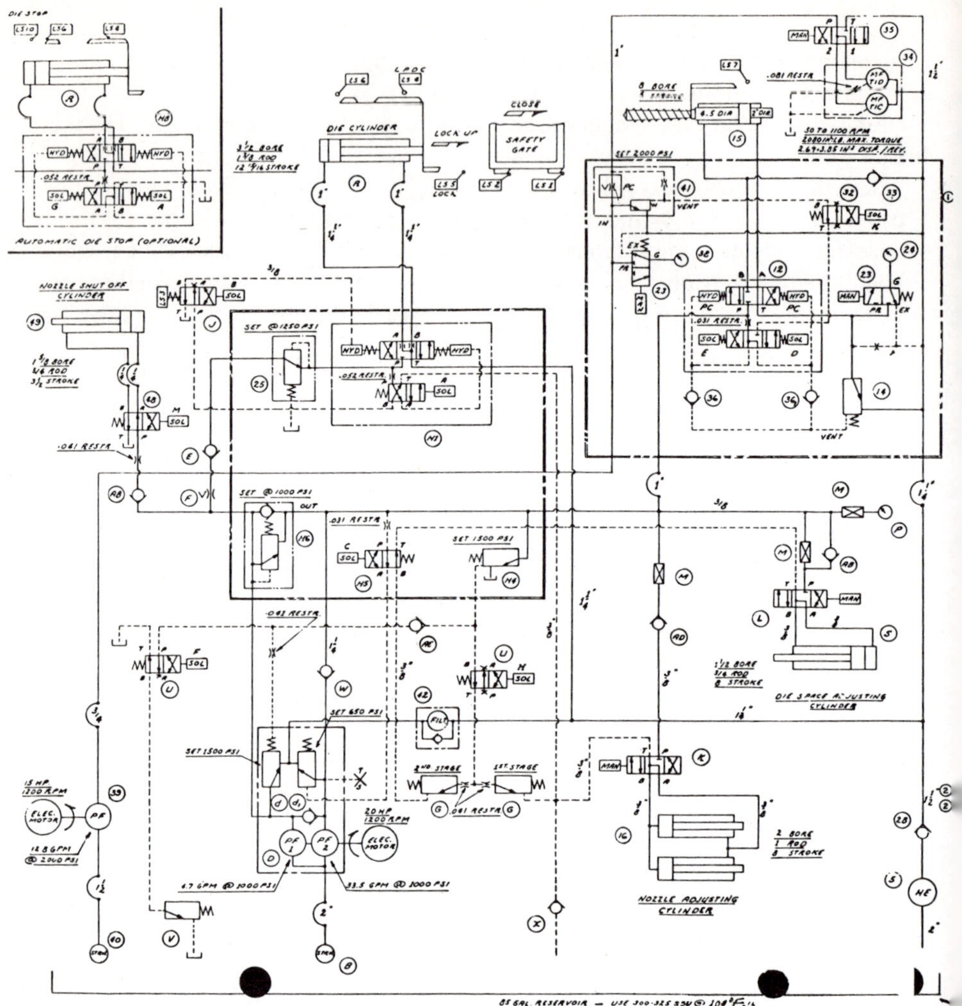

FIG. 8.41 Courtesy, Reed-Prentice Div., Package Machinery Co. (See pages 357 and 358 for General Description and Sequence of Operations.)

GENERAL DESCRIPTION

Pump (39) protected by strainer (40) delivers oil solely to fluid mtr. (34). Flow control valve (41) governs speed of mtr. (34), and also limits max. resultant pres. from (39).

Large volume (PF-2) and high pres. (PF-1) pumps, protected by strainer (B) provide oil for the remainder of the system. Filter (42) provides 25 Micron filtration.

Injection fluid mtr. and plunger back pres. are indicated on gages (P), (24) and (38), respectively. Gage (P) is protected by shut-off (M) and (24) and (38) by isolators (23).

Selector valve (35) directs oil from (39) to either or both cartridges within (34).

Check valves (X) and (28) allow servicing of components without draining the reservoir.

Pump (PF-2) unloads through valve (d_1) when system pres. exceeds its setting, providing sol. H of valve (U) is not energ. If H is energ., the pilot circuit of (d_1) is interrupted causing it to close so that max. pres. attained by oil from pump (PF-2) will be limited to setting of relief valve (d) or (H4) whichever is lower.

Max. Pres. resulting on (PF-1) is limited by relief valve (d). Check valve (w) prevents pump (PF-1) from unloading through (d_1).

Check valve within (H6) allows large pump (PF-2) to enter the die cylinder circuit, but prevents the high pres. pump (PF-1) from entering the plunger cylinder circuit except through valve (H6) at its set pres. Essentially, this valve allows (PF-1) to be effective at the plunger. During the "cure" when (PF-2) is unloaded, without any appreciable lowering of pres. at the die cylinder in spite of the fact that valve (H4) may be set to relieve at a low pres.

Throttle valve (F) is for die speed control, and check valves (B), (21) and (AD) minimize effect of pres. drops on cylinders (R), (49) and (16) respectively. Cooler (5) is provided as a means of regulating oil temperature between 100 and 115°F.

Cylinders (16) and manual selector valve (K) are for positioning the nozzle against or away from the sprue bushing. cyl. (s) and manual selector valve (L) are for die height adjustment and check valve (AB) prevents breaking the rod of cyl. (S) in the event that the dies were closed without the tie-bar nuts first being seated.

SEQUENCE OF OPERATIONS

(L.P.D.C. "ON")

1. Operator closes safety door to trip LSI and 2 and hold sol. B of safety interlock valve (J) energ.
2. Spool of interlock valve shifts to trip LS-3 and hold sol. C, H & A energ. This prevents pump (PF-2) from unloading thru valve (d_1), blocks 2nd stage pilot relief valve, blocks remote control of main relief valve (H4) and directs pilot oil via (J) to shift slave spool of (HI) so that combined del. of pump (d) goes to headend of cyl. (R).

   Dies start closing with max. pres. resulting on pump (PF-2) limited to setting of (d) or (H4) whichever is lower, and that resulting on (PF-1) limited by (d).
3. LS-4 is released to hold sol. F energ. and connect remote circuitry of relief valves (H4) and (d) to pilot relief valve (V) which is adjustable for low pressure.
4. If no obstruction prevents dies from "kissing" LS-4 is re-tripped to reset alarm timer TD1 and DE-energ. Sol. F which blocks remote circuitry of #3 above and dies lock-up at max. pres.
5. At lock-up LS-5 is tripped to hold sol. D Energ. and shift slave spool of (12) to direct combined delivery of pumps (d) to the head-end of cyl. (15). Plunger starts injecting and relief valve (14) is vented through (36_1).

FIG. 8.41 (*Continued*)

6. LS-7 is released to de-energ. sol. H so that injection pres. is limited to setting of 1st stage pilot relief valve (G).
7. T3 times out DE-energ. sol. C allowing large vol. pump (PF-2) to unload through (d_1) and plunger remains "IN" on pump (PF-1) at pres. limited by setting of 2nd stage pilot relief valve (G). Valve (H6) disallows pres. below its setting at die cyl. (R).
8. T1 times out DE-energ. sol. D and causing sol. E to be held energ. which shifts (12) to direct oil from pumps (D) to rod-end of (15) and start plunger back. Relief (14) is vented through (36).
9. TD2 times out DE-energizing solenoid (E) which pressure centers slave spool of (12). This blocks both the pres. and cyl.-port-to-rod-end of cyl. (15) and connects the head-end of (15) to tank through relief (14) which is now devented because both (36) and (36_1) are held seated by pilots pres. from (12). Also sol. K is held energ. which devents relief (41) to direct oil from (39) to fluid mtr. (34) causing it to rotate the screw plasticizer. As the screw rotates, it causes cyl. (15) to move back exhausting oil from its head-end with back pres. regulated by (14). At this time, "make-up" oil is supplied to the rod-end via check valve (33).
10. Cyl. (15) continues back until LS-7 is tripped which DE-energ. sol. K to again vent (41) and divert oil from pump (39) to tank, thus stopping mtr. (34).
11. T2 times out to DE-energ. sol. A and safety interlock sol. B thus opening dies and releasing LS-3.
12. Dies open completely to trip LS-6 and operator opens gate to release LS-1 and 2. End cycle.

FIG. 8.41 (*Continued*)

Several of the components in this schematic are circumscribed by a dashed line. This implies that all the components shown within the dashed line are included in the valve body. This is referred to as a "Component Enclosure."

An interesting circuit which eliminates the four-way operating valve is shown in Fig. 8.40. The reciprocating action of the cylinder is obtained by reversing the direction of the three-phase motor driving the gear pump, thus causing the oil to flow in the desired direction. This system is used to power a stripping machine.

The modern injection machine hydraulic circuit includes many of the features just discussed. A typical circuit of a late model screw-injection machine is shown in Fig. 8.41. A components list, general circuit description, and sequence of operations is included. Careful study will be valuable in gaining understanding of circuit principles.

SYSTEM RESERVOIR

The hydraulic system fluid reservoir and its accessory equipment is of fundamental importance in any hydraulic installation. Figure 8.42 depicts a reservoir which includes the features conforming to the Joint Industries Conference (J.I.C.) specifications. The tank is of welded steel construction with large removable endplates to facilitate cleaning. The bottom is either sloped or concave and

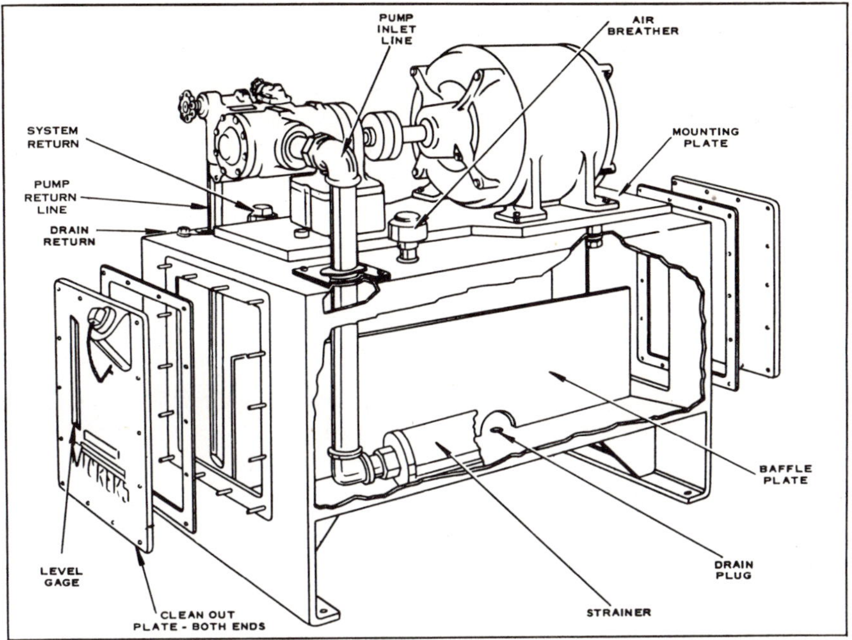

FIG. 8.42 Hydraulic system fluid reservoir. (*Courtesy Vickers, Inc.*)

fitted with a drain plug so that complete draining may be accomplished. A magnetic plug is often times used to trap steel and iron particles, and is highly recommended as a system safeguard.

The tank is also fitted with an oil level gauge and a capped filler equipped with a fine screen filter to prevent introduction of foreign matter when the tank is filled. The breather mounted on the top of the tank must contain a filter to clean the air entering the tank, and should be sized to maintain atmospheric pressure in the tank. As shown, the pump inlet is fitted with an inlet strainer and the inlet pipe is secured to the tank with a packed flange plate of such size that the pipe and filter can be removed from the tank for inspection and cleaning. Cleaning is essential for good service.

Suction and return lines must be placed near the bottom of the tank. Drain lines from valves usually enter the tank above liquid level to prevent back pressure on the valves.

The tank is also equipped with a longitudinal baffle plate which separates the suction and return side of the system. The baffle plate prevents direct recirculation of the fluid and permits foreign particles to settle to the tank bottom, thus helping maintain a uniform fluid temperature.

FILTERS AND STRAINERS

The Joint Industry Conference (J.I.C.) has defined the function of the strainer as "a device for the removal of solids from a fluid wherein the resistance to motion of such solids is in a straight line." The filter is defined as "a device for the removal of solids from a fluid wherein the resistancy to motion of such solids is in a tortuous path." Thus a strainer is generally constructed of fine 60 to 100-mesh wire screen wrapped around a metal supporting frame. They are usually installed on pump inlet lines to offer the least pressure drop and yet prevent damaging solids such as grit, sludge, rust, etc., from entering the system. The flow as defined is in a straight line through the wire screen, protecting the pump and other system elements from large particles.

Filters are made of many materials such as plastics, wood cellulose, felt, sintered metal, activated clay and fuller's earth. Elements such as activated clay and fuller's earth should not be used with fluids containing additives, since they will remove the additives as well as the impurities. The fluid flow through these units is in a random or "tortuous" path. Filters are sometimes installed in the pressure line, and thus subjected to full line pressure which is not desirable. More commonly they are used in the return line where the filter operates at substantially reduced pressure. In either case, its purpose is to remove damaging solids from the system.

HEAT EXCHANGERS

Efficient operation of any hydraulic system is to a large extent dependent on maintaining a reasonably low fluid temperature in the order of 100° to 140°F. Higher temperatures adversely affect the fluid viscosity, thus resulting in reduced pump efficiency, by-passing in valves, deterioration of seals, excess water, etc.

Unless the system components such as the reservoir, piping, etc., are capable of dissipating the heat generated by the operation, it is necessary to install a heat exchanger in the system. A typical device is shown in Fig. 8.43, and is installed in the low-pressure return-side of the system. This shell and tube heat exchanger is water cooled. In situations where the use of water is too expensive, air cooled units are used. These are similar to an automobile radiator with the oil flowing through the core and with a fan blowing air through it to effect the cooling.

HYDRAULIC FLUIDS

The basic function of the fluid in the hydraulic system is to transmit power from one point to another. In the modern hydraulic system, the fluid must also provide lubrication, corrosion prevention, and sealing to the system components. For plastics machinery applications, petroleum oils fulfill the requirements.

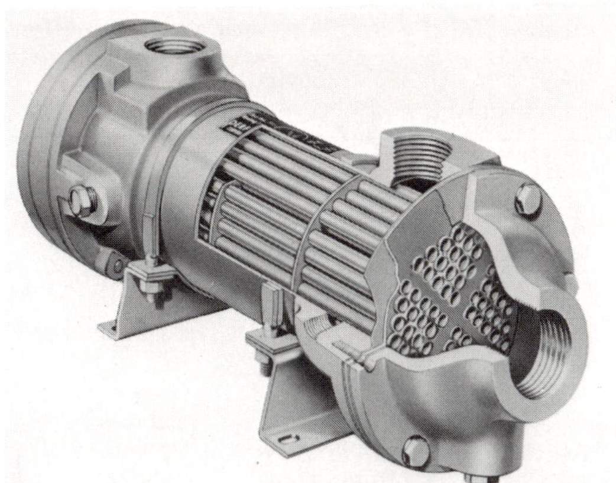

FIG. 8.43 The tube type heat exchanger is used to cool the hydraulic oil and maintain correct fluid temperature. (*Courtesy Perfex Corp.*)

Machine and pump manufacturers specify the oil to be used in the machine. However, an understanding of the specifications is important.

Briefly, the important considerations in a hydraulic fluid used in plastics processing machinery are: viscosity, rust, oxidation and wear-inhibiting qualities. The oils used in the majority of plastics-molding machinery are mineral (petroleum), oil-based and compounded with additives to provide the inhibiting qualities stated above. Quality performance requires regular oil checks.

The viscosity of hydraulic oils is most often stated as time in seconds for 66 cc of the oil under test to flow through a standard orifice under a standard head at a standard temperature of 100°F or 210°F. The apparatus used is the Saybolt University Viscosimeter, and the results are thus stated as "Saybolt Seconds, Universal" or "SSV."

The wear-inhibiting additive is perhaps the most important additive to the base oil. This is a recent development based on the addition of small percentages of dithiophosphates to the oil. Published test data show extraordinary reduction in wear of pump elements, and it can be safely assumed that the same would be true in wear reduction of valve spools and other system elements. The dithiophosphate also provides oxidation stability and rust-inhibiting properties to the oil.

In recent years the demand for fire-resistant fluids has increased, especially in diecasting machine applications where a considerable fire hazard exists. The types of fire-resistant fluids are water-in-oil emulsions, glycol-water, and phosphate ester or silicone-based synthetic fluids. The synthetic fluids are fire resistant because of their chemical nature. The emulsions and the glycol-water fluids get their fire resistance from the blanketing and cooling action of the steam that is

TABLE 8.1 COMPARISON OF FLUID PROPERTIES*

		Fire-Resistant Types		
	Petroleum Oil	Emulsions (Water-in-Oil)	Glycol-Water	Synthetic (Phosphate Ester)
Cost	X	2 to 3X	3 to 4X	5 to 6X
High-temperature service	Life shortened at bulk temperatures above 160 to 180°F	Excessive water evaporation above 150°F	Excessive water evaporation at elevated temperature.	Stable at relatively high temperature.
Low-temperature service	Fair to good	Poor	Good	Poor to fair
Oxidation stability	Good	Good	Good	Excellent
Rust protection	Good to Excellent	Good	Fair to good	Fair to good
Foaming	Excellent	Excellent	Good	Good
Viscosity index	Good	Excellent	Excellent	Poor
Effect on seals and gaskets	Oil-resistant materials required.	Compatible with seals used for petroleum oils; materials made from paper, fiber, leather not satisfactory.	Not compatible with some types of cork, paper, leather, and synthetic fiber materials.	Not compatible with seals used for petroleum oils; materials resistant to these fluids must be used.
Effect on paints and insulation	Oil resistant paints required. No specific problem with insulation.	Same as petroleum oil.	Not compatible with some paints.	High solvency for most paints and insulation.
Lubricity	Excellent	Good	Good	Excellent
Heat transfer	Good	Excellent	Excellent	Good
Compressibility	Moderate	Moderate	Moderate	Low

*Courtesy *Machine Design.*

formed when the fluid strikes a hot surface. Table 8.1 exhibits the properties of the various hydraulic fluids that have been discussed.

HYDRAULIC NOISE

Hydraulic systems can be a great source of noise contamination in the plastics molding plant. The more important sources of this noise derive from pumps, valves, hydraulic plumbing and electric drive motors.

Present day pumps have been designed to meet the Federal Noise Regulations but unless operation of the pump is correct, noise will develop. If air is introduced into the system through any source, for example, through the use of a poor grade of hydraulic oil without anti-foaming additives, noise will result. Oil temperatures must be held at from 100 to 140°F. High oil temperatures will drop the oil viscosity causing noise as well as excessive pump wear. If the coupling between the motor and the pump is worn, noise will result.

Chattering valves, because of hydraulic instability, are a noise source. A pilot operated relief valve is preferable to a direct operated relief valve as far as this characteristic is concerned. The piping system must be adequately sized to keep the oil velocities low (15 fps on pressure lines and 5 fps on suction lines). Sharp bends must be avoided. The oil reservoir must be properly sized and designed to permit free escape to air evolved from the oil. Baffles are required in the tank to preclude resonant conditions being set-up and the suction to the pump must be far removed from the return lines.

Tremendous hydraulic shocks can be generated when the fluid rapidly changes direction and velocity. This will occur when a solenoid operated directional valve functions. These shocks are deleterious to the system, placing substantial overloads on the piping system in addition to making a lot of disconcerting noise. A shock depressing valve installed in the system will negate this problem and protect the pump and the system.

9 THERMAL AND MACHINE CONTROL

All plastics processes rely on the application of heat, pressure and time to convert the material to its new form. This is true of both thermosets as well as thermoplastics. The essential difference lies in the fact that the thermosetting product can be removed from the mold at molding temperature, while the thermoplastics must be cooled in the mold so that they may retain their new shape.

HEATING METHODS FOR THERMOSETS

Methods for heating thermoset molds are:

(1)—Direct: by passing steam, superheated water, heat transfer fluids, etc., through cored passages drilled into the mold frame. Oftentimes the same effect is achieved by fitting electrical cartridge heaters into the cored holes in the frame. Gas flames directed against the mold are sometimes used to obtain extra high heat.

(2)—Indirect: by placing the mold on press platens heated in the same fashion as described in (1).

(3)—Mechanical Work: by use of an extrusion screw or elastic rotor. In any event, sufficient heat energy (Btu or calories) must be added to the molding material so that the resin can flow and polymerize during the molding cycle.

Thermosetting resins are heated in the mold; often they are preheated before molding. Preheating is used to speed the molding cycle, as discussed in Chapter 5. Several preheating methods are used: (a) hot plate, (b) oven, (c) infrared lamps,

(d) steam, (e) high-frequency dielectric. In limited molding operations, thermosetting preforms or powder may be preheated by placing the charge on a hot plate or in a small oven. Large charges are difficult to preheat by this method because the material is a good thermal insulator, and the interior of the charge remains cool while the surface of the charge, in contact with the heating plate, may become too hot and precure. Infrared lamps do a fairly satisfactory job of preheating small charges for slow molding cycles. Preforms or powder may be exposed to live steam in an oven for prewarming, and this is fast and effective for some work. Difficulty may be experienced by the moisture that is left in the compound and this system has a limited field of application. Best preheating comes from mechanical working of the compound in an extruder system or by the use of internal heat generated in the material by a high frequency preheater.

ELECTRONIC HEATERS FOR THERMOSETS

In high-frequency electronic preheating, the compound is placed between electrodes which are subjected to radio frequency energy in the order of 70 to 80 megacycles. The molecules of the resin between the electrodes are thus molecularly excited, causing frictional heat which brings the entire compound mass to the desired plastic state. The compound is removed from the preheater at the proper time and immediately placed in the hot mold. The additional heat to which the material is subjected during the molding cycle completes the polymerization. While the molding cycle progresses, the preheater is loaded with material for the succeeding cycle. Thus, preheating shortens the overall cycle in molding thermosetting materials. Because of the completely uniform heating obtained in electronic preheating, a more complete "cure" or polymerization is gained, resulting in best physical, chemical and electrical properties for the molded product. It is important to use hard preforms of uniform density to gain optimum preheat value.

Figure 9.1 illustrates a 5 KW preheating unit which is also equipped with a preform rotator to insure maximum uniformity of the preheat. Precise timing is controlled by a cycle timer.

THERMOPLASTICS

Thermoplastics are brought to a plastic state by means of an external heating device such as an electrically heated injection cylinder, or screw-injection unit. These are fully described in Chapter 4. The hot plastic mass is forced from this "melt plasticizer" through a suitable passageway into the mold cavity. Since the mass must lose heat energy to obtain a solid state, the mold is cool, as compared with the temperature of the injected plastics.

The temperature of the mold depends on the plastics material being used, the product and mold design. Some molds are best run at an elevated temperature,

FIG. 9.1 Five-KW dielectric preheater used to heat thermosets. Preforms are rotated during the preheat period to insure uniformity of temperature. (*Courtesy W. T. LaRose & Associates, Inc.*)

perhaps 250°F; others are run with chilled water or refrigerants as low as 20°F, and still others at intermediate higher levels.

Temperature control in injection molds is obtained by passing a controlled temperature fluid through passages drilled into the mold frame (or base), and through cavities and cores. Sometimes when a high-temperature mold is necessary, such as in molding F.E.P.-"Teflon," the mold is fitted with electrical cartridge heaters controlled by a suitable pyrometer.

The simplest and most commonly used cooling system is derived from passing tap or well water through the mold. The desired temperature is achieved by throttling the outlet water from the mold. Water from the mold is often wastefully dumped to the sewer.

In any event, the productivity of the mold is dependent on the rate at which heat energy is removed from the injected material. This is a function of the mold design as well as the coolant temperature and the rate and turbulence of the coolant flow.

Chillers

Lower temperatures to a practical value of 20°F and above are obtained by use of "beside-the-press" or central water-chilling systems. These devices depend on mechanical refrigerating or evaporative cooling units to bring the cooling fluid to the required low temperature. Details on their operation can be obtained from any mechanical engineering handbook. A "beside-the-press-unit" refrigerating unit is adjusted to provide the precise temperature requirements of the mold it is servicing.

In most molding shops the temperature requirements for the molds vary from one mold to another, as well as in a given mold. It may be necessary to run the mold force and the cavity at different temperatures. With a central system, the mold temperature is maintained in each mold by means of an "at the press" heat exchanger, which independently circulates fluid through the mold at its "set" level, using the fluid from the chiller to remove heat from the circulated fluid at the required rate. Any number of heat exchangers may be used in the system to obtain other temperatures than the basic output of the chiller. The chiller service may also be directly connected to the molds that require the lowest temperature fluid.

Beside-the-press and central chilling systems, a closed system should be located where all the fluid is recirculated. This is quite important when water cost and supply problems are considered.

Elevated Temperature Controllers

For cooling molds at temperatures above tap water and up to 250°F, recirculating water temperature controllers are used. Two systems are available—the additive and the closed.

The additive system consists of a small volume tank fitted with electrical resistance heaters, a high-velocity high-volume rotary pump to circulate the water through the system under pressure, and an adjustable modulating type temperature control valve to maintain a "set" temperature by adding cold tap water directly to the system and expelling hot water from it as required. A thermostat must also be installed to control the electrical heaters. Thermometers to indicate "in" and "out" water temperatures should also be included.

In operation at elevated temperatures, the heaters quickly bring the water to "set" temperature and, through the thermostatic control, add further heat only when required. If an override should occur, the modulating valve brings in cold water to reduce the circulating fluid to "set" temperature. When the unit is used for cooling only, the heaters are off; and again the modulating valve adds cold water and dumps hot water to maintain the "set" temperature. The additive system provides very close temperature control ($\pm \frac{1}{2}$°F), and fast action in changing from one temperature to another. The units are self-contained and readily portable. A typical unit is shown in Fig. 9.2. The disadvantage is that

FIG. 9.2 Elevated temperature controller for injection molds. (*Courtesy Mokon Division, Protective Closures Co., Inc.*)

chemical deposits can occur since raw tap water circulates in the system. Experience has shown that this is minimal since, when the unit is used as a heater, the system is closed. When a small volume of water is in the system, little deposition can take place. During cooling operation, precipitation is at a minimum under any circumstance.

The closed system consists of a tank fitted with electrical resistance heaters, a heat exchanger coil, and an adjustable thermostatic switch. A pump is also installed to circulate water in the system. In operation, the heaters bring the water to "set" temperature. If an override occurs, the thermostatic switch energizes a solenoid valve which allows tap water to circulate through the heat exchanger, thus reducing the tank water temperature. When set temperature is reached, the solenoid valve closes, stopping the tap-water flow. Cycling is continuous to hold "set" temperature.

The prime advantage of the closed system is that the circulating fluid can be treated to eliminate mold corrosion and chemical deposits such as obtained from untreated water. The disadvantages are that the system usually contains a large volume of water which takes considerable time to bring to a set temperature. Changes from one temperature to another are also time consuming. Cooling of the tank water is expensive, since the efficiency of removing heat through an exchanger is much lower than by adding cold water directly to the system.

Heating of the water is also more expensive, since higher wattage heaters are required to heat the large volume of water in the tank in a reasonable time.

DRYERS

Many of the thermoplastics require predrying to remove surface or included moisture to insure optimum physical or optical properties. Among these are polycarbonate, A.B.S. and F.E.P. Teflon.

Drying of these plastics is accomplished in trays in circulating air ovens fitted with desiccators or dehumidifiers. This is a batch-type process requiring the rapid emptying of trays into the machine hopper and reloading the trays with the next batch.

A more sophisticated and labor saving method is to use the dehumidifying hopper dryer which will dry properly and supply the material directly to the machine hopper as needed. Many of these devices are furnished with two dessicant supplies which allows one to be regenerated while the other continues to dry the material. This ideal arrangement provides a continuous supply of dried material.

The drying time and temperature are very important. Too low a temperature, of course, does not properly dry the material and one that is too high can agglomerate the material so that it cannot be fed into the process. The resin manufacturers' drying specifications must be followed carefully.

Process Temperature Control

Temperature control is the most important aspect of any plastics process. This control is accomplished by means of a sensing device called a thermocouple, a meter to express the value, and a relay system actuated by the meter to control the power to the electrical heaters.

Thermocouples. A thermocouple consists of two dissimilar metals welded together at one end. When the welded joint is heated, a small voltage is generated. This voltage is directly proportional to the temperature at the joint. The voltage generated is quite small, and is measured in millivolts.

There are four types of thermocouples in common use. These are copper-constantan whose range is to $600°F$, iron-constantan to $1600°F$, chromel-alumel to $2100°F$, and platinum rhodium-platinum whose upper limit is $2900°F$. The application of a particular type is of course dependent on the temperature range to which the thermocouple will be subjected. In plastics processing equipment, the choice is iron-constantan.

Location of Thermocouples. The location of the thermocouple on the injection cylinder or extruder barrel is of great importance for proper operation of the equipment. The usual location is shown in Fig. 9.3. The dashed line (2) in the graph shows the temperature at the thermocouple. The fluctuation is due to the large heat capacity stored in the barrel wall. Even though the thermocouple

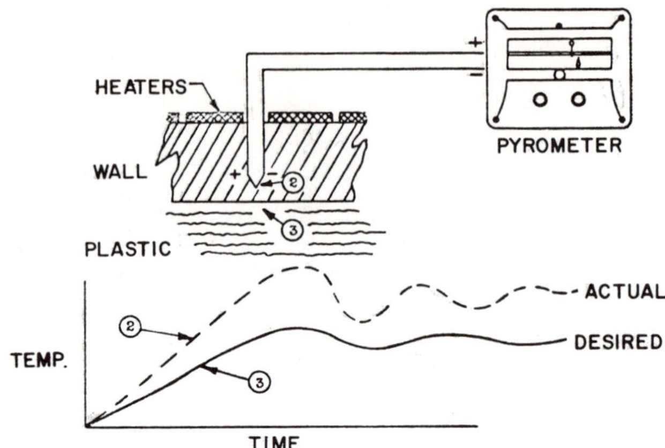

FIG. 9.3 Poor location of thermocouple on injection machine cylinder and resulting heater fluctuation. (*Courtesy West Instrument Co.*)

signals the control instrument to shut off the power to the heaters, the residual heat in the barrel causes the temperature to override. This is known as thermal inertia, and functions as well on the descending portion of the temperature curve. In other words, even though the thermocouple signals for heat, the temperature continues to fall although the heaters are on.

In one instance of record this phenomenon resulted in approximately $\pm 10°F$ variation in the plastics melt temperature. The product being molded varied in

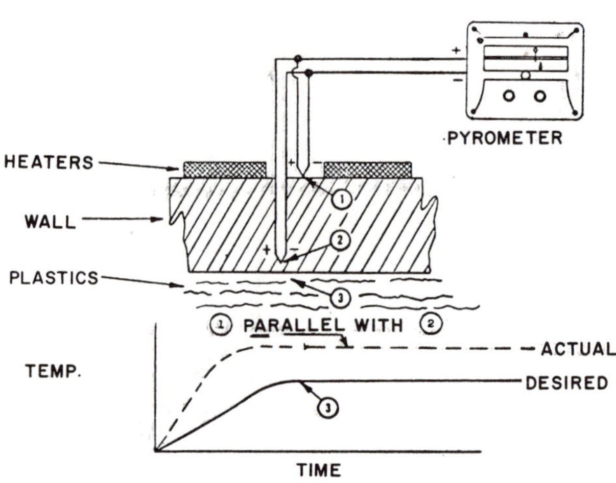

FIG. 9.4 Best thermocouple installation to minimize thermal variation or "hunting." (*Courtesy West Instrument Co.*)

size in direct relation to the temperature variation, and tolerances could not be maintained. The problem was solved by installing dual thermocouples, as suggested in an *SPE Journal* article, by R. K. West. The schematic of the installation is shown in Fig. 9.4. When connecting thermocouples and thermocouple lead wires to the instrument, polarity must be strictly observed or else the instrument will indicate false readings. Plus must always connect to plus, and minus to minus.

Pyrometer Temperature Controllers

The pyrometer temperature controller is an instrument used to measure the temperature of the system and to control its value. The instrument consists of a millivoltmeter which measures the temperature, and a relay system for controlling the heaters.

Pyrometer controllers are available as simple on-off, proportioning and stepless controllers of several types. The on-off type controller is not often used where precise control of a plastics process is desired; temperature under- and override are common problems with this type of instrument.

Proportioning Controllers. The proportioning pyrometer controller is in very general use on plastics processing machinery, such as molding machines and extruders. The purpose of the proportioning system is to "straighten out" the temperature control curve to achieve minimal over- and under ride of the set temperature. The effect is graphically shown in Fig. 9.5. All controllers of this type have three systems to accomplish the job and these are: an indicating system, a relay system, and a proportioning system. A brief description of one of several units on the market follows.

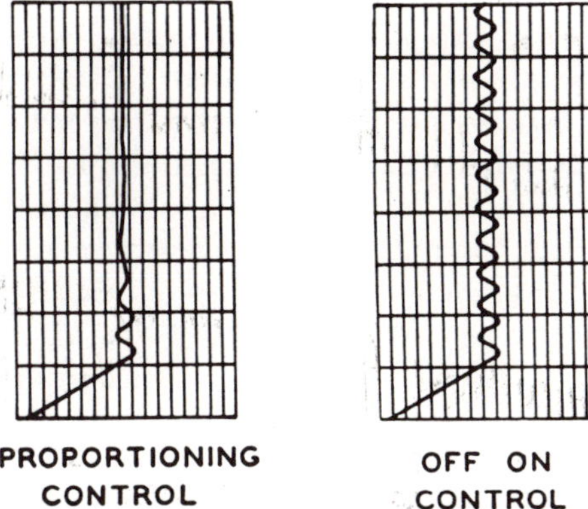

FIG. 9.5 The proportioning control system reduces thermal fluctuations that result when "on"-"off" control is used. (*Courtesy West Instrument Co.*)

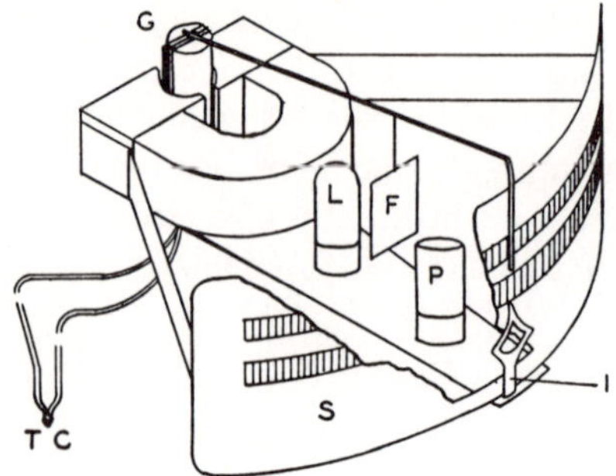

FIG. 9.6 Indicating portion of a proportioning controller. (*Courtesy West Instrument Corp.*)

Figure 9.6 shows the indicating portion of the instrument. Thermocouple "TC" is connected to the galvanometer "G." As "TC" is heated, the indicating pointer moves upscale. A manually adjusted index pointer "I" can be set at any desired point on scale "S." Mounted on the arm are a light source "L," and a photocell "P." As long as light from "L" strikes the photocell, sufficient current flows in the system to hold in a relay (not shown) to provide power to the heaters (usually through a contactor to carry the high heater load). However, when the opaque flag "F," attached to the pointer, interrupts the light to the photocell, the current flow is insufficient and the relay drops out. Power to the heaters is shut off.

The description given thus far is completely applicable to the operational function of the "on-off" controller as well as the proportioning instrument. The proportioning system depends on a rather complicated electronic circuit whose description is beyond the scope of this book. In the proportioning instrument, "the anticipatory action is accomplished by modulation of values within the electronic circuit when the instrument is "ON" and by reversing the procedure when the instrument is "OFF." For example, when the instrument is "ON" and calling for heat, the electronic circuit automatically lowers the point on the temperature scale where the relay will drop out and shut off the heat input. Now that the instrument is "OFF," the temperature indicating pointer starts down the scale due to cooling. The electronic circuit has changed and the point where the instrument will call for heat again is moving back up the temperature scale. The proportioning control action occurs sooner in each cycle until the true temperature indicating pointer becomes motionless, indicating a constant fixed temperature on most reasonably balanced thermal systems. The control system then continues to modulate and determine the required time "OFF" and "ON" to maintain this fixed temperature. The proportioning system has no moving parts or delicate filaments and is of permanent construction requiring no maintenance at any time." A typical proportioning controller is shown in Fig. 9.7.

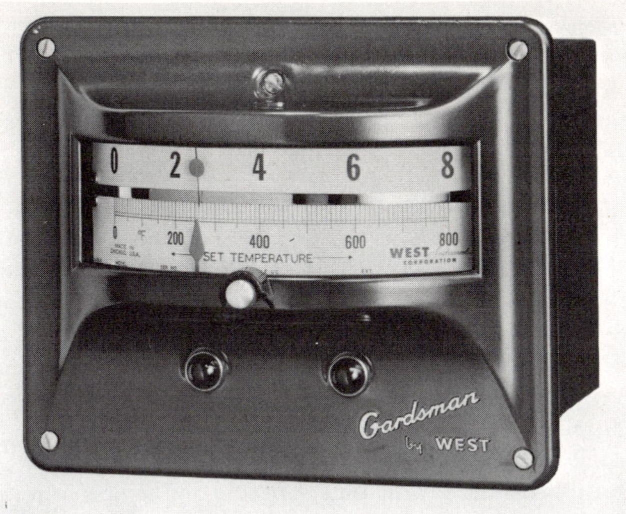

FIG. 9.7 Proportioning controller for plastics processing heat control. (*Courtesy West Instrument Corp.*)

Three-Position Temperature Controllers. In extruders, oftentimes the forward zone or zones will override the set point due to frictional heat. Air blowers or coolants, through coils wrapped around the barrel, are used to control these zones. A proportioning instrument fitted with an additional "ON-OFF" set point is used to control the cooling needs of the extrusion system. These instruments are sometimes called "Double Target Controllers."

Stepless Controllers. Stepless controllers are used where the ultimate in temperature control is necessary. These devices consist of a pyrometer control instrument used to vary the output of a power unit, which may in turn control a saturable core reactor for heavy power loads. The device operates on the principle of varying the power input (voltage) to the resistance heating units on the machine.

POST-MOLDING CONSIDERATIONS

Even though molding conditions may be carefully controlled and product and mold design expertly executed, molded-in strains will exist in the molded product. Proper conditioning of the product after molding can minimize these. The most common method used is to after-bake the product in an oven to allow the stresses to relieve themselves. This will also result in greater dimensional stability of the product. Parts, as removed from the mold, are oftentimes buried in dry sawdust and allowed to cool slowly to room temperature. In other instances, the parts, as ejected from the mold, are immediately submerged in a liquid bath where they are held and cooled in the bath. All methods, and there are others, have one goal—to relieve stresses. Specific methods vary in the large family of resins, and the material supplier should be consulted for applicable methods. A redesign for wall uniformity may help.

INJECTION PROCESS CONTROL

In Chapter 4 you were introduced to the principles of the injection process. Seven elements of the process were listed, all of which are important to successful practice. With close operator correction of variables as they occur, a high percentage of good quality parts will be yielded. But because of the scarcity of operators competent and willing to perform this function and the high cost of labor, other reliable means must be found. Significant progress in automatic control of the injection process has been achieved.

The injection process deals with time, pressure and temperature. To achieve *consistent* quality each of these parameters must be controlled accurately to fill the mold with exactly the same amount of polymer during each cycle. The sub-variables which most effect a drift from this situation are changes in the polymer melt viscosity and the rate of injection of the polymer into the mold cavity.

Material viscosity can vary as a result of barrel temperature variation, or change in the polymer itself. To fill the mold properly the pressure or temperature can be modified to compensate. The problem has been to observe and correct the drift soon enough to avoid poor quality production.

Equipment is now available which will automatically make the necessary adjustments as the variations occur. This is done by observing electronically the rate of injection during each cycle and comparing it to the set points (go/no-go) for acceptable production. If the rate drifts beyond the acceptable set points, the injection pressure or flow rate through feedback control is adjusted in small (positive or negative) increments to compensate. If a pre-set number of adjustment increments is exceeded, a signal bell, light, or if desired, machine shut-down will take place. The equipment to accomplish this is based on solid state devices.

The feedback equipment is only useful when operating the machine on an automatic cycle.

SOLID STATE CONTROL

Solid state control is a system of controlling the machine functions electronically through the use of low powered logic devices instead of relays and contactors.

In the conventional injection machine, process functions are controlled through limit switches, electro-mechanical temperature controllers and timers, solenoid operated valves, mechanically pre-set pressure control valves, etc. In the solid state controlled machine, these functions are taken over by transistors, diodes, silicon-controlled rectifiers, resistors, capacitors, servo-controlled pressure-control valves, solid state timers and temperature-control instruments. The question one might ask at this point is—why change? The answer is that the solid state system provides more precisely repetitive operating cycles with less set-up and down time for corrections or adjustments.

The solid-state control system simply replaces the relays with logic devices, the limit switches with linear transducers—usually potentiometers—with pressure being controlled by a direct current solenoid operated servo valve. Solid state temperature controllers with rate and integral feedback maintain temperatures within narrow limits.

Since there are no moving parts in the logic devices which take the place of the relays, "wear" in the order of billions of cycles as compared with the NEMA life ratings for relays of 50 million cycles is an outstanding improvement. The servo valves for pressure control modulates smoothly over the operating ranges, avoiding the hydraulic shocks common in the present system. This is an advantage in minimizing maintenance costs due to hydraulic leaks resulting from these shocks.

Heater barrel temperature control is also achieved through solid state instruments, and an explanation of these devices follows later in the chapter. Solid state timers are available in both analog or digital modes with the digital form considered more accurate and repeatable.

Solid-state control systems are available as part of the injection molding machine from most molding machine manufacturers, and are expected to become standard in the industry.

The parameters controlled by the solid state systems are held to much tighter tolerances. Timers are repeatably accurate to within $1/100$ second. Temperatures at the sensing point are held to $\pm 1°F$ with solid state controllers. Linear position of the injection screw as it determines shot volume and cushion is highly important and can be controlled to thousandths of an inch. All of this means that once a machine and job is properly tuned in it will continue to make good parts without readjustments. Scrap will be reduced.

Another real advantage lies in the repeatability of the parameters when next the job goes into production. In most shops the first time a job goes into production the set-up and tune-in are supervised by the engineer or molding room superintendent. When parts, satisfactory in all respects, dimension, finish, cycle, etc., are produced, the job goes into production. At this time all the settings are recorded on the set-up sheet and approved by engineering, manufacturing and quality control.

THE SOLID STATE SYSTEM*

GENERAL CONSIDERATIONS

Control Systems

Various components of plastics processing machinery must operate in some pre-determined sequence in order to produce the desired results. This is accomplished by the machine control system. The following discussion relates primarily to control systems for injection molding machines, but identical principles apply to extruders, blow molders, and other similar machinery.

Elements of a control system may be divided into three categories:

1. *Sensing elements* which provide information on the current status of various parts of the machine. These sensing devices may include limit switches or potentiometers to indicate position, thermocouples to detect temperatures, pressure switches or potentiometers to determine pressure levels, and similar devices.

2. *Decision Elements* On the basis of information derived from the sensing elements, the decision elements issue the proper sequence of commands to the work elements in accordance with a predetermined plan. The decision elements may be relays, timers, or solid-state logic devices to be discussed later.

3. *Work Elements* These are devices such as solenoid valves, contactors, and solid-state static switches controlling cylinders, motors, and other power devices on the machine in accordance with commands from the decision elements.

SOLID STATE LOGIC

When solid state logic is employed, the conventional *decision elements* (relays and electromechanical timers) are replaced by transistors, diodes, resistors, capacitors, and similar electronic components usually arranged as modules and mounted on plug-in circuit boards for easy servicing. Conventional *sensing elements* such as limit switches may be replaced by such devices as potentiometers, proximity switches, and photoelectric sensors. *Work elements* such as power relays may be replaced by static switches (bidirectional thyristors or triacs). Many installations are hybrid, i.e., the *decision elements* may be solid state while all or part of the sensing elements and work elements are conventional.

---

*This section was prepared by Mr. Philip A. Heist, Manufacturing Engineer, Revell, Inc., Venice, California.

ANATOMY OF SOLID STATE LOGIC

Gates

A basic building block of solid state logic is the gate. Essentially, the gate is an electronic switching device in which a small amount of power at the input can switch on or off a larger amount at the output. Unlike an amplifier, in which the output is *proportional* to the input, the gate is a binary device; the outputs and inputs are either fully on or off. In this respect the gate is like a relay, where the contacts are either closed or open depending on whether the control coil is energized or de-energized.

Just as the coils and contacts of a number of relays may be interconnected to perform logical operations, elements of electronic gates may be combined to perform analoguous functions. Logic gates are usually operated at low d.c. voltage levels, 12 volts being typical for industrial applications. Components such as resistors, diodes, and transistors are mounted on small circuit boards as modules which "plug-in" to sockets in a wiring frame, allowing for easy replacement in case of trouble. Because of the very compact nature of solid state elements, several gate combinations are usually installed in one card.

Schematic diagrams representing gate circuits are usually easier to follow than those representing relay circuitry.

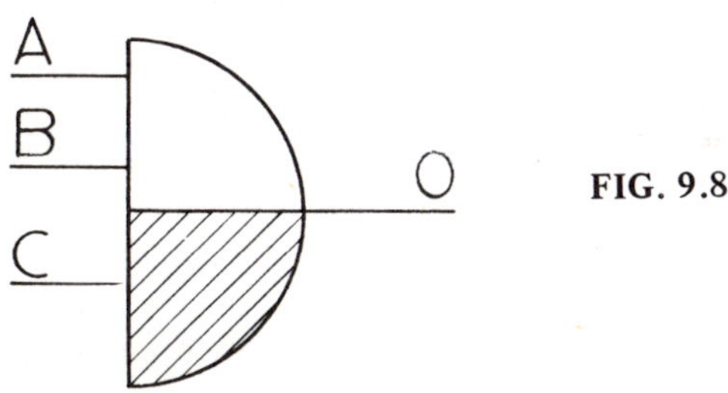

FIG. 9.8

As an example see Figure 9.8, which shows a symbol used by one manufacturer, (Solid Controls, Inc.), to represent an "OR gate with Inhibit." (Note: Because of the variety of logic schemes and gate arrangements used by various manufacturers, there is no widespread standardization of symbols.) In this example, if there is no input at A, B, or C, there will not be output at O. If there is no input at C (which connects into the shaded area indicating output inhibition), but there is signal present on A or B or both, then there will be an output signal at O. The relay circuitry for accomplishing exactly the same logical function is shown in Fig. 9.9. Note that in the solid state symbolism, associated inputs and outputs are shown together, whereas in the standard ladder diagram for relay logic inputs (relay coils) may be widely separated from outputs (relay contacts).

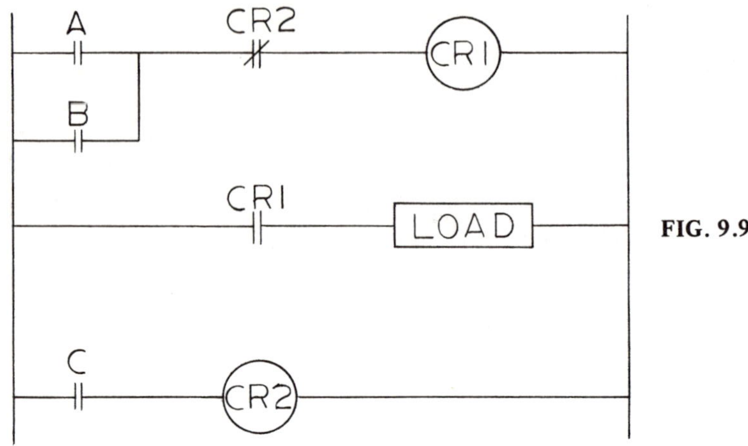

FIG. 9.9

Also note that in solid state symbolism the power supply connections to each gate are not shown. Only the inputs and outputs are represented, with a symbol of logic function between them. This makes it much easier to follow the logical sequence of events.

The "OR gate with Inhibit" shown in Fig. 9.8 is easily arranged to perform a number of functions usually accomplished by relays. Figure 9.10 shows a "holding relay" connection. If input B is pulsed momentarily, output from O will be fed back into input A, keeping the output energized until input C is energized. Equivalent relay circuitry is shown in Fig. 9.11.

Figure 9.12 shows connection as a "NOT" gate. Input A is continuously energized so that there is continuous output at O so long as input C is not energized, whereupon it will cease. Figure 9.13 gives the equivalent relay circuitry, a relay with a normally closed contact.

Where the output of the logic gate must operate one of the work elements such as a solenoid valve, motor contactor, or heater, the logic gate is used to trigger a power gate with adequate voltage and current output characteristics.

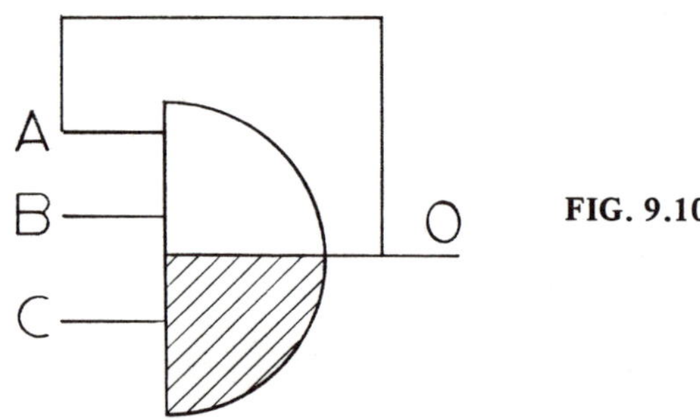

FIG. 9.10

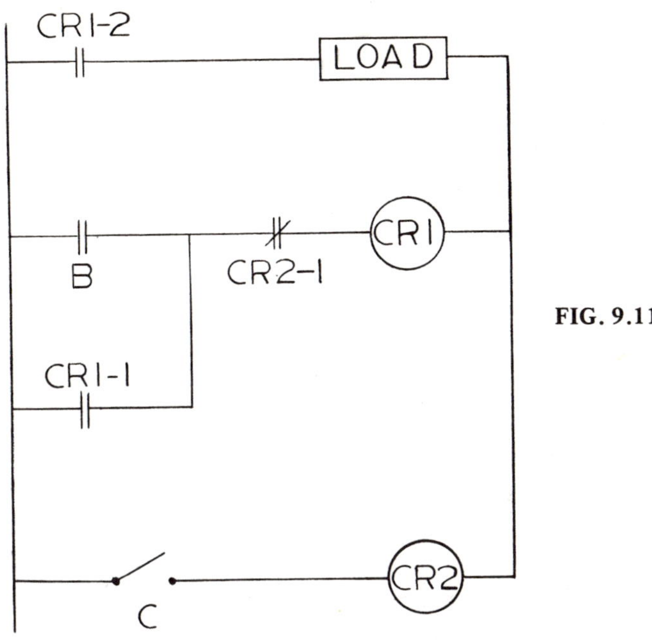

FIG. 9.11

Such gates are sometimes known by other names such as static switches, and usually consist of a solid state device such as a silicon controlled rectifier or triac with necessary triggering circuitry. Like the logic gates, they contain no moving parts, are fast in action, and compact in size.

Timers

Solid-state timing circuits may be fixed or adjustable. Usually the interval is based on the time taken to charge a capacitor to some predetermined voltage

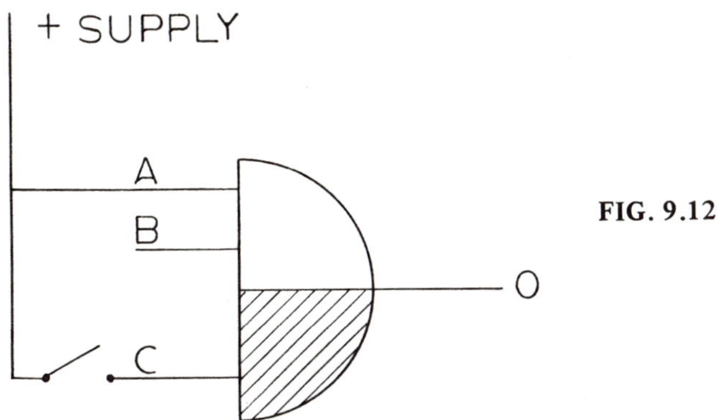

FIG. 9.12

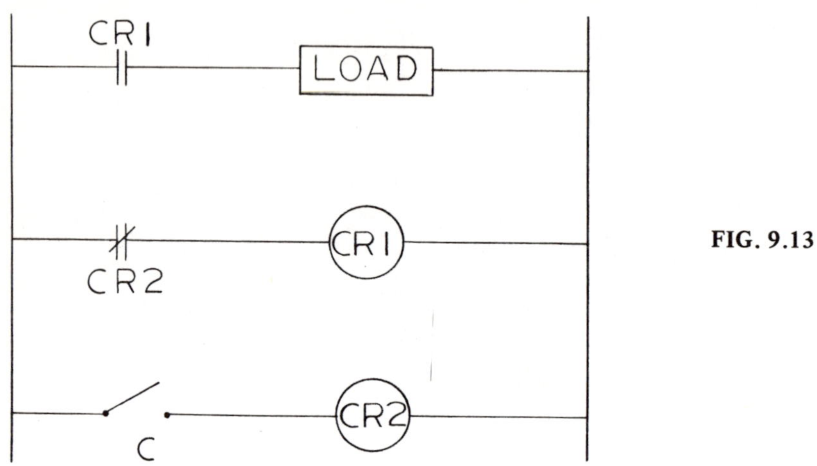

FIG. 9.13

through a calibrated resistor. When the critical voltage is reached, as determined by an electronic voltage sensor, the outputs change from OFF to ON or *vice versa* as required by the logic plan. Components are mounted on standard plug-in cards, except for the resistor in the adjustable types, which is mounted on the control panel for easy adjustment.

Figure 9.14(a) illustrates in symbolic form a timer with ON DELAY and OFF

	INITIAL	TIMING	TIMED OUT
INPUT — A	O	X	X
OUTPUT — B	O	X	O
OUTPUT — C	O	O	X

(a)

(b)

FIG. 9.14

DELAY outputs. Note again that power supply connections are not shown. The O and X notation gives the condition of the inputs and outputs under various circumstances. X represents ON and O represent OFF. Referring to the table (Fig. 9.14b) the first vertical column indicates that initially the input is OFF and so are both outputs. When a signal is applied to the input, the timing period begins. Reading down the second column, with input A ON, output B also goes ON but output C remains off. When the timing period is over, with input still applied, output B goes OFF and output C comes ON. When the input signal is withdrawn, the timer resets and returns to initial conditions.

The adjustable resistor which determines the timing interval may be a calibrated multiple turn potentiometer, or, more commonly, a string of precision fixed resistors switched by a bank of thumbwheel digital decade switches. Repeatability up to 0.1% may be expected from this type of control.

Linear Positioning

The control system for an injection molding machine makes important decisions depending on the linear position of machine elements. For example, at the clamp end, the points where low pressure is applied to the clamp cylinder depends on the position of the platen. Injection cannot be started until the clamp is fully closed. The shot volume also depends on the length of the injection stroke. These critical positions may be determined by means of limit switches and cams. With solid state controls better accuracy, repeatability, and setting convenience are obtainable.

A common method used to control the position of hydraulically driven machine elements is to mechanically couple the moving machine part (such as the moving platen) to an instrument potentiometer. This device serves to produce a low level d.c. control voltage which is exactly proportional to the linear position of the moving part, and which is known as the signal voltage. Another potentiometer calibrated for accurate manual adjustment is located in the control panel, and may be set to match the signal voltage at any desired point. This is the reference voltage.

To complete our control loop we need an electronic circuit known as a voltage comparator.

If the signal voltage is less than the reference voltage, the output from this device is zero. If the signal voltage exceeds the reference voltage, full output voltage is produced.

Figure 9.15 represents an application of these principles. Suppose that it is necessary to control a hydraulic cylinder so that it will extend to some pre-set position less than its full travel and then stop. The sense potentiometer is coupled to the piston rod (or the driven part) and as the piston starts from the retracted position the initial output of this pot is very low. The output from the reference pot, manually set to correspond to some extended position of the piston, is higher. Thus, no output is produced at the comparator. Signal is now

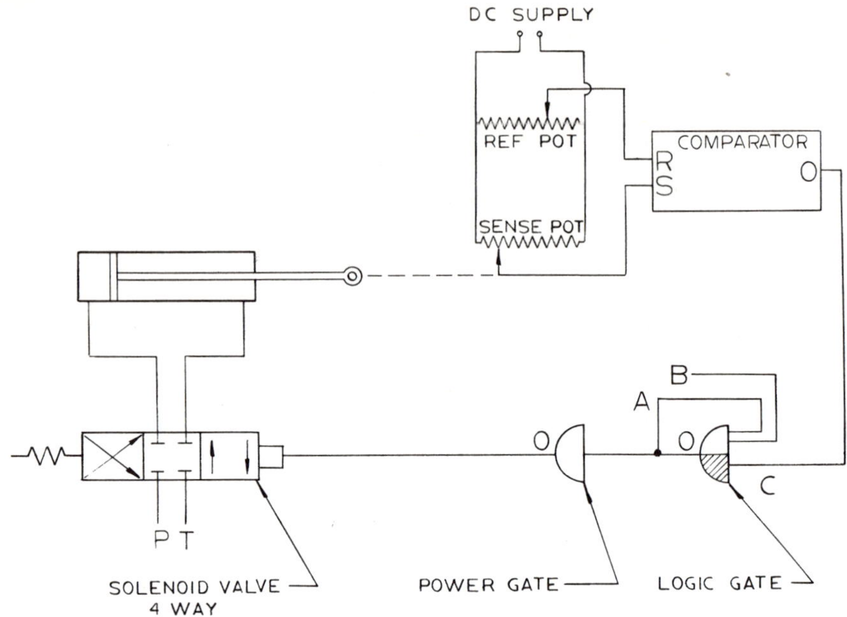

FIG. 9.15

applied to input B of the logic gate, sealing it through input A. The output operates the power gate which in turn energizes the solenoid valve. As the piston moves forward, voltage from the sense pot increases until it equals the set-point voltage, at which time output is produced at the comparator, inhibiting the logic gate, and stopping the action of the piston. On successive cycles the piston may be expected to stop within about 0.1% of its full travel distance from the same point. The same logic is used in controlling the reciprocating screw on the injection machine. It can be seen that the precise control of screw stroke thus obtained, will be reflected in accurate, repeatable shot volume, which is extremely important in the production of high quality plastics products.

Figure 9.15 is intended to illustrate the concept only and should not be interpreted to represent an actual machine circuit.

Electrohydraulic Valves

Most electrohydraulic valves used in plastics processing are binary devices, that is, their inputs (solenoid coils) are either "off" or "on" as are their outputs (valve spools). With solid state control, these are operated in the proper sequence from the low voltage logic system controlling power gates as previously discussed.

Pressure regulating and flow control valves are proportional devices. The output is proportional to the input, which in mechanical systems is usually set by a screw adjustment of spring pressure on a pilot valve.

With solid state control, the spring is replaced by a direct current solenoid operating a poppet valve against the pressure of the fluid to be controlled. This poppet valve serves as a pilot for the main control valve. The point at which fluid pressure exactly balances solenoid pressure (and thus the setting at which the system stabilizes) is a linear function of the solenoid current. Solenoid current in turn can be made directly proportional, through electronic means, to a setting of a calibrated potentiometer located on the control panel. A typical pressure control valve of this kind is shown in Fig. 9.16.

Advantages of the solid-state controlled proportional valves over their mechanical counterparts are:

(1) Better repeatability from one cycle to the next, very quick and precise action with no "hunting."

(2) Better repeatability from one mold setting to the next. With *mechanical adjustments*, once the optimum adjustment is found it is impossible to duplicate it accurately the next time the same mold is installed. The laborious "tuning up" process must be repeated by skilled personnel.

(3) Greater convenience for set-up, resulting in shorter mold-change times.

FIG. 9.16 Pressure control valve for solid state control. (*Courtesy Racine Hydraulics Inc.*)

Further discussion of electrohydraulic valves and pumps will be found in Chapter 8.

Temperature Controllers

Modern solid state temperature controllers are composed of *sensing elements*, usually thermocouples, *decision elements*, which are electronic circuits which assess the temperature conditions, and issue instructions to *work elements* which apply necessary heating or cooling as required. For most plastics processing applications we are considering the addition of heat to the plasticizing barrels of molding machines or extruders. Such heat is usually applied electrically. In some applications it is necessary to control the removal of heat by chilled water or other media. The control principles are the same.

Earliest temperature control devices were purely binary. The decision element determined only if the input from the sensing element was above or below a set point (off or on) and in response called on the work elements to go full off or full on. Because of the thermal inertia of processing machinery and the wide temperature gradients between sensing points (such as thermocouples), and heat generation areas (such as heater bands), wide temperature overshoots from the set points were encountered. Some attempts to improve this situation are discussed earlier in this Chapter.

In modern solid state technology, the voltage output of the sensing element (thermocouple) is compared by the decision element to a reference voltage corresponding to the desired temperature. From this difference is derived a control signal to the work element. In *proportional systems* the control signal is proportional to the temperature error, so that large corrections will be applied to large errors, decreasing as the set point is approached, and thus reducing overshoot. The disadvantages of a purely proportional system are:

(1) Slow response.
(2) Droop. A system may be adjusted to stabilize about a set point with constant thermal load. When this load is changed (as by operation of the extruder screw in a barrel heated by electrical bands) the system will then stabilize about some new temperature level offset from the desired reference level.

To correct droop, modern controllers offer a feature known as *integral control* (also called *automatic reset*). With this scheme, temperature variations above and below the set point are compared over a period of time. The control signal is altered to make the mean value of these fluctuations correspond with the selected set point.

Where extreme fluctuations in thermal load are encountered, *rate control* (also known as derivative control) may be added. This type of control is used when fast response to large load variations is required. With rate control, the proportional control band is narrowed, causing large corrections as the set point is approached. The rate system senses the rate of change of corrective action, and may reduce or even reverse the control signal prior to reaching the set point, thus preventing overshoots.

Proportional control with *automatic reset* has been found to be adequate for most plastics processing operations.

The work element must convert the level of the control signal to a proportional thermal input to the machine. Truly proportional control of large electrical currents is very expensive, so usually some form of time proportioning is employed. Power is switched on and off cyclically, the ratio of "on" time to "off" time during any cycle varying with the magnitude of the control signal. If the period of the cycle is small compared to the process speed, very smooth results may be obtained with very little thermal shock to the heating elements. Modern power gates, such as Triacs, permit high speed switching with short cycles impossible to achieve with relay control.

A modern solid state temperature controller is illustrated in Fig. 9.17. Note that the reference temperature is set on decade digital thumbwheel switches. The small meter indicates instantaneous deviations from the set-point tempera-

FIG. 9.17 Solid state temperature controller. (*Courtesy Barber Colman Co.*)

FIG. 9.18 Electronic components of solid state temperature controller. (*Courtesy Barber Colman Co.*)

ture. The electronic components are located on circuit boards behind the panel as shown in Fig. 9.18 and the entire assembly "plugs in" to a socket in the cabinet, permitting easy access for service or replacement.

10 THE LAMINATED AND REINFORCED PLASTICS

A laminated structure is formed from layers of materials bonded together into a unit body. For example, plywood is a laminated product formed from several thin wood sections glued together into an integral body. The plastics industry manufactures large quantities of laminated sheets, tubes, rods, and products using various materials such as paper, cloth, asbestos, wood cellulose, glass fabric, etc., bonded by the synthetic resins. Such products are traditionally known as laminated plastics when high molding pressures (1200 to 2500 psi) are used, and as reinforced plastics when low molding pressures are employed (0 to 1000 psi). Molding compounds, of resins mixed with fillers, are often called reinforced plastics.

THE LAMINATED PLASTICS

The electrical industry makes use of the electrical and thermal insulating properties, and the non-magnetic quality of the laminated plastics. Tiny pieces stamped from the laminated sheets are used in radio and electrical assemblies, while large tubes, six feet in diameter, and eight feet long, are used in large power transformers. These represent typical electrical uses of laminated phenolic materials. Copper-surfaced laminated sheets are widely used for etched circuits.

Mechanical uses for laminated plastics, as shown in Fig. 10.1, include such items as ball retainers for ball-bearing assemblies, automobile timing gears, cams, silent gears, clutches, fan blades, and the heavy-duty roll neck bearings used in

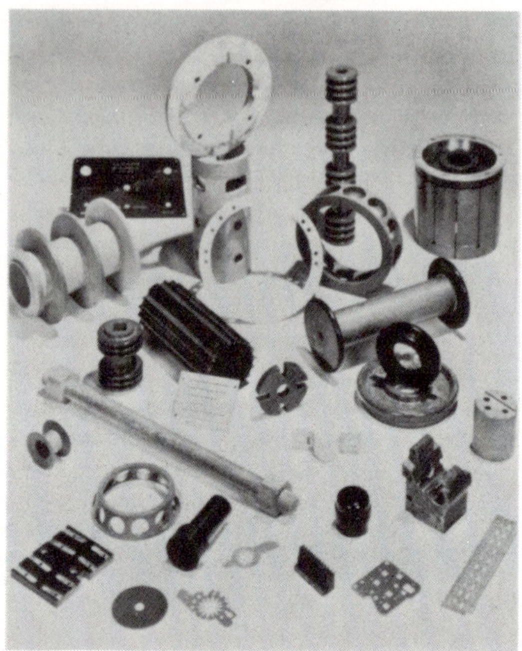

FIG. 10.1 Typical industrial laminated plastics products include high-speed precision ball retainers, pump bearings, cams, guides, rotor vanes, pulleys, clutch cones and other parts machined from laminated plastics. (*Courtesy Synthane Taylor Co.*)

the steel industry. Textile manufacturers use laminated shuttle box plates, picker blocks and shuttle runs.

Because of their resistance to chemicals, the laminated materials are used in the production of buckets for spinning rayon, and for plating barrels, plater's racks, frames and baskets. The paper mill industry uses laminated plastics because they resist corrosion and are lubricated by water. Toll covers, toe blocks, bearings, shake springs and doctor blades are typical paper mill applications.

In the decorative field, resin treated, selected wood surfaces, and colorful fabrics are used to produce decorative table tops which are wear-resistant and impervious to alcoholic products, and can not be stained by burning cigarettes. The urea resins are combined with pure cellulose and rag papers to produce light-diffusing panels for lighting fixtures. These panels are excellent for light transmission; they reduce glare, provide uniform distribution, and do not burn.

Resins for Laminated Materials

Resins for this purpose are generally used in a varnish form, or a water solution. The varnish is produced by dissolving the soluble and fusible form of the resin in alcohol. After this varnish is applied to the laminations, the treated sheets are pressed, causing the resin to flow and bond the adjacent sheets into a unit structure. The resin becomes very hard, producing a dense, strong and tough substance. Typical resins used are phenolic, melamine, and epoxy.

Some special-purpose varnishes, or impregnating resins, are used to make laminated materials having specific characteristics. Certain laminated applications call for materials which may be punched easily and cleanly, without surface cracks. In this case, various oils are added to the impregnating varnish to serve as a softening or plasticizing agent. This oil-modified varnish produces a sheet material which is soft and flexible, and which may be fabricated by punching into very intricate sections.

See Chapters 2 and 3 for more information on the plastics materials available for the laminated and reinforced plastics.

Filler Sheet Materials

Rag paper makes the best filler for lamination material because of its great toughness. Less costly materials are often used, however, since rag paper is very expensive. Alpha paper, which is produced from purified wood cellulose pulp, is used for surface sheets, where uniformity in appearance and good electrical properties are desired. Kraft paper is the least expensive of the available paper materials, and is extensively used. Kraft paper displays good mechanical and electrical properties. In every case, the paper selected is an absorbent type, made especially for laminated work.

The cloth fillers are made in a variety of weights, to give a wide range of properties. These cloth materials vary all the way from $1\frac{1}{2}$-ounce fabric, to a heavy 40-ounce canvas.

Sheets of asbestos paper and fabric produce a material having maximum temperature resistance, dimensional stability and resistance to certain chemicals. A glass fabric, woven from very fine filaments of glass, is used as a filler to produce a strong material with high resistance to impact, moisture and heat.

The finished laminated sheet is composed of inner core sheets, with surface sheets of better color and quality. Paper core sheets are often .011 inch thick. A high resin content produces the very hard, best quality product. The surface sheets are given a high resin content, and are usually made from purified wood cellulose, or alpha paper, .003 to .007-inch thick. The surface sheets are treated on one side only. For certain decorative purposes, colored and printed papers are used for the surface sheet. By this printing process, simulated oak, mahogany, marble and walnut, or other surface designs, are created.

Treating the Lamination Sheets with Resin

The paper, cloth and other lamination materials, are received in large rolls and passed through a machine which applies the resin and dries the treated sheet. In this machine, the paper dips down into the varnish, and then passes through a wringer or squeeze roll, which controls the resin content. From the squeeze roll, the paper goes through a drying oven, which evaporates the alcohol or water from the varnish. This oven also advances the cure of the resin so that it will cure quickly and correctly in the final pressing operation. All of these steps require continuous control and sampling, because the process is very critical.

These rolls of treated material are stored in rooms having a low moisture content, until they are ready for the presses.

Pressing Laminated Sheets

Huge hydraulic presses are used to provide the heat and pressure required to form the finished sheets. Both a temperature of 300 to 350°F, and a pressure of 1200 to 2500 psi, are used to "cure" the resins and bond the sheets. The steel platens between the press openings are drilled—this allows the circulation of hot water or steam, which provides the heat. The force applied by the press compresses the filler sheets, and causes the resin to flow. As the pressure and heat continue, the resin hardens, bonding the sheets in the highly compressed state created by the hydraulic pressure. The press platens are cooled before the pressure is removed. The RF plastics sheets are pressed similarly with polyester or epoxy resins, at low pressure.

Making Laminated Tubes and Rods

Two processes are used in making laminated tubing, producing either "rolled tubing" or "molded tubing." In the making of rolled tubing, the treated paper is passed over a heated roll and wound on a steel mandrel of the required diameter. This mandrel is centered between three rollers, which apply pressure as the tube is wrapped. The rolling process melts the resin and bonds the paper roll as it is wrapped on the mandrel. The mandrel, with its wrapped paper covering, is then cured in an oven for approximately 18 hours, at 275°F. Tubing with heavy wall requires a longer bake to fully polymerize or "cure" the resin. The final processing of these tubes is accomplished by stripping them from the mandrel, and grinding their outside surfaces to the correct diameter in a centerless grinder.

Molded tubes and rods are produced by wrapping the treated paper or cloth on mandrels, and placing the unit in a closed mold, as shown in Fig. 10.2, to which pressure and heat are then applied. The mandrel used for wrapping the rod filler material is small, and it is pulled out before inserting in the mold for compression into the solid rod shape. Molded tubes or rods have greater density since the resin is cured under greater pressure. However, they also have a seam or parting line where the mold closes, which may cause a line of weakness.

Laminated Nameplate Materials

Permanent and attractive nameplate laminated materials find many markets, as shown in Fig. 10.3. These nameplate materials have an inner core of one color,

FIG. 10.2 Laminated phenolic rods are often made by pressing the treated stock in heated molds. Other rods are made by a rolling operation.

for example white, with black face sheets. In making the nameplates, it is merely necessary to cut through the black face sheet with an engraving machine, and expose the white core. Printed face sheets may also be used, bonding the printed data as a permanently outer sheet on the laminate.

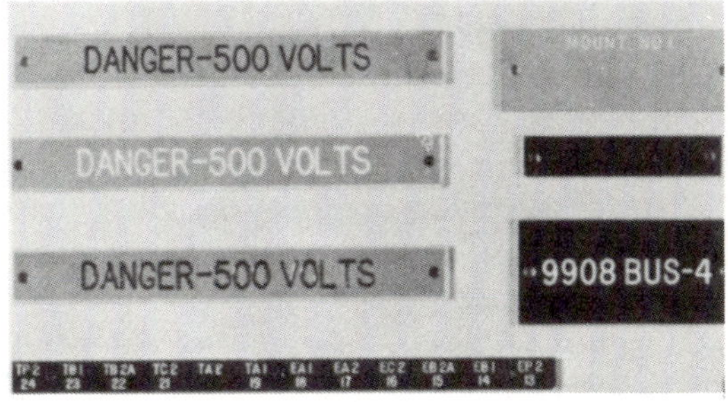

FIG. 10.3 Laminated nameplate stock is formed by laminating colored surface sheets over an inner core of another color material. The letters are then engraved through the surface sheet exposing the inner core. Multiple colors are obtained by painting the engraved letters.

Uses for Laminated Materials

Table 10.1 gives an idea of the variety of special materials which have been developed. The laminated products were originally made available only for electrical insulation purposes. Gear materials were subsequently developed and, from this start, the wide range of applications and materials have evolved.

The best electrical materials are very hard, and therefore difficult to punch. These glass-hard materials are used for many important applications in radio, as well as other high frequency and high voltage apparatus. One grade of material requires 62,000 volts to break through a $1/16$-inch thick piece. Parts to be made from this hard material should be designed very simply so that they may be fabricated without punching whenever possible.

The conventional electrical insulating pieces may be punched easily from the many punching grades of laminated sheet. Typical electrical insulating applications for laminated plastics are motor-slot wedges, terminal blocks, field-coil collars, washers, panels, barriers, tube sockets, coil forms, spool heads, bus-bar insulation, transformer insulation, spacers, relay cams, lighting fixture insulation, and automotive electrical circuitry. The fiber-faced laminates are often used for high voltage fuses where the fiber core is subjected to the high voltage flash; the laminated outer tube provides dielectric and moisture resistance. Vulcanized fiber is quite resistant to electrical arc.

The laminated materials make quiet heavy-duty gears because of their resilience. A gear train is usually quiet when there is a true rolling motion of one tooth over the mating tooth. In service, the vibration and surface impact of metal gear teeth ultimately distort the gear-tooth shape, and increase the noise and rate of tooth wear. The laminated gears absorb these shocks and retain the original tooth form, as well as a retention of the true rolling action between the gear teeth. The result is that properly designed and made laminated gears operate very quietly, and give long life under extremely trying conditions. The tooth deflection of laminated gears may be thirty times that of steel. This deflection absorbs the impact, and eliminates the noise and wear. The laminated gears should mesh with a metallic gear to provide best service and long life.

Any application of laminated or molded plastics material, which depends upon rubbing, rolling, or sliding surface movement, must have a hard material in contact with the plastics surface. Hardened steel and chrome-plated areas provide best contact surfaces for cam followers, and similar applications. Cams which are cut from laminated sheets, or molded from laminated or molded materials containing some graphite, give most unusual service. This material weighs only half as much as aluminum; the reduction in weight immensely improves the starting and stopping operations of large, high-speed cams.

An excellent use for laminated cloth materials is in the large, heavy-duty bearings used in the steel mills on the roll necks. These rolls press the steel into the various sections that are being rolled, subjecting the bearings to terrific

TABLE 10.1 PROPERTIES OF LAMINATED PLASTICS*

		PAPER REINFORCED GRADES							GLASS REINFORCED GRADES	
STANDARD SHEET	X X-115 X-116 Phenolic	XX XX-7 Phenolic	XPC Phenolic	XXX Phenolic	XP-240 X-235 Phenolic	XXP-241 XXP-271 Phenolic	XXXP Phenolic	XXXPC Phenolic	G-3 GF-3 G-3HT Phenolic	G-5 Melamine
GOVERNMENT SPECIFICATION	LP-509	LP-509	LP-509	LP-509	LP-509	LP-509	LP-509	LP-509	LP-509	....
NEMA DESIGNATION	X	XX	XPC	XXX	XP	XXP	XXXP	XXXPC	G-3	G-5
Flexural Strength, psi, ⅛" thk. Lengthwise Crosswise	25,000 22,000	15,000 14,000	10,000 8,000	13,500 11,800	13,000 11,000	14,000 12,000	12,000 10,500	12,000 10,500	20,000 18,000	44,000 38,000
Bonding Strength, lbs., ½" thk., Cond. A	700	800	....	950	....	....	....	....	850	1570
Izod Impact, ft-lbs/in. of notch, ⅛" Edgewise, Crosswise	0.50	0.35	0.55	0.35	0.50	0.40	0.30	....	5.50	5.50
Water Absorption, %, 24 hrs. 1/16" ⅛" ½"	6.0 3.3 1.1	2.0 1.3 0.55	5.5 3.0	1.40 0.95 0.45	3.60 2.20	1.80 1.10	1.00 0.75	0.75 0.55	2.70 2.00 1.50	2.70 2.00 1.50
Dielectric Breakdown, KV, 1/16" Parallel to Laminations Step x Step, Cond. A	....	40	....	50	40	60	60	60	....	23
Arc Resistance, Seconds, Cond. A	....	....	....	....	....	....	....	....	....	180
Dissipation Factor, 1 MHz, Cond. A	....	0.045	....	0.038	....	0.040	0.035	0.035	....	0.020
Dielectric Constant, 1 MHz, Cond. A	....	5.5	....	5.3	....	5.0	4.6	4.6	....	7.8
Tensile Strength, psi, Lengthwise Crosswise	20,000 16,000	16,000 13,000	10,500 8,500	15,000 12,000	12,000 9,000	11,000 8,500	12,400 9,500	12,400 9,500	23,000 20,000	37,000 30,000
Compressive Strength, psi Flatwise Edgewise	36,000 19,000	34,000 23,000	22,000	32,000 25,500	25,000	25,000	25,000	25,000	50,000 17,500	70,000 25,000
Modulus of Elasticity in Flexure Lengthwise, psi Crosswise, psi	1,800,000 1,300,000	1,400,000 1,100,000	1,000,000 800,000	1,300,000 1,000,000	1,200,000 900,000	900,000 700,000	1,000,000 700,000	1,000,000 700,000	1,500,000 1,200,000	1,700,000 1,500,000
Shear Strength, psi	12,000	11,000	....	10,000	8,000	11,000	11,000	11,000	18,000	20,000
Izod Impact, ft-lbs/in. of notch, ½" Flatwise, Crosswise	4.0	1.3	....	1.0	....	....	....	....	6.5	12.0
Rockwell Hardness, M Scale	110	105	75	110	95	100	105	105***	100	120
Specific Gravity	1.36	1.34	1.34	1.32	1.33	1.32	1.30	1.30	1.65	1.90
Insulation Resistance, Megohms Condition: 96 hrs, 90% RH, 95°F	....	60	....	1,000	....	500	20,000	50,000	....	100
Coeff. of Thermal Exp., Cm/Cm/°C	2×10^{-5}	2×10^{-5}	2×10^{-5}	2×10^{-5}	2×10^{-5}	2×10^{-5}	2×10^{-5}	2×10^{-5}	1.8×10^{-5}	1×10^{-5}
Thermal Conductivity, Calories/sec/Cm²/°C/Cm	7×10^{-4}	7×10^{-4}	7×10^{-4}	7×10^{-4}	7×10^{-4}	7×10^{-4}	7×10^{-4}	7×10^{-4}	....	12×10^{-4}
Max. Constant Operating Temp., °F	280	280	280	280	280	280	270	265	350	280
Thickness, Inches, Minimum Maximum	0.010 2	0.010 2	0.010 0.125	0.010 2	0.010 2	0.010 0.250	0.010 0.250	0.032 0.125	0.020 3	0.015 3
Standard Colors	Natural Black Choc	Natural Black	Natural	Natural Black	Natural Black Choc	Natural Black Choc	Natural Black	Natural	Natural Black	Natural White
Standard Finishes See Page 8 for Description	#4	#8	#4	#8	#4	#4	#4	#4	#4	#4

GUARANTEED PROPERTIES‡ spans the rows from GOVERNMENT SPECIFICATION through Dielectric Constant. TYPICAL PROPERTIES spans Tensile Strength through Max. Constant Operating Temp. GENERAL spans Thickness through Standard Finishes.

NOTE: Data listed above are NEMA values for those grades and properties established by NEMA.

‡Guaranteed maximum or minimum values as the case may be. Other values shown are typical values derived from a number of tests and are included for design information only.

*Cold punching
**GEC-500: 2000 lbs.; GEC-111: 1600 lbs.; FB600: 2000 lbs.
***XXXP-C—95
†GEC-500: 280; GEC-111: 320

*Courtesy Synthane-Taylor Corp.

TABLE 10.1 (Continued)

COTTON REINFORCED GRADES

STANDARD SHEET	C C-701 Phenolic	CE CB CE-6 Phenolic	CPF†† C-7 Phenolic	CCM CFBM Melamine	CYB Phenolic Graph- itized	L Phenolic	LE LB LE-6 LB-31**** Phenolic	LM Melamine	LYB Phenolic Graph- itized
GOVERNMENT SPECIFICATION	LP-509	LP-509	LP-511	LP-509	MIL-P-5431	LP-509	LP-509	LP-509	MIL-P-5431
NEMA DESIGNATION	C	CE	CF	—	—	L	LE	—	—
Flexural Strength, psi, 1/8" thk.									
Lengthwise	17,000	17,000	16,000	16,000	14,500	15,000	15,000	15,000	14,500
Crosswise	16,000	14,000	....	13,000	13,500	14,000	13,500	13,500	13,500
Bonding Strength, lbs., 1/2" thk., Cond. A	1,800	1,800	....	1,600	1,600	1,600	1,600	1,500	1,500
Izod Impact, ft-lbs/in. of notch, 1/8" Edgewise, Crosswise	1.90	1.40	1.5	0.50	1.40	1.10	1.00	0.60	1.00
Water Absorption, %, 24 hrs.		CE CB					LE LB		
1/16"	4.40	2.20 2.00	4.4	3.5	2.20	2.50	1.95 1.50	2.4	2.20
1/8"	2.50	1.60 1.50	2.5	2.5	1.60	1.60	1.30 1.20	1.2	1.60
1/2"	1.20	0.75 0.65	....	0.9	0.75	0.90	0.70 0.65	0.7	0.75
Dielectric Breakdown, KV, 1/16" Parallel to Laminations Step x Step, Cond. A	15	35	....	20	....	15	40	20	....
Arc Resistance, Seconds, Cond. A	....	....	....	100	....	....	....	100	....
Dissipation Factor, 1 MHz, Cond. A	....	....	....	0.050	....	....	0.055	0.050	....
Dielectric Constant, 1 MHz, Cond. A	....	....	....	7.0	....	....	5.8	7.0	....
Tensile Strength, psi, Lengthwise	10,000	9,000	7,500	10,000	7,000	13,000	12,000	12,000	7,000
Crosswise	8,000	7,000	....	7,000	5,000	9,000	8,500	8,000	5,000
Compressive Strength, psi									
Flatwise	37,000	39,000	30,000	38,000	37,000	35,000	37,000	37,000	35,000
Edgewise	23,500	24,500	....	24,000	24,000	23,500	25,000	25,000	23,500
Modulus of Elasticity in Flexure									
Lengthwise, psi	1,000,000	900,000	825,000	900,000	900,000	1,100,000	1,000,000	1,200,000	900,000
Crosswise, psi	900,000	800,000	....	800,000	800,000	850,000	850,000	900,000	800,000
Shear Strength, psi	12,000	11,000	....	13,000	11,000	12,000	11,500	11,000	11,000
Izod Impact, ft-lbs/in. of notch, 1/2" Flatwise, Crosswise	3.2	2.3	....	0.6	2.3	2.5	1.8	0.8	2.5
Rockwell Hardness, M Scale	103	105	....	110	95	105	105	115	95
Specific Gravity	1.36	1.33	1.35	1.40	1.34	1.35	1.33	1.45	1.34
Insulation Resistance, Megohms Condition: 96 hrs, 90% RH, 95°F	....	....	....	....	....	....	30	120	....
Coeff. of Thermal Exp., Cm/Cm/°C	2×10^{-5}	2×10^{-5}	....	2×10^{-5}	2×10^{-5}	2×10^{-5}	2×10^{-5}	2×10^{-5}	2×10^{-5}
Thermal Conductivity, Calories/sec/Cm²/°C/Cm	7×10^{-4}	7×10^{-4}	....	....	....	7×10^{-4}	7×10^{-4}	....	....
Max. Constant Operating Temp., °F	250	250	....	275	250	250	250	275	250
Thickness, Inches, Minimum	0.010	0.010	0.031	0.031	0.047	0.008	0.008	0.012	0.015
Maximum	10	10	0.250	2	4	6	6	2	2
Standard Colors	Natural Black	Natural Black	Natural Black	Natural	Black	Natural Black	Natural Black	Natural	Natural
Standard Finishes See Page 8 for Description	#4	#4	#4	#4	#4	#4	#4	#4	#4

Row-label group markers printed at left: **GUARANTEED PROPERTIES‡** (Flexural Strength through Dielectric Constant), **TYPICAL PROPERTIES** (Tensile Strength through Max. Constant Operating Temp.), **GENERAL** (Thickness through Standard Finishes).

NOTE: Data listed above are NEMA (National Electrical Manufacturers Association) values for those grades and properties established by NEMA.

‡Guaranteed maximum or minimum values as the case may be. Other values shown are typical values derived from a number of tests and are included for design information only.
§AA-HT retains 40% of mechanical strength after 500°F. for 100 hrs.
††Properties before post-forming.
****LB-31 exhibits superior wearing properties.

TABLE 10.1 (*Continued*)

ASBESTOS REINFORCED GRADES		OTHER GRADES		
A	AA AA-F AA-HT AAF-HT	EP-22	AAA	N-1 NS
Phenolic	Phenolic	Epoxy	Phenolic	Phenolic
LP-509	LP-509	MIL-P-22324		MIL-P-15047
A	AA	—	—	N-1
13,000	18,000	20,000	22,000	10,000
11,000	16,000	16,000	22,000	9,500
700	1,800	....	1,200	1,000
0.60	3.00	....	1.60	2.00
1.50	3.00	0.65	2.50	0.60
0.95	2.50	0.50	1.60	0.40
0.55	1.25	....	0.90	0.35
5	....	60	....	60
....	....	....	....	....
....	....	0.035	....	0.038
....	....	4.6		3.9
10,000	12,000	14,000	16,000	8,500
8,000	10,000	12,000	14,000	8,000
40,000	38,000	30,000	45,000	....
17,000	21,000	....	....	....
2,300,000	1,600,000	1,500,000	1,600,000	600,000
1,400,000	1,400,000	1,000,000	1,400,000	500,000
9,000	12,000	11,000	19,000	14,000
1.8	3.6	....	2.4	4.0
111	103	100	105	105
1.72	1.70	1.36	1.70	1.15
....	....	100,000	....	50,000
1.5×10^{-5}	1.5×10^{-5}	2×10^{-5}	1.5×10^{-5}	8×10^{-5}
5×10^{-4}	5×10^{-4}	....	....	....
300	310§	250	300§	165
0.025	0.062	0.031	0.020	0.020
2		0.750	4	1
Natural	Natural	Natural	Natural	Natural
#4	#4	*****	#4	#4

*****Special Release Finish.

punishment. This continuous hammering and pounding for a short period ruined the babbit and bronze bearings which they replaced. Their resiliency enables the laminated canvas bearings to withstand this beating. Laminated bearings have been known to run fifty times as long as metallic bearings. Cooling water is used for lubrication; the coefficient of friction is much lower than that of the metallic bearings with grease lubrication.

In the textile industry, the rayon buckets are made from laminated products. These buckets rotate at high speeds, therefore requiring higher tensile strength and lighter weight. The corrosive chemicals used in making the synthetic fabrics have little effect on the laminated materials that are available.

Clutches are frequently made from laminated materials because their light weight contributes to fast start and stop. The face of the laminated products provides a good clutch surface for many applications. Asbestos fillers are used when there is considerable clutch slip to withstand the heat that is generated. Laminated fan blades are very popular and useful because of their light weight and quietness of operation. All plastics materials are mechanically non-resonant, tending to dampen and reduce sound vibrations. Laminated separators are often

FIG. 10.4 Designed for fire hazard requirements, the laminated plastics panels shown here add beauty, warmth and minimized maintenance over very long periods. (*Courtesy Formica Corp.*)

used in structural applications to reduce the transmission of noise, vibration, heat, or electricity.

Laminated Materials for Decorative Purposes

A large quantity of laminated materials, as shown in Fig. 10.4, are used for table tops, counter tops, wall panels, doors, inlaid murals, windowsills, and other interior surfaces. These products provide beauty and long life—and are not affected by beverages or burning cigarettes. Their burn resistance is increased by the inclusion of thin-sheet aluminum foil below the surface sheet. This aluminum foil prevents localization of the heat, so that a cigarette lying on the surface will burn out without causing any damage.

Surface sheets for the decorative laminates employ printed fabrics, wood veneers, printed designs, melamine surface resins which can be brightly colored, and pieces of inlaid metals to produce attractive surface decorations. Some of the particularly attractive laminates utilize thin laminations of wood for the surface, gaining all the beauty of the wood, and the wear resistance of the resin bonded laminates.

Copper-Clad Laminates

The electrical and electronic industries use large quantities of copper-clad sheet for printed circuitry applications. For this purpose, thin sheets of copper are bonded to the surface of the laminate. The desired circuitry is printed on the

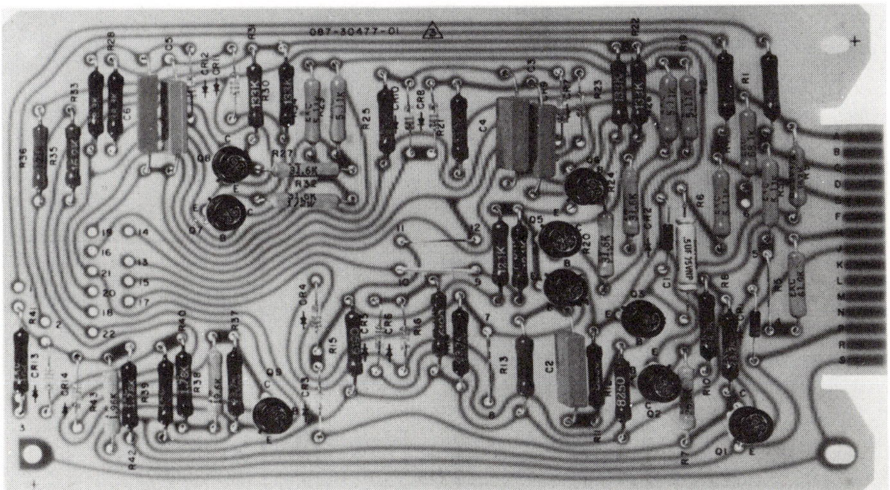

FIG. 10.5 This printed circuit was made by a photo-etching process. The entire sheet of laminated plastics is covered with copper foil. The desired circuit is printed on the copper with an ink that will resist an acid which will remove all unprinted areas. After the acid-etch to remove unwanted copper, the circuit remains and the resist ink is removed by a solvent. (*Courtesy Synthane Taylor Corp.*)

copper with an acid-resisting ink so that immersion in a suitable acid will eat away all of the copper, leaving only the desired circuitry, as shown in Fig. 10.5, on the surface of the laminate. The printed resist is then removed from the copper by its solvent. Printed circuitry materials are made from many substances, including paper-base phenolic and epoxy, glass-cloth, TFE-glass cloth, and other special combinations.

Another form of molded circuitry makes use of an unmolded resin-treated substrate, into which the copper circuitry is punched from the copper sheet. Subsequent curing produces flush-type circuitry, suitable for rotary switches, current carrying components, etc.

Many of the etched circuits are also produced on thermoplastics films such as teflon to gain flexibility for wrap arounds and interconnections. The flexible circuit shown in Fig. 7.5 is wrapped around the glass bonded mica substance.

THE REINFORCED PLASTICS (RP)

All of the statistics point to an accelerating growth rate for the reinforced plastics. The annual production is in excess of one billion pounds. This is expected to increase to two billion pounds by 1978. Ninety per cent of this volume is glassfiber reinforced. The largest growth areas include marine, auto-motive, furniture and construction products.

The reinforced plastics materials constitute a very large special section of the plastics industry. Typical products, depicted in Figs. 10.6 and 10.7, are auto-mobile bodies, missile and aircraft sections, boats, building panels, tanks and fishing rods, etc. The reinforced plastics are similar to the laminates in many ap-plications, differing primarily by their use of resins that do not require the very high molding pressures used in making the conventional laminated sheet mate-rials. These materials are also called composite plastics, or filled plastics. Rein-forcements include glass, boron, carbon, graphite, beryllia and alumina fibers, asbestos, woven and nonwoven textiles, and sisal. Fillers are often particles such as powders, pellets, spheres, and needles—granular, rather than fibrous in shape.

Resins used for the reinforced plastics include polyester, epoxy, phenolic, melamine, diallyl phthalate, vinyl ester and silicone. A large portion of this work is done with polyester resins. Low pressure molding procedures employed are predicated on the use of a catalyst, freshly mixed with the resin monomer and filler at the time of molding. The catalyst causes the hardening of the resin and filler combination, without external pressure being necessary. Pressure is often used to gain improved density, desirable surface textures, and faster cure. In the curing process, a soft gel is initially formed, and heat is produced by the chemical reaction taking place which converts the resin and filler into a hard integral mass, without liberation of volatile materials.

More than 90% of the products use thermosetting resins. Of the remaining 10%, approximately 30% are polypropylene, 18% nylon, 8% polyethylene, and 4% acetal thermoplastics.

FIG. 10.6 Fiber glass panels of reinforced plastics material are used for the roof of this large warehouse. (*Courtesy International Filon Producers Assn.*)

Resins for Reinforced Plastics (See also Chapters 2 and 3)

Polyesters. Leading resin for reinforced plastics, polyesters, are comparatively low in cost, and easy to mold at low temperatures. They have good mechanical, chemical, and electrical properties. Some types are fire retardant. Polyesters are used primarily with fibrous glass for radomes, aircraft parts, building panels, electrical components, motor components, and ducting. With special papers they are used for decorative and protective surfaces, wall board, furniture, etc.

Epoxy. Epoxy resins provide high mechanical and fatigue strength, excellent dimensional stability, corrosion resistance, and interlaminar bond, good electrical properties, and very low water absorption. Epoxy-glass and epoxy-paper combinations are widely used for printed circuits. Epoxy-glass is a major material for filament-wound structures, and is also used in ducting, aircraft and rocket parts.

Phenolic. Phenolic resins have high mechanical strength, excellent resistance to high temperature, good thermal insulation and electrical properties, and good chemical resistance. Low-cost phenolic is used with glass for high-strength interior aircraft parts, fishing rods and honeycombs, with asbestos for missile parts, with cotton and paper for safety helmets and electrical laminates and with high-silica glass or nylon for high temperature applications.

FIG. 10.7A Making the layup for the Corvette front upper-body section. The operators are pouring resin on the fiberglass in this view.

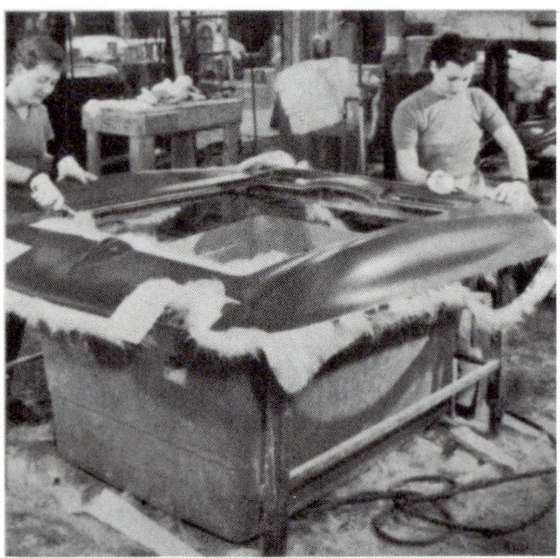

FIG. 10.7B The workers shown in this picture are removing the flash from a front upper section of the Corvette body that has just been molded of reinforced plastics. (*Courtesy Molded Fiber Glass Body Co.*)

Melamine. Melamine resins have excellent color range and retention, good abrasion resistance, good electrical properties, are resistant to alkalis, and are also flame resistant. Melamine is used with rayon-paper for molded-in decoration of dinnerware, closures and wall tiles, with paper for decorative surfaces on counter tops, furniture, and wall panels, and also with glass, for industrial laminates.

Silicone. Silicone resins provide the highest electrical properties available in reinforced plastics, and are the most heat stable. They retain strength and electrical characteristics under long exposure, from 500 to 1000°F. Silicone-glass is used for electrical insulation, hot-air ducts, and other aircraft and missile structures. Higher temperature resistance and thermal insulation is gained with asbestos and high-silica glass fillers, for rocket nozzles and missile motor housings.

Reinforcement (See also Chapter 1)

A variety of types of reinforcement are offered to meet the multiple product requirement, as shown in Table 10.2; glass is one of the most commonly used RP reinforcement materials. Continuous strand glass gives unidirectional reinforcement. Glass fabric essentially reinforces the object in two directions. Woven glass roving gives high strength and is lower in cost than conventional glass fabrics. Chopped glass strands give a random reinforcement. Reinforcing glass mats are lower in cost than fabric, and give random reinforcement. Surfacing glass mats give virtually no reinforcement, but give a decorative and smoother surface finish.

Glass Fiber. Glass fiber reinforcement gives outstanding strength-to-weight characteristics, high tensile strength, high modulus of elasticity and resiliency, and excellent dimensional stability. Glass fiber is the reinforcement most widely used in premixes (also called prepregs). It is used with all principal resins for such products as aircraft and missile parts, ducts, trays, electrical components, motor body parts, and building panels.

There are two basic forms of glass fiber: continuous filament and staple fiber. Both forms begin with the same manufacturing process. Various ingredients are mixed to make a batch of a specified formulation. The batch, in bulk or marble

TABLE 10.2 GENERAL PROPERTIES OF REINFORCED PLASTICS (ASTM TEST PROCEDURES)*

Material	Specific Gravity	TENSILE Strength 10^3 psi	TENSILE Modulus 10^6 psi	Compressive Strength 10^3 psi	Flexural Strength 10^3 psi	Izod Impact Strength ft. lb./in. notch	Heat Resistance Continuous °F	Arc Resistance Seconds
Polyester								
Glass cloth	1.5-2.1	30-70	1-3	25-50	40-90	5-30	300-350	60-120
Glass mat	1.3-2.3	20-25	½-2	15-50	25-40	2-10	300-350	120-180
Asbestos	1.6-1.9	30-60	1-3	30-50	50-70	2-8	300-450	100-140
Paper	1.2-1.5	6-14	½-1½	20-25	13-28	1-2	220-250	28-75
Cotton cloth	1.2-1.4	7-9	½-1½	23-24	13-18	1-4	230-250	70-85
Epoxy								
Glass cloth	1.9-2.0	20-60	2-4	50-70	70-100	11-26	330-500	100-110
Glass mat	1.8-2.0	14-30	1-3	30-38	20-26	8-15	330-500	110-125
Paper	1.4-1.5	10-19	½-1	20-28	19-24	½-1	260-300	30-100
Phenolic								
Glass cloth	1.8-2.0	40-60	1-3	35-40	65-95	10-35	350-500	20-130
Glass mat	1.7-1.9	5-20		17-26	10-60	8-16	350-500	40-150
Asbestos	1.7-1.9	40-65	2-5	45-55	50-90	1-6	350-600	120-200
Paper	1.3-1.4	8-20	1-2	20-40	10-30	⅓-1	225-250	Tracks
Cotton cloth	1.3-1.4	7-16	½-1½	30-44	14-30	½-3	225-250	Tracks
Nylon cloth	1.1-1.2	5-10	¼-½	28-36	9-22	2-4	150-165	Tracks
Silicone								
Glass cloth	1.6-1.9	10-35	1-2	25-46	10-38	5-13	400-700	150-250
Asbestos cloth	1.7-1.8	10-25	1-2	40-50	·12-20	6-9	450-730	150-300

**Plastics World*, Feb. 1966.

FIG. 10.8 The 58-foot diameter sectional fiberglass reinforced plastic cover on this reservoir is self-supporting, self-ventilating, and virtually maintenance free. (*Courtesy Durez Division, Hooker Chemical Corp. and Fiberglass Specialty Co.*)

form, is then fed into a furnace where it is melted at high temperature. Fine, precisely controlled filaments are drawn, or pulled rapidly, from streams of the molten glass. These filaments may be sized and then wound on packages for further fabrication.

The difference between continuous filament and staple filament is the manner in which they are drawn from the streams flowing from the furnace.

A continuous filament is an individual fiber of such length that it extends substantially throughout the length of any continuous filament strand. A strand is composed of many continuous filaments. The number of filaments depends upon the specific requirements. These continuous filaments are drawn from the molten glass at speeds in excess of two miles per minute.

A staple filament is an individual fiber (8 to 15 inches long). It is formed by jets of air which pull the glass filaments from the molten glass streams onto a revolving vacuum drum where they are gathered into a strand.

Both continuous and staple fibers can be fabricated into yarns and cords through conventional twisting, plying and cabling operations.

Asbestos. Asbestos felts provide maximum tensile and flexural strength, excellent thermal insulation, good abrasion resistance, as well as high impact strength. Asbestos papers also give low cost thermal insulation. Asbestos is used a good deal for missile and aircraft parts, its short-term heat and flame resistance con-

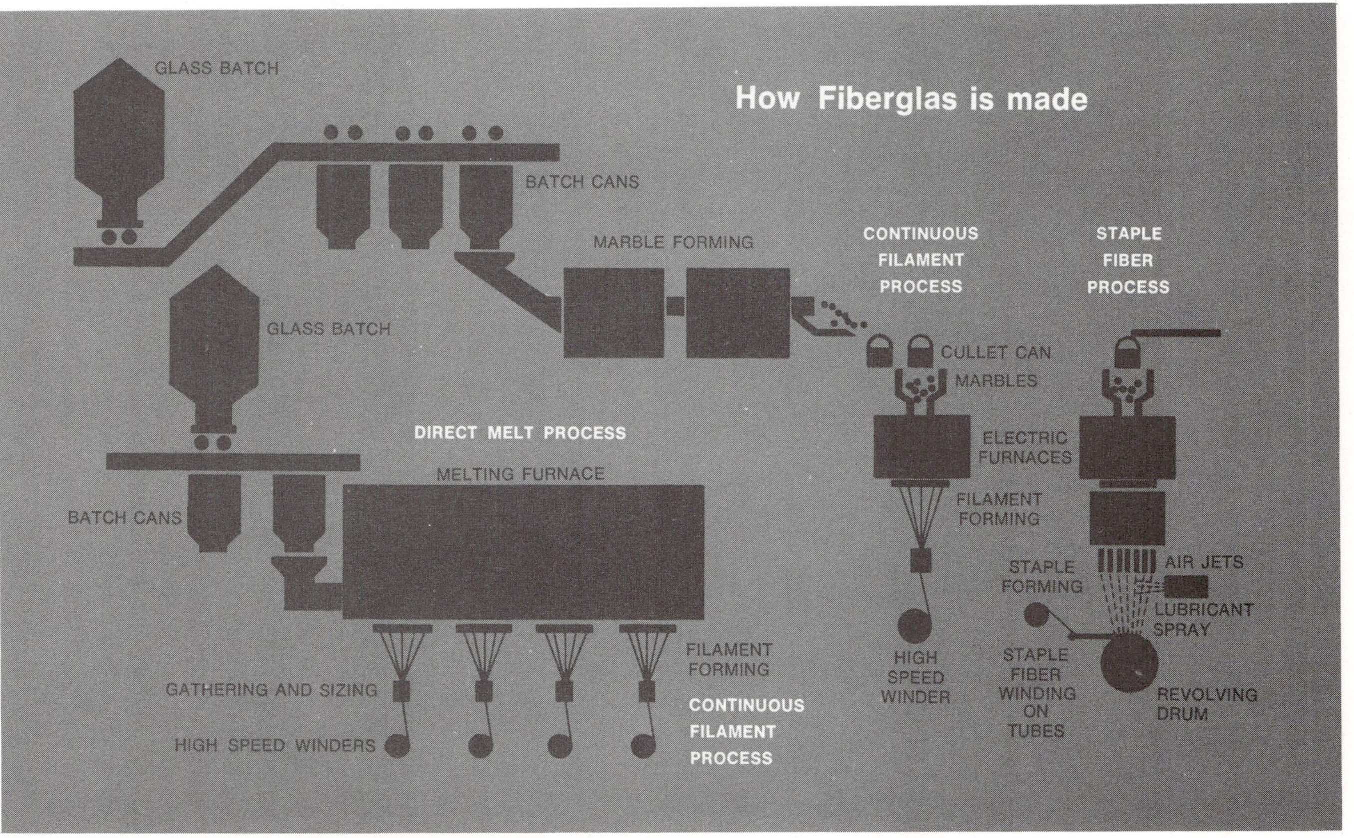

FIG. 10.9 How "Fiberglas" is made. (*Courtesy Owens Corning Fiberglas Corp.*)

sidered outstanding (6900°F for 60 seconds). Asbestos fabrics are used in flat laminates; asbestos paper is used for low-cost, non-structural insulation. Asbestos is, however, mainly used with phenolics and silicones.

Paper. Paper is the leading reinforcement for high-pressure laminating, and is often used for low-pressure reinforced plastics. It is inexpensive, prints well, and has adequate strength for many applications. Paper is used for circuit boards, and for such other products as decorative but hard-wearing laminates, furniture, and wall board. It is widely used for honeycomb structures. The main use of paper, however, is with phenolics and melamines.

Unusual Reinforcements. Such materials as high-silica glass, quartz, graphite, and other exotic reinforcements, are used to meet unusual high-temperature demands. Synthetic fibers find many special uses. Typical applications include high-silica glass as insulation in rockets, quartz for structures exposed to long term elevated temperatures, and graphite for abrasion-resistant moldings in rockets, as well as synthetics for flexible ducting. (See Chapter 2 for epoxy graphite compounds.)

Molds for Reinforced Plastics

Open and closed molds are used for this process and, since low pressures can be used, it is also possible to produce these products in simple plaster, concrete or wood molds. Open molds, male or female, are used for no-pressure molding, and they produce parts with one finished surface only. Closed molds provide products with inner and outer surfaces furnished with excellent detail.

Open Mold RP. Lowest cost molds are used in this process since pressures are low, and little strength is required. For contact molding by the open mold process, as shown in Fig. 10.11, polyester or epoxy resins are commonly used, and the mold surface is protected from sticking by a layer of celophane, wax or silicone-mold release. Layers of glass cloth, or other filler and resin loads, are placed in the mold and worked into intimate contact with the mold surfaces by the use of squeegees, rollers, or hand. If needed, additional layers are added to build up the proper thickness. Cure is achieved as a result of the hardener or catalyst that was added to the resin just prior to its use. Cure can be speeded up by heat. This process can be used for boats, auto-body repairs, swimming pools, ducts, sheets, tubes, and housings. Building and machinery repairs are made by this process also.

Pressure may also be used with open molds to improve quality, uniformity of density, and better finish on the open side. One procedure makes use of a vinyl acetate or cellophane bag that encloses the mold and it contents, as depicted in Fig. 10.12. Air is withdrawn from the bag by a vacuum pump, producing a uniform pressure on the product surface that is not in contact with the mold. The reverse of this process is pressure bag molding, wherein rubber sheeting is placed against the plastics lay-up, and then exposed to air or steam pressures as high as 50 psi, as shown in Fig. 10.13. This bag-pressure type mold may also be

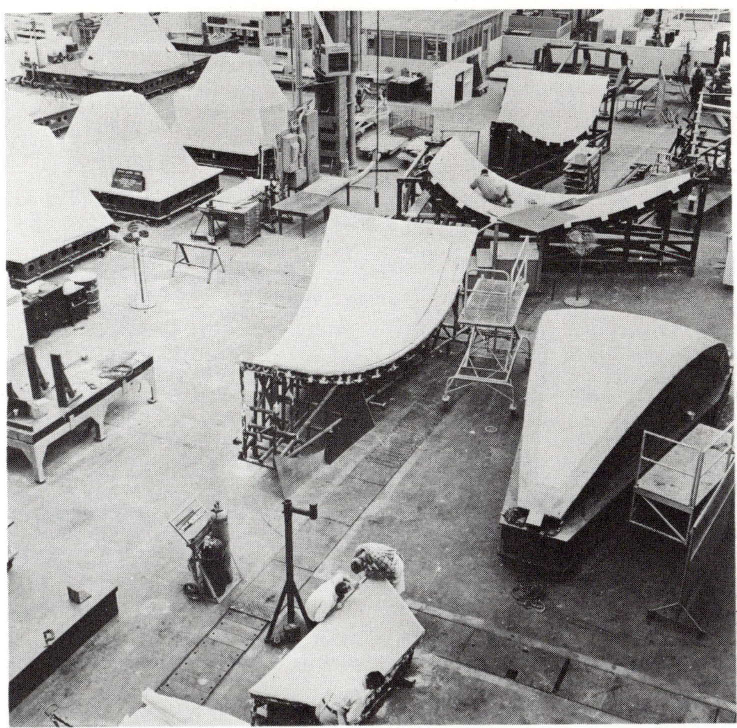

FIG. 10.10 Tooling for reinforced plastics molding. The molds shown here were used for making the Saturn space vehicles. They were made from a two component, white colored epoxy laminating resin system. (*Courtesy Furane Plastics, Inc.*)

CONTACT MOLDING

Resin is in contact with air. Lay-up normally cures at room temperature. Heat may accelerate cure. A smoother exposed side may be achieved by wiping on cellophane.

FIG. 10.11

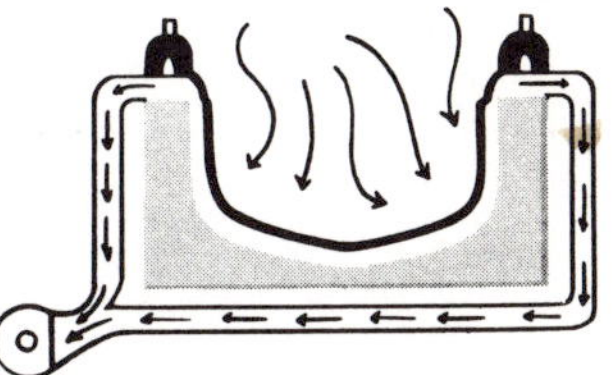

VACUUM BAG

Cellophane or polyvinyl acetate is placed over lay-up. Joints are sealed with plastic; vacuum is drawn. Resultant atmospheric pressure eliminates voids and forces out entrapped air and excess resin.

FIG. 10.12

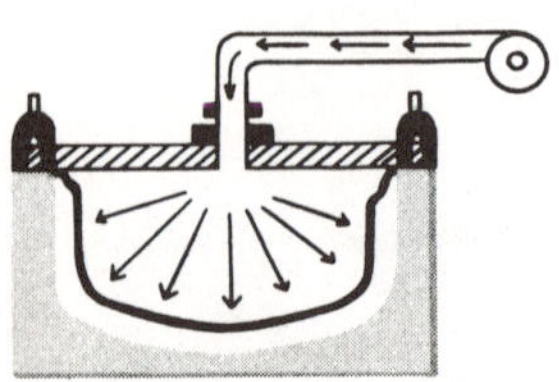

Tailored bag — normally rubber sheeting—is placed against lay-up. Air or steam pressure up to 50 psi is applied between pressure plate and bag.

FIG. 10.13

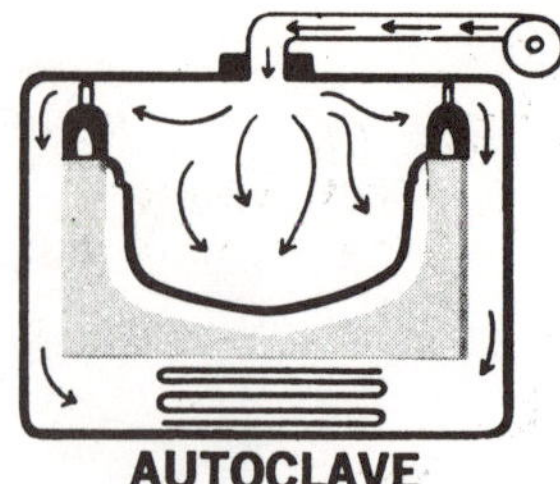

Modification of the pressure bag method: after lay-up, entire assembly is placed in steam autoclave at 50 to 100 psi. Additional pressure achieves higher glass loadings and improved removal of air.

FIG. 10.14

inserted in an autoclave, as seen in Fig. 10.14, where steam pressure builds up as high as 100 psi. This process permits maximum loading with glass, as well as greatest product density.

Spray-up RP. The spray-up RP molding process, as shown in Fig. 10.15, makes use of a multiple-headed gun which blasts chopped glass fibers, resin and catalyst simultaneously, from one of its three heads. This spray is directed by the molder to apply a uniform build-up over the entire mold surface. After the desired

SPRAY-UP

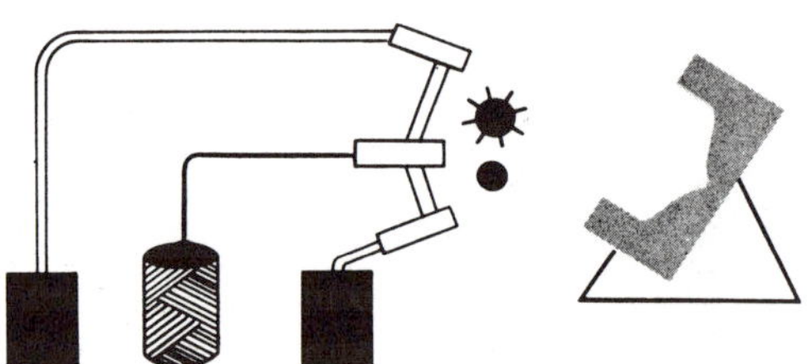

Fiberglas and resin are simultaneously deposited in a mold. Roving is fed through a chopper and ejected into a resin stream, which is directed at the mold by either of two spray systems: (1) A gun carries resin premixed with catalyst, another gun carries resin premixed with accelerator. (2) In a second system, ingredients are fed into a single gun mixing chamber ahead of the spray nozzle. By either method the resin mix precoats the strands and the merged spray is directed into the mold by the operator. The glass-resin mix is rolled by hand to remove air, lay down the fibers, and smooth the surface. Curing is similar to hand lay-up.

FIG. 10.15 Spray-up reinforced plastics.

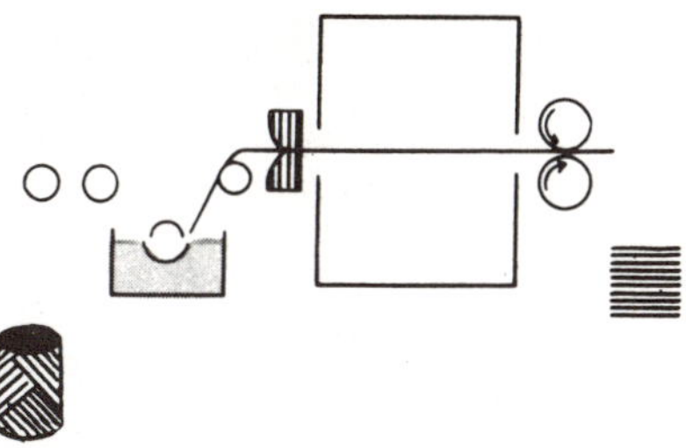

Continuous strand — in the form of roving — or other forms of reinforcement is impregnated in a resin bath and drawn through a die which sets the shape of the stock and controls the resin content. Final cure is effected in an oven through which the stock is drawn by a suitable pulling device.

FIG. 10.16 Continuous pultrusion reinforced plastics.

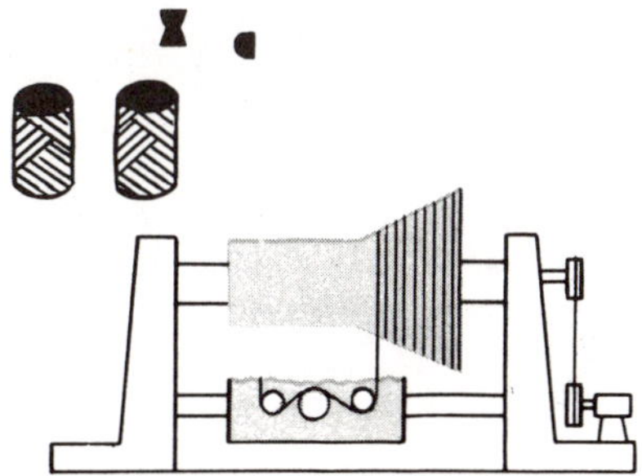

Filament winding uses continuous reinforcement to achieve efficient utilization of glass fiber strength. Roving or single strands are fed from a creel through a bath of resin and wound on suitably designed mandrel. Preimpregnated roving is also used. Special lathes lay down glass in a predetermined pattern to give max. strength in the directions required. When the right number of layers have been applied, the wound mandrel is cured at room temperature or in an oven.

FIG. 10.17 Filament winding process.

(FIGS. 10.8–10.23 *Courtesy Owens Corning Fiberglas Corp.*)

FIG. 10.17A The filament winding machine is used to produce very high strength tanks and products from the reinforced plastics.

thickness has been piled up, the exposed product area is smoothed by the use of rollers, etc., and curing is achieved by one of the aforementioned contact-molding processes.

This system is particularly suitable for tank linings, large bodies, pools, roofs, etc., and makes use of an inexpensive filler, glass roving. Labor costs are greatly reduced by this process.

Pultrusion Molding. Fishing rods, tubing, profile shapes, etc., are molded by the use of continuous strands of glass fiber that are impregnated in a catalytically accelerated resin bath, while being pulled along through a die that fixes the product shape. Cure is accelerated by passing through a heated zone, as illustrated in Fig. 10.16.

Filament Winding. Highest strength products are made by a process which winds continuous strands of glass fibers, or roving, on a core or mold of the desired shape. A mechanically controlled machine coats the filaments with resin, and directs their placement to develop uniform wall coverage and thick-

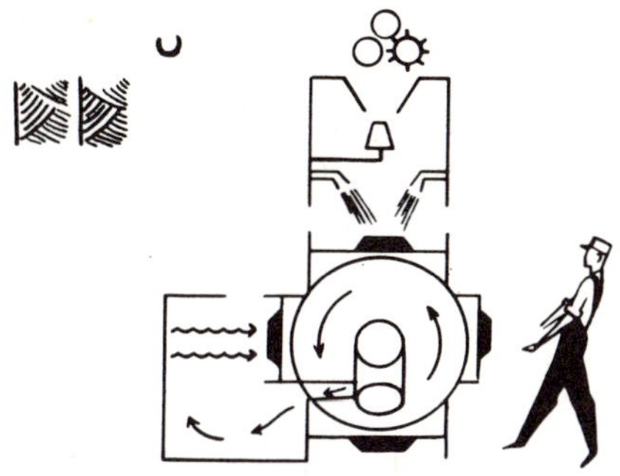

PLENUM CHAMBER

Roving is fed into a cutter on top of plenum chamber. Chopped strands are directed onto a spinning fiber distributor to separate chopped strands and distribute strands uniformly in plenum chamber. Falling strands are sucked onto preform screen. Resinous binder is sprayed on. Preform is positioned in a curing oven. New screen is indexed in plenum chamber for repeat cycle.

FIG. 10.18

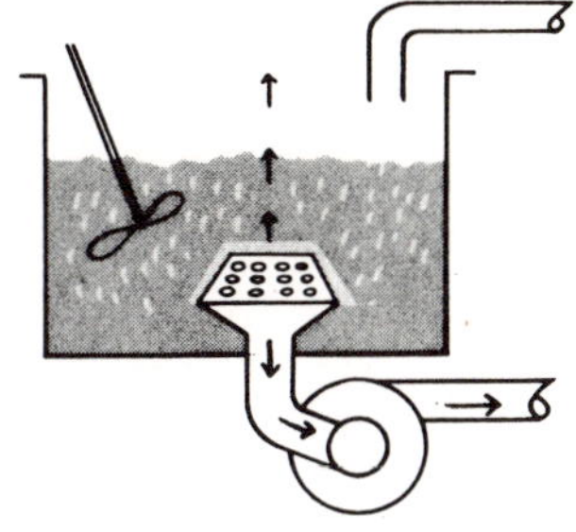

WATER SLURRY

Chopped strands are pre-impregnated with pigmented polyester resin and blended with cellulosic fiber in a water slurry. Water is exhausted through a contoured, perforated screen and glass fibers and cellulosic material are deposited on the surface. The wet preform is transferred to an oven where hot air is sucked through the preform. When dry, the preform is sufficiently strong to be handled and molded.

FIG. 10.19

ness, as illustrated in Figs. 10.17 and 10.17A. Highest strength pipe, rocket motor cases, gun barrels, and aircraft and missile bodies are produced by this process. Other gains include maximum strength to weight ratio, low-cost reinforcement materials, and best uniformity of properties. Collapsible or fusible mold cores are used to facilitate their removal after the product is wound and cured on such cores.

Many variations of these processes are employed in fully mechanized production systems for high-volume programs. Matched metal dies are usually used for such processes. For this work, precut shapes of resin-treated mat or fabric may be employed for simple shapes. More complex shapes are molded from preforms, composed of chopped strands suitably treated with resins, and formed into a load of the desired shape and distribution to be placed in the die. Some of these preforms are shaped by vacuum, as illustrated in Fig. 10.18, which sucks up a suitable layer of the desired piece on a shaped screen. Another preform process introduces the resin-impregnated strands in a water slurry. Suction pulls the water through a screen of correct shape, and leaves a preform built up on the screen which is then removed and dried, prior to loading in the mold. This is shown in Fig. 10.19.

Premixes. Resin, catalyst and filler are all blended together into a pre-mixed charge that may be molded by the conventional transfer injection and compression processes, using molding pressures from 1500 to 2000 psi; resins commonly used include all of the thermosetting resins for the production of switchgear, trays, housings, as well as structural and functional components. Similarly, the thermoplastics resins are reinforced with chopped glass fibers, asbestos, mica, etc., and are then molded by the conventional injection process. (BMC)

The reinforced plastics manufacturers often refer to the premixes as prepregs. For this purpose, prepregs often consist of a tack-free glass-mat reinforcement, a low viscosity polyester resin, fillers, color pigment and catalyst, all ready for molding in the "as-purchased" condition. Incorporation of MgO produces a delayed but controlled viscosity increase, after mixing, impregnation and thorough wetting of the glass fiber substrate. This thickening reaction allows use of optimum monomer resin levels, resulting in uniform, integral flow of reinforcement and resin to all parts of the mold cavity. This facilitates the use of faster curing, more reactive resins, and minimizes the segregation of resin with the resulting crazing problems.

Sheet molding compound (SMC) is extensively used for many of the reinforced plastics molded parts. Its usage simplifies the preparation and handling of materials for compression molding in matched die molds and for sheet making. In this process, the polyester is mixed with a thermoplastic copolymer additive that will reduce shrinkage during cure. Fillers are added at this point to thicken the mix which is then completed by the addition of catalysts and chopped glass fibers. The resin mix and chopped strands of glass are deposited evenly on a

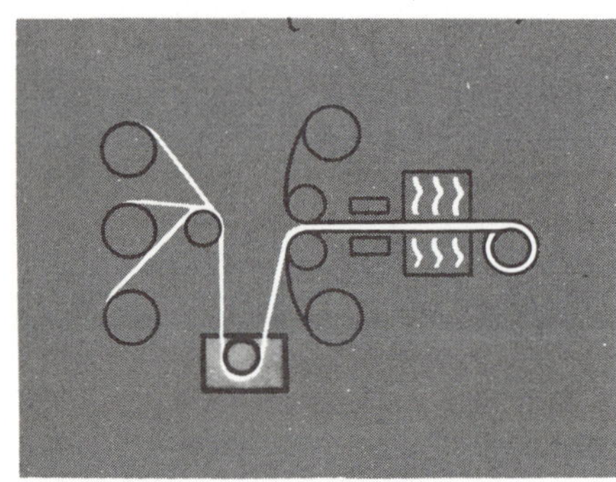

CONTINUOUS LAMINATING
Several layers of fabric or mat are passed through a resin die and brought together between cellophane covering sheets. Laminate thickness and resin content are controlled by squeeze rolls. Lay-up passes through a heating zone to cure resin.

FIG. 10.20

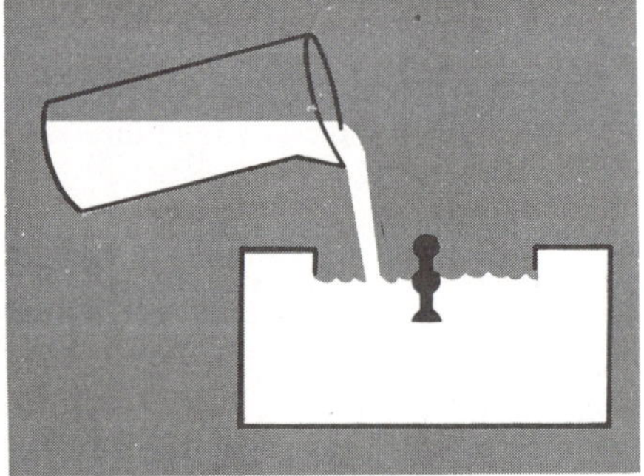

ENCAPSULATION
Milled fibers or short chopped strands are combined with catalyzed resin and poured into open blocks. Fibers decrease shrinkage and crazing and increase useful temperature range.

FIG. 10.21

MATCHED DIE MOLDING
This is the mass production method for
manufacturing Fiberglas Reinforced
Plastics. Mat, Fabric, or Preform
Reinforcement is combined with resin.
In the mold, matched metal dies under
pressure form and cure the part at
100-300 psi and 225 to 300°F. Depending
on part size and shape, cure cycles range
from one to five minutes.

FIG. 10.22

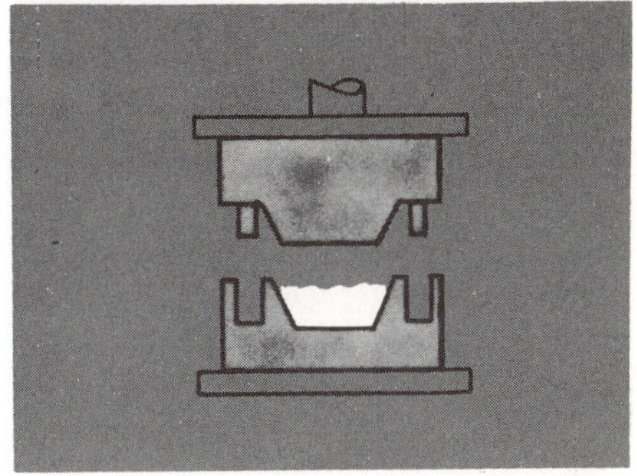

PREMIX/MOLDING COMPOUND (BMC)
Glass reinforcement is combined with
resin, pigment, filler and catalyst. This
mix is then placed in mold cavity under
heat (225°F to 300°F) and pressure from
100 to 1500 psi. Molding compounds
ready for use are sold by many
manufacturers.

FIG. 10.23

polyethylene film strip carried on a moving belt that moves the mat through wet-out rollers into a roll-up position. Other production methods for RF plastics are shown in Figs. 10.20–10.25.

A nylon based line of stampable thermoplastic materials* will offer better answers to many problems and open new markets. This line of materials is mineral-filled to provide high flexural modulus and easy formability. Long glass fiber reinforcement adds high heat distortion, thermal endurance and impact resistance. Composite sheets are provided in formulations and constructions to cover a range of physical properties. Precut blanks of this line of stampable thermoplastics may be formed into complex shapes (e.g. exterior automobile

*STX® Allied Chemical Company.

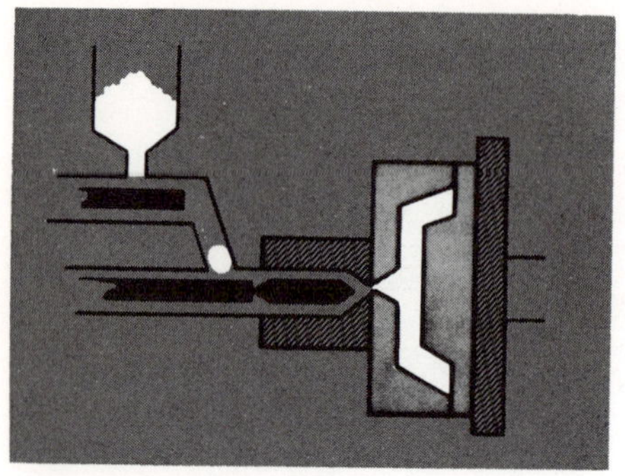

INJECTION MOLDING
This is a high production process using thermoplastic materials. A solid glass-resin compound is introduced into a heated chamber where it softens. The soft mass then is injected into the mold cavity where it solidifies into the part.

FIG. 10.24

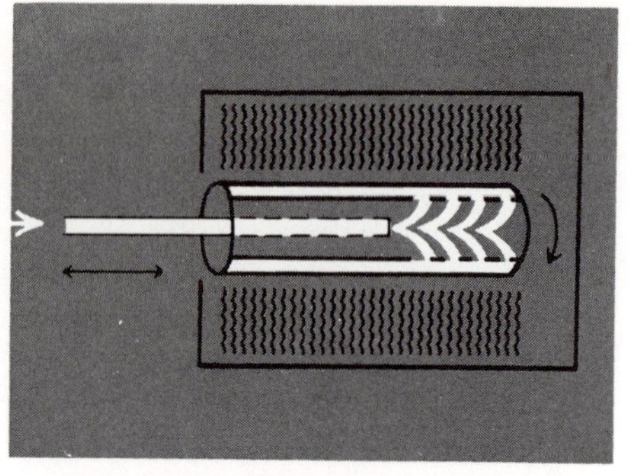

CENTRIFUGAL CASTING
Chopped strand mat is positioned inside a hollow assembly and placed in an oven. As assembly rotates, resin is introduced uniformly through reinforcement. Centrifugal force forms glass and resin to shape against walls of assembly.

FIG. 10.25

body components) on conventional metal stamping or compression molding equipment.

REINFORCED PLASTICS MARKET SUMMARY

Transportation. Auto bodies, trucks, cabs and bodies, trailers, cooling system components, component parts in buses, trucks, autos, motor scooters, insulated tank trucks; special purpose vehicles, farm equipment.

Construction. Structural shapes, patio roofing, panelling, siding and curtain wall materials, glazing and skylighting panels, concrete pouring forms, rooms.

Marine. Boats, ships and submarine hulls for recreation, Navy, commerce, and Coast Guard; shipboard applications such as fuel and water storage tanks, duct work, ventilation cowls; marker buoys, marinas, floating docks.

Materials Handling. Pharmaceutical trays, industrial tote trays and boxes, food processing and delivery trays, boxes, bins, piping and storage tanks.

Electrical. Applications in transformers, motors, generators, switchgear, and electronic equipment.

Sporting Goods. Fishing rods, archery bows and arrows, golf carts, golf clubs, hockey sticks, vaulting poles, shot gun barrels, water and snow skis, protective helmets (racing and baseball), swimming pools, playground equipment.

Seating. Luxury office seating, bowling alley settees, school and auditorium furniture, occasional chairs, lawn furniture, subway and bus seating, benches and arena seating.

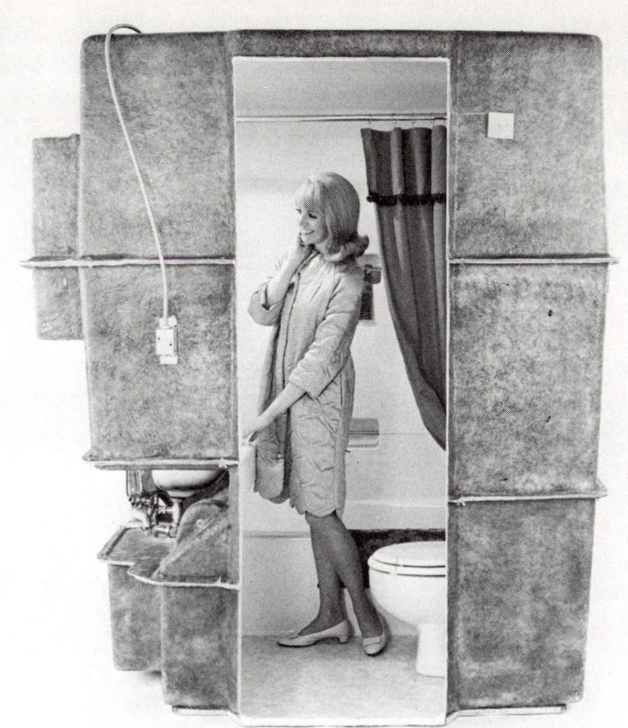

FIG. 10.26 Complete bathroom molded in sections of polyester glass is inexpensive and points to the future of the reinforced molded plastics in housing. (*Courtesy The Crane Co.*)

Corrosion Applications. Fuel and chemical storage tanks, fertilizer, and chemical storage hoppers, tanks, pipe, fume carrying ductwork, ventilator housings, pumps, stacks, cooling tower frames and grids, scrubbers, chemical and photographic processing equipment.

Protective Covers and Housings. Outboard engine and lawn mower housings, electrical equipment, computer and business equipment housings, photo equipment cases, machinery guards and mail boxes.

Appliance and Equipment. Consumer appliance components, dryer ducts, dishwasher lids, decorative panels, structural gussets, laundry tubs, water softener tanks and pipes, clothes washer tubs, gear housings, air conditioner components, fan housings and hand tool housings.

Aerospace and Military Market. Rocket motor cases, nozzles, nose cones, submarine fairwaters, and pressure hulls, radomes, wing tip tanks, helicopter rotor blades, pressure bottles and spheres, assault boats, liquified gas storage vessels, rifle stocks, sonar domes, aircraft lounge seats, engine cowlings, duct work and complete aircraft fusilage surfaces.

FIG. 10.27 This one piece front end for GMC truck is made of fiberglass polyester sheet molding compound and weighs only 70 pounds. (*Courtesy Owens Corning*)

Fabricating Laminated and Reinforced Plastics

The successful fabrication and machining of laminated plastics depend chiefly upon the use of sharp tools operated at proper speeds. Tools operate best at high speeds, except where the cut is so large that the tool overheats. In general, the cloth-based materials are easily machined, the paper-based materials are somewhat more difficult, and the glass and mineral-filled products are the most difficult of all.

Sawing. Circular saws for the paper and cloth-filled materials should be made of hard steel, just soft enough to permit filing. Diamond saws and tungsten carbide are much more appropriate for the glass and mineral-filled plastics. For the conventional laminates, the smoothness of the cut desired governs the number of teeth, as well as the amount of set. For extremely smooth cuts, the saws should be hollow ground, and used without set. The saw table should be adjusted so that the saw teeth will be just above the top of the laminated stock. The work should be fed slowly into the saw.

Drilling. Laminated materials should be drilled with carbide or high carbon steel drills without lubricant (it may be advantageous to use lubricant on holes over $3/16$-inch in diameter). Water or CO_2 jets may be used to cool the drill when desirable. Best results in drilling are obtained by the use of drills especially

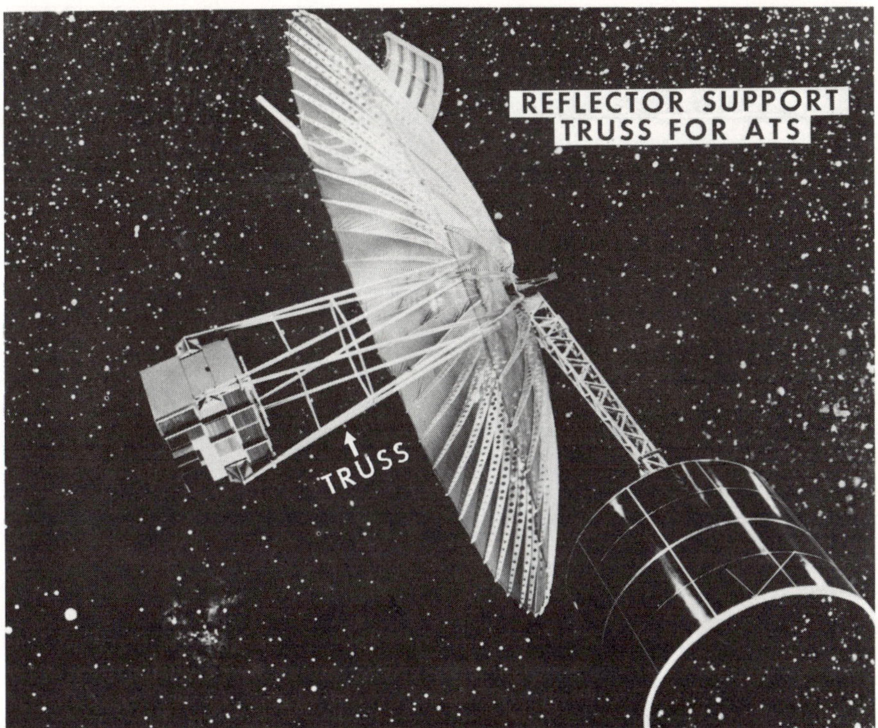

FIG. 10.28 This 14½ foot long support truss for this NASA communications satellite truss was made from an epoxy graphite compound. Graphite fibers were used to gain maximum strength, low density and very low thermal expansion. A 50% weight saving over modern high strength metals was realized. (*Courtesy Hercules, Inc.*)

designed for the drilling of laminated plastics. Such drills are provided with wider flutes and extra clearance on the edge of the flutes, in order to reduce friction and prevent overheating. Good results will be obtained from slate or marble drills, as well as ordinary twist drills, with the lips backed off and ground slightly off center. If ordinary twist drills are not ground off center, the holes will be true and smooth, but about .003-inch undersize. Tungsten-carbide spade drills, with water lubrication, are recommended for the glass-filled products. A pecking motion will give best results, since the frequent removal of the drill from the hole will remove chips, and minimize wear. In feeding the drill, care should be taken not to subject the material to unnecessary crowding, which would overheat the drill, thus splitting the work. Multiple drilling can be done with the multi-drill heads. A piece of similar material should be firmly clamped to the exit side to minimize the chipping when the drill comes through.

For drilling parallel to the laminations, the material should be firmly clamped in order to prevent splitting, and the drill then fed carefully into the stock to avoid undue strain. When comparatively large holes are to be drilled, a small hole, followed by a counterbore, should be used before using the large drill.

The speed of drilling varies according to the diameter of the hole, and should be as high as can be used without burning the tool. Coolants will speed up drilling, and are essential to production drilling of the mineral and glass-filled products.

Turning. The tools used for turning should be made of tungsten carbide, diamond, or high-speed steel. The conventional cloth and paper laminates may be machined in the manner used for soft steel, but at an increased speed. Tools should be ground to a diamond point, with plenty of clearance.

Tapping. The instructions for drilling also apply to tapping. Care must be taken to insure against splitting the laminations. Lubricants are advisable, and the stock should be firmly clamped in compression when drilling parallel to the laminations. It may be necessary to hand tap the most exacting jobs in the glass and mineral-filled materials.

Threading. For the paper and cloth laminates, the conventional methods for cutting threads in soft steel may be used. High cutting speeds should be employed. The glass and mineral-filled products may require grinding. When grinding is not essential, carbide and diamond cutters are preferred for accuracy and production work.

Milling. Good results will be obtained by the use of standard milling cutters, ground with a slight rake, and operated at high speed and a course feed. Proper clearance may be obtained by the use of side-milling cutters. However, milling

FIG. 10.29 The punch presses in this picture are stamping out small functional parts of laminated plastics. (*Courtesy Insulating Fabricators, Inc.*)

is not always practical with the glass and mineral-filled materials, and such products may require a grinding operation.

Punching. Most of the laminated materials may be punched into simple shapes and thin sections, provided good punching practice is used. This includes sharp dies, proper clearance between punch and die (.001 inch generally recommended), proper stripper plates, and proper preheating devices. An example may be seen in Fig. 10.29. For clean punchability in the thicker sections, the laminated sheets are often preheated in ovens or on hot plates. A variety of cold punching stocks are produced to give clean punchings at very high speed punching rates.

In good punching practice, the edges of the piece should not be closer to the edge of the strip than twice its thickness. Pierced holes should not be less than one and a half times the thickness of the material from the edge of the piece; the thickness of the material allowed between holes should be at least equal to the thickness of that material. Pierced holes should be no smaller in diameter than the thickness of the material—nor should they have square corners. For all practical purposes, it may not be possible to punch or shear the glass and mineral-reinforced plastics except in relatively thin sections.

Flycutting. Discs and gear blanks are frequently cut from sheet materials with flycutters, as shown in Fig. 10.30. The glass-filled materials do not cut easily

FIG. 10.30 Gear blanks and discs are often cut from the laminates by a flycutting operation.

FIG. 10.31 These pieces were post formed from flat laminated sheets after heating the special phenolic post-forming materials to a high temperature. Conventional metal-working dies and fixtures are used for this work. (*Courtesy Synthane Corp.*)

when employing this process, and grinding or turning may be necessary to make discs of these abrasive materials when close dimensional control is necessary. Diamond tools are essential for glass fillers.

Cementing. Epoxy adhesives are widely used for cementing the laminated plastics. Surfaces must be kept clean and free or oils, silicones, etc., when cementing operations are to be done.

Post-Formed Laminates. The laminated plastics may be formed into shapes such as those depicted in Fig. 10.31. These formed pieces retain the properties of the original flat sheets, and this process facilitates the production of numerous special shapes, with no high mold investment. The forming process makes use of high temperature preheats before the forming is done. Fixtures and dies, like those used for the metals, are widely used for the post-formed laminates. In one commonly used process, the laminated pieces are preheated by immersion in a molten lead bath for a brief period of time.

11 MOLDING MACHINE MAINTENANCE

While the title is "Molding Machine Maintenance" it is impossible to discuss all ramifications of the subject because of the many differences which exist between machines of different manufacture. The fundamentals which apply to all are worthy of serious study.

Plastics molding machines are generally hydraulically powered. To obtain best operating efficiency and long machine life, it is essential to keep the hydraulic fluid cool and clean. As discussed in Chap. 8, the fluid is cleaned by filters installed in the suction line, the return line, or both. The temperatures must be maintained as outlined in the machine manual. Cooling water supply to the oil cooler should be adjusted to maintain an oil temperature of 110 to 150°F. This cool oil will assure long life of the hydraulic components. Depending on the quality of the water supply, it will be necessary to disassemble the cooler and remove chemical deposits from the tubes, and perhaps, at intervals, some sludge on the oil side of the cooler. Regular maintenance is very important.

Greasing of platen and link bushings is equally important—follow the specifications of the machine manual. On machines equipped with automatic oiling systems, the reservoir must be checked frequently and refilled as required. Oil flow must be adjusted to provide adequate lubrication without flooding.

It is further suggested that all machines be equipped with elapsed hour timers, so that a regular maintenance procedure can be established. If properly applied, preventative procedures can keep breakdowns to a minimum during production. All machine components have a fixed life. By observing their hours of life, replacements can be made before breakdown. This applies to such elements as seals, limit switches, bushings, etc.

The machine and its surroundings must be kept clean at all times. At the close of the work week the machine should be wiped down. This should include working surfaces such as tie rods, toggle mechanism, hydraulic lines, etc. The machine should then be started and operated by the maintenance crew. Examination of the components under these clean conditions will show up leaks and other defects which can then be corrected.

Electrical connections to the heaters on the heating cylinders must also be checked at this time. Thermocouple leads should also be examined during this inspection, and repairs made on all elements as required. Burnout signals are desirable on all heaters; otherwise they need a check on each shift. The shift foreman should examine critically the operation of all machines during his shift and report his findings in writing at the conclusion. Such information is invaluable in heading off breakdowns. It is strongly recommended that spares such as heater bands, timers, limit switches, "O" rings, fuses, pump cartridges, etc., be kept in stock to cope with a breakdown, should one result. Consult with the machine maker on which spares are required or recommended. In any event, a maintenance program based on the recommendations of the machine manufacturer must be set up and *maintained*. The cost of repairs and rejects is much greater than the cost of preventative maintenance.

Molding Machine Check List

Some machine manufacturers furnish a maintenance check list with the machines. If not supplied, a check list should be prepared for them all. In many shops these lists are posted on the machine. A check list for a hypothetical injection machine might be similar to the one shown below.

CHECK LIST

Daily:

1. Check hydraulic tank level and oil temperature under operating conditions. Check position of indicator on "tell-tale" oil filter. Change filter as required.

2. Check and refill toggle mechanism lubricator. Check setting of the system. A small excess of oil should always be present in knuckle joints, tie bar, bushings, etc.

3. Remove dirt accumulations from all sliding surfaces such as tie bars, ways, guides, etc.

4. In some shops inspection and function of safety devices such as safety doors, guards, electrical mechanical and hydraulic interlocks are checked daily also.

5. Grease wheels on safety doors, if machine is operating on a semi-automatic cycle. If on an automatic cycle, this is unnecessary, weekly greasing should be adequate.

Weekly:

1. Wipe down and clean the entire machine. Operate the machine to find oil leaks and correct as required.

2. Check connections and wiring to heaters and thermocouples. Correct as required.

Annually:

1. Replace grease in ball-bearings of electric motors. Follow motor manufacturers' instructions.

2. Replace hydraulic oil. (Some shops use an auxiliary filtering system which at intervals pumps the hydraulic oil through a special filtering system to remove the finest contaminants. Experience with such systems, coupled with relatively low oil operating temperatures, has allowed hydraulic oil to be used for many years.)

Trouble shooting of press failures is a function of determining if the failure is mechanical, hydraulic or electrical. The important clues usually lie in what the machine was doing when the failure occurred.

Many hours have been lost in checking hydraulic failure when the problem lies in a simple fuse failure in a control circuit. If the press will operate manually after failure on an automatic cycle, the trouble obviously lies in a control circuit. Since automatic control is based on timers, these should be checked or replaced. Oftentimes the problem lies in loose connections. It is recommended that before any new press is placed in service, all electrical connections be checked for tightness. When any electrical failure is suspected, the tightness of electrical connections should be verified.

Mechanical failures are generally quite obvious—a broken toggle pin, etc. However, these are rare. Hydraulic failures are often elusive and require a careful examination of the facts and thoughtful reference to the hydraulic circuit diagram to pinpoint the source of trouble.

Mold Set-Ups

A mold is an expensive tool. Much mold and press damage can be traced to improper set-up. While basic differences exist between various makes of presses, some generalities exist. For example, in a hydraulically clamped press, clamping pressure on the mold is easily adjusted. With a toggle operated press, careful and painstaking adjustment is required.

A good set-up man will always study the set-up and provide all necessary tools and equipment such as clamps, bolts, washers, hoses for cooling lines, wrenches, etc., to accomplish expeditiously the task. In the better shops, the set-up man will arrive at the press with the mold and a well-equipped portable tool chest, so that he can cope with any problem which might arise.

In a toggle press, after the previous production mold is removed, the distance between closed platens is measured and the extended moveable die plate and link mechanism is adjusted to accommodate the mold to be installed. After the mold is attached to the platens, the press should be opened and closed while observing the link action. The tie rod nuts should then be *equally* adjusted so that when the mold clamps, there is a slight hesitation in the link action as it locks. Generally speaking, this setting should be adequate to provide sufficient clamping pressure. Always consult the machine manual for explicit instructions to obtain suitable clamping pressure. The penalty for overclamping is broken tie rods. Closing the mold on a sheet of paper will disclose any non-uniformity of clamp.

After mold installation, the knock-out system is installed. Since this set-up varies rather considerably from one make press to another, it is recommended that the machine instructions be followed. In any event, provide only sufficient stroke to completely eject the products. All mold stroke and ejection strokes must be adjusted to a minimum. This will help to obtain optimum production rates.

Mold temperature control hose lines are now attached to the mold. Depending on the job, the hoses may be connected to "city" water, a mold chiller (refrigeration system), steam, or a heater. Connections must be leak free. Initial temperature adjustments should be set at this time.

While the mold set-up is being made, heaters on the injection cylinder should be on and set at the proper temperature for the material to be molded. When the set-up is completed, material is fed to the cylinder, the nozzle retracted from the mold, and cylinder purging commences. When material passing through the nozzle is clean and smooth in appearance, the nozzle is seated in the sprue bushing, and test shots are then made.

On a plunger machine, material feed and plunger stroke settings are now established so that the cavities are filled out without packing. Similarly, settings of screw rpm, stroke and cushion are set in a screw-injection machine. The cycle, injection, dwell, cooling, mold open-time, etc., are simultaneously adjusted, thus providing an economical production time. After these initial adjustments are completed and production goes on stream, the set-up man or foreman should "sharpen-up" all elements of the cycle, including mold temperature, to establish the best cycle time consistent with acceptable product quality standards. A sprue grinder should be at the press so that sprues and runners can be reground immediately without contamination. The reground material is mixed with virgin material in fixed proportions, and fed into the machine hopper. Depending on the job, in some shops the runners and sprues are snipped into small pieces and fed into the cylinder, and are directly reprocessed.

The product should be examined by the operator, and on many jobs, checked with gauges to make certain that the quality does not fall short of the established standard. Inconsistencies should be immediately reported to the foreman or shift leader for correction. In some shops an inspector will pick production samples at regular intervals for check. His findings are promptly reported to the

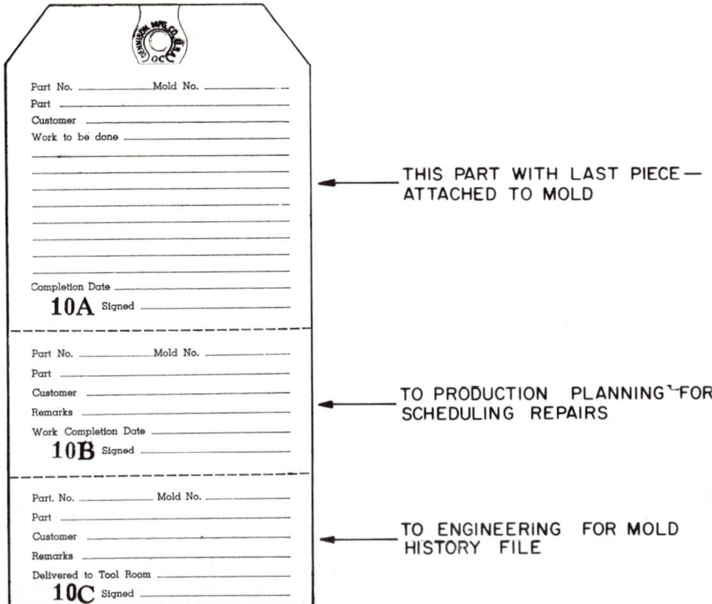

FIG. 11.1 Proper inspection of last production piece and listing of essential mold repairs is essential to economical production.

foreman, who initiates action as required. Every operating employee should be an "inspector." The earnings of his company and certainly, as a result, his own earnings depend on acceptable quality. Scrap earns no money!

Before removal of mold from press, a last piece sample should be inspected carefully and the mold condition noted. This last piece and a list of essential repairs to the mold should be attached thereto before it goes into storage. A duplicate of this report, Fig. 11.1, should go to the scheduling department to facilitate mold maintenance before it returns to production.

Trouble Shooting

Trouble shooting was considered an art for a long time. This is nonsense. Any person who teaches himself to visualize what is happening in the injection or screw cylinder, the nozzle, runner and the cavity can correct injection molding problems. The same is true for transfer and compression molding. If the extrudate from the nozzle is lumpy, showing unplasticized granules, obviously the barrel heat should be increased—but how much? A little at a time until the situation clears. Does the material freeze in the runner? Injection pressure may be too low. Perhaps the cavity does not fill out. The same reason applies, or the gate may be too small. Flash on the part—increase the clamp pressure, decrease the material temperature, or decrease the injection pressure—many things can be done, but *never* change more than one variable at a time. The logical course to follow is to observe the results, and only then proceed with the adjustment of another variable.

Many charts and data sheets are available from all material suppliers, which carefully detail molding difficulties and means for their correction. These should be consulted. With experience, trouble shooting becomes a "sixth sense," and this is only developed by observing the effect of the process variables on many jobs—time, temperature, and pressure are the basic machine variables. Material conditioning is a basic need.

On a very small scale, time and motion studies that organize material and insert handling, minimize walk or reach time, and prevent sequence errors will pay off—these savings are achieved by careful thought and planning at the set-up time.

FIG. 11.2 Automated compression molding. Rotary plastics molding presses are adapted for a wide range of automatic molding operations using thermosetting molding compound. They are extremely proficient for producing small plastic parts as used in the manufacture of electrical wiring devices, electrical appliances, automobile and airplane parts, bases for electronic tubes, small statues, toys, instrument parts, and many other parts. One operator can run six or eight presses. (*Courtesy New England Butt Co.*)

Quality Control Fundamentals

Quality control in most molding shops is thought of as physical inspection of the product. This is only a small part of the whole job of quality control. The task starts with the conception of the product. The product must be designed to fulfill its mechanical, electrical or chemical resistance requirements. These criteria determine the plastics material to be used, and this in turn establishes the electrical and physical characteristics of the product, such as tensile, compressive or impact strength, and dimensional tolerances. Product requirements must be completely specified to assure an acceptable quality level from the supplier, whether custom or proprietary. All quotations from material and insert suppliers must detail the quality they propose to supply, including exceptions to the specifications. There must be no misunderstanding as to what will be supplied.

To achieve the required quality standards, modern quality control systems must be employed. The usual method of inspection procedure simply removes defective parts from the production lot. The method provides no clue as to why

FIG. 11.3 Completely mechanized injection molding system for producing camera parts. (*Courtesy Eastman Kodak Co.*)

the defective parts occurred. Statistical quality control—"S.Q.C."—(and please do not let the word "statistical" deter from further study), is a process of quality control which is aligned with the production process. Statistical quality control requires the measurement of only a small number of samples taken from the process at intervals to ascertain whether the process is within limits or drifting beyond the limits prescribed. The data obtained is plotted on a control chart, which shows graphically whether or not the process is in good control. Such information indicates whether adjustments to the process should be made.

The general mechanics for economical quality control are too lengthy to detail in this book. Excellent books[1] are available. The booklet "S.Q.C.—A New Management Tool for the Plastics Industry," by Monsanto Chemical Co., is highly recommended as a primer on this subject.

FIG. 11.4 High volume injection molding plant with mechanized feed, control and take-away. (*Courtesy Ford Motor Co.*)

[1] "Quality Control for Plastics Engineers," Debing, L. M., Van Nostrand Reinhold Co., 1959; "Statistical Quality Control," Grant, E. L., McGraw-Hill Book Co., Inc.; "Quality Control Handbook," Juran, J. M., McGraw-Hill Book Co., Inc.

Automation

Automation is practiced in some form in even the smallest molding shop. If a press is set up and operated without an attendant, as shown in Fig. 11.2, automation is practiced. See also Chapters 4 and 5. Whenever the product design or production quantity warrants the expense, the mold must be designed to operate automatically. The product should be degated in the mold, and sprues and runners should be conducted to a separate receptacle. Blow molds are sometimes also arranged to deflash the product during the molding cycle. Direct labor is the largest cost item. The highest order of automation is often found in plants producing a product in millions, at the lowest cost possible. A particularly fine example of this lies in the Eastman Kodak Co. production of the "Kodapak" polystyrene cartridge for the "Instamatic" camera, and illustrated in Fig. 11.3. The cartridge is produced in a battery of twelve fully automatic 25-ounce reciprocating screw machines. Six machines make covers, and the other six make the bodies. All molds are eight cavity, and the molded parts drop onto conveyors under the machine. These inclined conveyors carry the product to conveyors situated high above the aisle between the machines. One of these carries the covers, and the other the bodies.

The parts are passed from the central conveyor belts into a group of vertical disk-type separators which separate runners from the parts. The runners proceed to a grinder, and the parts continue to a machine which assembles the covers and bodies. From here, the assemblies travel to spool assembly. Film spools are molded in a similar set-up. Molding material for all machines is supplied by a vacuum system. Reground material is automatically blended with virgin material in pre-set proportions.

Figure 11.4 illustrates another type of automated injection molding as used for automotive products.

One operator for each six machines checks all machines, and maintains an operating log for his shift. Samples from each cavity from each machine are checked in an electronic gauging machine every hour. Molding conditions are then corrected as required.

12 ENGINEERING AND ESTIMATING MOLDED PRODUCTS

In a bitterly competitive business with multiple variables in methods, materials, and processes, the estimator is required to do a value analysis on each project. The customer may specify the material, delineate the design, spell out his essential tolerances and finishing needs, and buy the part from someone who changes everything to provide greater values for the customer. Shown in Fig. 12.1 is a check list of data essential to a good quotation. Estimating to win requires more than a cost analysis of what the customer thinks he wants to get—it requires consideration of all the alternatives that can give better performance and more values for the price asked. The customer must however outline fully the product requirements so that a correct value analysis can be made.

Initial considerations are concerned with a general analysis of production needs, material specified, mold size, as well as molding method, with the logical alternatives being considered as the value analysis proceeds.

Most engineers use the weight calculation period for the product design analysis. Weight is estimated by breaking down each section of the part into a standard shape from which volume can be calculated and then converted into weight by gravity conversion tables. Volume of models can be determined by displacement. Weight calculations can often be helped by use of a planimeter. During these calculations, the engineer sees each section clearly and evaluates its moldability, tolerance control, flow considerations, warpage probability, air entrapment, etc. Each section of the part can present its own special molding problem which shows up clearly during this sectional analysis. The estimator concludes

```
Probable order size .................................................. Annual requirement ..............................................
Quote in lots of..........................................................................................................................
Material  ..............................................................................Desired mold size............. cavities
QUOTE:
..............1-cavity mold and economy sampling       ............ to drawing tolerances only
..............economical mold size for min. cost        ..............on "standard" tolerances
..............subj. to agreement on tolerances          ..............subj. agreement on design rv
Desirable alternatives to be covered by quote .......................................................................

.......................................................................................................................................

.......................................................................................................................................
Inserts to be furnished by molder ................................. Customer................................
SPECIAL REQUIREMENTS:
Electrical .................................................................. Temperature ..........................................
Mechanical ................................................................ Assembly................................................
Testing ..................................................................... Printing.................................................
Desired date to start production .............................. Probable order date ..............................
Competitive material ................................................ Why unsatisfactory ..............................

.......................................................................................................................................
Is customer experienced plastics user ? ...............................................................................
Desired price, when known .......................................................................................................
Quote to be made to customer by .................................................................... (Person)
```

Fig. 12.1   Desirable data for quotation preparation.

this study with an estimated weight and a series of notes on the negative factors, tolerances, wall thickness, undercuts, and essential design or material changes.

Material section considerations involve a study of many factors, as outlined in Chapter 7. The estimator must always remember that there may be a better material and a better way, and the engineer who finds the best way can have the best price and win the order. With the "best value" material selected, the molding method is also clarified. Next come the mold design considerations.

The estimator must then visualize the mold that will make the part. Where will the parting line be located for lowest mold cost and for best operation? This choice is often the lesser of the several evils that he sees in each design. How will side cores be pulled? Where must he have KO pins, and what minimum size will do a good job? Where will most heat or cooling be needed, and how can the mold be built to get the fastest cycle? What type of runner system and what kind of gate is economical? Can a self-degating mold be designed to give fully automatic operation? Will the inserts "float"? Are essential mold sections too fragile, where are radii essential, and what must be done to be sure there will be no lines that fail to knit? Every detail of the mold construction, mold material and operating cycle must be visualized at this point, preparatory to estimating part and mold cost. Mold size is dictated by volume needs, cost of mold, tolerances, available presses, and life expectancy.

At this time the estimator can stipulate the mold design details with adequate clarity and detail to facilitate preparation of a mold estimate by the mold maker. Failure to base the mold estimate on a good performing mold that will produce the part "in tolerance," on a good cycle with minimum maintenance, will cost his company a great deal of money. No one ever succeeded with cheap and poor molds. Extra hand work to compensate for poor mold details is impossibly expensive. The mold must be right, or all is lost.

With the type of mold and the mold size determined, the operational calculations come up for study.

Thermosetting products are generally more time-consuming and difficult to estimate because of their greater number of variables and hand operations. Preform size and cost can be the first step, followed by preheat time and loading procedure. Other time factors, then to be determined, may include insert and preform load time, walk time and wait time, rate of close, ejection procedure, and time. Valving of available press can be a real factor along with the speed of the press. Time required to break flash and insert piece in shrink fixture may need to be determined, along with gate removal or other post-molding operations to be done by the molder at the press. The number of molds that can be combined to share the molder's time is a real consideration. Sometimes calculations are necessary both ways to learn whether or not single, dual or triple press operation will give the most economical answer. A real job of industrial and plastics engineering has to be done by the estimator on each quote if he is to meet smart competition. Any one of the foregoing details can be as important in the winning price as the overhead.

The probability of unbalanced production from family molds must be carefully considered, and plans made to insure good pieces from all cavities.

For injection molding, the rate of fill and the hardening time are the principal factors to be estimated. When automatic degating or runnerless molds can not be used, the value of the sprue must be determined; some can be used almost 100%—others cannot. What is a safe reasonable estimate of the sprue's value? What fixtures are needed for degating, dimensional control and post-molding operations? Because of fast injection cycles, mechanized fixtures are mandatory—and they must be designed mentally with the details recorded for transmission to the person estimating these fixtures. Insert loading is a real problem; perhaps the insert loads can be pressed-in after molding. Thermoplastic parts are easily marred, especially when warm and fresh out of the mold. How must they be handled from press to carton?

Operational details and time cycles are often determined by simulating the motions essential to the operation while timing them with a stop watch. Records of similar jobs are often checked to compare results. Many estimators maintain graphs, tables, and charts of details to eliminate time-consuming calculations, and to insure repetitive accuracy in the selection of proven data. With several hundred materials, colors, quantity prices, gravity differentials, etc., the material maker's data sheets are constantly used in estimating.

Volume jobs require a careful estimate of packing cost. An over-costly carton can lose a job, and excess breakage in shipment can cause loss to the molder. Quantity of parts per carton, spacers, wrapping, etc., are all vital details that must be evaluated carefully.

How much inspection will be needed? What kind of inspection will insure economical quality control? Who will do it? Can the molder inspect his parts—and—will he do it? What tool wear is expected that must be included as an extra maintenance charge? How much time is necessary to set up and remove the mold? How much purge time and material will be required?

As each detail is calculated, it is recorded on a basic form where the final arithmetic can be done easily with a calculator, and then checked quickly by a supervisor. Thus we arrive at a selling price for the parts.

In most cases, the best mold price is gotten only by a "give and take" bargaining and exchange of engineering views with the mold maker. This is a simple process with a captive mold shop. In going outside for tool prices, the estimator must know where to go for best value on each type of job. Availability or historical success with certain types of jobs, knowledge of the special equipment in some tool shop that will cut costs, use of standard components, and many other similar details help get low tool costs. One sure way to get a poor mold estimate is to mail product prints to a mold maker and ask him to give you a price on an "x" cavity mold. Without an intimate discussion of details, the mold price quoted will include several safety or ignorance factors; one might also get a mold that will not operate on the estimated cycle, thus requiring many post-molding finishing operations. A good mold at minimum cost results only from intimate liaison between a plastics engineer who knows what he must have, and an experienced mold maker. Mold drawings must be checked and approved by the molder to make sure that the mold, if executed per drawing, will be suitable for its purpose. A copy of the drawing will also be essential for future mold change or repair consideration.

Estimating skills are achieved by continuing practice. The estimator must be informed of the results of each proposal and quote. He learns from his successes and failures. His company profits when he outsmarts the competition with skillful engineering, judicious choice of materials, excellent mold design, creative industrial engineering, and accurate calculations. The selection of the most economical material, design and molding process are vital price considerations. Figure 12.2 illustrates   typical quotation forms and estimates.

The plastics engineer has serious responsibilities that must follow the estimating. When a decision is reached to proceed with the product, the entire estimate must be reviewed, and final changes made to the product design—then the mold design is started.

The following summary of responsibilities usually assigned to the plastics engineer will be a helpful guide and check-list.

TYPICAL DUTIES OF PLASTICS ENGINEERS

Quotations and order processing work sheets must be complete in every detail, with all pertinent information pertaining to the part, material, operations, inserts,

COPIES: _____________ ESTIMATED BY _____________   QUOTE NO. _____________

CUSTOMER _____________   DATE _____________

ADDRESS _____________   CUST. INQUIRY _____________

_____________   PART NO. _____________ REV. _______

ATT. OF _____________ TITLE _____________   PART NAME ___INSULATOR___

TOOLING DESCRIP.	MOLD AND/OR TOOL CHARGE	EST'D. TOOL TIME, WEEKS	EST'D. WEEKLY MOLD OUTPUT
ONE (1) SINGLE CAVITY SEMI-AUTOMATIC PLUNGER MOLD.	$2300	13	3M

MATERIAL ___PHENOLIC GENERAL PURPOSE BLACK___ OR EQUIVALENT

PIECE PRICE

PER RELEASE:

5M	$ 52.45/C
2.5M	60.38/C
1M	73.11/C
500	105.88/C

INSERTS   ☐ T-A FURNISHES   ☐ CUST. FURNISHES WITH   OVER ORDER REQUIREMENTS

SPECIAL NOTES AND CONDITIONS:

XXXXXXXX 1. TERMS: MOLDS 50% CASH WITH ORDER BALANCE ON COMPLETION OF MOLDS. PRICES FOR MOLDS AND TOOLS ARE NET. MOLDED PIECES NET 30 DAYS, F.O.B. MOLD PLANT. TERMS HEREIN ARE SUBJECT TO APPROVAL OF OUR CREDIT DEPARTMENT UPON RECEIPT OF ORDER.

2. THIS QUOTATION PREDICATED ON ENGINEERING DISCUSSION AT TIME OF ENTERING ORDER TO REVIEW DESIGN FOR BETTER MOLD AND PLASTIC PART DESIGN WHICH WILL BE MUTUALLY BENEFICIAL TO CUSTOMER AND US.

3. WE WILL FURNISH THIS PART PREDICATED ON AN APPROXIMATE NET WEIGHT OF 14.5 GRAMS/PIECE DEGATED AND BULK PACKED IN SUITABLE CARTONS.

4. DUE TO THE UNCERTAINTY ON PLASTIC MATERIAL PRICES, WE RESERVE THE RIGHT TO ADJUST OUR PARTS PRICES, IF NECESSARY FOR MATERIALS (SUBJECT TO AVAILABILITY OF MATERIAL) AT TIME OF PLACEMENT OF ORDER. THE ABOVE PRICES ARE BASED ON A MATERIAL PRICE OF $.245/LB.

Fig. 12.2A Pricing and terms of sale details as given to customer.

finish, mold type and number of cavities, jigs, fixtures, size of press, and special press requirements. The ordering specification for materials must be prepared. All technical details of the mold and fixture design must be adequate to be understandable in 6 to 9 months, with no cost details overlooked. All details for

CAVITIES __1__ TYPE __INJECTION__ PRESS __13 oz__
O'ALL CYCLE __0' 55__ CURE ______ BENCH ______
HEATS/HR. __65__ PCS./HR. __65__ COST/LB. __.65 .85__
PC. WT. (GMS) __85__ WT. & LOSS __98__ LBS./C PCS. __22.__
PKG. DET. __PACK APPROX 12 PCS/#8 CARTON @ .23__

	PROD./HR.	$/C	$/C	$/C	$/C	$/C	$/C	$/C	$/C
MATERIAL			14.300	18.700					
PREFORMS									
PKG. MAT'L.			1.917	1.917					
TUMBLE									
WHEELABRATE									
MOLDING $ 9.75 /HR.			15.000	15.000					
/MIN.									
DEGATE									
INSPECTION									
SHIPPING LABOR									
TOTAL MATERIAL			16.217	20.617					
FINISHING LABOR									
MOLDING LABOR			15.000	15.000					
TOTAL MAT'L. & LABOR			31.217	35.617					
5 % REJECTS			1.561	1.781					
TOTAL MFG. COST			32.778	37.398					
MATERIAL 10 %			1.622	2.062					
FINISHING %									
S.P. LESS LABOR M.U.			34.400	39.460					

QUAN.	HRS.	%	MOLDING LABOR MARK UP						
15M	230	.240		3.600				
10M	153	.377		5.655				
5M	76	.570			8.550			

QUAN.	HRS.	%	SELLING PRICE						
15M			39.140	38.000				
10M			41.257	40.055				
5M				49.450				

Fig. 12.2B Quotation worksheet as developed by product engineer.

the quotation, including cost of mold, essential molding, finishing and inspection fixtures, special set-up expense, insert cost (if furnished), along with the estimated cost of the pieces in the specified quantities, should be prepared. When design changes are necessary, a marked print or sketch should be prepared, show-

CAVITIES: 1 TYPE: T.S PLUNGER PRESS: ____

O'ALL CYCLE: 2' 0 CURE: 1' 0 BENCH: ____

HEATS/HR.: 30 PCS./HR.: 30 COST LB.: .25

PC. WT. (GMS): 15 WT. & LOSS: 29 LBS./C PCS.: 6.4

PKG. DET.: BULK PACK 650 PCS/ #1 CARTON @.16

	PROD./HR.	$/C	$/C	$/C	$/C	$/C	$/C	$/C	$/C	$/C
MATERIAL				1.600						
PREFORMS										
PKG. MAT'L.				.025						
TUMBLE										
WHEELABRATE 7.50	15M			.050						
MOLDING $ 10.50 HR.				35.000						
DEGATED MIN.										
THREADED DETAILS 2 IN 2 OUT IN MOLD										
INSPECTION										
SHIPPING LABOR										
TOTAL MATERIAL				1.625						
FINISHING LABOR				.050						
MOLDING LABOR				35.000						
TOTAL MAT'L. & LABOR				36.675						
5 % REJECTS				1.834						
TOTAL MFG. COST				38.509						
MATERIAL 10 %				.163						
FINISHING %										
S.P. LESS LABOR M.U.				38.672						

QUAN.	HRS.	%		MOLDING LABOR MARK UP						
5M	166	.350		12.250						
2.5M	83	.570		19.950						
1M	33	.923		32.305						
500	16	1.832		64.120						

QUAN.	HRS.	%		SELLING PRICE						
5M				52.450	50.922					
2.5M				60.381	58.622					
1M				73.106	70.977					
500				105.876	102.792					

Fig. 12.2C Thermosetting quotation work sheet.

ing the essential deviations from the print. Particular attention must be given to the tolerances specified, and any special notes concerning finishing, MIL-specifications, and inspection features with essential deviations, clearly described.

When an order is received to proceed on a new part from one of these quotations, it should be processed as follows by the responsible plastics engineer:

(1) Recheck the quotation and the customer's order to be sure that what the

COPIES: _______________ ESTIMATED BY _______________   QUOTE NO. _______________

CUSTOMER _________________________________   DATE _______________

ADDRESS _________________________________   CUST. INQUIRY _______________

_________________________________   PART NO. _______________ REV. _______

ATT. OF _______________ TITLE _______________   PART NAME _______________

TOOLING DESCRIP. _________________________________   MOLD AND/OR TOOL CHARGE | EST'D. TOOL TIME, WEEKS | EST'D. WEEKLY MOLD OUTPUT

MATERIAL _________________________________ OR EQUIVALENT

PIECE PRICE

PER RELEASE:

INSERTS [ ] T-A FURNISHES [ ] CUST. FURNISHES WITH   OVER ORDER REQUIREMENTS

SPECIAL NOTES AND CONDITIONS:

2. T-A WILL _________________________________

Fig. 12.2D

customer expects to get has been quoted. This check should include:

Dimensional tolerances and inspection needs.
Essential product changes.
Material, insert and color demands.
Mold delivery expected and its capacity.

Molding, finishing and inspection fixtures.
Shipping requirements, special packing or cartons.

If change negotiations with customer are indicated, they must be initiated immediately and confirmed before mold and customer's order are processed. When necessary, a copy of the customer's authorization to change his order must be solicited and received in writing.

On new jobs, a request to the customer will be made by the plastics engineer that one copy of the drawing for internal distribution go to the files, molding, tool shop, finishing department, plastics engineer and inspection.

(2) As the first step in mold procurement, the plastics engineer should prepare a product drawing or detailed instruction to the mold builder. This drawing or instruction may be a marked copy of the customer's drawing or a newly prepared drawing to show the part as you propose to manufacture it. This drawing should show all essential changes, tolerances, finishing data, radii and sharp corner locations, holes to be tapped, drilled, etc., after molding, as specified in the quotation. When necessary to get approval, the customer should return an initialed copy of the drawing as proof of his approval. All subsequent design changes, etc., should be handled in a similar manner, with absolute records in the files for every change. Each change or special instruction pertaining to any processing, inspection, finishing or shipping detail, should be written up as a change notice with copies to all departments.

The plastics engineer and the production manager should consult on the "preliminary manufacturing" specification, which informs the foreman and the mold sampler of the essential methods and desirable sampling data, with copies to all departments. The "preliminary specification" is the criterion by which all departments should be guided until the samples are approved, the first run completed, and the "master specification" issued.

(3) The production scheduler will establish a delivery promise-date, after a review of the mold maker's reliability and delivery promise, press availability, material, and insert and special fixture deliveries. No production should be made without written approval of samples.

(4) When the tool and fixture drawings are completed, they should be checked by the plastics engineer and the production manager, for comments and approval. Every new mold design should be checked by two product engineers, and the production manager as well. If major changes are suggested, revised drawings must be prepared and resubmitted to the customer for his approval. The toolshop foreman should look for mold-design factors that are known to have high maintenance cost. It is the production manager's job to make sure that the mold is designed for continuous production, and that it will meet the rates established for the job, with the available operators.

(5) Upon receipt of the mold, it must be checked thoroughly by the tool and

ENGINEERING CHECK LIST

TO: (Product Engineer)   DATE ________________________

     We have received from ________________________ their order #________________________

for ________________________ dated ________________________ at selling price of ________________________

with expected delivery of ________________________.

     Please follow and take necessary action for satisfactory economical completion.

Note ☐ marks for functions.

TASK	ACTION REQUIRED	DATE COMPLETED
☐ Review Quote		_____________
☐ Review Part Print		_____________
☐ Solicit Tool Bids		_____________
☐ Confer with Chief Engineer		_____________
☐ Confer with ________________		_____________
☐ Establish Tool Delivery		_____________
☐ Issue Tool Order		_____________
☐ Issue Purchase Order		_____________
☐ Issue Sample Order		_____________
☐ Return Copy Cust's Order to Prod.		_____________
☐ Notify Cust. Approx. Mold Del.		_____________
☐ Advise Prod. Mat'l Req'd.		_____________
☐ Advise Prod. Sp. Packing Req'd.		_____________
☐ Advise Prod. Inserts Req'd.		_____________
☐ Cust. Clearance on Parting Lines, Design Changes, Ejector, Trademark and Gate Location, etc.		_____________
☐ Follow and Review Tool Designs		_____________
☐ Review Mold Designs with Mfg.		_____________
☐ Advise Prod. Equipment Changes		_____________
☐ Fixtures Req'd.		_____________
☐ Follow Mold Progress ☐1 ☐2 ☐3 ☐4 ☐5 ☐6		
☐ Mold Checked Prior to Shipment		_____________
☐ Mold Rec'd ________ Date ________________.		
☐ Mold Sampled ________ Date ________________.		
☐ Samples Checked ________ Date ________________.		
☐ Samples Submitted ________ Date ________________.		
☐ Samples Rejected ________ Date ________________.		
☐ Action Req'd.		_____________
☐ Cust. Apprv'l Rec'd		_____________
☐ Operation Sheets Made and Issued ☐ Molding ☐ Finishing ☐ Inspection ☐ Packaging		
☐ Released to Production		_____________
☐ Final Review of Quote with Actual Operations: ☐ Profit ☐ Loss		_____________
☐ Final Review with Chief Engineer		_____________
☐ Obsolete Old Prints		_____________
☐ File Active Prints		_____________
☐ Completion Date		_____________
☐ Miscellaneous		_____________

Fig. 12.3 Product engineer's check list.

plastics engineer, tool-room supervisor, and the production manager, who will issue orders for essential changes, additions, or other supplemental work. The production scheduler should issue instructions for mold installation when installation has been approved by the foregoing persons, and the press is available for sampling. Instructions for sampling should also be given by the production manager, in conference with the plastics engineer.

(6) Samples for submittal to a customer will be finished in the finishing department by an average operator under the instructions of the plastics engineer so that they represent the average kind of work that will be delivered in production. Samples are then checked by production, inspection and engineering before submittal to customer, thus insuring that production standards are established. Hand "manicured" samples should never be submitted for approval.

If the sampling and finishing discloses cost errors that must be corrected, all parties must be notified immediately. Customers must be informed of potential cost increases at the earliest possible date. When formal sample approval is received, a shop order should then be issued, informing all parties and release production. Any special and subsequent comments, suggestions, or criticisms must be included in the shop order.

(7) When the order is completed, the production manager must make sure that the "master specification" is in accordance with the facts of production. This will include mention of special inspection procedures, sources of loss, scrap utilization, essential changes in manufacturing, mold defect allowances, inspection, finishing or packaging procedures.

It is the job of the tool engineer, plastics engineer and the chief inspector or the production manager to request the stopping of the job at any time they believe continuance will result in loss.

The foreman should then check each mold at the end of its run, listing all essential mold repair and change work that should be done. When questionable, the plastics engineer can be called in to check on the work and the mold change needs. The mold should be tagged with a list of essential repairs before it goes into storage as shown in Fig. 11.1. Such work will be authorized by the production scheduler in anticipation of the next run. Chronic tool breakage and repair work will be brought to the attention of the tool engineer, who will recommend reaction.

An engineering check list is shown in Fig. 12.3, which lists the duties of the plastics engineer in processing an order for a new mold and molded part.

GLOSSARY[*]

A

A-Stage—An early stage in the reaction of a thermosetting resin in which the material is still soluble in certain liquids and fusible. See also *B-* and *C-Stage*.

Ablative Plastics—This description applies to a material which absorbs heat (while part of it is being consumed by heat) through a decomposition process known as pyrolysis, which takes place in the near surface layer exposed to heat.

ABS—See *Acrylonitrile Butadiene Styrene*.

Absolute Viscosity—Of a fluid, the tangential force on unit area of either of two parallel planes at unit distance apart when the space between the planes is filled with the fluid in question and one of the planes moves with unit differential velocity in its own plane.

The C.G.S. unit for absolute (or dynamic) viscosity is the poise (dyne-sec./sq. cm.). The centipoise (0.01 poise) if often used.

Accelerator—A substance that hastens a reaction, particularly one which speeds up the vulcanization of rubber. Also known as *Promoter*.

Accumulator—A device that stores energy by means of gravity forces, springs, or the compression of gases.

Accumulator—A term used also with reference to blow molding equipment which designates an auxiliary ram extruder which is used to provide extremely fast parison delivery. The accumulator cylinder is filled with plasticated melt coming from the extruder between parison deliveries or "shots" and is stored or "accumulated" until the plunger is required to deliver the next parison.

Acetal Resins—The molecular structure of the polymer is that of a linear acetal, consisting of unbranched polyoxymethylene chains.

*Revised from "Modern Plastics Encyclopedia," and Phillips Petroleum Co. originals.

Acrylic Ester—An ester of acrylic acid, or of a structural derivative of acrylic acid, e.g., methyl methacrylate.

Acrylic Resin—A synthetic resin prepared from acrylic acid or from a derivative of acrylic acid.

Acrylonitrile—A monomer with the structure (CH_2:CHCN). It is most useful in copolymers. Its copolymer with butadiene is nitrile rubber, and several copolymers with styrene exist that are tougher than polystyrene. It is also used as a synthetic fiber and as a chemical intermediate.

Acrylonitrile-Butadiene-Styrene (abbreviated ABS)—Acrylonitrile and styrene liquids and butadiene gas are polymerized together in a variety of ratios to produce the family of ABS resins.

Adiabatic—An adjective used to describe a process or transformation in which no heat is added to or allowed to escape from the system under consideration. It is used, somewhat incorrectly, to describe a mode of extrusion in which no external heat is added to the extruder although heat may be removed by cooling to keep the output temperature of the melt passing through the extruder constant. The heat input in such a process is developed by the screw as its mechanical energy is converted to thermal energy.

Additives—Products that are combined with resins and polymers as extenders or modifiers.

Adhesion Promoter—A coating which is applied to the substrate before it is extrusion coated with the plastics and which improves the adhesion of of the plastics to the substrate.

Adhesive—A substance which applied as an intermediate is capable of holding materials together by surface attachment.

Adsorption—The adhesion of the molecules of gases, dissolved substances, or liquids in more or less concentrated form to the surfaces of solids or liquids with which they are in contact.

Aging—The change of a material with time under defined environmental conditions, leading to improvement or deterioration of properties.

Aging, Artificial—The accelerated testing of plastics specimens to determine their changes in properties. Carried out over a short period of time, such tests are indicative of what may be expected of a material under service conditions over extended periods. Typical investigations include those for dimensional stability; the effect of immersion in water, chemicals and solvents; light stability and resistance to fatigue.

Air-Assist Forming—A method of thermoforming (q.v.) in which air flow or air pressure is employed to partially preform the sheet immediately prior to the final pulldown onto the mold using vacuum.

Air Gap—In extrusion coating, the distance from the die opening to the nip formed by the pressure roll and the chill roll.

Air Ring—A circular manifold used to distribute an even flow of the cooling medium, air, onto a hollow tubular form passing through the center of the ring. In blown tubing, the air cools the tubing uniformly to provide uniform film thickness.

Air-Slip Forming—A variation of snapback forming in which the male mold is enclosed in a box in such a way that when the mold moves forward toward the hot plastics, air is trapped between the mold and the plastics sheet. As the mold advances, the plastics is kept away from it by the air cushion formed as described above, until the full travel of the mold is reached, at which point a vacuum is applied, destroying the cushion and forming the part against the plug.

Aliphatic Hydrocarbons—Saturated hydrocarbons having an open chain structure. Familiar examples: gasoline and propane.

Alkyd Resin—Alkyd resins as used for coating materials contain drying oils and phthalate plus other acids and diols. Alkyds as used for molding compounds are unsaturated polyester resins plus maleic or fumaric acid.

Alkyl—A general term for monovalent aliphatic hydrocarbon radicals.

Allyl Diglycol Carbonate (ADC)—A thermosetting crystal clear plastics with exceptional scratch resistance; used for goggles, scratch-resistant glazing, etc.

Alloy—Composite material made up by blending polymers or copolymers with other polymers or elastomers under selected conditions, e.g., styrene-acrylonitrile copolymer resins blended with butadiene-acrylonitrile rubbers.

Allyl Resin—A synthetic resin formed by the polymerization of chemical compounds containing the group $CH_2 = CH - CH_2 -$. The principal commercial allyl resin is a casting material that yields allyl carbonate polymer.

Alpha-Cellulose—A very pure cellulose prepared by special chemical treatment.

Alpha Nylon—A crystalline structure nylon.

Amino—Indicates the presence of an $-NH_2$ or $-NH$ group.

Amino Plastics—Urea and melamine formaldehyde.

Amorphous Phase—Devoid of crystallinity—no definite order. At processing temperatures, the plastics is normally in the amorphous state.

Angle Press—A hydraulic molding press equipped with horizontal and vertical rams, and specially designed for the production of complex moldings containing deep undercuts.

Aniline—$C_6H_5NH_2$. An important organic base made by reacting chlorobenzene with aqueous ammonia in the presence of a catalyst. It is used in the production of aniline formaldehyde resins, q.v., and in the manufacture of certain rubber accelerators and antioxidants.

Aniline Formaldehyde Resins—Members of the aminoplastics family made by the condensation of formaldehyde and aniline in an acid solution. The resins are thermoplastic and are used to a limited extent in the production of molded and laminated insulating materials. Products made from these resins have high dielectric strength and good chemical resistance.

Annealing—A process of holding a material at a temperature near, but below, its melting point, the objective being to permit stress relaxation without distortion of shape. It is often used on molded articles to relieve stresses set up by flow into the mold.

Antioxidant—Substance which prevents or slows down oxidation of material exposed to air.

Antistatic Agents—Methods of minimizing static electricity in plastics materials. Such agents are of two basic types: (1) metallic devices which come into contact with the plastics and conduct the static to earth. Such devices give complete neutralization at the time, but because they do not modify the surface of the material it can become prone to further static during subsequent handling; (2) chemical additives which, mixed with the compound during processing, give a reasonable degree of protection to the finished products.

ARC Extinguishing Plastics (ExarcII)—A thermoplastic which gives off a gas that has arc extinguishing properties under flash conditions; it can be injection or extrusion molded.

Arc Resistance—Time required for a given electrical current to render the surface of a material conductive because of carbonization by the arc flame.

Armor—A solid or braided metal jacket for imparting maximum abrasion resistance to the completed cable. Braided armor is sometimes used in lieu of solid armor for improved flexibility.

Aromatic Hydrocarbons—Hydrocarbons derived from or characterized by presence of unsaturated resonant ring structures.

Aromatic Polyamides—A highly heat resisting "paper" used for electrical insulation.

Atactic—A chain of molecules in which the position of the methyl groups is more or less random.

Autoclave—1) Closed strong vessel for conducting chemical reactions under high pressure; 2) in low-pressure laminating, a round or cylindrical container in which heat and gas pressure can be applied to resin-impregnated paper or fabric positioned in layers over a mold.

Autoclave Molding—Modification of the pressure bag method for molding reinforced plastics. After lay-up, entire assembly is placed in steam autoclave at 50 to 100 psi. Additional pressure achieves higher reinforcement loadings and improved removal of air.

Automatic Mold—A mold for injection or compression molding that repeatedly goes through the entire cycle, including ejection, without human assistance.

Average Molecular Weight (viscosity method)—The molecular weight of polymeric materials determined by the viscosity of the polymer in solution at a specific temperature. This gives an average molecular weight of the molecular chains in the polymer independent of specific chain length. Falls between weight average and number average molecular weight.

B

B-Stage—An intermediate stage in the reaction of the thermosetting resin in which the material softens when heated and swells in contact with certain liquids but does not entirely fuse or dissolve. Resins in thermosetting molding compounds are usually in this stage. See also *A-Stage* and *C-Stage*.

Backing Plate—In injection molding, a plate used as a support for the cavity blocks, guide pins, bushings, etc.

Back Pressure—The viscosity resistance of a material to continued flow when a mold is closing. In extrusion, the resistance to the forward flow of molten material.

Baffle—A device used to restrict or divert the passage of fluid through a pipe line or channel. In hydraulic systems the device, which often consists of a disc with a small central perforation, restricts the flow of hydraulic fluid in a high pressure line. A common location for the disc is in a joint in the line. When applied to molds, the term is indicative of a plug or similar device located in a steam or water channel in the mold and designed to divert and restrict the flow to a desired path.

Bag Molding—A method of applying pressure during bonding or molding, in which a flexible cover, usually in connection with a rigid die or mold, exerts pressure on the material being molded, through the application of air pressure or drawing of a vacuum.

Bakelite—The proprietary name for phenolic and other plastics materials produced by Union Carbide Corp., but often used indiscriminately to describe any phenolic molding material or molding. The name is derived from that of Dr. Leo Hendrik Baekeland (1863-1944), a Belgian who did the work on the synthesis of phenolic resins and their commercial development in the early 1900's.

Balanced Runner—Balance runner systems are designed to place each cavity at an equal distance from the sprue.

Banbury—An apparatus for compounding materials composed of a pair of contra-rotating rotors which masticate the materials to form a homogeneous blend. This is an internal type mixer which produces excellent mixing.

Barrier Resin—Nitrile based compounds developed for soft drinks.

Bead—The annular ring below the threads of a bottle finish.

Benzene Ring—The basic structure of benzene, the most important aromatic chemical. It is an unsaturated, resonant 6-carbon ring having three double bonds. One or more of the 6 hydrogen atoms of benzene may be replaced by other atoms or groups.

Beta Gage (or beta-ray gage)—A gage consisting of two facing elements, a B-ray-emitting source and a B-ray detector. When a sheet material is passed between the elements, some of the B-rays are absorbed, the percent absorbed being a measure of the areal density or the thickness of the sheet.

Bio-degradable—Such products will decompose under natural environmental conditions—time and nature.

Blanking—The cutting of flat sheet stock to shape by striking it sharply with a punch while it is supported on a mating die. Punch presses are used. Also called *Die Cutting*.

Bleed—To give up color when in contact with water or a solvent; undesired movement of certain materials in a plastic (e.g., plasticizers in vinyl) to the surface of the finished article or into an adjacent material. Also called *Migration*.

Blister—A raised area on the surface of a molding caused by the pressure of gases inside it on its incompletely hardened surface.

Blocking—An undesired adhesion between touching layers of a material, such as occurs under moderate pressure during storage or use.

Bloom—See *Lubricant Bloom*.

Blow Molding—A method of fabrication in which a parison (hollow tube) is forced into the shape of the mold cavity by internal air pressure.

Blow Pressure—The air pressure used to form a hollow part by flow molding.

Blow Rate—The speed at which the air enters the parison during the blow molding cycle.

Blowing Agents—See *Foaming Agents*.

Blowup Ratio—In blow molding, the ratio of the mold cavity diameter to the parison diameter. In blown tubing (film), the ratio of the final tube diameter (before gusseting, if any) to the original die diameter.

Blown Tubing—A thermoplastic film which is produced by extruding a tube, applying a slight internal pressure to the tube to expand it while still molten and subsequent cooling to set the tube. The tube is then flattened through guides and wound up flat on rolls. The size of blown tubing is determined by the flat width in inches as wound rather than by the diameter as in the case of rigid types of tubing.

Blueing—A mold blemish in the form of a blue oxide film which occurs on the polished surface of a mold as a result of the use of abnormally high mold temperatures.

Blunt Thread Start—A thread design where the start of the thread has been squared off for exact locating of the container in a printing or labeling machine. See *Lug*.

Booster—A booster or intensifier uses a large quantity of low pressure fluid to produce a small quantity of high pressure fluid.

Boss—Protuberance on a plastic part designed to add strength, to facilitate alignment during assemlby, to provide for fastenings, etc.

Boston Round—A particular shape of container; cross section as well as shoulders are round.

Bottom Blow—A specific type of blow molding machine which forms hollow articles by injecting the blowing air into the parison from the bottom of the mold.

Bottom Plate—Part of the mold which contains the heal radius and the push-up.

Branched—In molecular structure of polymers (as opposed to *Linear*), refers to side chains attached to the main chain. Side chains may be long or short.

Breakdown Voltage—The voltage required, under specific conditions, to cause the failure of an insulating material. See *Dielectric Strength*.

Breaker Plate—A perforated plate located at the rear end of an extruder head. It often supports the screens that prevent foreign particles from entering the die.

Breathing—The opening and closing of a mold to allow gases to escape early in the molding cycle. Also called *degassing*. When referring to plastic sheeting, "breathing" indicates permeability to air.

Bubbler Mold Cooling (Injection Molding)—A method of cooling an injection mold in which a stream of cooling liquid flows continuously into a cooling cavity equipped with a cooling cavity equipped with a coolant outlet normally positioned at the end opposite the inlet. Uniform cooling can be achieved in this manner.

Bulk Density—The mass per unit volume of a molding powder as determined in a reasonably large volume. The recommended test method is ASTM D1182-54.

Bulk Factor—Ratio of the volume of loose molding powder to the volume of the same weight of resin after molding.

Burning Rate—A term describing the tendency of plastics articles to burn at given temperatures. Certain plastics, such as those based on shellac, burn readily at comparatively low temperatures. Others will melt or disintegrate without actually burning, or will burn only if exposed to direct flame. These latter are often referred to as self-extinguishing.

Bushing (Extrusion)—The outer ring of any type of a circular tubing or pipe die which forms the outer surface of the tube or pipe.

Butadiene—$CH_2:CH \cdot CH:CH_2$. A gas, insoluble in water but soluble in alcohol and ether, obtained from the cracking of petroleum, from coal tar benzene or foam acetylene produced from coke and lime. It is widely used in the formation of copolymers with styrene, acrylonitrile, vinyl chloride and other monomeric substances, where it imparts flexibility to the subsequent moldings.

Butadiene Styrene—A thermoplastic polymer used for film and sheet.

Butt-Fusion—A method of joining pipe, sheet, or other similar forms of a thermoplastic resin wherein the ends of the two pieces to be joined are heated to the molten state and then rapidly pressed together to form a homogeneous bond.

Buttress Thread—A type of threading in which the thread sides terminate abruptly in threading gradually tapering down to the neck finish. Designed to withstand maximum force in one direction only. Cross section of thread is triangular.

C

C-Stage—The final stage in the reactions of a thermosetting resin in which the material is relatively insoluble and infusible. Thermosetting resins in fully cured plastics are in this stage. See *A-Stage* and *B-Stage*.

Cable—A standard conductor; or a group of solid or standard conductors laid together but insulated from one another.

Calender (*v.*)—To prepare sheets of material by pressure between two or more counter-rotating rolls. (*n.*)—The machine performing this operation.

Caprolactam—A cylic amidetype compound, containing 6 carbon atoms. When the ring is opened, caprolactam is polymerizable into a nylon resin known as type-6 nylon or polycaprolactam.

Carbon Black—A black pigment produced by the incomplete burning of natural gas or oil. It is widely used as a filler, particularly in the rubber industry. Because it possesses useful ultraviolet protective properties, it is also much used in polyethylene compounds intended for such applications as cold water piping and black agricultural sheet.

Casein—A protein material precipitated from skimmed milk by the action of either rennet or dilute acid. Rennet casein finds its main application in the manufacture of plastics. Acid casein is a raw material used in a number of industries including the manufacture of adhesives.

Cast—1) to form a "plastics" object by pouring a fluid monomer-polymer solution into an open mold where it finishes polymerizing. 2) forming plastics film and sheet by pouring the liquid resin onto a moving belt or by precipitation in a chemical bath.

Casting (*n.*)—The finished product of a casting operation; should not be used for *molding,* q.v.

Casting Area—The moldable area of a thermoplastic in square inches for a given thickness and under a given set of injection molding conditions. Casting area is a measure of flow under actual molding conditions where flow is unrestricted by cavity boundaries.

Catalyst—A substance which markedly speeds up the cure of a compound when added in minor quantity as compared to the amounts of primary reactants. See *Hardener, Inhibitor, Promoter.*

Cavity—Depression in a mold made by casting, machining, hobbing, or a combination of these methods; depending on number of such depressions, molds are designated as *Single-Cavity* or *Multi-Cavity.*

Cellular Plastics—See *Foamed Plastics.*

Celluloid—A thermoplastic material made by the intimate blending of cellulose nitrate, q.v., with camphor. Alcohol is normally employed as a volatile solvent to assist plasticization, and is subsequently removed.

Cellulose—A natural high polymeric carbohydrate found in most plants; the main constituent of dried woods, jute, flax, hemp, ramie, etc. Cotton is almost pure cellulose.

Cellulose Acetate—An acetic acid ester of cellulose. It is obtained by the action, under rigidly controlled conditions, of acetic acid and acetic andydride on purified cellulose usually obtained from cotton linters. All three available hydroxyl groups in each glucose unit of the cellulose can be acetylated but in the material normally used for plastics it is usual to acetylate fully and then to lower the acetyl value (expressed as acetic acid) to 52-56% by partial hydrolysis. When compounded with suitable plasticizers it gives a rough thermoplastic material.

Cellulose Acetate Butyrate—An ester of cellulose made by the action of a mixture of acetic and butyric acids and their anhydrides on purified cellulose. It is used in the manufacture of plastics which are similar in general properties to cellulose acetate but are tougher and have better moisture resistance and dimensional stability.

Cellulose Ester—A derivative of cellulose in which the free hydroxyl groups attached to the cellulose chain have been replaced wholly or in part by acidic groups, e.g., nitrate, acetate, or stearate groups. Esterification is effected by the use of a mixture of an acid with its anhydride in the presence of a catalyst, such as sulfuric acid. Mixed esters of cellulose, e.g., cellulose acetate butyrate, are prepared by the use of mixed acids and mixed anhydrides. Esters and mixed esters, a wide range of which is known, differ in their compatibility with plasticizers, in molding properties, and in physical characteristics. These esters and mixed esters are used in the manufacture of thermoplastic molding compositions.

Cellulose Nitrate (Nitrocellulose)—A nitric acid ester of cellulose manufactured by the action of a mixture of sulfuric acid and nitric acid on cellulose, such as purified cotton linters. The type of cellulose nitrate used for celluloid manufacture usually contains 10.8-11.1% of nitrogen. The latter figure is the nitrogen content of the dinitrate.

Cellulose Propionate—An ester of cellulose made by the action of propionic acid and its anhydride on purified cellulose. It is used as the basis of a thermoplastic molding material.

Cellulose Triacetate—A cellulosic material made by reacting purified cellulose with acetic anhydride in the presence of a catalyst. It is used in the form of film and fibers. Films and sheet are cast from clear solutions on to "drums" with highly polished surfaces. The film, which is of excellent clarity, has high tensile strength, and good heat resistance and dimensional stability. Applications include book jackets, magnetic recording tapes, and various types of packaging. Cellulose triacetate sheet has somewhat similar properties to those of the film and is used to make such articles as safety goggles, map wallets and transparent covers of many kinds.

Center Gated Mold—An injection mold wherein the cavity is filled with resin through an orifice interconnecting the nozzle and the center of the cavity area. Normally, this orifice is located at the bottom of the cavity when forming items such as containers, tumblers, bowls, etc.

Centrifugal Casting—A method of forming thermoplastic resins in which the granular resin is placed in a rotatable container, heated to a molten condition by the transfer of heat through the walls of the container, and rotated so that the centrifugal force induced will force the molten resin to conform to the configuration of the interior surface of the container. Used to fabricate large diameter pipes and similar cylindrical items.

Ceramoplastic—See *Glass Bonded Mica*.

Chain Length—See *Degree of Polymerization*.

Charge—The measurement or weight of material used to load a mold at one time or during one cycle.

Chase—An enclosure of any shape, used to: a) shrink-fit parts of a mold cavity in place; b) prevent spreading or distortion in hobbing; c) enclose an assembly of two or more parts of a split cavity block.

Chelating Agent—A substance, which through chemical means, will form complexes with various ions and remove them from exerting an influence on the system under study.

Chill Roll—A cored roll, usually temperature controlled with circulating water, which cools the web before winding. For chill roll (cast) film, the surface of the roll is highly polished. In extrusion coating, either a polished or a matte surface may be used depending on the surface desired on the finished coating.

Chill Roll Extrusion (or cast film extrusion)—The extruded film is cooled while being drawn around two or more highly polished chill rolls cored for water cooling for exact temperature control.

Chlorinated Polyether—The polymer is obtained from pentaerythritol by preparing a chlorinated oxetane and polymerizing it to a polyether by means of opening the ring structure.

Chlorinated Polyethylene—A modifier for vinyl and polyethylene sheets.

Choked Neck—Narrowed or constricted opening in the neck of a container.

Chromium Plating—An electrolytic process that deposits a hard film of chromium metal onto working surfaces of other metals where resistance to corrosion, abrasion, and/or erosion is needed.

CIL (Flow Test)—A method of determining the rheology or flow properties of thermoplastic resins developed by Canadian Industries Limited. In this test, the amount of the molten resin which is forced through a specified size orifice per unit time when a specified, variable force is applied gives a relative indication of the flow properties of various resins.

Clamping Plate—A plate fitted to a mold and used to fasten the mold to a molding machine.

Clamping Pressure—In injection molding and in transfer molding, the pressure which is applied to the mold to keep it closed, in opposition to the fluid pressure of the compressed molding material.

Clarifier—An additive that increases the transparency of a material.

Clearance—A controlled distance by which one part of an object is kept separated from another part.

Coathanger Die—This is one basic type of slot die for extrusion that is shaped internally like a coat hanger to gain better distribution across the full width of a sheet extrusion.

Coating—See specific type of coating such as curtain, extrusion, kiss-roll spray.

Coating Weight—The weight of coating per unit area. In the United States usually "per ream," i.e., 500 sheets 24″ × 36″ (3000 sq. ft.), but sometimes 1000 sq. ft.

Coefficient of Expansion—The fractional change in length (sometimes volume, specified) of a material for a unit change in temperature. Values for plastics range from 0.01 to 0.2 mils/in., °C.

Coextrusion—A process for extruding two or more materials in a single sheet or film. Two identical polymers may be laminated by coextrusion.

Cold Flow—See *Creep*.

Cold Molding—A procedure in which a composition is shaped at room temperature and cured by subsequent baking.

Cold Slug—The first material to enter an injection mold; so called because in passing through sprue orifice it is cooled below the effective molding temperature.

Cold Slug Well—Space provided directly opposite the sprue opening in an injection mold to trap the cold slug.

Cold Stretch—Pulling operation, usually on extruded filaments, to improve tensile properties.

Collapse—Contraction of the walls of a container, e.g., upon cooling, leading to a permanent indentation.

Compression Mold—A mold which is open when the material is introduced and which shapes the material by heat and by the pressure of closing.

Compression Molding—A technique of thermoset molding in which the molding compound (generally preheated) is placed in the open mold cavity, mold is closed, and heat and pressure (in the form of a downward moving ram) are applied until the material has cured.

Compression Ratio—In an extruder screw, the ratio of volume available in the first flight at the hopper to the to the last flight at the end of the screw.

Compressive Strength—Crushing load at the failure of a specimen divided by the original sectional area of the specimen.

Concentricity—For a container, the shape in which various cross sections have a common center.

Condensation—A chemical reaction in which two or more molecules combine with the separation of water or some other simple substance. If a polymer is formed, the condensation process is called *Polycondensation*. See also *Polymerization*.

Condensation Resin—A resin formed by polycondensation, e.g., the alkyd, phenol-aldehyde, and urea form-aldehyde resins.

Conditioning—The subjection of a material to a stipulated treatment so that it will respond in a uniform way to subsequent testing or processing. The term is frequently used to refer to the treatment given to specimens before testing.

Conductor—A wire, or combination of wires not insulated from each other, suitable for carrying electricity.

Contact Pressure Resins—Liquid resins which thicken or resinify on heating and, when used for bonding laminates, require little or no pressure.

Continuous Thread—A spiral, protruding finish on the neck of a container to hold a screwtype closure.

Continuous Tube Process—A blow-molding process that uses a continuous extrusion of tubing to feed into the blow molds as they clamp in sequence.

Convergent Die—A die in which the internal channels leading to the orifice are converging (only applicable to dies for hollow bodies).

Conversion Polymerization—The degree to which a monomer has been reacted with itself and other monomers when related to the original quantities.

Conveyor—A mechanical device to transport material from one point to another, often continuously.

Cooling Channels—Channels or passageways located within the body of a mold through which a cooling medium can be circulated to control temperature on the mold surface.

Cooling Fixture—Block of metal or wood holding the shape of a molded piece which is used to maintain the proper shape or dimensional accuracy of a molding after it is removed from the mold until it is cool enough to retain its shape without further appreciable distortion. Also known as *Shrink Fixture*.

Copolymer—See *Polymer*

Core—1) The central member of a sandwich construction (can be honeycomb material, foamed plastic, or solid sheet) to which the faces of the sandwich are attached; the central member of a plywood assembly. 2) A channel in a mold for circulation of heat-transfer media. 3) Part of a complex mold that molds undercut parts. Cores are usually withdrawn to one side before the main sections of the mold open. Also called *Core Pin*.

Corona Resistance—A current passing through a conductor induces a surrounding electrostatic field. When voids exist in the insulation near the conductor, the high voltage electrostatic field may ionize and rapidly accelerate some of the air molecules in the void. These ions can then collide with the other molecules ionizing them, and thereby "eating" a hole in the insulation. Resistance to this process is called corona resistance.

Cover—In wire coating, a coating whose primary purpose is to "weatherproof" or to prevent casual grounding (such as contact with a wet tree branch), or to otherwise protect a conductor.

Crazing—Fine cracks which may extend in a network on or under the surface or through a layer of a plastic material.

Creep—The dimensional change with time of a material under load, following the initial instantaneous elastic deformation. Creep at room temperature is sometimes called *Cold Flow*.

Crosshead (extrusion)—A device generally employed in wire coating which is attached to the discharge end of the extruder cylinder, designed to facilitate extruding material at an angle. Normally, this is a 90 degree angle to the longitudinal axis of the screw.

Cross-Linking—Applied to polymer molecules, the setting-up of chemical links between the molecular chains. When extensive, as in most thermosetting resins, cross-linking makes one infusible supermolecule of all the chains.

Crystallinity—A state of molecular structure in some resins which denotes uniformity and compactness of the molecular chains forming the polymer. Normally can be attributed to the formation of solid crystals having a definite geometric form.

Cull—Material remaining in a transfer chamber after mold has been filled. Unless there is a slight excess in the charge, the operator cannot be sure cavity is filled. Charge is generally regulated to control thickness of cull.

Cure—To change the physical properties of a material by chemical reaction, which may be condensation, polymerization, or vulcanization; usually accomplished by the action of heat and catalysts, alone or in combination, with or without pressure.

Curing Temperature—Temperature at which a cast, molded, or extruded product, a resin-impregnated reinforcing material, an adhesive, etc., is subjected to curing.

Curing Time (Molding Time)—In the molding of thermosetting plastics, the interval of time between the instant of cessation of relative movement between the moving parts of a mold and the instant that pressure is released.

Curling—A condition in which the parison curls upwards and outwards, sticking to the outer face of the die ring. Balance of temperatures between die and mandrel will normally relieve this problem.

Curtain Coating—A method of coating which may be employed with low viscosity resins or solutions, suspensions, or emulsions of resins in which the substrate to be coated is passed through and perpendicular to a freely falling liquid "curtain" (or "waterfall"). The flow rate of the falling liquid and the linear speed of the substrate passing through the curtain are coordinated in accordance with the thickness of coating desired.

Curvature—A condition in which the parison is not straight, but somewhat bending and shifting to one side, leading to a deviation from the vertical direction of extrusion. Centering of ring and mandrel can often relieve this defect.

Cut-Off—The line where the two halves of a compression mold come together; also called *Flash Groove* or *Pinch-off*.

Cycle—The complete, repeating sequence of operations in a process or part of a process. In molding, the *cycle time* is the period, or elapsed time, between a certain point in one cycle and the same point in the next.

Cylindrical—Refers to the shape of a container which has a circular cross section parallel to the minor axis and a rectangular cross section parallel to the major axis.

D

Dash-Pot—A device used in hydraulic systems for damping down vibration. It consists of a piston attached to the part to be damped and fitted into a vessel containing fluid or air. It absorbs shocks by reducing the rate of change in the momentum of moving parts of machinery.

Daylight Opening—Clearance between two platens of a press in the open position.

Debossed—An indent or cut in design or lettering of a surface.

Deckle Rod—A small rod, or similar device, inserted at each end of the extrusion coating die which is used to adjust the length of the die opening.

Decorative Sheet—A laminated plastics sheet used for decorative purposes in which the color and/or surface pattern is an integral part of the sheet.

Deflashing—Covers the range of finishing techniques used to remove the flash (excess, unwanted material) on a plastic molding.

Degassing—See *Breathing*.

Degree of Polymerization (DP)—The number of structural units or mers in the "average" polymer molecule in a particular sample. In most plastics the DP must reach several thousand if worthwhile physical properties are to be had.

Delamination—The separation of the layers in a laminate caused by the failure of the adhesive.

Deliquescent—Capable of attracting moisture from the air.

Denier—The weight (in grams) of 9000 meters of synthetic fiber in the form of continuous filament.

Density—Weight per unit volume of a substance, expressed in grams per cubic centimeter, pounds per cubic foot, etc.

Desiccant—Substance which can be used for drying purposes because of its affinity for water.

Destaticization—Treating plastics materials to minimize their accumulation of static electricity and, consequently, the amount of dust picked up by the plastics because of such charges.

Detergents—Substances with a high surface activity. Similar in general to soaps; made synthetically to a large extent.

Die Blades—Deformable member(s) attached to a die body which determine the slot opening and which are adjusted to produce uniform thickness across the film or sheet produced.

Die Cutting—1) Blanking (q.v.); 2) Cutting shapes from sheet stock by striking it sharply with a shaped knife edge known as a "steel-rule die." Clicking and Dinking are other names for die cutting of this kind.

Die Gap—The distance between the metal faces forming the die opening.

Dielectric—Insulating material. In radio-frequency preheating, dielectric may refer specifically to the material which is being heated.

Dielectric Constant—Normally the relative dielectric constant; for practical purposes, the ratio of the capacitance of an assembly of two electrodes separated solely by a plastics insulating material to its capacitance when the electrodes are separated by air (ASTM D150-59T).

Dielectric Heating (Electronic heating)—The plastic to be heated forms the dielectric of a condenser to which is applied a high-frequency (20 to 80 mc.) voltage. Dielectric loss in the material is the basis. Process used for sealing vinyl films and preheating thermoset molding compounds.

Dielectric Strength—The electric voltage gradient at which an insulating material is broken down or "arced through," in volts per mil of thickness.

Die Lines—Vertical marks on the parison caused by damage of die parts or contamination.

Die Swell Ratio—The ratio of the outer parison diameter (or parison thickness) to the outer diameter of the die (or die gap). Die swell ratio is influenced by polymer type, head construction, land length, extrusion speed, and temperature.

Dimensional Stability—Ability of plastics parts to retain the precise shape in which it was molded fabricated, or cast.

Dimer—A substance (comprising molecules) formed from two molecules of a monomer.

Dip Coating—Applying a plastics coating by dipping the article to be coated into a tank of melted resin or plastisol, then chilling the adhering melt.

Direct Gate—A direct gate is large and at the end of the sprue.

Discoloration—Any change from the original color, often caused by overheating, light exposure, irradiation, or chemical attack.

Dispersion—A more or less uniform distribution of one substance in another. The breakdown of large particles into small is also called peptization; the converse is called aggregation.

Dissipation Factor—See *Power Factor*.

Divergent Die—A die in which the internal channels leading to the orifice are diverging (applicable only to dies for hollow bodies).

Double-Shot Molding—A means of turning out two-color parts in thermoplastics materials by successive molding operations.

Draft—The degree of taper of a side wall or the angle of clearance designed to facilitate removal of parts from a mold.

Drape Assist Frame—In sheet thermoforming, a frame (made up of anything from thin wires to thick bars) shaped to the peripheries of the depressed areas of the mold and suspended above the sheet to be formed. During forming, the assist frame drops down, drawing the sheet tightly into the mold and thereby preventing webbing between high areas of the mold and permitting closer spacing in multiple molds.

Drape Forming—Method of forming thermoplastic sheet in which the sheet is clamped into a movable frame, heated, and draped over high points of a male mold. Vacuum is then pulled to complete the forming operation.

Draw Down Ratio—The ratio of the thickness of the die opening to the final thickness of the product.

Drawing—The process of stretching a thermoplastic sheet or rod to reduce its cross-sectional area.

Dry Coloring—Method commonly used by fabricators for coloring plastics by tumble blending uncolored particles of the plastic material with selected dyes and pigments.

Dry Strength—The strength of an adhesive joint determined immediately after drying under specified conditions or after a period of conditioning in the standard laboratory atmosphere. See *Wet Strength*.

Ductility—The extent to which a solid material can be drawn into a thinner cross section.

Dwell—A pause in the application of pressure to a mold, made just before the mold is completely closed, to allow the escape of gas from the molding material.

Dyes—Synthetic or natural organic chemicals that are soluble in most common solvents. Characterized by good transparency, high tinctorial strength, and low specific gravity.

E

Ecology—That branch of biology that deals with the relations between living organisms and their environment.

Effective Thread Turns—The number of full 360° turns on a threaded closure that are actually in contact with the neck thread.

Ejector Pin (or sleeve)—A pin or thin plate that is driven into a mold cavity from the rear as the mold opens, forcing out the finished piece. Also *Knockout Pin*.

Ejector Return Pins—Projections that push the ejector assembly back as the mold closes; also called *Surface Pins* and *Return Pins*.

Ejector Rod—Bar that actuates the ejector assembly when mold is opened.

Elastic Deformation—The part of the deformation of an object under load which is recoverable when the load is removed.

Elastic Rotor—A form of screwless extrusion where melt is acheived by mechanical work between rotating members.

Elastomer—A material which at room temperature stretches under low stress to at least twice its length and snaps back to the original length upon release of stress. See also *Rubber*.

Electronic Treating—A method of oxidizing a film of polyethylene to render it printable by passing the film between the electrodes and subjecting it to a high voltage corona discharge.

Electroformed Molds—A mold made by electroplating metal on the reverse pattern on the cavity. Molten steel may be then sprayed on the back of the mold to increase its strength.

Electrostatic Bed—See *Fluidized Bed coating*.

Electrostatic Powder Coating—Dry powder with a high voltage charge is directed to a metal object where it adheres and is bonded later by fusing.

Elongation—The fractional increase in length of material stressed in tension.

Embossing—Techniques used to create depressions of a specific pattern in plastics film and sheeting.

Emulsifier—A substance which is surface active and allows two immiscible liquids to remain intimately mixed and stable.

Emulsion—A suspension of fine droplets of one liquid in another.

Encapsulating—Enclosing an article (usually an electronic component or the like) in a closed envelope of plastics.

Engraved-Roll (or Gravure) Coating—The amount of coating applied to the web is metered by the depth of the over-all engraved pattern in a print roll. This process is frequently modified by interposing a resilient offset roll between the engraved roll and the web.

Entrance Angle—Maximum angle at which the molten material enters the land area of the die, measured from the center line of the mandrel.

Environmental Stress Cracking (ESC)—The susceptibility of a thermoplastic article to crack or craze formation under the influence of certain chemicals and stress.

Epoxy Resins—Based on ethylene oxide, its derivatives or homologs, epoxy resins form straight-chain thermoplastics and thermosetting resins, e.g., by the condensation of bisphenol and epichlorohydrin.

Ester—The reaction product of an alcohol and an acid.

Ethylene Ethyl Acrylate—A copolymer with elastomeric qualities of vulcanized rubber.

Ethylene Tetrafluoroethylene Copolymer (ETFE)—A moldable variety of teflon (TFE).

Ethylene-Vinyl Acetate—Copolymers from these two monomers form a new class of plastics materials. They retain many of the properties of polyethylene, but have considerably increased flexibility for their density—elongation and impact resistance are also increased.

Exotherm—1) The temperature/time curve of a chemical reaction giving off heat, particularly the polymerization of casting resins. 2) The amount of heat given off. The term has not been standardized with respect to sample size, ambient temperature, degree of mixing, etc.

Expanded Plastics—See *Foamed Plastics.*

Extender—A substance, generally having some adhesive action, added to a plastics composition to reduce the amount of the primary resin required per unit area.

Extrudate—The product or material delivered by an extruder, such as film, pipe, the coating on wire, etc.

Extrusion—The compacting of a plastics material and the forcing of it through an orifice in more or less continuous fashion.

Extrusion Coating—The resin is coated on a substrate by extruding a thin film of molten resin and pressing it onto or into the substrates, or both, without the use of an adhesive.

F

Fabricate—To work a material into a finished form by machining, forming, or other operation or to make flexible film or sheeting into end-products by sewing, cutting, sealing, or other operation.

Fadeometer—An apparatus for determining the resistance of resins and other materials to fading. This apparatus accelerates the fading by subjecting the article to high intensity ultraviolet rays of approximately the same wave length as those found in sunlight.

False Neck—A neck construction which is additional to the neck finish of a container and which is only intended to facilitate the blow molding operation. Afterwards the false neck part is removed from the container.

Family Molds—Two or more dissimilar mold cavities that are run concurrently in the same mold frame.

Fan Gate—The fan gate permits inflow of material through a wide but shallow gate.

Fatty Acid—An organic acid obtained by the hydrolysis (saponification) of natural fats and oils, e.g., stearic and palmitic acids. These acids are monobasic, may or may not have some double bonds, contain 16 or more C atoms.

Fault—An electrical short circuit or leakage path to ground or from phase to phase inadvertently created.

FDA—Food and Drug Administration.

Female—In molding practice, the indented half of a mold designed to receive the male half.

Fiber—This term usually refers to relatively short lengths of very small cross-sections of various materials. Fibers can be made by chopping filaments (converting). Staple fibers may be ½ to a few inches in length and usually 1 to 5 denier (½ to 1 mil in diameter).

Filament—A variety of fiber characterized by extreme length, which permits its use in yarn with little or no twist and usually without the spinning operation required for fibers.

Filament Winding—Roving or single strands of glass, metal, or other reinforcement are wound in a predetermined pattern onto a suitable mandrel. The pattern is so designed as to give maximum strength in the directions required. The strands can either be run from a creel through a resin bath before winding or preimpregnated materials can be used. When the right number of layers have been applied, the wound mandrel is cured at room temperatures or in an oven.

Fill-and-Wipe—Parts are molded with depressed designs; after application of paint, surplus is wiped off, leaving paint remaining only in depressed areas.

Fill Point—The level to which a container must be filled to furnish a designated quantity of the content.

Filler—An inert substance added to a plastic to make it less costly. Fillers may also improve physical properties, particularly hardness, stiffness, and impact strength. The particles are usually small, in contrast to those of reinforcements (q.v.); but there is some overlap between the functions of the two.

Fillet—A rounded filling of the internal angle between two surfaces of a plastic molding.

Film—An optional term for sheeting having a nominal thickness not greater than 0.010 inch.

Fin—The web of material remaining in holes or openings in a molded part which must be removed in finishing.

Fines—Very small particles (usually under 200 mesh) accompanying larger grains, usually of molding powder.

Finish—The plastic forming the opening of a container shaped to accommodate a specific closure. Also, the ultimate surface structure of an article.

Finish Insert—A removable part of a blow mold to form a specific neck finish of a plastics bottle. Sometimes called *Neck Insert.*

Fish Eye—A fault in transparent or translucent plastics materials, such as film or sheet, appearing as a small globular mass and caused by incomplete blending of the mass with surrounding material.

Fitment—A device used as a part of a closure assembly to accomplish certain purposes such as a dropper sprinkler, powder shaker, etc.

Flake—Used to denote the dry, unplasticized base of cellulosic plastics.

Flame Retardant Resin—A resin which is compounded with certain chemicals to reduce or eliminate its tendency to burn. For polyethylene and similar resins, chemicals such as antimony trioxide and chlorinated paraffins are useful.

Flame Spraying—Method of applying a plastics coating in which finely powdered fragments of the plastics, together with suitable fluxes, are projected through a cone of flame onto a surface.

Flame Treating—A method of rendering inert thermoplastic objects receptive to inks, lacquers, paints, adhesives, etc., in which the object is bathed in an open flame to promote oxidation of the surface of the article.

Flammability—Measure of the extent to which a material will support combustion.

Flash—Extra plastics attached to a molding along the parting line; it must be removed before the part can be considered finished.

Flash Gate—A long, shallow rectangular gate.

Flash Line—A raised line appearing on the surface of a molding and formed at the junction of mold faces.

Flash Mold—A mold designed to permit excess molding material to escape during closing.

Flash Point—The lowest temperature at which a combustible liquid will give off a flammable vapor that will burn momentarily.

Flexible Molds—Molds made of rubber or elastomeric plastics used for casting plastics. They can be stretched to remove cured pieces with undercuts.

Flexibilizer—An additive that makes a resin or rubber more flexible, i.e., less stiff. Also a plasticizer.

Flexographic Printing—A rubber roll, partially immersed in an ink fountain, transfers the ink to a fine, screen-lined steel roller which, in turn, deposits a thin layer of ink on the printing plate.

Flexural Strength—The strength of a material in bending, expressed as the tensile stress of the outermost fibers of a bent test sample at the instant of failure. With plastics, this value is usually higher than the straight tensile strength.

Floating Platen—A platen located between the main head and the press table in a multidaylight press and capable of being moved independently of them.

Flock—Short fibers of cotton, etc., used as fillers, q.v., for molding materials.

Flow—A qualitative description of the fluidity of a plastic material during the process of molding.

Flow Line (weld line)—A mark on a molded piece made by the meeting of two flow fronts during molding.

Flow Marks—Wavy surface appearance of an object molded from thermoplastic resins caused by improper flow of the resin into the mold.

Fluidized Bed Coating—A method of applying a coating of a thermoplastic resin to an article in which the heated article is immersed in a dense-phase fluidized bed of powdered resin and thereafter heated in an oven to provide a smooth, pin-hole-free coating.

Fluorescent Pigments—By absorbing unwanted wavelengths of light and converting them into light of desired wavelengths, these colors seem to possess an actual glow of their own.

Fluorinated Ethylene Propylene (FEP)—A member of the fluorocarbons, q.v., family of plastics, it is a copolymer of tetrafluoroethylene and hexafluoropropylene, possessing most of the properties of polytetrafluoroethylene (PTFE), q.v., and also having a melt viscosity low enough to permit conventional thermoplastic processing. Available in pellet form for molding and extrusion, and as dispersions for spray or dip-coating processes.

Fluorine (F)—The most reactive nonmetallic element. A pale yellow gas which is both corrosive and poisonous, it reacts vigorously with most oxidizable substances at room temperature and forms fluorides. It is used in the production of metallic and other fluorides, some of which are used to introduce fluorine into organic compounds, i.e., the fluorocarbons, q.v.

Fluorocarbons—The family of plastics including polytetrafluoroethylene (PTFE); polychlorotrifluoroethylene (PCTFE); polyvinylidene and fluorinated ethylene propylene (FEP), q.v. They are characterized by properties including good thermal and chemical resistance and non-adhesiveness, and possess a low dissipation factor and low dielectric constant. Depending upon which of the fluorocarbons is used, they are available as molding materials, extrusion materials, dispersions, film or tape.

Flux—1) An additive to a plastics composition during processing to improve its flow. For example, coumarone-indene resins are used as a flux during the milling of vinyl polymers. 2) Indicating a state of fluidity.

Foam Molding—A molding process whereby heat softened plastics containing a foaming agent is injection molded into a cavity where it hardens producing a product having a solid contiguous skin with a foam core. See also *Structural Foam* and *Sandwich Molding*.

Foamed Plastics—Resins in sponge form. The sponge may be flexible or rigid, the cells closed or interconnected, the density anything from that of the solid parent resin down to, in some cases, 2 lb./cu. foot. Compressive strength of rigid foams is fair, making them useful as core materials for sandwich structures. Both types are good heat barriers.

Foaming Agents—Chemicals added to plastics and rubbers that generate inert gases on heating, causing the resin to assume a cellular structure.

Foam-in-Place—Refers to the deposition of foams which requires that the foaming machine be brought to the work which is "in place" as opposed to bringing the work to the foaming machine.

Foil Decorating—Molding paper, textile, or plastic foils printed with compatible inks directly into a plastic part so that the foil is visible below the surface of the part as integral decoration.

Foot—The widest portion of a bottle's bottom.

Force Plate—The plate that carries the plunger or force plug of a mold and guide pins or bushings. Since it is usually drilled for steam or water lines, it is also called the *Steam Plate*.

Force Plug—The portion of a mold that enters the cavity block and exerts pressure on the molding compound, designated as *Top Force* or *Bottom Force* by position in the assembly; also called *Plunger* or *Piston*.

Forging—Fabrication of thermoplastics by impact or hydraulic pressure.

Formaldehyde (HCHO)—A colorless gas (usually employed as a solution in water) which possesses a suffocating, pungent odor. It is derived from the oxidation of methanol or low-boiling petroleum gases such as methane, ethane, propane and butane. It is widely used in the production of phenol formaldehyde (phenolic), urea formaldehyde (urea), and melamine formaldehyde (melamine) resins.

Friction Calendering—A process whereby an elastomeric compound is forced into the interstices of woven or cord fabrics while passing through the rolls of calender.

Friction Welding—A method of welding thermoplastics materials whereby the heat necessary to soften the components is provided by friction.

Frost Line—In the extrusion of polyethylene, lay-flat film, a ring-shaped zone located at the point where the film reaches its final diameter. This zone is characterized by a "frosty" appearance to the film caused by the film temperature falling below the softening range of the resin.

Frothing—Technique for applying urethane foam in which blowing agents or tiny air bubbles are introduced under pressure into the liquid mixture of foam ingredients.

Furane Resins—Dark colored, thermosetting resins available primarily as liquids ranging from low-viscosity polymers to thick, heavy syrups.

Furfural Resin—A dark-colored synthetic resin of the thermosetting variety obtained by the condensation of furfural with phenol or its homologues. It is used in the manufacture of molding materials, adhesives and impregnating varnishes. Properties include high resistance to acids and alkalis.

Fuse—In plastisol molding, to heat the plastisol to the temperature at which it becomes a single homogeneous phase. In this sense, *Cure* is the same as *Fuse*.

G

Gate—In injection and transfer molding, the orifice through which the melt enters the cavity. Sometimes the gate has the same cross-section as the runner leading to it; often, it it severely restricted.

Gel (*n.*)—In polyethylene, a small amorphous resin particle which differs from its surroundings by being of higher molecular weight and/or crosslinked, so that its processing characteristics differ from the surrounding resin to such a degree that it is not easily dispersed in the surrounding resin. A gel is readily discernible in thin films.

Gel Coat—A thin, outer layer of resin, sometimes containing pigment, applied to a reinforced plastics molding as a cosmetic.

Glass-Bonded Mica—A moldable thermoplastic material that uses glass as a binder and mica as a filler. Also known as *Ceramoplastic*.

Glass Thread—A type of threading in which the thread sides gradually taper down to the neck finish. Cross section of threads are semicircular.

Glitter (or Flitter or Spangles)—A group of special decorative materials consisting of flakes large enough so that each separate flake produces a plainly visible sparkle or reflection. They are incorporated directly into the plastic during compounding.

Gloss—The shine or lustre of the surface of a material.

Graduated—Molded in scale to indicate content level in container.

Graft Copolymers—A chain of one type of polymer to which side chains of a different type are attached or grafted (i.e., polymerizing butadiene and styrene monomer at the same time).

Granular Structure—Nonuniform appearance of finished plastics material due to retention of, or incomplete fusion of, particles of composition, either within the mass or on the surface.

Gravure Printing—Depositing ink on plastic film or sheeting or product from depressions of a specific depth, pattern, and spacing, which have been either mechanically or chemically engraved into a printing cylinder.

Grit Blasted—A surface treatment of a mold in which steel grit or sand materials are blown to the walls of the cavity to produce a roughened surface. Air escape from mold is improved and special appearance of molded article is often obtained by this method.

Guide Pins—Devices that maintain proper alignment of force plug and cavity as mold closes.

Gum—An amorphous substance or mixture which, at ordinary temperatures, is either a very viscous liquid or a solid which softens gradually on heating, and which either swells in water or is soluble in it. Natural gums, obtained from the cell walls of plants, are carbohydrates or carbohydrate derivatives of intermediate molecular weight.

Gusset—A tuck placed in each side of a tube of blown tubing as produced to provide a convenient square or rectangular package, similar to that of the familiar brown paper bag or sack, in subsequent packaging.

Gutta-Percha—A rubber-like material obtained from the leaves and bark of certain tropical trees. Sometimes used for the insulation of electrical wiring, and for transmission belting and various adhesives.

H

Hand Mold—A mold that is taken out of the press after each shot for removal of part.

Hardener—A substance or mixture of substances added to plastics composition, or an additive to promote or control the curing reaction by taking part in it. The term is also used to designate a substance added to control the degree of hardness of the cured film. See also *Catalyst*.

Hardness—The resistance of a plastics material to compression and indentation. Among the most important methods of testing this property are Brinell hardness, Rockwell hardness and Shore hardness, q.v.

Haze—The degree of cloudiness in a plastics material.

Head—The end section of a blow molding machine (in a general extruder) in which the melt is transformed into a hollow parison.

Head Space—The space between the fill level of a container and the sealing plane.

Heat-Distortion Point—The temperature at which a standard test bar (ASTM D 648) deflects 0.010 in. under a stated load of either 66 or 264 p.s.i.

Heat-Sealing—A method of joining plastics films by simultaneous application of heat and pressure to areas in contact. Heat may be supplied conductively or dielectrically.

Heating Chamber—In molding, that part of the machine in which the cold feed is reduced to a hot melt. Also *Heating Cylinder*.

Heel—The part of a container between the bottom bearing surface and the side wall.

Heel Radius—The degree of curvature at the extreme bottom end of a container extending upward from the bearing surface. Also called *Base Radius*.

Heel Tap—A bottle imperfection resulting from one area of the bottom having excess thickness.

Hexachlorobutadiene—A plastics that is used to form ultra thin pinhole-free films on metals.

H.F. Preheating—See *Dielectric Heating*.

High-Load Melt Index—The rate of flow of a molten resin through a 0.0825 inch orifice when subjected to a force of 21,600 grams at 190 C. See *Melt Index*.

High Polymer—A macromolecular substance which, as indicated by the term "polymer" and by the name (e.g., polyvinyl chloride) and formula (e.g., CH_2CHCL) by which it is identified, consists of molecules which are (at least approximately) multiples of the low molecular unit.

High-Pressure Laminates—Laminates molded and cured at pressures not lower than 1000 p.s.i. and more commonly in the range of 1200 to 2000 p.s.i.

Hob—A master model in hardened steel used to sink the shape of a mold into a soft steel block.

Hobbing—Forming multiple mold cavities by forcing a *hob* (q.v.) into soft steel (or beryllium-copper) cavity blanks.

Homopolymer—A polymer, consisting of (neglecting the ends, branch junctions, and other minor irregularities) a single type of repeating unit.

Honeycomb—Manufactured product consisting of sheet metal or a resin impregnated sheet material (paper, fibrous glass, etc.) which has been formed into hexagonal-shaped cells. Used as core material for sandwich constructions.

Hopper—Conical feed reservoir into which molding powder is loaded and from which it falls into a molding machine or extruder, sometimes through a metering device.

Hopper Dryer—A combination feeding and drying device for extrusion and injection molding of thermoplastics. Hot air flows upward through the hopper containing the feed pellets.

Hopper Loader—A curved pipe through which molding powders are pneumatically conveyed from shipping drums to machine hoppers.

Hot Gas Welding—A technique of joining thermoplastic materials (usually sheet) whereby the materials are softened by a jet of hot air from a welding torch, and joined together at the softened points. Generally a thin rod of the same material is used to fill and consolidate the gap.

Hot-Runner Mold—A mold in which the runners are insulated from the chilled cavities and are kept hot. Parting line is at gate of cavity, runners are in separate plate(s), so they are not, as is the case usually, ejected with the piece.

Hot-Stamping—Engraving operation for marking plastics in which roll leaf is stamped with heated metal dies onto the face of the plastics. Ink compounds can also be used. By means of felt rolls, ink is applied to type and by means of heat and pressure, type is impressed into the material, leaving the marking compound in the indentation.

Hydraulic—A system in which energy is transferred from one place to another by means of compression and flow of a fluid (e.g., water, oil).

Hydrocarbon Resin—A family of thermoplastic compounds based on liquid butadiene-sytrene copolymers, polybutadiene with various high molecular weight hydrocarbon polymers such as polyethylene and elastomers.

Hydrogenation—Chemical process whereby hydrogen is introduced into a compound.

Hydrolysis—Chemical decomposition of a substance involving the addition of water.

Hydrophilic—A substance with an affinity for water.

Hydrophobic—A substance with an aversion to water.

Hygroscopic—Tending to absorb moisture.

I

Immiscible—Descriptive of two or more fluids which are not mutually soluble.

Impact Bar (Specimen)—A test specimen of specified dimensions which is utilized to determine the relative resistance to plastics to fracture by shock.

Impact Resistance—Relative susceptibility of plastics to fracture by shock, e.g., as indicated by the energy expended by a standard pendulum type impact machine in breaking a standard specimen in one blow.

Impact Strength—1) The ability of a material to withstand shock loading. 2) The work done in fracturing, under shock loading, a specified test specimen in a specified manner.

Impregnation—The process of thoroughly soaking a material such as wood, paper or fabric, with a synthetic resin so that the resin gets within the body of the material. The process is usually carried out in an impregnator.

Impulse Sealing—A heat sealing technique in which a pulse of intense thermal energy is applied to the sealing area for a very short time, followed immediately by cooling. It is usually accomplished by using an RF heated metal bar which is cored for water cooling or is of such a mass that it will cool rapidly at ambient temperatures.

Infra-Red—Part of the electromagnetic spectrum between the visible light range and the radar range. Radiant heat is in this range, and infra-red heaters are much used in sheet thermoforming.

Inhibitor—A substance that slows down chemical reaction. Inhibitors are sometimes used in certain types of monomers and resins to prolong storage life.

Injection Blow Molding—A blow molding process in which the parison to be blown is formed by injection molding.

Injection Mold—A mold into which a plasticated material is introduced from an exterior heating cylinder.

Injection Molding—A molding procedure whereby a heat softened thermoplastic material is forced from a plasticizing device into a relatively cool mold cavity for hardening. Thermosetting resins are also injection molded by heat softening the material in a plasticizing device and then forcing the material into a hot mold for the cure-hardening. See *Transfer Molding*.

Injection Molding Cycle—The complete time cycle of operation utilized in injection molding of an object including injection; die close and die open time.

Injection Pressure—The pressure on the face of the injection ram at which molding material is injected into a mold. It is usually expressed in p.s.i.

Injection Ram—The ram which applies pressure to the plunger in the process of injection molding.

Inorganic Pigments—Natural or synthetic metallic oxides, sulfides, and other salts, calcined during processing at 1200 to 2100°F. They are outstanding in heat- and light-stability, weather resistance, and migration resistance.

In Place Process—A blow molding process that makes use of a manifold with an orifice leading to each blow mold.

Insert—An integral part of a plastics molding consisting of metal or other material which may be molded into position or may be pressed into the molding after the molding is completed.

In Situ Foaming—The technique of depositing a foamable plastics (prior to foaming) into the place where it is intended that foaming shall take place. An example is the placing of foamable plastics into cavity brickwork to provide insulation. After being positioned, the liquid mix foams to fill the cavity. See also *Foamed Plastics*.

Insulated Runner Mold—See *Hot Runner Mold*.

Insulation—A coating of a dielectric or essentially non-conducting material whose purpose it is to prevent the transmission of electricity.

Insulation Resistance—The electrical resistance of an insulating material to a direct voltage. It is determined by mesauring the leakage of current which flows through the insulation.

Interlock—A safety device designed to insure that a piece of apparatus will not operate until certain precautions have been taken.

Internal Mixers—Mixing machines using the principle of cylindrical containers in which the materials are deformed by rotating blades or rotors. The containers and rotors are cored so that they can be heated or cooled to control the temperature of a batch. These mixers are extensively used in the compounding of plastics and rubber materials and have the inherent advantage of keeping dust and fume hazards to a minimum.

Intrinsic Viscosity—The intrinsic viscosity of a polymer is the limiting value at infinite dilution of the ratio of the specific viscosity of the polymer solution to its concentration in mols per liter.

$$n_1 = C \xrightarrow{\text{Lim}} O \ \frac{n^{sp}}{C}$$

where n_1 = intrinsic viscosity, n^{sp} = specific viscosity, c = concentration in moles per liter. Intrinsic viscosity is usually estimated by determining the specific viscosity at several low concentrations and extrapolating the values of $\frac{n^{sp}}{C}$ to c = 0. The concentration is expressed in terms of the repeating unit. In the case of polystyrene the repeating unit is $—CH_2—CH(C_6H_5)—$ and has a molecular weight of 104.

Introfaction—The change in fluidity and wetting properties of an im-pregnating material, produced by the addition of an introfier, q.v.

Introfier—A chemical which will convert a colloidal solution into a molecular one. See also *Introfaction*.

Ion Exchange Resins—Small granular or beadlike particles containing acidic or basic groups, which will trade ions with salts in solutions. Generally used for softening and purifying water.

Ionomer—Ionized carboxyl groups used for film lamination in the coextrusion process.

Ionomer Resins—A new polymer which has ethylene as its major component, but containing both covalent and ionic bonds. The polymer exhibits very strong interchain ionic forces. The anions hang from the hydrocarbon chain and the cations are metallic—sodium, potassium, magnesium. These resins have many of the same features as polyethylene plus high transparency, tenacity, resilience and increased resistance to oils, greases and solvents. Fabrication is carried out as with polyethylene.

Irradiation (Atomic)—As applied to plastics, refers to bombardment with a variety of subatomic particles, generally alpha-, beta-, or gamma-rays. Atomic irradiation has been used to initiate polymerization and copolymerization of plastics and in some cases to bring about changes in the physical properties of a plastic material.

Isinglass—A white, tasteless gelatin derived from the bladder of fishes, usually the sturgeon. It is used as an adhesive and clarifying agent.

Isocyanate Resins—Most applications for this resin are based on its combination with polyols (e.g., polyesters, polyethers, etc.). During this reaction, the reactants are joined through the formation of the urethane linkage—and hence this field of technology is generally known as urethane chemistry.

Isotactic—A chain of unsymmetrical molecules combined head to tail with their methyl groups occupying the same relative position in space along the chain.

Izod Impact Test—A test designed to determine the resistance of a plastics material to a shock loading. It involves the notching of a specimen, which is then placed in the jaws of the machine and struck with a weighted pendulum. See also *Impact Strength*.

J

Jacket—A tough sheath to protect an insulated wire or cable, or to permanently group two or more insulated wires or cables.

Jet Molding—Processing technique characterized by the fact that most of the heat is applied to the material as it passes through the nozzle or jet, rather than in a heating cylinder as is done in conventional processes.

Jet Spinning—For most purposes similar to melt spinning. Hot gas jet spinning uses a directed blast or jet of hot gas to "pull" molten polymer from a die lip and extend it into fine fibers.

Jetting—Turbulent flow of resin from an undersize gate or thin section into a thicker mold section, as opposed to lamular flow of material progressing radially from a gate to the extremities of the cavity.

Jig—Tool for holding component parts of an assembly during the manu-facturing process, or for holding other tools. Also called a *Fixture*.

Jute—Blast fiber obtained from the stems of several species of the plant Corchorus found mainly in India and Pakistan. Used as a filler, q.v., for plastics molding materials, and more recently as a reinforcement for polyester resins in the fabrication of reinforced plastics.

K

Keyhole Die—See *Slot Extrusion Die*.

Kirksite—An alloy of aluminum and zinc used for the construction of blow molds; it imparts high degree of heat conductivity to the mold.

Kiss-Roll Coating—This roll arrangement carries a metered film of coating to the web; at the line of web contact, it is split with part remaining on the roll, the remainder of the coating adhering to the web.

Knife Coating—A method of coating a substrate (usually paper or fabric) in which the substrate, in the form of a continuous moving web, is coated with a material whose thickness is controlled by an adjustable knife or bar set at a suitable angle to the substrate. In the plastics industry PVC formulations are widely used in this work and curing is effected by passing the coated substrate into a special oven, usually heated by infrared lamps or convected air. There are a number of variations of this basic technique and they vary according to the type of product required.

Knit Lines—See *Weld Mark*.

Knockout Pin—A device for knocking a cured piece from a mold. Also called *Ejector Pin*.

Knot Tenacity (Knot Strength)—The tenacity in grams per denier of a yarn where an overhand knot is put into the filament of yarn being pulled to show up sensitivity to compressive or shearing forces.

Kraft Paper—Paper made from sulfate wood pulp.

L

Label Panel—The plain portion of a decorated container set up for application of labels.

Lacquer—Solution of natural or synthetic resins, etc., in readily evaporating solvents, which is used as a protective coating.

Laminar Flow—Laminar flow of thermoplastic resins in a mold is accompanied by solidification of the layer in contact with the mold surface that acts as an insulating tube through which material flows to fill the remainder of the cavity. This type of flow is essential to duplication of the mold surface.

Laminated Plastics (Synthetic Resin-Bonded Laminate, Laminate)—A plastics material consisting of superimposed layers of a synthetic resin-impregnated or -coated filler which have been bonded together, usually by means of heat and pressure, to form a single piece.

Laminated Wood—A high-pressure bonded wood product composed of layers of wood with resin as the laminating agent. The term *Plywood* covers a form of laminated wood in which successive layers of veneer are ordinarily cross laminated, the core of which may be veneer or sawn lumber in one piece or several pieces.

Land—1) The horizontal bearing surface of a semipositive or flash mold by which excess material escapes. See *Cut-off*. 2) The bearing surface along the top of the flights of a screw in a screw extruder; 3) the surface of an extrusion die parallel to the direction of melt flow.

Landed Plunger Mold—See *Semi-Positive Mold*.

Landed Positive Mold—See *Semi-Positive Mold*.

Latex—An aqueous dispersion of a rubber or resin solid.

Lay-up—(*n.*) As used in reinforced plastics, the reinforcing material placed in position in the mold; also the resin-impregnated reinforcement. (*v.*)—The process of placing the reinforcing material in position in the mold.

Leach—To extract a soluble component from a mixture by the process of percolation.

Leaker—Any condition of the finish where the normal sealing device or closure will not retain the air or liquid content of the bottle.

Light-Resistance—The ability of a plastics material to resist fading after exposure to sunlight or ultraviolet light. Nearly all plastics tend to darken under these conditions.

L/D Ratio—A term used to define an extrusion screw which denotes the ratio of the screw length to the screw diameter.

Linear Molecule—A long chain molecule as contrasted to one having many side chains or branches.

Linters—Short fibers that adhere to the cotton seed after ginning. Used in rayon manufacture, as fillers for plastics, and as a base for the manufacture of cellulosic plastics.

Lip—The extreme outer edge of the top of a container intended to facilitate pouring.

Litharge—PbO. An oxide of lead used as an inorganic accelerator, as a vulcanizing agent for neoprene, and as an ingredient of paints.

Loading Shoe Mold—A variation of the flash type mold which introduces an intermediate plate to provide greater loading space for high bulk thermoset compounds.

Loading Tray (Charging Tray)—A device in the form of a specially designed tray which is used to load the charge simultaneously into each cavity of a multi-cavity mold by the withdrawal of a sliding bottom from the tray.

Loop Tenacity (Loop Strength)—The tenacity or strength value obtained by pulling two loops, as two links in a chain, against each other to demonstrate the susceptibility that a yarn, cord or rope has for cutting or crushing itself.

Loose Detail Mold—A mold with parts that come out with the piece.

Loss Factor—The product of the power factor and the dielectric constant.

Low Pressure Laminates—In general, laminates molded and cured in the range of pressures from 400 p.s.i. down to and including pressures obtained by the mere contact of the plies.

Lubricant Bloom—An irregular, cloudy, greasy film on a plastic surface.

Lubricants—Additives used to improve the processability and appearance of plastics products.

Lug—An indentation or raised portion of the surface of a container,

provided to control automatic (multicolor) decorating operations.

Luminescent Pigments—Special pigments available to produce striking effects in the dark. Basically there are two types: one is activated by ultra-violet radiation, producing very strong luminescence and, consequently, very eye-catching effects; the other type, known as phosphorescent pigments, does not require any separate source of radiation.

M

Macerate—(*v.*) To chop or shred fabric for use as a filler for molding resin. (*n.*) The molding compound obtained when so filled.

Machine Shot Capacity—Refers to the maximum weight of thermoplastic resin which can be displaced or injected by the injection ram in a single stroke.

Macromolecule—The large ("giant") molecules which make up the high polymers.

Mandrel—1) The core around which paper, fabric, or resin-impregnated fibrous glass is wound to form pipes or tubes. 2) In extrusion, the central finger of a pipe or tubing die.

Manifold—A term used mainly with reference to blow molding and sometimes with injection molding equipment. It refers to the distribution or piping system which takes the single channel flow output of the extruder or injection cylinder and divides it to feed several blow molding heads or injection nozzles.

Masterbatch—A plastics compound which includes a high concentration of an additive or additives. Masterbatches are designed for use in appropriate quantities with the basic resin or mix so that the correct end concentration is achieved. For example, color masterbatches for a variety of plastics are extensively used as they provide a clean and convenient method of obtaining accurate color shades.

Mat—A randomly distributed felt of glass fibers used in reinforced plastics lay-up molding.

Matched Metal Molding—Method of molding reinforced plastics between two close-fitting metal molds mounted in a hydraulic press.

Material Well—Space provided in a compression or transfer mold to care for bulk factor.

Melamine Formaldehyde Resin—Classified as a synthetic resin derived from the reaction of melamine (2,4,6-triamino-1,3,5-triazine) with formaldehyde or its polymers.

Melt Extractor—Usually refers to a type of injection machine torpedo but could refer to any type of device which is placed in a plasticating system for the purpose of separating fully plasticated melt from partially molten pellets and material. It thus insures a fully plasticated discharge of melt from the plasticating system.

Melt Fracture—An instability in the melt flow through a die starting at the entry to the die. It leads to surface irregularities on the finished article like a regular helix or irregularly-spaced ripples.

Melt Index—The amount, in grams, of a thermoplastic resin which can be forced through a 0.0825 inch orifice when subjected to 2160 gms. force in 10 minutes at $190^{\circ}C$.

Melt Instability—An instability in the melt flow through a die starting at the land of the die. It leads to the same surface irregularities on the finished part as melt fracture.

Melt Strength—The strength of the plastics while in the molten state.

Meniscus—The free surface of a liquid in a container, for example, water in contact with air confined in a capillary tube. The meniscus may be convex, e.g., mercury vs. air in glass, or concave, e.g., water vs. air in glass.

Mer—The repeating structural unit of any high polymer.

Metalizing—Applying a thin coating of metal to a nonmetallic surface. May be done by chemical deposition or by exposing the surface to vaporized metal in a vacuum chamber.

Metallic Pigments—A class of pigments consisting of thin opaque aluminum flakes (made by ball milling either a disintergrated aluminum foil or a rough metal powder and then polishing to obtain a flat, brilliant surface on each particle) or copper alloy flakes (known as bronze pigments). Incorporated into plastics, they produce unusual silvery and other metal-like effects.

Metering Screw—An extrusion screw which has a shallow constant depth, and constant pitch section over, usually, the last 3 to 4 flights.

Methacrylonitrile—A vinyl nitrile compound somewhat similar to ABS.

Methyl Methacrylate—
$CH_2CCH_3COOCH_3$. A colorless, volatile liquid derived from acetone cyanohydrin, methanol and dilute sulphuric acid, and use in the production of acrylic resins, q.v.

Methylpentene (TPX)—A thermoplastic material based on 4 methylpentene 1. Low gravity 0.83, 90% optical transmission and 390°F. thermal range plus excellent electrical properties.

Migration of Plasticizer—Loss of plasticizer from an elastomeric plastic compound with subsequent absorption by an adjacent medium of lower plasticizer concentration.

Mn (Number-average molecular weight)—The total weight of all molecules divided by the total number of molecules.

Modified—Containing ingredients such as fillers, pigments or other additives, that help to vary the physical properties of a plastics material. An example is oil modified resin.

Modulus of Elasticity—The ratio of stress to strain in a material that is elastically deformed.

Moisture Vapor Transmission—The rate at which water vapor permeates through a plastic film or wall at a specified temperature and relative humidity.

Mold (*v.*)—To shape plastics parts or finished articles by heat and pressure. (*n.*)—1) The cavity or matrix into which the plastics composition is placed and from which it takes its form. 2) The assembly of all the parts that function collectively in the molding process.

Mold Efficiency—In a multimold blowing system the percentage of the total turnaround time of the mold actually required for forming, cooling and ejection of the container.

Molding Cycle—1) The period of time occupied by the complete sequence of operations on a molding press requisite for the production of one set of moldings. 2) The operations necessary to produce a set of moldings without reference to the time taken.

Molding Powder—Plastics material in varying stages of granulation, and and comprising resin, filler, pigments, plasticizers, and other ingredients, ready for use in the molding operation.

Molding Pressure—The pressure applied to the ram of an injection machine or press to force the softened plastics completely to fill the mold cavities.

Molding Shrinkage (Mold Shrinkage, Shrinkage, Contraction)—The difference in dimensions, expressed in inches per inch, between a molding and the mold cavity in which it was molded, both the mold and the molding being at normal room temperature when measured.

Mold Release—*See Parting Agent.*

Mold Seam—A vertical line formed at the point of contact of the mold halves. The prominence of the line depends on the accuracy with which the mating mold halves are matched. See *Parting Line.*

Molecular Weight Distribution—The ratio of the weight average molecular weight to the number average molecular weight gives an indication of the distribution.

Monofilament (Monofil)—A single filament of indefinite length. Monofilaments are generally produced by extrusion. Their outstanding uses are in the fabrication of bristles, surgical sutures, fishing leaders, tennis-racquet strings, screen materials, ropes and nets; the finer monofilaments are woven and knitted on textile machinery.

Monomer—A relatively simple compound which can react to form a polymer. See also *Polymer*.

Mounting Plate—The part of the molding unit to which the mold is attached.

Movable Platen—The large back platen of an injection molding machine to which the back half of the mold is secured during operation. This platen is moved either by a hydraulic ram or a toggle mechanism.

Multi-Cavity Mold—A mold with two or more mold impressions, i.e., a mold which produces more than one molding per molding cycle.

Multifilament Yarn—The multifilament yarn is composed of a multitude of fine continuous filaments, often 5 to 100 individual filaments, usually with some twist in the yarn to facilitate handling. Multifilament yarn sizes are described in denier and range from 5–10 denier up to a few hundred denier. The larger deniers, even in the thousands, are usually obtained by plying smaller yarns together. Individual filaments in a multifilament yarn are usually about 1 to 5 denier (which is about ½ mil to 1 mil diameter in Marlex polyethylene).

Multi-Orifice Die—See *Spinneret*.

Multiple Gate Mold—Many small gates into the side of restricted flow areas where fragile mold sections necessitate a minimum and distributed material flow.

Multiple Head Machine—A (blow molding) machine in which the plastics melt prepared by the extruder is divided into a multiplicity of separate streams (parisons) each giving ultimately a finished item.

Mw (Weight-average molecular weight) —The sum of the total weights of molecules of each size multiplied by their respective weights divided by the total weight of all molecules.

N

Neck—The part of a container where the shoulder cross section area decreases to form the finish.

Neck Bead—A protruding circle on a container at the point where the neck meets the finish, the diameter of which usually equals the outside diameter of the closure.

Neck-in—In extrusion coating, the difference between the width of the extruded web as it leaves the die and the width of the coating on the substrate.

Neck Insert—Part of the mold assembly which forms the neck and finish. Sometimes called *Neck Ring*.

Neck Ring Process—In this blow molding process, the blownware is formed neck down. The neck portion of the blow mold has a hollow tube to form the inside of the neck and admit blow air. The neck ring starts at the orifice and is filled with the material first and then drops down as the parison forms into its final position and is clamped as the mold closes and blow starts.

Needle Blow—A specific blow molding technique where the blowing air is injected into the hollow article through a sharpened hollow needle which pierces the parison.

Nest Plate—A retainer plate with a depressed area for cavity blocks used in mold construction.

Nip—The "V" formed where the pressure roll contacts the chill roll.

Nitrile Barrier Resins—See *Barrier Resins.*

Nonpolar—Having no concentrations of electrical charge on a molecular scale, thus, incapable of significant dielectric loss. Examples among resins are polystyrene and polyethylene.

Nonrigid Plastic—A non-rigid plastic is one which has a stiffness or apparent modulus of elasticity of not over 50,000 p.s.i. at 25°C. when determined according to ASTM test procedure D747-43 T.

Notch Sensitivity—The extent to which the sensitivity of a material to fracture is increased by the presence of a surface in homogeneity such as a notch, a sudden change in section, a crack, or a scratch. Low notch sensitivity is usually associated with ductile materials, and high notch sensitivity with brittle materials.

Novolac—A phenolic-aldehyde resin which, unless a source of methylene groups is added, remains permanently thermoplastic. See also *Resinoid* and *Thermoplastic.*

Nozzle—The hollow cored metal nose screwed into the extrusion end of *a*) the heating cylinder of an injection machine or *b*) a transfer chamber where this is a separate structure. A nozzle is designed to form, under pressure, a seal between the heating cylinder or the transfer chamber and the mold. The front end of the nozzle may be either flat or spherical in shape.

Nutating Motor—A hydraulic motor which offers high torque and low speed without gear trains.

Nylon—The generic name for all synthetic fiber-forming polyamides; they can be formed into monofilaments and yarns characterized by great toughness, strength and elasticity, high melting point, and good resistance to water and chemicals. The material is widely used for bristles in industrial and domestic brushes, and for many textile applications; it is also used in injection molding gears, bearings, combs, etc.

O

Oblong—A particular shape. A container which has a rectangular cross section perpendicular to the major axis.

Off Center—Any condition where the finish opening is not centered over the bottom of the container. Also, the condition where the mandrel is not concentric with the ring of the blowing head.

Offset—A printing technique in which ink is transferred from a bath onto the raised surface of the printing plate by rollers. Subsequently, the printing plates transfer the ink to the object to be printed.

Oil-Soluble Resin—Resin which at moderate temperatures will dissolve in, disperse in, or react with, drying oils to give a homogeneous film of modified characteristics.

Olefins—A group of unsaturated hydrocarbons of the general formula C_nH_{2n}, and named after the corresponding paraffins by the addition of "ene" or "ylene" to the stem. Examples are ethylene and propylene.

Oleo Resins—Semi-solid mixtures of the resin and essential oil of the plant from which they exude, and sometimes referred to as balsams. Oleoresinous materials also consist of products of drying oils and natural or synthetic resins.

One Shot Molding—In the urethane foam field, indicates a system whereby the isocyanate, polyol, catalyst, and other additives are mixed together directly and a foam is produced immediately (as distinguished from *prepolymer*).

Opaque—Descriptive of a material or substance which will not transmit light. Opposite of transparent, q.v. Materials which are neither opaque nor transparent are sometimes described as semi-opaque, but are more properly classified as translucent, q.v.

Orange-Peel—Said of moldings that have unintentionally rough surfaces.

Organic Pigments—Characterized by good brightness and brilliance. They are divided into toners and lakes. Toners, in turn, are divided into insoluble organic toners and lake toners. The insoluble organic toners are usually free from salt-forming groups. Lake toners are practically pure, water-insoluble heavy metal salts of dyes without the fillers or substrates of ordinary lakes. Lakes, which are not as strong as lake toners, are water-insoluble heavy metal salts or other dye complexes precipitated upon or admixed with a base or filler.

Organosol—A vinyl or nylon dispersion, the liquid phase of which contains one or more organic solvents. See also *Plastisol*.

Orientation—The alignment of the crystalline structure in polymeric materials so as to produce a highly uniform structure. Can be accomplished by cold drawing or stretching during fabrication.

Orifice—The opening in the extruder die formed by the orifice bushing (ring) and mandrel.

Orifice Bushing—The outer part of the die in an extruder head.

Out-of-Round—A plastics container manufacturing variance in which a round container, when formed, does not remain round.

Oval—A particular shape. A container which has an egg-shaped cross section perpendicular to the major axis.

Overcoating—In extrusion coating, the practice of extruding a web beyond the edge of the substrate web.

Overflow Capacity—The capacity of a container to the top of the finish or to the point of overflow.

Overlay Sheet (Surfacing Mat)—A non-woven fibrous mat (either in glass, synthetic fiber, etc.) used as the top layer in a cloth or mat lay-up to provide a smoother finish or minimize the appearance of the fibrous pattern.

Oxidation—The addition of oxygen to a compound or the reduction of hydrogen.

P

Paneling—Distortion of a container occurring during aging or storage, caused by the development of a reduced pressure inside the container.

Parallels—1) Spacers placed between the steam plate and press platen to prevent the middle section of the mold from bending under pressure. 2) Pressure pads or spacers between the steam plates of a mold to control height when closed and to prevent crushing the parts of the mold when the land area is inadequate.

Parison—The hollow plastic tube from which a container, toy, etc. is blow molded.

Parison Swell—In blow molding, the ratio of the cross-sectional area of the parison to the cross-sectional area of the die opening.

Parting Agent—A lubricant, often wax, used to coat a mold cavity to prevent the molded piece from sticking to it, and thus to facilitate its removal from the mold. Also called *Release Agent.*

Parting Line—Mark on a molding or casting where halves of mold met in closing.

Partitioned Mold Cooling—A large diameter hole drilled into the mold (usually the core) and partitioned by a metal plate extending to near the bottom end of the channel. Water is introduced near the top of one side of the partition and removed on the other side.

Parylene—Poly-para-xylene is used in ultra thin films for capacitor dielectrics and as a pore free coating. Films are formed by heating a monomer and condensing it on a cool surface.

Pearlescent Pigments—A class of pigments consisting of particles that are essentially transparent crystals of a high refractive index. The optical effect is one of partial reflection from the two sides of each flake. When reflections from parallel plates reinforce each other, the result is a silvery luster. Effects possible range from brilliant highlighting to moderate enhancement of the normal surface gloss.

Perfluoroalkoxy Teflon (PFA)—A melt extrudable variety of TFE.

Permanent Set—The increase in length, expressed in a percentage of the original length, by which an elastic material fails to return to original length after being stressed for a standard period of time.

Permeability—1) The passage or diffusion of a gas, vapor, liquid, or solid through a barrier without physically or chemically affecting it. 2) The rate of such passage.

pH—An expression of the degree of acidity or alkalinity of a substance. Neutrality is pH_7—acid solutions being under 7 and alkaline solutions over 7. pH meters are commercially available for accurate readings.

Phenolic Resin—A synthetic resin produced by the condensation of an aromatic alcohol with an aldehyde, particularly of phenol with formaldehyde. Phenolic resins form the basis of thermosetting molding materials, laminated sheet, and stoving varnishes. They are also used as impregnating agents and as components of paints, varnishes, lacquers, and adhesives.

Phenoxy Resins—A high molecular weight thermoplastic polyester resin based on bisphenol-A and epichorohydrin. Recently developed in the United States, the material is available in grades suitable for molding, extrusion, coating and adhesives, q.v.

Phosphonitrilic—A flame resistant thermosetting resin.

Phthalate Esters—A main group of plasticizers, q.v., produced by the direct action of alcohol on phthalic anhydride. The phthalates are the most widely used of all plasticizers, and are generally characterized by moderate cost, good stability, and good all-round properties.

Phthalocyanine Pigments—Organic pigments, q.v., of extremely stable chemical configuration resulting in very good fastness properties. These properties are enhanced by the formation of the copper complex which is the phthalocyanine blue most used. The introduction of chlorine atoms into the molecule of blue gives the well-known phthalocyanine green, also usually in the form of copper complex.

Pill—See *Preform*.

Pinch-off—A raised edge around the cavity in the mold which seals off the part and separates the excess material as the mold closes.

Pinch-off Blades—The part of the mold which compresses the parison to effect sealing of the parison prior to blowing and to permit easy removal and cooling of flash.

Pinch-off Land—The width of pinch-off blade which effects sealing of the parison.

Pinch-off Tail—The bottom of the parison that is pinched off when the mold closes.

Pinch Tube Process—A basic blow molding process in which the extruder drops a tube between mold halves that is pinched off when the mold closes.

Pinhole—A very small hole in the extruded resin coating.

Pinpoint Gate—A restricted orifice of 0.030 inches or less in diameter through which molten resin flows into a mold cavity.

Pipe Train—A term used in extrusion of pipe which denotes the entire equipment assembly used to fabricate the pipe, e.g., extruder, die, cooling bath, haul-off and cutter.

Pitch—The distance from any point on the flight of a screw line to the corresponding point on an adjacent flight, measured parallel to the axis of the screw line or threading.

Plastic—(a) Pliable and capable of being shaped by pressure. Plastic is incorrectly used as the generic word for the industry and its products.

Plastics—(*n.*) A generic term for the industry and its products which is properly used only as a plural word. The plastics products include polymeric substances, natural or synthetic; exclude the rubbers.

Plastic Deformation—A change in dimensions of an object under load that is not recovered when the load is removed; opposed to elastic deformation.

Plastics Tooling—Tools, e.g., dies, jigs, fixtures, etc., for the metal forming trades constructed of plastics, generally laminates or casting materials.

Plasticate—To soften by heating or kneading. Synonyms are: plastify, flux, and, (imprecisely) plasticize (q.v.).

Plasticity—The quality of being able to be shaped by plastic flow.

Plasticize—To soften a material and make it plastic or moldable, either by means of a plasticizer or the application of heat.

Plasticizer—Chemical agent added to plastic compositions to make them softer and more flexible.

Plasticorder or Plastigraph—A laboratory device used to predict the performance of a plastics material by measurement of temperature, viscosity and shear rate relationships.

Plastigel—A plastisol exhibiting gel-like flow properties.

Plastisols—Mixtures of resins and plasticizers which can be molded, cast, or converted to continuous films by the application of heat. If the mixtures contain volatile thinners also, they are known as *Organosols*.

Plastometer—An instrument for determining the flow properties of a thermoplastic resin by forcing the molten resin through a die or orifice of specific size at a specified temperature and pressure.

Plate Dispersion Plug—Two perforated plates held together with a connecting rod which are placed in the nozzle of an injection molding machine to aid in dispersing a colorant in a resin as it flows through the orifices in the plates.

Platens—The mounting plates of a press to which the entire mold assembly is bolted.

Platform Blowing—A special technique for blowing large parts. To prevent excessive sag of the heavy parison the machine employs a table which after rising to meet the parison at the die descends with the parison but at a slightly lower rate than the parison extrusion speed.

Plug-and-Ring—Method of sheet forming in which a plug, functioning as a male mold, is forced into a heated plastic sheet held in place by a clamping ring.

Plug Forming—A thermoforming process in which a plug or male mold is used to partially preform the part before forming is completed using vacuum or pressure.

Plug Gate—See *Tunnel Gate*.

Plunger—See *Force Plug*.

Pneumatic—A system in which energy is transferred by compression, flow and expansion of air.

Pock Marks—Irregular indentations on the surface of a blown container caused by insufficient contact of the blown parison with the mold surface. They are due to low blow pressure, air gas entrapment or moisture condensation on mold surface.

Poise—The unit of viscosity, q.v., expressed as one dyne per second per square centimeter.

Polar—See *Nonpolar*.

Polishing Roll (s)—A roll or series of rolls, which have a highly polished chrome plated surface, that are utilized to produce a smooth surface on sheet as it is extruded.

Polyacrylate—A thermoplastic resin made by the polymerization of an acrylic compound such as methyl methacrylate.

Polyallomers—
Crystalline polymers produced from two or more olefin monomers.

Polyamide—A polymer in which the structural units are linked by amide or thio-amide groupings. Many polyamides are fiber-forming.

Polyamide-Imide—A high temperature resin used for magnet wire insulation, laminates and adhesives.

Polyaminobismaleimide Resins (PABMS)—Highly stable resins are adaptable to many fillers and are processed like phenolics and epoxies.

Polyaryl Ether—A $300°$F. heat deflection thermoplastic.

Polyaryl Sulfone—A very highly heat and are resistant thermoplastic.

Polybenzimidazole (PBI)—A highly oxidative stable thermosetting resin used for sophisticated aerospace products.

Polybenzothiazoles—A thermosetting laminating resin with higher thermal and oxidative stability than the polyimides.

Polybutadiene—A thermosetting resin for high heat distortion and low moisture absorption.

Polybutadiene Glycol—A thermosetting high performance resin used in some laminates.

Polybutylene (PB)—A thermoplastic with creep, craze and stress crack resistance.

Polyblends—Colloquial term generally used in the styrene field to apply to mechanical mixtures of polystyrene and rubber.

Polycarbonate Resins—Polymers derived from the direct reaction between aromatic and aliphatic dihydroxy compounds with phosgene or by the ester exchange reaction with appropriate phosgene-derived precursors.

Polycarboranesiloxane—An elastomeric resinous and liquid plastics for high heat products.

Polycondensation—See *Condensation*.

Polyester—A resin formed by the reaction between a dibasic acid and a dihydroxy alcohol, both organic. Modification with multi-functional acids and/or bases and some unsaturated reactants permit cross-linking to thermosetting resins. Polyesters modified with fatty acids are called *Alkyds*.

Polyester-Imide—A high temperature magnet wire insulation and similar to the aromatic polyimides.

Polyester Reinforced Urethane—A poromeric material which may have a urethane impregnation or a silicone coating and used for shoe uppers and industrial leathers.

Polyethylene—A thermoplastic material composed by polymers of ethylene. It is normally a translucent, tough, waxy solid which is unaffected by water and by a large range of chemicals.

Polyimide Resins—A new group of resins recently introduced in the United States. The material is an aromatic polyimide made by reacting pyromellitic dianhydride with aromatic diamines. The polymer is characterized by the fact that is has rings of four carbon atoms tightly bound together. This resin has greater resistance to heat than any other unfilled organic material yet discovered.

Polyisobutylene—The polymerization product of isobutylene. It varies in consistency from a viscous liquid to a rubber-like solid with corresponding variation in molecular weight from 1000 to 400,000.

Polyliner—A perforated longitudinally ribbed sleeve that fits inside the cylinder of an injection molding machine; used as a replacement for conventional injection cylinder torpedos.

Polymer—A high-molecular-weight organic compound, natural or synthetic, whose structure can be represented by a repeated small unit, the *mer*; e.g., polyethylene, rubber, cellulose. Synthetic polymers are formed by addition or condensation polymerization of monomers. If two or more monomers are involved, a copolymer is obtained. Some polymers are eslastomers, some plastics.

Polymerization—A chemical reaction in which the molecules of a monomer are linked together to form large molecules whose molecular weight is a multiple of that of the original substance. When two or more monomers are involved, the process is called copolymerization or heteropolymerization. See also *Degree of, Condensation,* and *Polymer*.

Polymethyl Methacrylate—A thermoplastic material composed of polymers of methyl methacrylate. It is a transparent solid with exceptional optical properties and good resistance to water. It is obtainable in the form of sheets, granules, solutions, and emulsions. It is extensively used for aircraft domes, lighting fixtures, decorative articles, etc.; it is also used in optical instruments and surgical appliances. Lucite and Plexiglas are typical brands.

Polyphenylene Oxide—Presently made commerically as a polyether of 2, 6-dimethyl-phenol via an oxidative coupling process by means of air or pure oxygen in the presence of a copper-amine complex catalyst. These resins have a useful temperature range from less than $-275°F$ to $375°F$ with intermittent use up to $400°F$ possible.

Polyphenylene Sulfide (PPS)—A high heat deflection thermoplastic with excellent chemical resistance.

Poly-P-Xylene—See *Parylene*.

Polypropylene—A tough, lightweight rigid plastics made by the polymerization of high-purity propylene gas in the presence of an organo-metallic catalyst at relatively low pressures and temperatures.

Polystyrene—A water-white thermoplastic produced by the polymerization of styrene (vinyl benzene). The electrical insulating properties of polystyrene are outstandingly good and the material is relatively unaffected by moisture. In particular the power loss factor is extremely low over the frequency range 10^3–10^8 c.p.s.

Polysulfone—A chemical linkage of isopropylidene, ether and sulfone provides this thermoplastic material with exceptional high temperature and low creep properties. It has arc resistance, good electricals, is self extinguishing and may be molded and extruded.

Polyterephthalate (PTMT)—A thermoplastic with a balance of properties comparable with acetal and nylon resins.

Polyterpene Resins—Thermoplastic resins obtained by the polymerization of turpentine in the presence of catalysts. These resins are used in the manufacture of adhesives, coatings, and varnishes, and in food packaging. They are compatible with waxes, natural and synthetic rubbers, and polyethylene.

Polytetrafluoroethylene (PTFE) Resins—Members of the fluorocarbons, q.v., family of plastics made by the polymerization of tetrafluoroethylene. PTFE is characterized by its extreme inertness to chemicals, very high thermal stability and low frictional properties. Among the applications for these materials are bearings, fuel hoses, gaskets and tapes, and coatings for metal and fabric.

Polyurethane Resins—A family of resins produced by reacting diisocyanate with organic compounds containing two or more active hydrogens to form polymers having free isocyanate groups. These groups, under the influence of heat or certain catalysts, will react with each other, or with water, glycols, etc., to form a thermosetting material.

Polyvinyl Acetal—A member of the family of vinyl plastics, q.v., polyvinyl acetal is the general name for resins produced from a condensation of polyvinyl alcohol with an aldehyde. There are three main groups: polyvinyl acetal itself; polyvinyl butyral, and polyvinyl formal, q.v. Polyvinyl acetal resins are thermoplastics which can be processed by casting, extruding, molding and coating, but their main uses are in adhesives, lacquers, coatings and films.

Polyvinyl Acetate—A thermoplastic material composed of polymers of vinyl acetate in the form of a colorless solid. It is obtainable in the form of granules, solutions, latices, and pastes, and is used extensively in adhesives, for paper and fabric coatings, and in bases for inks and lacquers.

Polyvinyl Alcohol—A thermoplastic material composed of polymers of the hypothetical vinyl alcohol. Usually a colorless solid, insoluble in most organic solvents and oils, but soluble in water when the content of hydroxy groups in the polymer is sufficiently high.

The product is normally granular. It is obtained by the partial hydrolysis or by the complete hydrolysis of polyvinyl esters, usually by the complete hydrolysis of polyvinyl acetate. It is mainly used for adhesives and coatings.

Polyvinyl Butyral—A thermoplastic material derived from a polyvinyl ester in which some or all of the acid groups have been replaced by hydroxyl groups and some or all of these hydroxyl groups replaced by butyral groups by reaction with butyraldehyde. It is a colorless flexible tough solid.

It is used primarily in interlayers for laminated safety glass.

Polyvinyl Carbazole—A thermoplastic resin, brown in color, obtained by reacting acetylene with carbazole. The resin has excellent electrical properties and good heat and chemical resistance. It is used as an impregnant for paper capacitors.

Polyvinyl Chloride (PVC)—A thermoplastic material composed of polymers of vinyl chloride; a colorless solid with outstanding resistance to water, alcohols, and concentrated acids and alkalies. It is obtainable in the form of granules, solutions, latices, and pastes. Compounded with plasticizers it yields a flexible material superior to rubber in aging properties. It is widely used for cable and wire coverings, in chemical plants, and in the manufacture of protective garments.

Polyvinyl Chloride Acetate—A thermoplastic material composed of copolymers of vinyl chloride and vinyl acetate; a colorless solid with good resistance to water, and concentrated acids and alkalies.

It is obtainable in the form of granules, solutions, and emulsions. Compounded with plasticizers it yields a flexible material superior to rubber in aging properties. It is widely used for cable and wire coverings, in chemical plants, and in protective garments.

Polyvinyl Formal—One of the groups of polyvinyl acetal resins, q.v., made by the condensation of formaldehyde in the presence of polyvinyl alcohol. It is used mainly in combination with cresylic phenolics, for wire coatings and for impregnations, but can also be molded, extruded or cast. It is resistant to greases and oils.

Polyvinylidene Chloride—A thermoplastic material composed of polymers of vinylidene chloride (1, 1-dichloroethylene). It is a white powder with softening temperature at $185–200°C$. The material is also supplied as a copolymer with acrylonitrile or vinyl chloride, giving products which range from the soft flexible type to the rigid type. Also known as saran.

Polyvinylidene Fluoride Resins—This recent member of the fluorocarbons,

q.v. family of plastics is a homo-
polymer of vinylidene fluoride. It is
supplied as powders and pellets for
molding and extrusion and in solu-
tion form for casting. The resin has
good tensile and compressive strength
and high impact strength. Among
anticipated applications are chemical
equipment such as gaskets, impellers
and other pump parts, and packaging
uses such as drum linings and pro-
tective coatings.

Porous Molds—Molds which are made
up of bonded or fused aggregate
(powdered metal, coarse pellets, etc.)
in such a manner that the resulting
mass contains numerous open inter-
stices of regular or irregular size
through which either air or liquids
may pass through the mass of the
mold.

Positive Mold—A mold designed to
trap all the molding material when it
closes.

Postforming—The forming, bending, or
shaping of fully cured, C-stage ther-
moset laminates that have been
heated to make them flexible. On
cooling, the formed laminate retains
the contours and shape of the mold
over which it has been formed.

Pot Life—See *Working Life*.

Potting—Similar to *Encapsulating*
(q.v.), except that steps are taken to
insure complete penetration of all
the voids in the object before the
resin polymerizes.

Pour Out Finish—A container finish
with an undercut below the top,
designed to facilitate pouring without
dripping.

Powder Molding—General term used to
denote several techniques for produc-
ing objects of varying sizes and shapes

by melting polyethylene powder,
usually against the inside of a mold.
The techniques vary as to whether
the molds are stationary (e.g., as in
variations on slush molding tech-
niques) or rotating (e.g., as in varia-
tions on rotational molding).

Power Factor—In a perfect condenser,
the current leads the voltage by 90°.
When a loss takes place in the insula-
tion, the absorbed current, which
produces heat, throws the 90° rela-
tionship out according to the propor-
tion of current absorbed by the
dielectric. The power factor is the
cosine of the angle between voltage
applied and the current resulting.
Measurements are usually made at
million-cycle frequencies.

P-oxybenzol—Thermosetting high tem-
perature polymer fabricated by pres-
sure sintering.

Preform—(*n.*) A compressed tablet or
biscuit of plastic composition used
for efficiency in handling and accu-
racy in weighing materials. (*v.*)—To
make plastic molding powder into
pellets or tablets.

Preheating—The heating of a compound
prior to molding or casting in order
to facilitate the operation or to re-
duce the molding cycle.

Preheat Roll—In extrusion coating, a
heated roll installed between the
pressure roll and unwind roll whose
purpose is to heat the substrate be-
fore it is coated.

Preimpregnation—The practice of mix-
ing resin and reinforcement before
shipping it to the molder.

Premix—In reinforced plastics molding,
the material made by "do-it-your-
selfers," molders, or end-users who
purchase polyester or phenolic resin,
reinforcement, fillers, etc., separately

and mix the reinforced molding compounds on their own premises.

Preplastication—Technique of premelting injection molding powders in a separate chamber, then transferring the melt to the injection cylinder. Device used for preplastication is commonly known as a preplasticizer. See *Plasticate.*

Preplasticize (Preplasticizer)—See *Preplastication.*

Prepolymer Molding—In the urethane foam field, indicates a system whereby a portion of the polyol is pre-reacted with the isocyanate to form a liquid propolymer with a viscosity range suitable for pumping or metering. This component is supplied to end-users with a second premixed blend of additional polyol, catalyst, blowing agent, etc. When the two components are mixed together, foaming occurs. (See *One-Shot Molding.*)

Prepreg—A term generally used in reinforced plastics to mean the reinforcing material containing or combined with the full complement of resin before molding.

Preprinting—In sheet thermoforming, the distorted printing of sheets before they are formed. During forming the print assumes its proper proportions.

Press Polish—A finish for sheet stock produced by contact, under heat and pressure, with a very smooth metal which gives the plastics a high sheen.

Pressure Forming—A thermoforming process wherein pressure is used to push the sheet to be formed against the mold surface as opposed to using a vacuum to suck the sheet flat against the mold.

Pressure Pads—Reinforcements of hardened steel distributed around the dead areas in the faces of a mold to help the land absorb the final pressure of closing without collapsing.

Pressure Roll—In extrusion coating, the roll which with the chill roll applies pressure to the substrate and the molten extruded web.

Pressure Sensitive Adhesive—An adhesive which develops maximum bonding power by applying only a light pressure.

Primary Plasticizer—Has sufficient affinity to the polymer or resin so that it is considered compatible and therefore it may be used as the sole plasticizer.

Printed Circuit—An electrical or electronic circuit produced mainly from copper clad laminates.

Printing of Plastics—Methods of printing plastics materials, particularly thermoplastic film and sheet, have developed side by side with the growth of usage of the materials, and are today an important part of finishing techniques. Basically, the printing processes used are the same as in other industries, but the adaptation of machinery and development of special inks have been a constant necessity, particularly as new plastics materials have arrived, each with its own problems of surface decoration. Among the printing processes commonly used are gravure, flexographic, inlay (or valley) and silk screen.

Profile Die—An extrusion die for the production of continuous shapes, excepting tubes and sheets, is called a profile die.

Programming—The extrusion of a parison which differs in thickness in the length direction in order to equalize wall thickness of the blown container. It can be done with a pneumatic or hydraulic device which activates the mandrel shaft and adjusts the mandrel position during parison extrusion (parison programmer, controller, or variator). It can also be done by varying extrusion speed on accumulator-type blow molding machines.

Projected Area—The total parting line area of all moldings, runners, sprues, vents or culls in a mold.

Promotor—A chemical, itself a feeble catalyst, that greatly increases the activity of a given catalyst.

Prototype Mold—A simplified mold construction often made from a light metal casting alloy or from an epoxy resin in order to obtain information for the final mold and/or part design.

Pulp—A form of cellulose obtained from wood or other vegetable matter by prolonged cooking with chemicals.

Pulp Molding—Process by which a resin-impregnated pulp material is preformed by application of a vacuum and subsequently oven cured or molded.

Purge Time—Time allowance used in estimating to record the purging cost.

Purging—Cleaning one color or type of material from the cylinder of an injection molding machine or extruder by forcing it out with the new color or material to be used in subsequent production. Purging materials are also available.

Push Up—The bottom contour of a plastics container designed in such a manner as to allow an even bearing surface on the outside edge and to prevent the bottle from rocking.

PVC—See *Polyvinyl Chloride.*

Pyrrones—A combination of polyimides and polybenzimidazoles which forms a 900°F material.

Q

Quench (Thermoplastics)—A process of shock cooling thermoplastic materials from the molten state.

Quench Bath—The cooling medium used to quench molten thermoplastic materials to the solid state.

Quench-Tank Extrusion—The extruded film is cooled in a quench-water bath.

R

Ra—The average deviation of a surface from a mean line through the surface texture as determined arithmetically.

Radio Frequency (R.F.) Preheating—A method of preheating used for molding materials to facilitate the molding operation or reduce the molding cycle. The frequencies most commonly used are between 10 and 100 Mc/sec.

Radio Frequency Welding—A method of welding thermoplastics using a radio frequency field to apply the necessary heat. Also known as high frequency welding.

Ram Travel—The distance the ram moves in filling the mold, in either injection or transfer welding.

Rayon—The generic term for fibers, staple, and continuous filament yarns composed of regenerated cellulose, q.v., but also frequently used to describe fibers obtained from cellulose acetate or cellulose triacetate. Rayon fibers are similar in chemical structure to natural cellulose fibers (e.g., cotton) except that the synthetic fiber contains shorter polymer units. Most rayon is made by the viscose process.

Ream—Usually 500 sheets, 24″ X 36″ of industrial paper. Sometimes expressed as 3000 sq. ft.

Recessed Panel—A container design in which the flat area for labeling is indented or recessed. Also see *Label Panel.*

Reciprocating Screw—An extruder system in which the screw when rotating is pushed backwards by the molten polymer which collects in front of the screw. When sufficient material has been collected, the screw moves forward and forces the material through the head and die at a high speed.

Reciproscrew®—An extrusion screw used for plasticizing the incoming melt and so designed to serve also as a plunger to push the melt into the mold.

Rectangular—A particular shape of a container which has right angles with adjacent sides of unequal dimensions.

Recycle—Ground material from flash and trimmings which after mixing with a certain amount of virgin material is fed back into the molding machine.

Reentrant Mold—A reentrant blow mold is one where the waist of the bottle forms an undercut in the cavity halves.

Regenerated Cellulose (Cellophane)—A transparent cellulose plastics material made by mixing cellulose xanthate with a dilute sodium hydroxide solution to form a viscose. Regeneration is carried out by extruding the viscose, in sheet form, into an acid bath to create regenerated cellulose. The material is very widely used as a packaging and overwrapping material of exceptional clarity. The film also has good electrical properties and is resistant to oils and greases. Included among recent application is the use of the material as a release agent in reinforced plastics moldings.

Reinforced Molding Compound—Compound supplied by raw material produced in the form of ready-to-use materials; as distinguished from pre-mix (q.v.).

Reinforcement—A strong inert material bound into plastics to improve its strength, stiffness, and impact resistance. Reinforcements are usually long fibers of glass, sisal, cotton, etc.—in woven or nonwoven form. To be effective, the reinforcing material must form a strong adhesive bond with the resin.

Relative Humidity—Ratio of the quantity of water vapor present in the air to the quantity which would saturate it at any given temperature.

Relative Viscosity—The relative viscosity of a polymer in solution is the ratio of the absolute viscosities of the solution (of stated concentration) and of the pure solvent at the same temperature.

$$n_r = \frac{n}{n_0}$$

where n_r = relative viscosity, n = absolute viscosity of polymer solution, n_0 = absolute viscosity of pure solvent.

Release Agent—See *Parting Agent.*

Relief Angle—The angle of the cutaway portion of the pinch-off blade measured from a line parallel to the pinch-off land.

Resiliency—Ability to quickly regain an original shape after being strained or distorted.

Resin—Any of a class of solid or semi-solid organic products of natural or synthetic origin, generally of high molecular weight with no definite melting point. Most resins are polymers (q.v.).

Resinoid—Any of the class of thermosetting synthetic resins, either in their initial temporarily fusible state or in their final infusible state. Compare with *Thermosetting*.

Resin Pocket—An apparent accumulation of excess resin in a small, localized section visible on cut edges of molded surfaces.

Resistivity—The ability of a material to resist passage of electrical current either through its bulk or on a surface. The unit of volume resistivity is the ohm-cm., of surface resistivity, the ohm.

Restricted Gate—A very small orifice between runner and cavity in an injection or transfer mold. When the piece is ejected, this gate breaks cleanly, simplifying separation of runner from piece.

Restrictor Ring—A ring-shaped part protruding from the torpedo surface which provides increase of pressure in the mold to improve, e.g., welding of two streams.

Retainer Plate—The plate on which demountable pieces, such as mold cavities, ejector pins, guide pins, and bushings are mounted during molding; usually drilled for steam or water.

Retarder—See *Inhibitor*.

Reverse-Roll Coating—The coating is premetered between rolls and then wiped off on the web. The amount of coating is controlled by the metering gap and also by the speed of rotation of the coating roll.

Rib—A reinforcing member of a fabricated or molded part.

Rigid PVC—Polyvinyl chloride or a polyvinyl chloride/acetate copolymer characterized by a relatively high degree of hardness; it may be formulated with or without a small percentage of plasticizer.

Rigid Resin—One having a modulus high enough to be of practical importance, e.g., 10,000 psi or greater.

Ring Gate—An annular ring at the point of material entry serves as a gate or runner on the end of thin wall round sections.

RMS—An obsolete measure of flatness now replaced by Ra.

Rocker—A plastic container with a bulged or deformed bottom, causing rocking of the container in the upright position.

Rockwell Hardness—A common method of testing a plastics material for resistance to indentation in which a diamond or steel ball, under pressure, is used to pierce the test specimen. The load used is expressed in kilograms and a 10-kilogram weight is first applied and the degree of penetration noted. The so-called major load (60 to 150 kilograms) is next applied and a second reading obtained. The hardness is then calculated as the difference between the two loads and expressed with nine different prefix letters to denote the type of penetrator used and the weight applied as the major load.

Roller Coating—Used for applying paints to raised designed or letters.

Roll Mill—Two rolls placed in close relationship to one another used to admix a plastics material with other substances. The rolls turn at different speeds to produce a shearing action to the materials being compounded.

Rosin—A resin obtained as a residue in the distillation of crude turpentine from the sap of the pine tree (gum rosin) or from an extract of the stumps and other parts of the tree (wood rosin).

Rotating Spreader—A type of injection torpedo which consists of a finned torpedo which is rotated by a shaft extending through a tubular cross-section injection ram behind it.

Rotational Casting (or molding)—A method used to make hollow articles from plastisols and latices. Plastisol is charged into hollow mold capable of being rotated in one or two planes. The hot mold fuses the plastisol into a gel after the rotation has caused it to cover all surfaces. The mold is then chilled and the product stripped out.

Round Square—Particular shape of a container which has sides of equal width with well-rounded corners and shoulders.

Roving—A form of fibrous glass in which spun strands are woven into a tubular rope. The number of strands is variable but 60 is usual. Chopped roving is commonly used in preforming.

RTP—Glass reinforced thermoplastics.

Rubber—An elastomer capable of rapid elastic recovery after being stretched to at least twice its length at temperatures from 0 to 150°F at any humidity. Specifically, Hevea or natural rubber, the standard of comparison for elastomers.

Runner—In an injection or transfer mold, the channel, usually circular, that connects the spure with the gate to the cavity.

S

Sag—The extension locally (often near the die face) of the parison during extrusion by gravitational forces. This causes necking-down of the parison. Also refers to the flow of a molten sheet in a thermoforming operation.

SAN—Styrene acrylonitrile thermoplastic copolymer with good stiffness, scratch chemical and stress crack resistance.

Sandwich Constructions—Panels composed of a lightweight core material—honeycomb, foamed plastic, etc., q.v.,—to which two relatively thin, dense, high strength faces or skins are adhered.

Sandwich Heating—A method of heating a thermoplastic sheet prior to forming which consists of heating both sides of the sheet simultaneously.

Sandwich Molding—A molding process where two different materials are injected consecutively into a mold cavity to produce products having surfaces of one plastics with desirable characteristics and with a core of another material with its desired characteristics.

Saturated Compounds—Organic compounds which do not contain double or triple bonds and thus cannot add on elements or compounds.

Scar—A characteristic mark on plastics containers which is confined mostly to the bottom. It is caused by the pinch-off operation and is often referred to as the length of the pinch-off.

Scrap—Any product of a molding operation that is not part of the primary product. In thermosetting molding, this includes flash, culls, runners, and is not reusable as a molding compound. Injection molding and extrusion scrap (runners, rejected parts, sprues, etc.) can usually be reground and remolded.

Screw Plasticating Injection Molding—A technique in which the plastic is converted from pellets to a viscous melt by means of an exturder screw which is an integral part of the molding machine. Machines are either single stage (in which plastication and injection are done by the same cylinder) or double stage in which the material is plasticated in one cylinder and then fed to a second for injection into a mold.

Sealing Plane—The plane on the inside of a bottle cap along the sealing surface.

Sealing Surface—The surface of the finish of the container on which the closure forms the seal.

Secondary Plasticizer (or Extender Plasticizer)—Has insufficient affinity for the resin to be compatible as the sole plasticizer and must be blended with a primary plasticizer. The secondary acts as a diluent with respect to the primary and the primary-secondary blend has less affinity for the resin than does the primary alone.

Segregation—A close succession of parallel, relatively narrow and sharply defined, wavy lines of color on the surface of plastics which differ in shade from surrounding areas, and create the impression that the components have separated.

Self-Extinguishing—A somewhat loosely-used term describing the ability of a material to cease burning once the source of flame has been removed.

Semi-Automatic Molding Machine—A modeling machine in which only part of the operation is controlled by the direct action of a human. The automatic part of the operation is controlled by the machine according to a predetermined program.

Semipositive Mold—A mold which allows a small amount of excess material to escape when it is closed. See *Flash Mold, Positive Mold.*

Set—To convert a liquid resin or adhesive into a solid state by curing (q.v.) or by evaporation of solvent or suspending medium or by gelling.

Setting Temperature—The temperature to which a liquid resin, an adhesive, or products or assemblies involving either is subjected to set the resin or adhesive.

Setting Time—The period of time during which a molded or extruded product, an assembly, etc., is subjected to heat and/or pressure to set the resin or adhesive.

Shark Skin—A surface irregularity in the form of finely-spaced sharp ridges caused by a relaxation effect of the melt at the die exit.

Shear Rate—The overall velocity over the cross section of a channel with which molten polymer layers are gliding along each other or along the wall in laminar flow.

$$\frac{\text{shear}}{\text{rate}} = \frac{\text{velocity}}{\text{clearance}} = \frac{\text{cm/sec.}}{\text{cm}} = \frac{1}{\text{sec.}}$$

Shear Strength—a) The ability of a material to withstand shear stress. b) The stress at which a material fails in shear.

Shear Stress—The stress developing in a polymer melt when the layers in a cross section are gliding along each other or along the wall of the channel (in laminar flow).

$$\text{shear stress} = \frac{\text{force}}{\text{area sheared}} = \text{psi}$$

Sheet (Thermoplastic)—A flat section of a thermoplastic resin with the length considerably greater than the width and 10 mils or greater in thickness.

Sheeter Lines—Parallel scratches or projecting ridges distributed over a considerable area of a plastic sheet.

Sheet Molding Compound (SMC)—A thermoplastic copolymer added to a polyester, reinforced with glass fibers and used in sheet form as a molding compound.

Sheet Train—The entire assembly necessary to produce sheet which includes extruder, die, polish rolls, conveyor, draw rolls, cutter and stacker.

Shelf Life—See *Storage Life.*

Shore Hardness—A method of determining the hardness of a plastics material using a scleroscope. This device consists of a small conical hammer fitted with a diamond point and acting in a glass tube. The hammer is made to strike the material under test and the degree of rebound is noted on a graduated scale. Generally, the harder the material the greater will be the rebound.

Short or Short Shot—In molding, failure to fill the mold completely.

Shot—The yield from one complete molding cycle, including scrap.

Shot Capacity—The maximum weight of material which an accumulator can push out with one forward stroke of the ram.

Shoulder—The wide portion of a bottle below the finish.

Shrinkage—Contraction of a container upon cooling.

Shrink Fixture—See *Cooling Fixture.*

Shrink Wrapping—A technique of packaging in which the strains in a plastics film are released by raising the temperature of the film thus causing it to shrink over the package. These shrink characteristics are built into the film during its manufacture by stretching it under controlled temperatures to produce orientation, q.v., of the molecules. Upon cooling, the film retains its stretched condition, but reverts toward its original dimensions when it is heated. Shrink film gives good protection to the products packaged and has excellent clarity.

Shuttle Press—Molding machines with a reciprocating platen to permit alternate injection into two molds.

Siamese Blow—A colloquial term applied to the technique of blowing two or more parts of a product in a single blow and then cutting them apart.

Side Bars—Loose pieces used to carry one or more molding pins, and operated from outside the mold.

Side Draw Pins—Projections used to core a hole in a direction other than the line of closing of a mold, and which must be withdrawn before the part is ejected from the mold.

Silicone—One of the family of polymeric materials in which the recurring chemical group contains silicon and oxygen atoms as links in the main chain. At present these compounds are derived from silica (sand) and methyl chloride. The various forms obtainable are characterized by their resistance to heat.

Silicones are used in the following applications: *a*) Greases for lubrication. *b*) Rubber-like sheeting for gaskets, etc. *c*) Heat-stable fluids and compounds for waterproofing, insulating, etc. *d*) Thermosetting insulating varnishes and resins for both coating and laminating.

Silk Screen Printing (Screen process decorating)—This printing method, in its basic form, involves laying a pattern of an insoluble material, in outline, on a finely woven fabric, so that when ink is drawn across it, it is able to pass through the screen only in the desired areas.

Single Cavity Mold—A mold having only one cavity in the body of the mold, as opposed to a multiple cavity mold or family mold which have numerous cavities.

Sink Mark—A shallow depression or dimple on the surface of an injection molded part due to collapsing of the surface following local internal shrinkage after the gate seals. May also be an incipient short shot.

Sinking a Mold—See *Hobbing*.

Sintering—In forming articles from fusible powders, e.g., nylon, the process of holding the pressed-powder article at a temperature just below its melting point for about ½ hour. Particles are fused (sintered) together, but the mass, as a whole, does not melt.

Sizing (*n*.)—The process of applying a material to a surface to fill pores and thus reduce the absorption of the subsequently applied adhesive or coating or to otherwise modify the surface. Also, the surface treatment applied to glass fibers used in reinforced plastics. The material used is sometimes called *Size*.

Slip Additive—A modifer that acts as an internal lubricant which exudes to the surface of the plastics during and immediately after processing. In other words, a nonvisible coating blooms to the surface to provide the necessary lubricity to reduce coefficient of friction and thereby improve slip characteristics.

Slip Forming—Sheet forming technique in which some of the plastics sheet material is allowed to slip through the mechanically operated clamping rings during a stretch-forming operation.

Slip-Plane—Plane within transparent material visible in reflected light, due to poor welding and shrinkage on cooling.

Slot-Extrusion—A method of extruding film sheet in which the molten thermoplastic compound is forced through a straight slot.

Slurry Preforming—Method of preparing reinforced plastics preforms by wet processing techniques similar to those used in the pulp molding (q.v.) industry.

Slush Molding—Method for casting thermoplastics, in which the resin in liquid form is poured into a hot mold where a viscous skin forms. The excess slush is drained off, the mold is cooled, and the molding stripped out.

Snap-Back Forming—Sheet forming technique in which an extended heated plastic sheet is allowed to contract over a male form shaped to the desired contours.

Solid Phase Forming—See *Forging*.

Solid State Control—Electronic devices that will measure and compensate for all variables in the processing of plastics.

Solution—Homogeneous mixture of two or more components, such as a gas dissolved in a gas or liquid, or a solid in a liquid.

Solvent—Any substance, usually a liquid, which dissolves other substances.

Solvent Molding—Process for forming thermoplastic articles by dipping a male mold in a solution or dispersion of the resin and drawing off the solvent to leave a layer of plastics film adhering to the mold.

Spacer Cable—A system of 2-15kv primary power distribution in which three partially insulated or covered phase wires and a high-strength messenger-ground wire are mounted in plastics or ceramic insulating spacers.

Spanishing—A method of depositing ink in the valleys of embossed plastics film.

SPE—Society of Plastics Engineers.

Specific Gravity—The density (mass per unit volume) of any material divided by that of water at a standard temperature, usually 4°C. Since water's density is nearly 1.00 g./cc., density in g./cc. and specific gravity are numerically nearly equal.

Specific Heat—The amount of heat required to raise a specified mass by one unit of a specified temperature.

Specific Viscosity—The specific viscosity of a polymer is the relative viscosity of a solution of known concentration of the polymer minus one. It is usually determined for a low concentration of the polymer (0.5 g. per 100 ml. of solution or less).

$$n^{SP} = \frac{n - n^0}{n^0} = n^r - 1,$$

where n^{SP} = specific viscosity, n^r = relative viscosity.

SPI—Society of the Plastics Industry.

SPI Tolerances—A presentation of workable tolerance values that can be achieved in molding various materials.

Spider—1) In a molding press, that part of an ejector mechanism which operates the ejector pins. 2) In extrusion, a term used to denote the membranes supporting a mandrel within the head/die assembly.

Spider or Spoke Gate—Round parts with a central opening are often gated from runners extending as spokes of a wheel from the central sprue.

Spider Lines—Vertical marks on the parison (container) caused by improper welding of several melt flow fronts formed by the legs with which the torpedo is fixed in the extruder head.

Spinneret—A type of extrusion die, i.e., a metal plate with many tiny holes, through which a plastic melt is forced to make fine fibers and filaments. Filaments may be hardened by cooling in air, water, etc., or by chemical action.

Spinning—Process of making fibers by forcing plastic melt through spinneret.

Spin Welding—A process of fusing two objects together by forcing them together while one of the pair is spinning, until frictional heat melts the interface. Spinning is then stopped and pressure held until they are frozen together.

Spiral Flow Test—A method for determining the flow properties of a thermoplastic resin in which the resin flows along the path of a spiral cavity. The length of the material which flows into the cavity and its weight gives a relative indication of the flow properties of the resin.

Spiral Mold Cooling—A method of cooling injection molds or similar molds wherein the cooling medium flows through a spiral cavity in the body of the mold. In injection molds, the cooling medium is introduced at the center of the spiral, near the sprue section, as more heat is localized in this section.

Split-Ring Mold—A mold in which a split cavity block is assembled in a chase to permit the forming of undercuts in a molded piece. These parts are ejected from the mold and then separated from the piece.

Spray Coating—Usually accomplished on continuous webs by a set of reciprocating spray nozzles traveling laterally across the web as it moves.

Sprayed Metal Molds—Mold made by spraying molten metal onto a master until a shell of predetermined thickness is achieved. Shell is then removed and backed up with plaster, cement, casting resin, or other suitable material. Used primarily as a mold in sheet-forming processes.

Spray-up—Covers a number of techniques in which a spray gun is used as the processing tool. In reinforced plastics, for example, fibrous glass and resin can be simultaneously deposited in a mold. In essence, roving is fed through a chopper and ejected into a resin stream which is directed at the mold by either of two spray systems. In foamed plastics, very fast-reacting urethane foams or epoxy foams are fed in liquid streams to the gun and sprayed

on the surface. On contact, the liquid starts to foam.

Spreader—A streamlined metal block placed in the path of flow of the plastics material in the heating cylinder of extruders and injection molding machines to spread it into thin layers, thus forcing it into intimate contact with the heating areas.

Spring Box Mold—This compression mold type has a spacing fork that is removed after partial compression to prevent loss of bottom loaded inserts or fine mold details.

Sprue—Feed opening provided in the injection or transfer mold; also the slug formed at this hole. *Spur* is a shop term for the sprue slug.

Sprue Bushing—A hardened steel insert in an injection mold which contains the tapered sprue hole and has a suitable seat for the nozzle of the injection cylinder. Sometimes called an *Adapter*.

Sprue Gate—A passageway through which molten resin flows from the nozzle to the mold cavity.

Sprue Lock—In injection molding, a portion of the plastic composition which is held in the cold slug well by an undercut; used to pull the sprue out of the bushing as the mold is opened. The sprue lock itself is pushed out of the mold by an ejector pin. When the undercut occurs on the cavity block retainer plate, this pin is called the *Sprue Ejector Pin*.

Sprueless Mold—Sprueless runner systems are designed with the nozzle extending as close as possible to the main runner.

Square Box Mold—Expanded styrene molds often use a common steam chamber or square box.

Stablizer—An ingredient used in the formulation of some plastics, especially elastomers, to assist in maintaining the physical and chemical properties of the compounded materials at their initial values throughout the processing and service life of the material.

Stampable Thermoplastics —Nylon-based, mineral-filled thermoplastics suitable for fabrication in metal stamping or compression presses.

Staple—Refers to textile fibers of a short length, usually ½ to 3″, for natural fibers and sometimes larger for synthetics.

Stationary Platen—The plate of a molding machine to which the front plate of the mold is secured during operation. This platen does not move during normal operation.

Steam Molding (expandable polystyrene)—Used to mold parts from pre-expanded beads of polystyrene using steam as a source of heat to expand the blowing agent in the material. The steam in most cases is contacted intimately with the beads directly or may be used indirectly to heat mold surfaces which are in contact with the beads.

Steam Plate—Mounting plate for molds, cored for circulation of steam.

Stereospecific Plastics—Implies a specific or definite order of arrangement of molecules in space. This ordered regularity of the molecules in contrast to the branched or random arrangement found in other plastics permits close packing of the molecules and leads to high *crystallinity* (i.e., as in polypropylene).

Stitching—The progressive welding of thermoplastic materials by successive applications of two small mechanically operated electrodes, connected to the output terminals of a radio frequency generator, using a mechanism similar to that of a normal sewing machine.

Storage Life—The period of time during which a liquid resin or packaged adhesive can be stored under specified temperature conditions and remain suitable for use. Storage life is sometimes called *Shelf Life.*

Straight-Sided Round—A round bottle with straight side walls from shoulder to base.

Stretch Forming—A plastics sheet forming technique in which the heated thermoplastic sheet is stretched over a mold and subsequently cooled.

Striation—Rippling of thick parisons, caused by a local orientation effect in the melt by the spider legs.

Stripper-Plate—A plate that strips a molded piece from core pins or force plugs. The stripper-plate is set into operation by the opening of the mold.

Structural Foam—This product has a rigid cellular core and a solid integral skin.

Subcavity Gang Mold—A group of cavities located in a common loading well.

Submarine Gate—A type of edge gate where the opening from the runner into the mold is located below the parting line or mold surface as opposed to conventional edge gating where the opening is machined into the surface of the mold. With submarine gates, the item is broken from the runner system on ejection from the mold. See *Tunnel Gate.*

Surface Resistivity—The electrical resistance between opposite edges of a unit square of insulating material. It is commonly expressed in ohms. (Also covered in ASTM D257-54T.)

Surface Treating—Any method of treating a polyolefin so as to alter the surface and render it receptive to inks, paints, lacquers, and adhesives such as chemical, flame, and electronic treating.

Surging—Unstable pressure build-up in an extruder leading to variable throughput and waviness of the parison.

Suspension—A mixture of fine particles of any solid with a liquid or gas. The particles are called the *disperse phase*, the suspending medium is called the *continuous phase*.

Sweating—Exudation of small drops of liquid, usually a plasticizer or softener, on the surface of a plastic part.

Syndiotactic—A chain of molecules in which the methyl groups alternate regularly on opposite sides of the chain.

Synergism—A term used to describe the use of two or more stabilizers in an organic material where the combination of such stabilizers improves the stability to a great extent than could be expected from the additive effect of each stabilizer.

T

Tab Gated—A small removable tab of approximately the same thickness as the mold item, usually located perpendicular to the item. The tab is used as a site for edge gate location, usually on items with large flat areas.

Tack—Stickiness of an adhesive, measurable as the force required to separate an adherent from it by viscous or plastic flow of the adhesive.

Tack Range—The period of time in which an adhesive will remain in the tacky-dry condition after application to an adherent, under specified conditions of temperature and humidity.

Tapered Cylinder—Refers to a particular shape of a container in which the circular cross section at the top is smaller in diameter than that at the bottom, or vice versa.

Tapping—Cutting threads in the walls of a circular hole.

T-Die—A term used to denote a center-fed, slot extrusion die for film which in combination with the die adapter resembles an inverted T.

Tenacity (*gpd*)—The term generally used in yarn manufacture and textile engineering to denote the strength of a yarn or of a filament for its given size. Numerically it is the grams of breaking force per denier unit of yarn or filament size; grams per denier, gpd. The yarn is usually pulled at the rate of 12 in./min. Tenacity equals breaking strength (gms) divided by denier. (Tenacity, gpd) (Spec. Gravity) 12,800 = (Tensile Strength, psi).

Tensile Bar (Specimen)—A compression or injection molded specimen of specified dimensions which is used to determine the tensile properties of a material.

Tensile Strength—The pulling stress, in psi, required to break a given specimen. Area used in computing strength is usually the original, rather than the necked-down area.

Therimage—A trademark for a decorating process for plastic which transfers the image of a label or decoration to the object under the influence of heat and light pressure.

Thermal Conductivity—Ability of a material to conduct heat; physical constant for quantity of heat that passes through unit cube of a substance in unit of time when difference in temperature of two faces is 1°.

Thermal Expansion (Coefficient of)—The fractional change in length (sometimes volume, specified) of a material for a unit change in temperature. Values of plastics range from 0.01 to 0.2 mils/in., °C.

Thermal Stress Cracking (TSC)—Crazing and cracking of some thermoplastic resins which results from overexposure to elevated temperatures.

Thermoforming—Any process of forming thermoplastic sheet which consists of heating the sheet and pulling it down onto a mold surface.

Thermoforms—The product which results from a thermoforming operation.

Thermoplastic—Capable of being repeatedly softened by heat and hardened by cooling.—A material that will repeatedly soften when heated and hardened when cooled. Typical of the thermoplastics family are the styrene polymers and copolymers, acrylics, cellulosics, polyethylenes, vinyls, nylons, and the various fluorocarbon materials.

Thermoplastic Polyester—An injection moldable polyester compound.

Thermoplastic Rubber—A rubberlike polyolefin that can be processed like conventional thermoplastics.

Thermoset—A material that will undergo or has undergone a chemical reaction by the action of heat, catalysts, ultra-violet light, etc., leading to a relatively infusible state. Typical of the plastics in the thermo-setting family are the aminos (melamine and urea), most polyesters, alkyds, epoxies, and phenolics.

Thixotropic—Said of materials that are gel-like at rest but fluid when agitated. Liquids containing suspended solids are apt to be thixotropic. Thixotropy is desirable in paints.

Thread Contour—The shape or type of thread design as observed in a cross section, along the major axis, i.e., flat-headed, square, round, etc.

Thread Plug—A part of a mold that shapes an internal thread and must be unscrewed from the finished piece.

Three Plate Mold—A third or intermediate movable plate is often used in injection molds to permit center or offset gating of each cavity.

Tie Bars—Bars which provide structural rigidity to the clamping mechanism often used to guide platen movement.

Toggle Action—A mechanism which exerts pressure developed by the application of force on a knee joint. It is used as a method of closing presses and also serves to apply pressure at the same time.

Tolerance—A specified allowance for deviations in weighing, measuring, etc., or for deviations from the standard dimensions or weight.

Top Blow—A specific type of blow molding machine which forms hollow articles by injecting the blowing into the parison at the top of the mold.

Torpedo (or Spreader)—A streamlined metal block placed in the path of flow of the plastics materials in the heating cylinder of exturders and injection molding machines to spread it into thin layers, thus forcing it into intimate contact with heating areas.

Torsion—Stress caused by twisting a material.

Tracking—A phenomenon wherein a high voltage source current creates a leakage or fault path across the surface of an insulating material by slowly but steadily forming a carbonized path.

Transfer Molding—A method of molding thermosetting materials, in which the plastics is first softened by heat and pressure in a transfer chamber, then forced by high pressure through suitable sprues, runners, and gates into closed mold for final curing.

Translucent—Descriptive of a material or substance capable of transmitting some light, but not clear enough to be seen through.

Transparent—Descriptive of a material or substance capable of a high degree of light transmission (e.g., glass). Some polypropylene films and acrylic moldings are outstanding in this respect.

Trapped Air Process—For blowing closed objects, the bottom pinch is conventional and after blowing the parison, sliding pinchers close off the top, forming a sealed air inflatted product.

Tree Wire—A special type of power line wire designed for installation in wooded areas.

T-Type Die—See *Slot Extrusion.*

Tumbleblasting—A finishing procedure for thermosetting parts where the pieces are exposed to a blast of walnut shell granules while tumbling to remove flash.

Tumbling—Finishing operation for small plastics articles by which gates, flash, and fins are removed and/or surfaces are polished by rotating them in a barrel together with wooden pegs, sawdust, and polishing compounds.

Tunnel Gate—Is similar to vertical tab gating except that a round tapered plug gate is used.

Turning Table—A rotating table or wheel carrying various molds in a multi-mold single parison blow molding operation.

Two Level Mold—Placement of one cavity of a mold above another instead of side by side to reduce clamping force is often used for large area parts such as picture frames.

U

Ultimate Strength—Term used to describe the maximum unit stress a material will withstand when subjected to an applied load in a compression, tension, or shear test.

Ultrasonic Sealing—A film sealing method in which sealing is accomplished through the application of vibratory mechanical pressure at ultrasonic frequencies (20 to 40 kc.). Electrical energy is converted to ultrasonic vibrations through the use of either a magnetostrictive or piezoelectric transducer. The vibratory pressures at the film interface in the sealing area develop localized heat losses which melt the plastic surfaces effesting the seal.

Ultraviolet—Zone of invisible radiations beyond the violet end of the spectrum of visible radiations. Since UV wavelengths are shorter than the visible, their photons have more energy, enough to initiate some chemical reactions and to degrade most plastics.

Undercut—(*a.*) Having a protuberance or indentation that impedes withdrawal from a two-piece, rigid mold. Flexible materials can be ejected intact even with slight undercuts. (*n.*) Any such protuberance or indentation; depends also on design of mold.

Unicellular—With foamed plastics, each cell an isolated unit. Equals "closed-cell."

Unit Mold—A simple mold which comprises only a single cavity without further mold devices to be used for production of sample containers having a shape which is difficult to blow.

Unsaturated Compounds—Any compound having more than one bond between two adjacent atoms, usually carbon atoms, and capable of adding other atoms at that point to reduce it to a single bond.

Urea Formaldehyde Resin (Urea Resin) —A synthetic resin derived from the reaction of urea (carbamide) with formaldehyde or its polymers.

Urethane—See description under *Isocyanate Resins*.

UV—Ultraviolet.

UV Stabilizer (Ultraviolet)—Any chemical compound which, when admixed with thermoplastic resin, selectively absorbs UV rays.

V

"V" Bar Mold—Two mold halves shaped in the form of a V which clamp when pressed into a "V" opening are commonly used for injection and transfer molding.

Vacuum Forming—Method of sheet forming in which the plastics sheet is clamped in a stationary frame, heated, and drawn down by a vacuum into a mold. In a loose sense, it is sometimes used to refer to all sheet forming techniques, including *Drape Forming* (q.v.), involving the use of vacuum and stationary molds.

Vacuum Metalizing—Process in which surfaces are thinly coated with metal by exposing them to the vapor of metal that has been evaporated under vacuum (one millionth of normal atmospheric pressure).

Valley Printing—Ink is applied to the high points of an embossing roll and subsequently deposited in what becomes the valleys of the embossed plastic material.

Vehicle—The liquid medium in which pigments, etc., are dispersed in coatings such as paint, q.v., and which enable the coating to be applied.

Vent—In a mold, a shallow channel or minute hole cut in the cavity to allow air to escape as the material enters.

Venturi Dispersion Plug—A plate having an orifice with a conical relief drilled therein which is fitted in the nozzle of an injection molding machine to aid in the dispersion of colorants in a resin.

Vertical Flash Ring—The clearance between the force plug and the vertical wall of the cavity in a positive or semi-positive mold; also the ring of excess material which escapes from the cavity into this clearance space.

Vinyl Resin—A synthetic resin formed by the polymerization of chemical compounds containing the group $CH^2 = CH—$. In particular, polyvinyl chloride, acetate, alcohol, and butyral, are referred to. (Though most addition polymers are within the above definition, it is seldom applied to any but the ones listed.)

Vinylidene Chloride—One of the very inert thermoplastics with excellent drug and meat packaging properties. It offers low water vapor and low odor and food flavoring transmission.

Vinylidene Fluoride—A thermoplastic chemical and UV resistant material with a thermal range from $-80°F$ to $300°F$. It is non-drip self-extinguishing, may be molded by all processes and is the basis for 30 year paints.

Viscosity—Internal friction or resistance to flow of a liquid. The constant ratio of shearing stress to rate of shear. In liquids for which this ratio is a function of stress, the term "apparent viscosity" is defined as this ratio.

Volume Resistivity (Specific Insulation Resistance)—The electrical resistance between opposite faces of a 1-cm. cube of insulating material. It is measured under prescribed conditions using a direct current potential after a specified time of electrification. It is commonly expressed in ohm-centimeters. The recommended test is ASTM D257-54T.

Vulcanization—The chemical reaction which induces extensive changes in the physical properties of a rubber and which is brought about by reacting the rubber with sulphur and/or other suitable agents. The changes in physical properties include decreased plastic flow, reduced surface tackiness, increased elasticity, much greater tensile strength, and considerably less solubility. More recently, certain thermoplastics, e.g., polyethylene, have been formulated to be vulcanizable. Cross-linking is encouraged, thereby giving resistance to deformation or flow above the melting point.

W

Waist—The central portion of a container which has a smaller cross section than the adjacent areas.

Wait Time—Time allowance used in estimating when operator is idle and waiting for the machine.

Walk Time—Time allowance used in estimating to compensate the operator for moving between presses.

Warpage—Dimensional distortion in a plastics object after molding.

Weatherometer—An instrument which is utilized to subject articles to accelerated weathering conditions, e.g., rich UV source and water spray.

Web—A thin sheet in process in a machine. The molten web is that which issues from the die. The substrate web is the substrate being coated.

Weld Lines—A mark on a container caused by incomplete fusion of two streams of molten polymer. See *Spider Lines*.

Weld Mark (also Flow Line)—A mark on a molded plastics piece made by the meeting of two flow fronts during the molding operation.

Welding—Joining thermoplastic pieces by one of several heat-softening processes. In hot-gas welding, the material is heated by a jet of hot air or inert gas directed from a welding "torch" onto the area of contact of the surfaces which are being welded. Welding operations to which this method is applied normally require the use of a filler rod. In *spin-welding* (q.v.), the heat is generated by friction. Welding also includes heat sealing and the terms are synonymous in some foreign countries, including Britain.

Wet Strength—The strength of paper when saturated with water, especially used in discussions of processes whereby the strength of paper is increased by the addition, in manufacture, of plastics resins. Also, the strength of an adhesive joint determined immediately after removal from a liquid in which it has been immersed under specified conditions of time, temperature, and pressure.

Wheelabrating—See *Tumbleblasting.*

Window—A defect in a thermoplastics film, sheet or molding, caused by the incomplete "plasticization" of a piece of material during processing. It appears as a globule in an otherwise blended mass. See also *Fish Eye.*

Wiper—Automatic molds often use a wiper to remove molded parts and runner after each shot.

Wire Train—The entire assembly which is utilized to produce a resin-coated wire which normally consists of an extruder, a crosshead and a die, cooling means, and feed and take-up spools for the wire.

Wood Model—A model of a product made from wood to assist in the design of a container.

Working Life—The period of time during which a liquid resin or adhesive, after mixing with catalyst, solvent, or other compounding ingredients, remains usable.

Y

Yield Value (Yield Strength)—The lowest stress at which a material undergoes plastic deformation. Below this stress, the material is elastic; above it, viscous.

Young's Modulus of Elasticity—The modulus of elasticity in tension. The ratio of stress in a material subjected to deformation.

APPENDIX

APPLICATION AND DESIGN PROPERTIES OF PLASTICS

The design technology of plastics including prediction of performance and endurance is quite different from that of metals and other traditional materials. Of critical importance to the designer of plastics parts is the fact that the tests for the engineering properties of plastics are different, and so is the interpretation of data produced by the traditional tests such as tensile strength, stress-strain modulus and Izod impact. These standard properties, commonly seen in data sheets and other publications are not suitable for material selection or design of parts because they do not adequately account for the effect of time under load, response to sustained temperature exposure, or relate to performance where the complexity of the applied stress does not permit direct application to design. The fundamental engineering properties required to overcome these limitations include strength and rigidity as a function of time, temperature and stress (creep rupture and creep), as well as realistic measure of strength at high rates of loading (impact) and repeated loading (fatigue). (Application and Design Properties of Plastics—Malcolm N. Riddell, Allied Chemical Corp.)

(Tables from the 1974 Materials Selector, Materials Engineering annual reference guide to the properties and characteristics of all engineering materials, Van Nostrand Reinhold Company.)

TABLE A-1 PLASTICS—Including Reinforced and Filled Types

ABS (Acrylonitrile Butadiene Styrene)

Type ➤	ASTM	Medium Impact	Very High Impact	Low Temp Impact	Heat Resistant	Self-extinguishing (No PVC)	Transparent
PHYSICAL PROPERTIES							
Specific gravity	D792	1.05-1.07	1.01-1.06	1.02-1.04	1.06-1.08	1.21	1.07
Refractive index, nD	D542	—	—	—	—	—	1.538
Transparency (visible light), %	—	Opaque	Opaque	Opaque	Opaque	Opaque	85
MECHANICAL PROPERTIES							
Tensile	D638						
Strength, 1000 psi (yield)	—	6.3-8.0	4.5-6.0	4.0-6.0	7.0-8.0	6.0	5.6
Elongation, % (ultimate)		5-20	20-50	30-200	20	20	—
Modulus, 1000 psi		330-400	200-310	200-310	350-420	320	290
Flexural	D790						
Strength, 1000 psi		9.9-11.8	6.0-9.8	5-8	11.0-12.0	10.0	9.95
Modulus, 1000 psi		350-400	200-320	200-320	350-420	330	300
Compressive str, 1,000 psi (2% offset)	D695	10.5-11.0	—	—	9.3-11.0	—	7.0
Impact strength							
Izod (notched), lb/in.	D256	2.0-4.0	5.0-7.0	6-10	2.0-4.0	4.0	5.3
Gardner, ft-lb		—	—	—	—	—	—
Hardness							
Rockwell	PD785	R108-115	R85-105	R75-95	R107-116	R100	R100
Shore	PD785	—	—	—	—	—	—
Barcol		—	—	—	—	—	—
Bending fatigue str (Woehler), 1,000 psi at 10^7 cycles	D671	—	—	—	—	—	—
Abrasion res (Taber, CS-17 wheel), mg/1000 cycles	D1044	—	—	—	—	—	83.8
Coef of static friction	D1894						
Against self		—	—	—	—	—	0.501
Against steel		—	—	—	—	—	0.281
ELECTRICAL PROPERTIES							
Vol res, ohm-cm	D257	2.7×10^{16}	$1\text{-}4 \times 10^{16}$	$1\text{-}4 \times 10^{16}$	$1\text{-}5 \times 10^{16}$	—	—
Dielec str, v/mil (dry)	D149						
Short-time		385	300-375	300-415	360-400	—	—
Step-by-step		—	—	—	—	—	—
Dielec constant (dry)	D150						
60 Hz		2.8-3.2	2.8-3.5	2.5-3.5	2.7-3.5	—	—
10^6 Hz		2.75-3.0	2.4-3.0	2.4-3.0	2.8-3.2	—	—
Dissip factor (dry)	D150						
60 Hz		0.003-0.006	0.005-0.010	0.005-0.01	0.030-0.040	—	—
10^6 Hz	D495	0.008-0.009	0.008-0.016	0.008-0.016	0.005-0.015	—	—
Arc res, sec		—	—	—	—	—	—
THERMAL PROPERTIES							
Ther cond, Btu/hr/sq ft/°F/ft (Cenco-Fitch)	C177	0.08-0.18	0.01-0.14	0.08-0.14	0.12-0.20	—	—
Coef of ther exp, 10^{-6}/°F	D696	3.2-4.8	5.0-6.0	5.0-6.0	3.0-4.0	5.3	0.46
Specific heat, Btu/lb/°F	—					—	—
Heat deflection temp °F	D648						
At 66 psi		—	—	—	—	198	195
At 264 psi		185-223	180-218	185-224	220-245	180	183
Brittleness temp, °F	D746	—	—	—	—	—	—
Max temp for continuous use (no load), °F	—	—	—	—	—	150	—
APPLICABLE PROCESSING METHODS	Injection mldg, extrusion, thermofrmg, blow mldg, foam mldg.						Injection mldg.
CHEMICAL RESISTANCE	Highly res to aqueous acids, alkalis, salts. Res to concentrated phosphoric and hydrochloric acids, alcohols, and animal, vegetable and mineral oils. Disintegrated by concentrated sulfuric or nitric acids. Soluble in esters, ketones, ethylene dichloride						
USES	Pipe, appliance housings, housewares, lawn and garden equipment, chrome plated parts, highway safety devices, extruded profiles, shoe heels, fume hoods and ducts, toys, office equipment; also available as formable sheet for such uses as cases, luggage, refrigerator linings						

ᵃ For a bar 2½ x ½ x ¼ in.

Acetals

Material →	ASTM	Homopolymer[c]			Copolymer[d]		
		Standard	20% Glass Reinforced	22% TFE Filled	Standard	25% Glass Reinforced	High Flow
PHYSICAL PROPERTIES	ASTM						
Specific Gravity	D792	1.425	1.56	1.54	1.410	1.61	1.410
Ther Cond, Btu/hr/sq ft/°F/ ft	—	0.13	—	—	0.16	—	1.6
Coef of Ther Exp, 10^{-5} per °F	D696	4.5	2.0-4.5	4.5	4.7	2.2-4.7	4.7
Specific Heat, Btu/lb/°F	—	0.35	—	—	0.35	—	0.35
Refractive Index, n_D	D542	Opaque	Opaque	Opaque	Opaque	Opaque	Opaque
Water Absorption (24 hr), %	D570	0.25	0.25	0.20	0.22	0.29	0.22
MECHANICAL PROPERTIES							
Tensile Strength, 1000 psi	D638						
Ultimate	—	10.0	8.5	6.9	8.8	18.5	8.8
Yield	—	10.0	—	—	8.8	18.5	8.8
Elongation, %	D638			12			
Ultimate	—	25	7	—	60-75	3	40
Yield	—	12	—	—	12	3	12
Mod of Elast in Ten, 10^5 psi	D638	5.2	—	—	4.1	12.5	4.13
Flex Strength, 1000 psi	D790	14.1	—	—	13	28	13
Mod of Elast in Flex, 10^5 psi	D790	4.1	8.8	4.0	3.75	11	3.75
Impact Str (Izod, notched), ft-lb/in.	D638	1.4	0.8	0.7	1.3	1.8	1.0
Compr Str (1%), 1000 psi	D695	5.2	5.2	4.5	4.5	—	4.5
Hardness (Rockwell)	D785	M94	M90	M78	M80	M79	M80
Coef of Static Frict (against steel)	—	0.1-0.3	0.1-0.3	0.05-0.15	0.15	0.15	0.15
Abr Res (Taber, CS-17), mg/1000 cyc	D1044	14-20	33	9	14	40	14
ELECTRICAL PROPERTIES							
Volume Resistivity, ohm-cm	D257	1×10^{15}	5×10^{14}	—	1×10^{14}	1.2×10^{14}	1.0×10^{14}
Dielectric Str (short-time), vpm	D149	500	500	—	500	580	500
Dielectric Constant	D150						
60 cycles	—	3.7	4.0	—	3.7 @ 100	3.9 @ 100	3.7 @ 100
10^6 cycles	—	3.7	4.0	—	3.7	3.9	3.7
Dissipation Factor	D150						
60 cycles	—	0.0048	0.0047	—	0.001 @ 100	0.003 @ 100	0.001 @ 100
10^6 cycles	—	0.0048	0.0036	—	0.006	0.006	0.006
Arc Resistance, sec	D495	129[b]	188	—	240	136	240
HEAT RESISTANCE							
Max Rec Service Temp, F	—	195	195	195	220	220	220
Deflection Temp, F	D648						
66 psi	—	338	345	329	316	331	316
264 psi	—	255	3.5	212	230	325	230
APPLICABLE PROCESSING METHODS		Injection molding, extrusion, rotational molding, blow molding			Injection molding, extrusion, rotational molding, blow molding		
CHEMICAL RESISTANCE		Excellent res to most organic solvents, including aliphatic and aromatic hydrocarbons. Not rec for use with strong acids and alkalis	Same as standard homopolymer	Same as standard homopolymer	Excellent res to strong alkalis. Most organic solvents including alcohols, ketones, esters, aliphatic and aromatic hydrocarbons and glycols do not seriously alter properties. Not recommended for use in strong mineral acids and oxidizing reagents.		
USES		Appliance parts, gears, bushings, aerosol bottles, auto, plumbing, textile, consumer uses	Same as homopolymer. Where high stiffness and dimensional stability are required	Same as homopolymer. Where low friction and high resistance to wear are required	Appliance parts, gears, bushings, aerosol bottles, various automotive, plumbing, textile machinery and consumer products.		

a 10% deformation.   b 15-mil specimen.   c "Delrin" is most common tradename.   d "Celcon" is most common tradename.

TABLE A-1 *Continued*
Acrylics

Type →	ASTM	Cast Sheets, Rods		Moldings		
		General Purpose Type I[a]	General Purpose Type II[a]	Grades 5, 6, 8[b]	High Impact Grade	Modified (XT acrylic)
PHYSICAL PROPERTIES	ASTM					
Specific Gravity	D792	1.17-1.19	1.18-1.20	1.18-1.19	1.12-1.16	1.10-1.12
Ther Cond, Btu/hr/sq ft/°F/ft	c	0.12	0.12	0.12	0.12	0.13
Coef of Ther Exp, 10^{-6} per °F	D696	4.5	4.5	3-4	4-6	4.4-4.5
Spec Ht, Btu/lb/°F	—	0.35	0.35	0.35	0.34	0.33
Refractive Index	D542	1.485-1.500	1.485-1.495	1.489-1.493	—	1.51
Transmittance (luminous, 0.125 in.), %	D791	91-92	91-92	>92	—	86-88
Haze, %	D672	1-2	1-2	<3	—	—
Water Absorption (24 hr), %	D570	0.3-0.4	0.2-0.4	0.3-0.4	0.2-0.3	0.3
MECHANICAL PROPERTIES						
Mod of Elast in Tension, 10^5 psi	D638	3.5-4.5	4.0-5.0	3.5-5.0	2.3-3.3	3.7-4.3
Ten Str, 1000 psi	D638	6-9	8-10	9.5-10.5	5.5-8.0	7.0-8.0
Elong (in 2 in.), %	D638	2-7	2-7	3-5	>25	12-30
Hardness (Rockwell)	D785	M80-90	M96-102	M80-103	L60-94	M45-68
Impact Str (Izod notched), ft-lb/in.	D256	0.4	0.4	0.2-0.4	0.8-2.3	1-2
Mod of Elast in Flex, 10^5 psi	D790	3.5-4.5	4.0-5.0	3.5-5.0	2.8-3.6	3.5-4.0
Flex Str, 1000 psi	D790	12-14	15-17	15-16	8.7-12.0	11-13
Compr Yld Str (0.1% offset), 1000 psi	D695	12-14	14-18	14.5-17	7.3-12.0	9.5-11.5
ELECTRICAL PROPERTIES						
Vol Res, ohm-cm	D257	$>10^{15}$	$>10^{15}$	$>10^{14}$	2.0×10^{16}	$>10^{16}$
Dielec Str (short time), v/mil	D149	450-530	450-500	400	400-500	400-500
Dielec Const						
60 Cycles	D150	3.5-4.5	3.5-4.5	3.5-3.9	3.5-3.9	—
10^6 Cycles	D150	2.7-3.2	2.7-3.2	2.7-2.9	2.5-3.0	2.78-2.86
Dissip Factor						
60 Cycles	D150	0.05-0.06	0.05-0.06	0.04-0.06	0.03-0.04	0.026-0.029
10^6 Cycles	D150	0.02-0.03	0.02-0.03	0.02-0.03	0.01-0.02	0.022-0.025
Arc Resistance, sec	. . . . .	No track	No track	No track	No track	No track
APPLICABLE PROCESSING METHODS		Thermoforming, casting		Injection molding, extrusion, thermoforming, blow molding		Injection molding, extrusion, thermoforming, blow molding
HEAT RESISTANCE						
Max Recommended Svc Temp, F		140-160	180-200	155-190	—	160
Heat Dist Temp, F		150-180	190-225	166-250[d]	169-205	195 (264 ps)
CHEMICAL RESISTANCE		Resists weak alkalis, acids and aliphatic hydrocarbons. Attacked by esters, ketones, aromatic hydrocarbons, chlorinated hydrocarbons and concentrated acids				
USES		Transparent aircraft enclosures, radio and television parts, lighting, drafting equipment, signs		Decorative and functional automotive parts, reflectors, protective goggle lenses, radio and television parts, appliances	Shoe heels, control knobs, business machine and piano keys, pump parts, sprinkler heads, tool handles	Packaging, lenses containers, shields

[a] ASTM D702. [b] Range includes typical values for Grades 5, 6, and 8, and may be superior to minimum or maximum requirements for these grades as detailed in ASTM D788. [c] Cenco-Fitch. [d] D788 specified values for Grades 5, 6, and 8: 149 F, 162 F, 183 F respectively.

Alkyds and Thermoset Carbonate

Material →	ASTM	Allyl Diglycol Carbonate	Alkyds			
			Putty (encapsulating)	Rope (general purpose)	Granular (high speed molding)	Glass Reinforced (heavy duty parts)
PHYSICAL PROPERTIES						
Specific Gravity	D792	1.32	2.05–2.15	2.20–2.22	2.21–2.24	2.02–2.10
Ther Cond, Btu/hr/sq ft/°F/in.	—	1.45	0.35–0.60	0.35–0.60	0.35–0.60	0.20–0.30
Coef of Ther Exp, per °F	D696	6×10^{-5}	$1\text{-}3 \times 10^{-5}$	$1\text{-}3 \times 10^{-5}$	$1\text{-}3 \times 10^{-5}$	$1\text{-}3 \times 10^{-5}$
Specific Heat, Btu/lb/°F	—	0.3	—	—	—	—
Water Absorption (24 hr), %	D570	0.20	0.10–0.15	0.05–0.08	0.08–0.12	0.07–0.10
Transparency (visible light), %	—	89–92	Opaque	Opaque	Opaque	Opaque
Refractive Index, n_D	D542	1.50	—	—	—	—
MECHANICAL PROPERTIES						
Tensile Strength, 1000 psi	D638	5–6	4–5	7–8	3–4	5–9
Tensile Mod, 10^5 psi	D638	3.0	20–27	19–20	24–29	20–25
Elongation, %	D638	—	—	—	—	—
Impact Str (Izod notched), ft-lb/in.	D256	0.2–0.4	0.25–0.35	2.2	0.30–0.35	8–12
Flex Strength, 1000 psi	D790	—	8–11	19–20	7–10	12–17
Mod of Elast in Flex, 10^5 psi	D790	2.5–3.3	—	22–27	22–27	22–28
Compr Strength, 1000 psi	D690	22.5	20–25	28	16–20	24–30
Hardness (Barcol)	D785	M95–M100 (Rockwell)	60–70	70–75	60–70	70–80
ELECTRICAL PROPERTIES						
Volume Resistivity, ohm-cm	D257	4×10^{14}	10^{14}	10^{14}	$1 \times 10^{14}\text{–}1 \times 10^{15}$	10^{14}
Dielectric Str (step-by-step), v/mil	D149	290	300–350	290	300–350	300–350
Dielectric Constant	D150					
60 cycles		4.4	5.4–5.9	7.4	5.7–6.3	5.2–6.0
10^6 cycles		3.5-3.8	4.5–4.7	6.8	4.8–5.1	4.5–5.0
Dissipation Factor	D150					
60 cycles		0.03-0.04	0.030–0.045	0.019	0.030–0.040	0.02–0.03
10^6 cycles		0.1-0.2	0.016–0.020	0.023	0.017–0.020	0.015–0.022
Arc Resistance, sec	D495	.185	180	180	180	180
APPLICABLE PROCESSING METHODS		Casting	Injection molding, compression molding, transfer molding			
HEAT RESISTANCE						
Max Rec Service Temp, F	—	212	250	300	300	300
Deflection Temp (264 psi), F	D648	—	350–400	>400	350–400	>400
CHEMICAL RESISTANCE		Resists nearly all solvents including acetone, benzene and gasoline, and practically all chemicals except highly oxiding acids	Resistant to weak acids; attacked by alkalis; practically unattacked by organic liquids such as alcohols, hydrocarbons and fatty acids			
USES		Aircraft windows, lenses, marine glazing, vending machine windows, slides, watch crystals, safety windows	Encapsulation of resistors, coils and small electronic parts	Molding of tube bases and sockets, connectors, tuning devices, electrical instrument parts, switches and relays. Parts for transformers, motor controllers and automotive ignition systems		Heavy duty circuit breaker and switchgear, stand-off insulators, electrical motor brush holders and end plates

TABLE A-1 *Continued*

Alloys—ABS/polycarbonate, ABS/PVC, Acrylic/PVC, ABS/Polysulfone, ABS/Polyurethane

Material →		ABS/ Polycarbonate	ABS/PVC (rigid)	Acylic/ PVC	ABS/Poly- sulfone (Polyaryl ether)	ABS/Poly- urethane
PHYSICAL PROPERTIES	ASTM					
Specific Gravity	D792	1.19	1.21	1.28-1.35	1.14	1.04-1.07
Refractive Index, nD	D542	—	—	—	—	—
Transparency (visible light), %	D1003	Opaque	Opaque	Translucent to opaque	Opaque	Opaque
MECHANICAL PROPERTIES						
Tensile	D638					
Str (yield), 1,000 psi		8.0	6.0	6.5-7.0	7.5	3.7-4.5
Elong (ult), %		—	—	75-150	25-90	120-200
Modulus, 1,000 psi		370	330	335-370	320	160-220
Flexural	D790					
Str, 1,000 psi		13.7	10.2	10.7-11	11	5.3-7.1
Modulus, 1,000 psi		380	340	380-400	300	150-210
Compressive str, 1,000 psi (2% offset)	D695	—	7.4	8.4	—	3.6-4.8
Impact str						
Izod (notched), ft-lb/in.	D256	10.5	12.5	12-15	8.0-10.0	8.4-10.1
Gardner, lb/in.		—	—	—	—	—
Hardness						
Rockwell	D785	R117	R102	R110-105	R117	R70-82
Shore	D785	—	—	—	—	—
Bending fatigue str (Woehler), 1,000 psi	D671	—	—	—	—	—
Abrasion res (Taber, CS-17 wheel, 1000g)	D1044					
mg/100 cycles		—	—	40	—	30-35
Coef of friction	D1894					
Against self		—	—	—	0.28	—
Against steel		—	—	—	0.31	—
ELECTRICAL PROPERTIES						
Vol res, ohm-cm	D257	10^{15} x 10^{16}	—	1.6 x 10^{16}	1.5 x 10^{16}	—
Dielec str, v/mil (dry)	D149					
Short-time		1250-1550	600	430-670	430	—
Step-by-step		—	—	647	—	—
Dielec constant (dry)	D150					
60 Hz		3.08	—	3.33-3.86	3.14	—
10^6 Hz		3.2-3.6	—	3.06-3.44	3.10	—
Dissip factor (dry)	D150					
60 Hz		0.019-0.021	—	0.016	0.006	—
10^6 Hz		0.020	—	0.017	0.007	—
Arc res, sec	D495	—	—	42-80	180	—
THERMAL PROPERTIES						
Ther cond, Btu/hr/sq ft/°F/in.	C177	0.95	—	1.01	2.07	—
Coef of ther exp, 10^{-5}/°F	D696	37	4.6	3.5-4.4	—	11.6-12.1
Specific heat, Btu/lb/°F		0.3	—	0.293	0.35	—
Heat deflection, °F	D648					
At 66 psi		238	—	177-205	320	214-216
At 264 psi		220	—	160-185	300	201-207
Brittleness temp, °F	D746	—	—	−46	—	—
Max temp for continuous use (no load), F	—	190-200	—	—	—	—
CHEMICAL AND ENVIRONMENTAL RESISTANCE						
Water absorption, %	D570					
In 24 hr		—	0.12	0.06-0.09	—	0.35-0.41
Saturation		—	—	—	—	—
Weathering		Slight effect	Slight effect	Resist	Embrittles	Slight effect
Acids						
Weak		Resist	Resist	Resist	Resist	Resist
Strong		Attack	Attack	Attack	Resist	Attack
Alkalis						
Weak		Resist	Resist	Resist	Resist	Resist
Strong		Resist	Resist	Resist	Resist	Resist
Organic solvents		Attack	Attack	Attack	Attack	Attack
Fuel		—	—	—	—	Attack
Oil and grease		—	—	—	—	—
METHODS OF PROCESSING		Inj mldg	Inj mldg, extrusion, thermoforming	Extrusion, thermoforming	Inj mldg, extrusion	Inj mldg, extrusion, thermoforming

Cellulose Acetate Butyrate and Cellulose Acetate Propionate

ASTM Grade[a] →	ASTM	Cellulose Acetate Butyrate			Cellulose Acetate Propionate		
		H4	MH	S2	1	3	6
PHYSICAL PROPERTIES							
Specific Gravity	D792	1.22	1.18±1.20	1.15–1.18	1.22	1.20–1.21	1.19
Ther Cond, Btu /hr /sq ft /°F /ft	C177	0.10–0.19	0.10–0.19	0.10–0.19	0.10–0.19	0.10–0.19	0.10–0.19
Coef of Ther Exp, 10^{-5} per °F	D696	$(6–9) \times 10^{-5}$	$(6–9) \times 10^{-5}$	$(6–9) \times 10^{-5}$	$(6–9) \times 10^{-5}$	$(6–9) \times 10^{-5}$	$(6–9) \times 10^{-5}$
Refractive Index	D543	1.46–1.49	1.46–1.49	1.46–1.49	1.46–1.49	1.46–1.49	1.46–1.49
Spec Ht, Btu /lb /°F		0.3–0.4	0.3–0.4	0.3–0.4	0.3–0.4	0.3–0.4	0.3–0.4
Luminous Transmittance, %	D791	75–92	80–92	85–95	80–92	80–92	80–92
Haze, %	D672	2–5	2–5	2–5	2–5	2–5	2–5
Water Absorption (24 hr), %	D570	2.0	1.3–1.6	0.9–1.3	1.6–2.0	1.3–1.8	1.6
Flammability, ipm	D635	0.5–1.5	0.5–1.5	0.5–1.5	0.5–1.5	0.5–1.5	0.5–1.5
MECHANICAL PROPERTIES							
Ten Str at Fracture, 1000 psi	D638	6.9	5.0–6.0	3.0–4.0	5.9–6.5	5.1–5.9	4.0
Hardness (Rockwell R)	D785	114	80–100	23–42	100–109	92–96	57
Impact Str (Izod), ft-lb /in. of notch	D256	3.0	4.4–6.9	7.5–10.0	1.7–2.7	3.5–5.6	9.4
Modulus of Elast in Flex, 10^5 psi	D747	1.80	1.20–1.40	0.70–0.90	1.7–1.8	1.45–1.55	1.1
Flex Str at Yield, 1000 psi	D790	9.0	5.6–6.7	2.5–3.95	6.8–7.9	5.6–6.2	—
Compr Str at Yield, 1000 psi	D695	8.8	5.3–7.1	2.6–4.3	6.2–7.3	4.9–5.8	—
ELECTRICAL PROPERTIES							
Vol Res, ohm-cm	D257	$10^{11}–10^{14}$	$10^{11}–10^{14}$	$10^{11}–10^{14}$	$10^{11}–10^{14}$	$10^{11}–10^{14}$	$10^{11}–10^{14}$
Dielec Str (short-time), v /mil	D149	250–400	250–400	250–400	300–450	300–450	300–450
Dielec Const							
60 Cycles	D150	3.5–6.4	3.5–6.4	3.5–6.4	3.7–4.0	3.7–4.0	3.7–4.0
10^6 Cycles	D150	3.2–6.2	3 2–6.2	3.2–6.2	3.4–3.7	3.4–3.7	3.7–3.4
Dissip Factor							
60 Cycles	D150	0.01–0.04	0.01–0.04	0.01–0.04	0.01–0.04	0.01–0.04	0.01–0.04
10^6 Cycles	D150	0.02–0.05	0.02–0.05	0.02–0.05	0.02–0.05	0.02–0.05	0.02–0.05
APPLICABLE PROCESSING METHODS		Injection molding, extrusion, thermoforming, rotational molding, blow molding			Injection molding, extrusion, thermoforming, rotational molding, blow molding		
HEAT RESISTANCE Heat Deflection Temp, °F							
66 psi		222	171–184	136–147	191–201	169–187	163
264 psi		196	146–160	118–130	163–173	141–157	129
CHEMICAL RESISTANCE		Unaffected by 3% sulfuric, 5% acetic, 10% hydrochloric and oleic acids; discolored by 10% nitric acid. Unaffected by 1% sodium hydroxide and 2% sodium carbonate; slightly softened by 10% sodium hydroxide and discolored by 10% ammonium hydroxide; Unaffected by white gasoline, but swollen or dissolved by ethyl alcohol, acetone, ethyl acetate, ethylene dichloride, carbon tetrachloride and toluene. Unaffected by water, salt water and 3% hydrogen peroxide.					
USES		Vacuum-formed outdoor signs and molded letters, blister packaging, TV and radio knobs, handles, pipe, pens, optical parts, containers			Telephones, steering wheels, blister packaging, toothbrushes, pens, knobs, containers, optical parts		

[a] According to ASTM D707–63 and D1562–60, respectively.

TABLE A-1 *Continued*

Diallyl Phthalates

Type ➡		Orlon-Filled	Dacron-Filled	Asbestos-Filled	Glass Fiber-Filled
PHYSICAL PROPERTIES	ASTM				
Specific Gravity	D792	1.31-1.35	1.40-1.65	1.50-1.96	1.55-1.85
Coef of Ther Exp, per °F	D696	5.0×10^{-5}	5.2×10^{-5}	4.0×10^{-5}	$2.2\text{-}2.6 \times 10^{-5}$
Water Abs (122 F, 48 hr), %	—	0.2-0.5	0.2-0.5	0.4-0.7	0.2-0.4
MECHANICAL PROPERTIES					
Mod of Elast in Tension, psi[a]	D638	6×10^5	—	12×10^5	—
Ten Str, psi	D638	4500-6000	4600-5500	4000-6500	5500-9500
Hardness (Rockwell)	D785	M108	—	M107	M108
Impact Str (Izod notched), ft-lb/in.	D256	0.5-1.2	1.7-4.5	0.30-0.50	0.5-15.0
Flex Str, 1000 psi	D790	7.5-10.5	9-11.5	8-10	10-18
Compr Str, 1000 psi	D695	20-25	20-30	18-25	25
ELECTRICAL PROPERTIES					
Dielec Str, v/mil					
Short Time (dry)	D149	400	376-390	350-450	350-430
Short Time (wet[b])	D149	375	360-391	300-400	300-420
Step-by-Step (dry)	D149	350	350-374	300-400	300-420
Step-by-Step (wet[b])	D149	325	350-361	250-350	275-420
Dielec Breakdown, kv					
Short Time (dry)	—	65-75	65	55-80	63-70
Short Time (wet[b])	—	60-65	60	55	45-65
Step-by-Step (dry)	—	55-60	60	38-70	55-65
Step-by-Step (wet[b])	—	46-60	55	39-60	45-65
Dissip Factor[c]					
Dry	D150	0.023, 0.015	0.008, 0.015	0.05, 0.03	0.01, 0.015
Wet[d]	D150	0.045, 0.040	0.009, 0.017	0.154, 0.050	0.012, 0.020
Dielec Const[e]					
Dry	D150	3.9, 3.3	3.8, 3.6	5.2, 4.5	4.5, 4.2
Wet[d]	D150	4.1, 3.4	3.9, 3.7	6.5, 4.8	4.6, 4.4
Vol Res, megohm-cm[d]	D257	60,000-6,000,000	100-25,000	100-5000	10,000-50,000
Surface Res, megohms[d]	D257	25,000-2,500,000	500-25,000	100-5000	10,000-100,000
Arc Resistance, sec	D495	85-115	105-125	125-140	125-135
APPLICABLE PROCESSING METHODS		Injection molding, compression molding, transfer molding, layup molding			
HEAT RESISTANCE					
Max Recommended Svc Temp, F	—	300	300-370	350-450	400-450
Heat Dist Temp, F	D648	240-266	270-290	300-350	350-500
CHEMICAL RESISTANCE		Unaffected by weak acids and alkalis and organic solvents. Slightly affected by strong acids and alkalis			
USES		Molding compounds—connectors, potentiometers, plugboards, housings, appliance fixtures, resistors, insulators, etc. Prepregs—radomes, aircraft leading edges, housings, nose cones, air ducts, etc. Laminates—decorative sheets for surfacing real and grain-printed wood and fabrics, etc.			

[a] Conditioned 48 hr at 122 F. [b] Tested after 48-hr immersion in water at 122 F. [c] Values given for frequencies of 1 kc and 1 mc, in that order. [d] Conditioned 30 days at 100% RH and 158 F. [e] Flame-resistant type is available. [f] 480 hr, 257 F.

Epoxies

Type →	ASTM	Standard Epoxies (Diglycidyl Ether of Bisphenol A)					
		Cast Rigid[a]	Cast Flexible[b]	Molded[c]	Unidirectional Laminate	High Strength Laminate[e]	Filament Wound Composite[f]
PHYSICAL PROPERTIES							
Specific Gravity	D792	1.15	1.14–1.18	1.80–2.0	1.8	1.84	2.18–2.17
Ther Cond, Btu/hr/sq ft/ °F/ft	D325	0.1–0.3	—	0.1–0.5	—	2.35	—
Coef of Ther Exp, 10^{-5} per °F	D696	3.3	3-5	1–2	$3.3–4.8 \times 10^{-6}$	$3.3–4.8 \times 10^{-6}$	2-6
Specific Heat, Btu/lb/ °F	—	0.4–0.5	—	—	—	0.21	0.24
Water Absorp (24 hr), %	D570	0.1–0.2	0.4–1.0	0.3–0.8	0.05–0.07	0.05	0.05–0.07
Transparency (visible light), %	—	90	85	Opaque	Opaque	Opaque	Opaque
Refractive Index, n_D	D542	1.61	1.61	—	—	—	—
MECHANICAL PROPERTIES							
Tensile Str, 1000 psi	D638	9.5–11.5	1.4–7.6	8–11	50–58	160	230–240 (Hoop)
Tensile Mod, 10^5 psi	D638	4.5	0.5–2.5	—	33–36	57–58	72–64
Elongation, %	D638	4.4	1.5–60	—	—	—	—
Impact Str (Izod notched), ft-lb/in.	D256	0.2–0.5	0.3–2.0	0.4–0.5	12–15	60–61 (edgewise)	—
Flexural Str, 1000 psi	D790	14–18	1.2–12.7	19–22	80–90	165–177	180–170
Mod of Elast in Flex, 10^5 psi	D790	4.5–5.4	0.36–3.9	15–25	36–39	53–55	69–75
Compr Strength, 1000 psi	D695	16.5–24	—	34–38	50–60	80–90 (edgewise)	—
Hardness (Rockwell)	D785	106M	50–100M	75–80 (Barcol)	115–117M	70–72 (Barcol)	98–120M
ELECTRICAL PROPERTIES							
Vol Resist, ohm-cm	D257	6.1×10^{15}	9.1×10^8—6.7×10^9	$1–5 \times 10^{15}$	—	$6.6 \times 10^7–10^9$	—
Dielectric Str (step-by-step), v/mil	D149	>400	400–410	360–400	450–550	650–750	—
Dielectric Constant	D150						
60 cycles		4.02	4.43–4.79	4.4–5.4	5.3–5.4	—	—
10^6 cycles		3.42	2.78–3.52	4.1–4.6	4.7–4.8	4.8–5.2	—
Dissipation Factor	D150						
60 cycles		0.0074	0.0048–0.0380	0.011–0.018	0.004–0.006	—	—
10^6 cycles		0.032	0.0369–0.0622	0.013–0.020	0.024–0.026	0.010–0.017	—
Arc Resistance, sec	D495	100	75–98	135–190	130–180	—	—
APPLICABLE PROCESSING METHODS		Casting	Casting	Injection, compression and transfer molding	Layup molding	Layup molding	Filament winding
HEAT RESISTANCE							
Max Rec Svc Temp, F	—	175–190	100–125	<400	250–350	250–350	250–350
Heat Deflection Temp (264 psi), F	D648	230	90–155	340–400	—	—	280–295
CHEMICAL RESISTANCE		Highly res to water and strong alkaline environments; less res to sulfuric and acetic acids, and oxidizing agents					
USES		Potting and encapsulation of electronic components, precision castings, tools and dies, patching compounds		Electrical moldings, such as condens-ers, switch plates, con-nector plugs, re-sistor bobbins and wirewound resistors, molded coils, relay assemblies	High strength parts, such as laminated tools for metal forming, aircraft structural parts, pipe, leaf or coil springs, high strength electrical or chemical resistant parts		Rocket motor cases, chemical tanks, pipe, pressure bottles, high strength tubing, shotgun barrels, missile bodies

[a]13 phr of TETA curing agent. [b]30-80 phr of flexible curing agent. [c]Mineral-glass reinforced. [d]23 phr aromatic amine curing agent, 12 plies E-181 glass cloth with Volan A finish. [e]36% resin, 64% unidirectional nonwoven glass fiber reinforcement. [f]NOL rings made with 12 end E-HTS glass; 15 phr metaphenylenediamine curing agent.

TABLE A-1 *Continued*

Epoxies

Type →	ASTM	Epoxy Novolacs			High Performance Resins (Cycloaliphatic Diepoxides)	
		Cast, Rigid[g]	Molded[h]	Glass Cloth Laminate[i]	Cast, Rigid[j]	Glass Cloth Laminate[k]
PHYSICAL PROPERTIES	ASTM					
Specific Gravity	D792	1.24	1.7	1.97	1.22	1.97
Ther Cond, Btu/hr/sq ft/°f/ft	—	—	—	—	—	—
Coef of Ther Exp, per °F x 10^{-6}	D696	—	1.7–2.2	—	1.6–3.0	—
Specific Heat	—	—	—	—	—	—
Water Absorption (24 hr), %	D570	—	0.11–0.2	0.04–0.06	0.1–0.7[l]	—
Transparency (visible light), %	—	—	Opaque	Opaque	—	Opaque
MECHANICAL PROPERTIES						
Tensile Strength, 1000 psi	D638	8–12	5.2–5.3	50–52	9.6–12.0	59.2
Tensile Modulus, 10^5 psi	D638	4–5	—	32–33	4.8–5.0	27.5
Elongation, %	D638	2–5	—	—	2.2–4.8	—
Impact Str (Izod notched), ft-lb/in	D256	0.5	0.3–0.5	13–17	—	—
Flexural Strength, 1000 psi	D790	11–16	10–12	70–72	12–13	84–89
Mod of Elast in Flex, 10^5 psi	D790	4–5	—	28–31	4.4–4.8	32–35
Compr Strength, 1000 psi	D695	17–19	22–26	67–71	30–50	48–57
Hardness (Rockwell)	D785	107–112[m]	94–96D[n]	75–80	—	—
ELECTRICAL PROPERTIES						
Volume Resistivity, ohm-cm	D257	2.10 x 10^{14}	1.4–5.5 x 10^{14}	—	>10^{16}	—
Dielectric Str (step-by-step), v/mil	D149	—	280–400	—	444 (short time)	—
Dielectric Constant	D150					
60 cycles		3.96–4.02	4.7–5.7	—	3.34–3.39	4.41–4.43
10^6 cycles		3.53–3.58	4.3–4.8	5.1[m]	—	—
Dissipation Factor	D150					
60 cycles		0.0055–0.0074	0.0071–0.025	—	0.001–0.007	—
10^6 cycles		0.029–0.028	—	0.015[m]	—	—
Arc Resistance, sec	D495	—	180–185	—	120	—
APPLICABLE PROCESSING METHODS		Casting	Injection molding, compression molding, transfer molding	Reinforced layup molding	Casting	Reinforced layup molding
HEAT RESISTANCE						
Max Rec Svc Temp, F	—	450	450–500	450–500	450–500	450–500
Heat Deflection Temp (264 psi), F	D648	300–400	300–425	—	300–525	—
CHEMICAL RESISTANCE		Res water and strong alkalis; more res to sulfuric and acetic acid and oxidizing agents than standard epoxy systems			Outstanding weather res compared to other epoxy systems. Highly res to water, strong alkaline environments; less res to sulfuric and acetic acids, oxidizing agents	
USES		Impregnation and potting requiring high heat res; adhesives	Electrical and electronic encapsulation designed for high temp	High temp tooling, structural laminates, ablatives	Encapsulation, impregnation and potting req outstanding arc and tracking res	Electrical laminates req outstanding weather res

[g]28 phr methylene dianiline; cure—16 hr at 130 F, 2 hr at 257 F, 2 hr at 347 F. [h]Mineral-filled proprietary compounds. [i]12 plies glass cloth with Volan A finish, 26–27% resin; cure–20 min at 383-400 F and contact pressure, plus 24 min at 419 F post cure. [j]12 phr of hexahydrophthalic and anhydride, 12 phr sodium alcoholate accelerator; cure–24 phr at 250 F and post cure of 3 phr at 400 F. [k]100 phr resin, 85 parts anhydride curing agent, 181 Volan glass cloth, [l]1 hr at 212 F. [m]1 mc. [n]Durometer.

Fluorocarbons

Type →	ASTM	Polytrifluoro-chloroethylene (PTFCE)	Polytetrafluoro-ethylene (PTFE)	Ceramic-Reinforced (PTFE)	Fluorinated Ethylene Pro-pylene (FEP)	Polyvinylidene-fluoride (PVF$_2$)	ETFE & ECTFE Std.	ETFE & ECTFE Glass reinf.	PFA
PHYSICAL PROPERTIES									
Specific Gravity	D792	2.10-2.15	2.1-2.3	2.2-2.4	2.14-2.17	1.77	1.68	1.86	2.12-2.17
Ther Cond, Btu/hr/sq ft/°F/ft	[c]	0.145	0.14	—	0.12	0.14	—	—	—
Coef of Ther Exp, per °F x 10^{-5}	D696	3.88	5.5	1.7-2.0	8.3-10.5	8.5	14	1.7	6.7-11.1
Refractive Index	D542	1.43	1.35	—	1.34	1.42	1.44	—	—
Specific Heat, Btu/lb/°F		0.22	0.25	—	0.28	0.33	—	—	—
Transmittance (luminous), %	D791	80-92	—	—	—	—	0.7-0.8	—	—
Water Absorption (24 hr), %	D570	0.00	0.01	>.2	0.01	0.03	—	0.01	0.03
MECHANICAL PROPERTIES									
Mod of Elast in Compr, psi	D638	1.8 x 10^5	0.70-0.90 x 10^5	1.5-2.0 x 10^5	0.6-0.8 x 10^5	1.7-2 x 10^5	—	—	—
Mod of Elast in Tension, psi	D638	1.9-3.0 x 10^5	0.38-0.65 x 10^5	1.5-2.0 x 10^5	0.5-0.7 x 10^5	1.7-2 x 10^5	2.4 x 10^5	11 x 10^5	—
Ten Str, 1000 psi	D638	4.6-5.7	2.5-6.5	.75-2.5	2.5-3.5	7.2-8.6	4.0-4.5	12.0	4.3
Elongation (in 2 in.), %	D638	125-175	250-350	10-200	250-330	200-300	150-200WIA	9	200
Hardness (Rockwell)	D785	R110-115	52D	R35-55	58D	R110	R95	—	D60
Abrasion Res, gm/cycle	[d]	0.0080	—	—	—	0.0006-0.0012	0.005	—	—
Impact Str (Izod notched), ft-lb/in.	D256	3.50-3.62	2.5-4.0	—	No break	3.8	No break	7	—
Mod of Elast in Flexure, psi	D747	2.0-2.5 x 10^5	0.6 x 10^5	4.64 x 10^5	0.8 x 10^5	—	2.4 x 10^5	9.5 x 10^5	1.0 x 10^5
Flex Str (0.1% offset), 1000 psi	D790	3.5	—	—	—	2	—	15	—
Compr Str (0.1% offset), 1000 psi	D695	2.0	0.7-1.8	1.4-1.8	1.6	12.8-14.2	—	—	—
ELECTRICAL PROPERTIES									
Volume Resistivity, ohm-cm	D257	10^{18}	>10^{18}	10^{15}	>2 x 10^{18}	5 x 10^{14}	10^{16}	—	>10^{18}
Dielec Str (short time), v/mil	D149	530-600	400-500	300-400	500-600	260	490	—	2000
Dielectric Constant									
60 Cycles	D150	2.6-2.7	2.1	2.9-3.6	2.1	10.0	2.6	—	2.1
10^6 Cycles	D150	2.30-2.37	2.1	2.9-3.6	2.1	7.5	2.5	—	2.1
Dissipation Factor									
60 Cycles	D150	0.02	0.0002	.0005-.0015	0.0003	0.050	0.0007	—	0.0002
10^6 Cycles	D150	0.007-0.010	0.0002	.0005-.0015	0.0003	0.184	0.009	—	0.0003
Arc Resistance, sec		>360	>200	—	>165	50-60	135	—	—
APPLICABLE PROCESS-ING METHODS		Compression mldg, isotactic pressing	Compression molding, isostactic molding		Inject molding, extrus compres molding	Inject molding, extrus, compres molding	Inject, rotational, blow & transfer molding; extrus; thermoforming; foam		Injection mldg, extrusion, blow mldg
HEAT RESISTANCE									
Max Rec Svc Temp, F		380	550	450-500	400	340	300-355	—	500
Heat Dist Temp, F									
66 psi	D648	196-291	—	350-480	—	300	220-240	—	—
264 psi	D648	151-178	—	170-220	—	232	160-170	285	—
CHEMICAL RESISTANCE		High res to corrosive chemical & most organic solvent	Inert to most chemicals and solvents with exception of alkali metals. Halogenated solvents at high temperatures and pressure have some effect			Res to most acids and bases except fuming sulfuric	Res to most acids & bases	Res to most acids & bases	Same as PTFE
USES		Chemical pipes pump parts, cables, tank linings, connectors, connector inserts, valve diaphragms, insulation	Chemical pipes, valves and liners, gaskets, packings, pump bearings and impellers, electrical equip, anti-adhesive coatings	Bearings, bushings, wear surfaces, elec insulators, gaskets, packings, valve seats in corrosive conditions	Electronic instruments, valve linings, laminates, corrosion resistant and non-adhesive coatings	Seals; chemical pipe and fittings, gaskets; electrical jackets and primary insulation; finishes			PTFE uses req more ease of processing

[a] A proprietary material consisting of polyetrafluorocthylene and special constituents designed to improve TFE's mechanical and thermal properties while retaining its electrical and chemical characteristics. [b] Range covers properties for compounds containing from 10-25% glass. [c] Cenco-Fitch. [d] Federal Spec L-P-406A No. 1092.1. [e] From 73 to 500 F.

TABLE A-1 *Continued*

Foams—Rigid (no surface skin)

Type→		ABS	Cellulose Acetate	Epoxy	Syntactic Epoxy[a]		Phenolic	Poly-ethylene	Poly-propylene
Density, pcf→		31	6-8	5-8	36	42	2-4	34	5
	ASTM								
Ther Cond, Btu/hr/sq ft/°F/in.	177	0.58	0.31-0.32	0.24-0.28	4.56	—	0.20-0.22	0.92	0.27
Coef of Ther Exp, per °F x 10⁻⁵	D696	9.7	2.5	2.3	4.5	—	0.5	4.18	—
Water Absorption, % vol.	C272	0.6	13-17	—	—	1.5	<3	0.22	—
Dielectric Constant at 10^6 cps	—	—	1.59	1.10-1.12	2.0	1.55	—	—	1.48	—
Dissipation Factor at 10^6 cps	—	0.007	0.003	0.005	0.01	—	—	0.0003	—
Max Rec Service Temp, F	—	200	350	400-500	300	300	270	195	230
Tensile Str, psi	D1623	1400	170	50-200	3300	4600	20-55	1000	170
Ultimate Ten Elong, %	D1623	—	—	—	—	—	—	—	—
Mod of Elast in Tension, 1000 psi	D1623	2.4	—	—	—	610	—	—	—
Compr Str, psi (10%)	D1621	—	125-150	60-90	9600	13,400	20-90	800	55
Mod of Elast in Compr, 1000 psi	D1621	—	5.5-13	2.1-6.5	373	480	—	—	1.2
Flex Str, psi	D790	2.4	150	200-800	3800	6000	25-65	1900	230
Mod of Elast in Flex, 1000 psi	D790	9	5.5	2.5-6	—	—	—	88	9.6
Shear Str, psi	C273	—	140	—	3800	4400	15-30	—	—
Mod of Elast in Shear, 1000 psi	C273	—	—	—	—	—	0.4-0.75	—	—
Hardness (Shore D)	—	60	—	—	80-85	—	—	—	—

[a]Glass microsphere-filled epoxy.

Type→		Polyvinyl Chloride	Polystyrene (expanded)		Urea	Urethane		
Density, pcf→		3	2	6	0.8-1.2	2-3	4-7	18-25
	ASTM							
Ther Cond, Btu/hr/sq ft/°F/in.	177	0.15-0.20	0.20-0.28	0.20-0.25	0.18-0.21	0.11-0.23	0.15-0.28	0.29-0.52
Coef of Ther Exp, per °F x 10⁻⁵	D696	2	2.7-4	2.7-4	—	3-4	4	4
Water Absorption, % vol	C272	0.1	<0.1	<0.1	—	3-4	1.5-2	0.2
Dielectric Constant at 10^6 cps	—	—	1.02-1.24	—	—	—	—	—
Dissipation Factor at 10^6 cps	—	—	<0.0005	—	—	—	—	—
Max Rec Service Temp, F	—	180	175	175	120	200-250	250-300	300-400
Tensile Str, psi	D1623	100-200	50-55	120	—	20-70	90-250	700-1300
Ultimate Ten Elong, %	D1623	5-20	5	2	—	—	—	—
Mod of Elast in Tension, 1000 psi	D1623	3-4	740	6100	—	—	—	—
Compr Str, psi (10%)	D1621	70-100	25-30	100-150	5	20-50	65-275	1200-2000
Mod of Elast in Compr, 1000 psi	D1621	3-4	0.55-2	3-6	—	0.3-0.6	1.5-4.5	10-40
Flex Str, psi	D790	120-160	55-75	200-300	17	60-100	200-350	700-2000
Mod of Elast in Flex, 1000 psi	D790	3-4	1.3-3.8	5-15	0.7	0.8-0.9	0.8-15	12-100
Shear Str, psi	C273	60-80	35	150	—	20-30	60-130	7600
Mod of Elast in Shear, 1000 psi	C273	2-2.5	1.15-1.6	3	—	0.17-0.21	0.5-1.5	3-9

Plastics and Rubber Foams — Flexible

Type, Density[a]→	Polyethylene (cellular)			Silicone (foam-in-place)	Urethane (prefoamed)		Vinyl (open cell)
	2.2	33[b]	41[c]	10	1.5-2.0	4.0-5.0	4 and up
Max Rec Service Temp, F	160	—	—	500	—	—	—
Elec Res, microhm-cm	1.5×10^{17}	—	—	3.8×10^{18}	—	—	—
Dielec Str (short time), v/mil	55	220	220	50	—	—	—
Dielec Const, 10^6 cycles	1.05[e]	1.5	1.84	1.17[d]	—	—	—
Dissip Factor, 10^6 cycles	0.0002[e]	0.0004	0.0006	0.001[d]	—	—	—
Tensile Strength, psi	20-30	600	2100	5	12-20	18-20	10-200
Elongation, %	60	300	350	40	150-250	100-125	75-300
Tear Strength, lb/in.	6	—	—	—	2.0-3.0	1.0-2.0	—
Compression Set (22 hr at 158 F), %	15	—	—	4	4-8	2-9	15
Compression-Deflection, psi							
Stress for 25% Defl	8.5[f]	—	—	1	0.35	0.9-2.2	3-500
Stress for 50% Defl	17.5	—	—	3	0.45	—	—

[a]Density in lb/cu ft. [b]Based on low-density polymer; typical expanded insulation (about 46% gas by vol). [c]Based on high-density polymer; typical expanded insulation (about 30% gas by vol.) [d]10^5 cycles. [e]At 10^9 cps. [f]Recovery at same temperature as compression.

Foams—Rigid (structural, integral skin type) [a]

	ASTM	ABS 50	Polycarbonate 50	Polyethylene 37.5	Polypropylene 37.5	Noryl 50	Polystyrene 43.5
Ther Cond, Btu/hr/sq ft/°F/in.	177	—	—	—	—	—	—
Coef of Ther Exp, per °F x 10^{-5}	D696	—	—	6.7	5.2	—	5.0
Water Absorption, % vol	C272	—	—	—	—	—	—
Heat Deflection (264 psi), F	D648	180	270	94	132	205	176
Max Rec Service Temp, F	—	—	—	—	—	—	—
Tensile Str, psi	D1623	2700	5500	1300	2100	3300	1800
Ultimate Ten Elong, %	D1623	—	—	—	—	—	—
Mod of Elast in Tension, 1000 psi	D1623	—	—	—	—	—	—
Compr Str, psi (10%)	D1621	1000	7500	—	—	5500	—
Mod of Elast in Compr, 1000 psi	D1621	—	—	—	—	—	—
Flex Str, psi	D790	3700	10,000	2700	3200	6000	4500
Mod of Elast in Flex, 1000 psi	D790	125	300	120	120	240	210
Shear Str, psi	C273	—	—	—	—	—	—
Mod of Elast in Shear, 1000 psi	C273	—	—	—	—	—	—
Hardness (Shore D)	—	—	—	54	62	—	74
Impact Str (Izod, unnotched), ft-lb	—	—	—	2.5	1.25	—	2.3

				Glass fiber reinforced types			
	ASTM	Polyurethane 34.5	Polyester 56	Polypropylene 45.5	Nylon 54	Polystyrene 52.5	ABS 52.5
Ther Cond, Btu/hr/sq ft/°F/in.	177	—	—	—	—	—	—
Coef of Ther Exp, per °F x 10^{-5}	D696	—	—	—	—	—	—
Water Absorption, % vol	C272	—	—	—	—	—	—
Heat Deflection (264 psi), F	D648	150	400 (66 psi)	162	390	190	210
Max Rec Service Temp, F	—	—	—	—	—	—	—
Tensile Str, psi	D1623	—	8550	3000	10,000	5000	7000
Ultimate Ten Elong, %	D1623	—	—	—	—	—	—
Mod of Elast in Tension, 1000 psi	D1623	—	—	—	—	—	—
Compr Str, psi (10%)	D1621	1600	9300	—	—	—	—
Mod of Elast in Compr, 1000 psi	D1621	—	—	—	—	—	—
Flex Str, psi	D790	4200	17,000	6000	16,000	8500	12,000
Mod of Elast in Flex, 1000 psi	D790	150	700	400	650	750	750
Shear Str, psi	C273	—	—	—	—	—	—
Mod of Elast in Shear, 1000 psi	C273	—	—	—	—	—	—
Hardness (Shore D)	—	70	—	—	—	—	—
Impact Str (Izod, unnotched), ft-lb	—	—	—	3.5	3.2	1.5	3.5

[a] Also samples 0.25 in. thick unless otherwise shown.

TABLE A-1 *Continued*
Nylons

Type→	ASTM	Type 6 — General Purpose[a]	Type 6 — Glass Fiber (30%) Reinforced[a]	Type 6 — Cast	Type 6 — Flexible[a] Copolymers	Type 8[b]	Type 11	Type 12	Transparent
PHYSICAL PROPERTIES									
Specific Gravity	D792	1.14	1.37	1.15	1.12-1.14	1.09	1.04	1.01	1.12
Ther Cond., Btu/hr/sq ft °F/in.	—	1.2	1.2-1.7	1.2-1.7	—	—	1.5	1.7	—
Coef of Ther Exp., 10^{-5} per °F	D696	4.8	1.2	4.4	—	—	5.5	7.2	2.8
Specific Heat, Btu/lb/°F.	—	0.4	—	0.4	—	0.4	0.58	0.28	—
Refractive Index, n_D	D542	—	—	—	—	—	—	—	1.566
Water Absorption (24 hr), %	D570	1.7-1.8	1.3	0.6	0.8-1.4	9.5	0.4	0.25	0.41
MECHANICAL PROPERTIES									
Tensile Strength (2 in./min)[d]	D638								
Ultimate		9.5-12.5	21-23	12.8	7.5-10.0	—	—	7.1-8.5	—
Yield		8.5-12.5	—	12.8	7.5-10.0	3.9	8.5	5.5-6.5	9.8
Elongation (2 in./min), %	D638								
Ultimate		30-220	2-4	20	200-320	400	100-120	120-350	130
Yield		—	—	5	—	—	—	5.8	—
Mod of Elast in Tension, 10^5 psi	D638	—	10-12	5.4	—	0.3	1.78-1.85	1.7-2.1	4.05
Flex Strength, 1000 psi	D790	Unbreak.	26-34	16.5	3.4-16.4	—	—	—	13.26
Mod of Elast in Flex, 10^5 psi	D790	1.4-3.7	10-12	5.05	0.92-3.2	0.4	1.51	—	3.86
Impact Str (Izod notched), ft-lb/in.	D256	0.8-1.2	3-2.3	1.2	1.5-19	>16	3.3-3.6	1.2-4.2	—
Compr Strength (1% offset)[d]	D695	9.7	19-20	14	—	—	—	—	3.39
Hardness (Rockwell)	D785	R118-R120	R121	R116	R72-R119	—	R100-R108	R106	M93
Coef of Dyn Frict	—	—	—	0.32[c]	—	—	—	—	—
Abrasion Res (Taber, CS-17), mg/1000 cycles	D1044	5	—	2.7	—	—	—	—	21
ELECTRICAL PROPERTIES									
Volume Resistivity, ohm-cm	D257	4.5×10^{15}	$2.8\text{-}10^{14}$ to 1.5×10^{15}	2.6×10^{14}	—	1.5×10^{11}	2×10^{13}	$10^{14}\text{-}10^{15}$	$>5 \times 10^{15}$
Dielectric Str (short time), v/mil	D149	385	400-450	380	440	340	425	840	670
Dielectric Constant	D150								
60 cycles		4.0-5.3	4.6-5.6	4.0	3.2-4.0	9.3	3.3(10^3cps)	3.6(10^3cps)	3.99
10^6 cycles		3.6-3.8	3.9-5.4	3.3	3.0-3.6	4.0	—	—	—
Dissipation Factor	D150								
60 cycles		0.06-0.014	0.022-0.008	0.015	0.007-0.010	0.19	0.03	0.04(10^3cps)	0.028
10^6 cycles		0.03-0.04	0.019-0.015	0.05	0.010-.0.015	0.08	0.02	—	—
Arc Resistance, sec	D495	—	92-81	—	—	—	—	—	120
HEAT RESISTANCE									
Max Rec Service Temp, F	—	250-300[f]	250-300	250-300	175-200	—	212-250	175-230	—
Deflection Temp, F	D648								
66 psi		360	425-428	420	260-350	129	302	—	284
264 psi		155-160	420-419	410	115-130	—	131	—	256
APPLICABLE PROCESSING METHODS		Injection molding, extrusion, rotational molding, blow molding		Rotational molding	Extrusion, injection molding	Extrusion, injection molding		Extrusion, injection molding	Injection molding, blow mold, extru, compr molding
CHEMICAL RESISTANCE		Resists esters, ketones, alkalis, weak acids, alcohols and common solvents. Not resistant to conc mineral acids	Res most organic chem, such as alcohols, ketones, hydrocarbons and chlor solv. Att by str acids, phenols, str oxidizing agents	Res esters, ketones, alkalis, weak acids, alcohols and common solvents. Not res to conc mineral acids	Exc res to aqueous alkalis, aliphatic and aromatic hydrocarbons, ether and mineral oils; poor-good res to dil min acids, alcohols, arom acids	Res alkalis, petroleum products and common organic solvents. Not res to phenols and conc acids and oxidants	Res alkalis, petroleum products and common organic solvents. Not res to phenols and conc acids and oxidants	Res weak acids, alkali, strong alkali, oil, greases. Attacked by strong acids alcohols and org solvents	
USES		Bearings, gears, bushings, coil forms, brush backs, rod tubing, tape	Gen purpose type 6 parts requiring greater stiffness and dimen stab	Bearings, wearplates, bushings, gears, rollers, stock shapes	Parts requiring high impact strength or flexibility	Molded parts requiring flexibility and chemical res	Elec insulation and other nylon uses where low mois absorp is needed	Filament, rod, tubing, sheet, moldings req dim stability and low moist absorp	Lenses, containers, gauges, fuel tanks

[a] Dry, as-molded properties. [b] Non-cross-linked; can be cross-linked. [c] Dynamic, no lubrication, nylon to steel. 1000 psi. [e] 0.4, self-ext to slow burn. [f] Heat stabilized.

Nylons

Type →	ASTM	6/6 Nylon				6/10 Nylon		6/12 Nylon	Mineral reinf nylon
		General Purpose Molding[a]	Glass Fiber Reinforced[b]	Glass Fiber, Molybdenum Disulfide Filled[c]	General Purpose Extrusion[a]	General Purpose[a]	Glass Fiber (30%) Reinforced[c]		
PHYSICAL PROPERTIES	ASTM								
Specific Gravity	D792	1.13–1.15,—	1.37, 1.47	1.37–1.41	1.13, 1.15	1.07–1.09, —	1.30	1.06–1.08	1.47
Ther Cond, Btu/hr/sq ft/°F/in	—	1.7,—	1.5, 3.3	—	1.7, —	1.5	3.5	1.5	—
Coef of Ther Exp, 10^{-5} per °F	D696	4.5,—	2.1, 1.4	1.75	—	5	2.5	5.0	2.7–5.0
Specific Heat, Btu/lb/°F	—	0.3–0.5	—	—	0.3–0.5	0.3–0.5	—	0.3–0.5	—
Refractive Index, n_D	D542	Transluc	Opaque	Opaque	Opaque	Opaque	Opaque	Translucent	Opaque
Water Absorption (24 hr), %	D570	1.5,—	0.9, 0.8	0.5–0.7	1.5	0.4	0.2	0.4	0.5–0.8
Coef of Static Frict (against self)	—	0.04–0.13,—	—	—	—	—	—	—	0.23
MECHANICAL PROPERTIES									
Tensile Strength, 1000 psi	D638								
Ultimate		11.8, 11.2	25, 30	19–22	12.6, 8.6	8.5, 7.1	19	8.8, 8.8	—
Yield		11.8, 8.5	—	—	12.6, 8.6	8.5, 7.1	—	8.8, 7.4	9–10
Elongation, %	D638								
Ultimate		60, 300	1.8, 2.2	3	90, 240	85, 220	1.9	150, 340	10–25
Yield		5, 25	—	—	5, 30	5, 30	—	7, 40	—
Mod of Elast in Tension, 10^5 psi	D638	4.75, 3.85	14, 20	—	—	2.8–3.0, —	—	—	5.0
Flex Strength, 1000 psi	D790	Unbreak.	26, 35	26–28	—	8	23	—	12–16
Mod of Elast in Flex, 10^5 psi	D790	410, 175	10, 18	11–13	4.1, 1.75	2.8, 1.6	8.5	2.9, 1.8	3.3–6.0
Impact Str (Izod notched), ft-lb/in	D638	1.0, 2.0	2.5, 3.4[e]	—	1.3, —	0.6, 1.6	3.4	1.0, 1.4	1.0–1.5
Compr Strength (1%), 1000 psi	D695	4.9,—	20, 24	—	4.9 (1%), —	3.0 (1%), —	18	2.4, —	—
Hardness (Rockwell)	D785	R118, R108	E60, E80	M95–100	R118–108	R111	E40–50	R114, —	R119–121
Abrasion Res (Taber CS-17, 1000g), mg/1000 cycles	D1044	3–5, 6–8	—	—	—, 3–5	—	—	—, 5.7	12–30
ELECTRICAL PROPERTIES									
Volume Resistivity, ohm-cm	D257	$10^{14} - 10^{15}$	5.5 x 10^{15}, 2.6 x 10^{15},	—	10^{15}	10^{15}	—	10^{15}, 10^{12}	10^{16}
Dielectric Str (short time), v/mil	D149	385	400, 480	300–400	—	470	—	—	280–485
Dielectric Constant	D150								
60 cycles		4.0, —	4.0, 4.4	—	—	3.9, —	—	4.0, 6.0	—
10^6 cycles		3.6, —	3.5, 4.1	—	—	3.5, —	—	3.5, 4.0	—
Dissipation Factor	D150								
60 cycles		0.014, 0.04	0.018, 0.009	—	—	0.04, —	—	—	—
10^6 cycles		0.04, —	0.017, 0.018	—	—	—	—	0.02, 0.03	—
Arc Resistance, sec	D495	120	148, 100	135	120	120	—	—	115
HEAT RESISTANCE									
Max Rec Service Temp, F	—	250-300[d]	250–300[d]	250–300[d]	250-300[d]	225–300[d]	250–300[d]	350	—
Deflection Temp, F	D648								
66 psi		470	507, 509	—	470	300	430	330	400
264 psi		220	495, 500	—	220	135	420	180	300
CHEMICAL RESISTANCE		Inert to most organic chemicals such as esters, ketones, alcohols and hydrocarbons. Resist alkalis and salt solutions, but att by phenols, formic acid, strong mineral acids and strong oxidizing agents							
APPLICABLE PROCESSING METHODS		Injection molding			Extrusion	Injection molding		Injection molding, blow molding, extrusion	Injection molding
USES		Bearings, gears, bushings, coil forms, brush backs, rod, tubing	Mech parts where lubrication is undesirable or diff		Tubing, rod, pipe, sheeting, laminations	Jacketing for wire and cable, special molded parts			Elec housings and mech. parts

[a] Where two values are given, first is for dry, as-molded material, and second for moisture equilibrium in air; single value pertains to dry material. [b] First value for 30% glass fiber and second for 40%. All values at moisture equilibrium. [c] 30% glass fiber. [d] Heat stabilized for max heat resistance. [e] ¼ in.

TABLE A-1 *Continued*

Phenolics

Type and Filler →	ASTM	General—Woodflour and Flock	Shock—Paper, Flock or Pulp	High Shock—Chopped Fabric or Cord	Very High Shock—Glass Fiber
PHYSICAL PROPERTIES					
Specific Gravity	D792	1.32–1.46	1.34–1.46	1.36–1.43	1.75–1.90
Ther Cond, Btu/hr/sq ft/°F/ft	C177	0.097–0.3	0.1–0.16	0.097–0.170	0.20
Coef of Ther Exp. 10^{-5} per °F	D696	1.66–2.50	1.6–2.3	1.60–2.22	0.88
Spec Ht, Btu/lb/°F		0.35–0.40	—	0.30–0.35	0.28–0.32
Water Absorption (24 hr), %	D570	0.3–0.8	0.4–1.5	0.4–1.75	0.1–1.0
MECHANICAL PROPERTIES					
Mod of Elast in Tension, 10^5 psi	D638	8–13	8–12	9–14	30–33
Ten Str, 1000 psi	D638, D651	5.0–8.5	5.0–8.5	5–9	5–10
Elong (in 2 in.), %	D638	0.4–0.8	—	0.37–0.57	0.2
Hardness (Rockwell)	D785	E85–100	E85–95	E80–90	E50–70
Impact Str (Izod notched), ft-lb/in.	D256	0.24–0.50	0.4–1.0	0.6–8.0	10–33
Mod of Elast in Flex, 10^5 psi	D790	8–12	8–12	9–13	30–33
Flex Str, 1000 psi	D790	8.5–12	8.0–11.5	8–15	10–45
Compr Str, 1000 psi	D695	22–36	24–35	15–30	17–30
ELECTRICAL PROPERTIES					
Vol Res, ohm-cm	D257	10^9–10^{13}	1–50×10^{11}	$>10^{10}$	7–10×10^{12}
Dielec Str (short time), v/mil	D149	200–425	250–350	200–350	200–370
Dielec Const					
60 Cycles	D150	5.0–9.0	5.6–11.0	6.5–15.0	7.1–7.2
10^6 Cycles	D150	4.0–7.0	4.5–7.0	4.5–7.0	4.6–6.6
Dissip Factor					
60 Cycles	D150	0.05–0.30	0.08–0.35	0.08–0.45	0.02–0.03
10^6 Cycles	D150	0.03–0.07	0.03–0.07	0.03–0.09	0.02
Arc Resistance, sec	D495	5–60	5–60	5–60	60
APPLICABLE PROCESSING METHODS		Injection molding, compression molding, transfer molding			
HEAT RESISTANCE					
Max Rec Svc Temp, F		300–350	300	250–300	350–450
Deflection Temp, F	D648	260–360	290–340	250–340	600
CHEMICAL RESISTANCE		Severely attacked by strong acids and strong alkalis. Effects of dilute acids, alkalis and and organic solvents vary with the reagent. Chemical resistance varies with the particular formulation and not all materials of a type are equally resistant			
USES		Mechanical applications include pulleys, wheels, motor housings, handles. Electrical uses include coil forms, ignition parts, condenser housings, fuse blocks, instrument panels. Thermal applications include handles, appliance connector plugs. Chemical uses include photographic development tanks, rayon spinning buckets and parts, milking machine cups. Decorative uses include radio and television cabinets, handles, knobs, buttons			

BLE A-1 *Continued*

Polyester—Thermoplastic　　　　　　　　　　　　　　　Phenolics

| Type → | ASTM | Injection Moldings | | | | | |
| | | Polybutylene Terephthalates | | | Polytetramethylene Terephthalate | | |
		General Purpose Grade	Glass-Reinforced Grades	Glass-Reinforced Flame Retardant	General Purpose Grade	Glass-Reinforced Grade	Asbestos-Filled Grade
PHYSICAL PROPERTIES	ASTM						
Specific Gravity	D792	1.31	1.52	1.58	1.31	1.45	1.46
Specific Heat, Btu/lb/°F	—	—	—	—	0.36-0.55	—	—
Ther Cond, Btu/hr/sq ft/°F/in.	C177	1.1	1.3	1.3	—	—	—
Coef of Ther Exp, 10^{-5} in./in./°F	D696	5.3	2.7-3.3	3.5	4.9-13.0[a]	—	—
Vol Resistivity, 10^{16} ohm-cm	D257	4	3.2-3.3	3.4	2×10^{15}	—	3×10^{14}
Dielectric Str (short time), v/mil	D149	590	560-750	750	420-540	—	580
Dielectric Constant	D150	3.1-3.3	3.7-4.2	3.7-3.8	3.16	—	3.5-4.2
Dissipation Factor (to 10^3 Hz)	D150	0.002	0.002-0.003	0.002	0.023	—	0.015
Arc Resistance, sec	D495	190	130	80	125	—	108
Water Absorption (24 hr), %	D570	0.08	0.06-0.07	0.07	0.09	0.07	0.1
MECHANICAL PROPERTIES							
Tensile Str, 10^3 psi	D638	8	17-17.3	17	8.2	14	12
F			7			—	—
F			5.5			—	—
Elongation, %	D638	300	1-5	5	250	<5	<5
Flex Str, 10^3 psi	D790	12.8	22-24	23	12	19	19
Flex Modulus, 10^6 psi	D790	0.34	1.2-1.5	1.2	3.3	8.7	9.0
F		—	0.63	—	—	—	—
F		—	0.53	—	—	—	—
Compressive Str, 10^3 psi	D695	13	16-18	18	—	—	—
Shear Str, 10^3 psi	D732	7.7	8.9	9	—	—	—
Impact Str (Izod), ft-lb/in.	D256						
Notched		1.2	1.3-2.2	1.8	1	1	0.5
Unnotched		—	4.2-15	15	—	7	6.0
Hardness, Rockwell	D785	R117	R118-M90	R119	R117	R117-M85	M85
Deflection Temp, F	D648						
psi		310	420	420	302	—	—
psi		130	415-416	415	122	380	330
Endurance Limit (10^7 cyc), 10^3 psi	D671	2.85	5	5	—	—	—
%	D674	1.1 (1000 psi)	0.44 (4000 psi)	0.44 (4000 psi)	—	—	—
Abrasion Resistance[c], mg/10^3 cyc	D1044	6.5	9-50	11	—	—	—
Coef of Friction	D1894						
		0.17	0.16	0.16	—	—	—
		0.13	0.14	0.14	—	—	—

PROCESSING METHODS

Injection molding, extrusion, thermoforming, foam molding

RESISTANCE

Resistant to aliphatic hydrocarbons, gasoline, carbon tetrachloride, perchloroethylene, oils, fats, alcohols, glycols, ethers, high molecular weight esters and ketones, dilute acids and bases, detergents, most aqueous salt sol, at ambient temp. Attacked by strong acids and bases. Resistant to potable water at ambient temp. Prolonged use in water above 150 F not recommended. Swollen by ethylene dichloride, low molecular weight ketones and substituted aromatic compounds

Gears, bearings, valves, pump parts, fittings, rollers, cams, bushings, electronic parts, textile machinery parts, tape cassettes, fasteners

own are at approximately room temperature unless otherwise noted. [a] For 32-302 F range. [b] Underwriters' Lab spec. [c] Taper abrasion; CS-17 wheel,

---

Phenolics

Type and Filler →	ASTM	Arc Resistant—Mineral	Rubber Phenolic—Woodflour or Flock	Rubber Phenolic—Chopped Fabric	Rubber Phenolic—Asbestos
PHYSICAL PROPERTIES	ASTM				
Specific Gravity	D792	1.6-3.0	1.24-1.35	1.30-1.35	1.60-1.65
Ther Cond, Btu/hr/sq ft/°F/ft	C177	0.24-0.34	0.12	0.05	0.04
Coef of Ther Exp, 10^{-5} per °F	D696	—	0.83-2.20	1.7	2.2
Spec Ht, Btu/lb/°F		0.28-0.32	0.33	—	—
Water Absorption (24 hr), %	D570	0.2	0.5-2.0	0.5-2.0	0.10-0.50
MECHANICAL PROPERTIES					
Mod of Elast in Tension, 10^5 psi	D638	10-30	4-6	3.5-6	5-9
Ten Str, 1000 psi	D638, D651	6	4.5-9	3.5	4
Elong (in 2 in.), %	D638	—	0.75-2.25	—	—
Hardness (Rockwell)	D785	E80-90	M40-90	M57	M50
Impact Str (Izod notched), ft-lb/in.	D256	0.32	0.34-1.0	2.0-2.3	0.3-0.4
Mod of Elast in Flex, 10^5 psi	D790	10-30	4-6	3.5	5.0
Flex Str, 1000 psi	D790	10	7-12	7	7
Compr Str, 1000 psi	D695	20	12-20	10-15	10-20
ELECTRICAL PROPERTIES					
Vol Res, ohm-cm	D257	6×10^{12}	10^8-10^{11}	10^{11}	10^{11}
Dielec Str (short time), v/mil	D149	380	250-375	250	350
Dielec Const					
60 Cycles	D150	7.4	9-16	15	15
10^6 Cycles	D150	5.0	5	5	5
Dissip Factor					
60 Cycles	D150	0.13-0.16	0.15-0.60	0.5	0.15
10^6 Cycles	D150	0.10	0.1-0.2	0.09	0.13
Arc Resistance, sec	D495	180	7-20	10-20	5-20

APPLICABLE PROCESSING METHODS

Injection molding, compression molding, transfer molding, foam molding, reinforced layup molding, laminating

HEAT RESISTANCE

Max Rec Svc Temp, F		400	212-300	212-225	225-360
Deflection Temp, F	D648	335	220-270	220-280	250-300

CHEMICAL RESISTANCE

Severely attacked by strong acids and strong alkalis. Effects of dilute acids, alkalis and organic solvents vary with the reagent. Chemical resistance varies with the particular formulation and not all materials of a type are equally resistant

USES

Mechanical applications include pulleys, wheels, motor housings, handles. Electrical uses include coil forms, ignition parts, condenser housings, fuse blocks, instrument panels. Thermal applications include handles, appliance connector plugs. Chemical uses include photographic development tanks, rayon spinning buckets and parts, milking machine cups. Decorative uses include radio and television cabinets, handles, knobs, buttons

TABLE A-1 *Continued*

Phenylene Oxides, Polysulfones, Polyarylsulfone

Material →	ASTM	Phenylene Oxides (Noryl)			Polysulfones		Polyaryl-sulfone
		SE-100	SE-1	Glass Fiber Reinforced[a]	Standard	Glass Fiber Reinforced[b]	
PHYSICAL PROPERTIES	ASTM						
Specific Gravity	D792	1.10	1.06	1.21, 1.27	1.24	1.41, 1.55	1.36
Ther Cond, Btu/hr/sq ft/°F/in.	C177	1.10	1.5	1.15, 1.1	1.8	—	1.1
Coef of Ther Exp, 10^{-5} per °F	D696	3.8	3.3	2.0, 1.4	3.1	1.6, 1.2	2.6
Specific Heat, Btu/lb/°F	—	—	—	—	0.24	—	—
Refractive Index, n_D	D542	Opaque	Opaque	Opaque	1.63	Opaque	—
Water Absorption (24 hr), %	D570	0.07	0.07	0.06	0.22	0.22, 0.18	1.8
MECHANICAL PROPERTIES							
Tensile Strength, 1000 psi	D638						
Ultimate		—	—	—	—	—	13
Yield		7.8	9.6	14.5, 17.0	10.2	17, 19	8
Elongation, %	—						
Ultimate		50	60	4-6	50-100	—	15-20
Yield		—	—	—	5.6	2, 1.6	13
Mod of Elast in Tension, 10^5 psi	D638	3.8	3.55	9.25, 13.3	3.6	10.9, 14.9	3.7
Flex strength, 1000 psi	D790	12.8	13.5	20.5, 22	15.4	25, 28	17.2
Mod of Elast in Flex, 10^5 psi	D790	3.6	3.6	7.4, 10.4	3.9	12, 15.5	4.0
Impact Str (Izod notched), ft-lb/in.	D638	5.0	5.0	2.3	1.3	1.8, 2.0	5.0
Compr Strength, 1000 psi	D695	12	16.4	17.6, 17.9	13.9	—	17.8
Hardness (Rockwell)	D785	R115	R119	L106, L108	R120	M84	M110
Coef of Static Frict (against self)	—	—	—	—	0.67	—	0.1-0.3
Abrasion Res (Taber, CS-17), mg/1000 cycles	—	100	20	35	20	—	40
ELECTRICAL PROPERTIES							
Volume Resistivity ohm-cm	D257	10^{17}	10^{17}	10^{17}	5×10^{16}	10^{17}	3.2×10^{16}
Dielectric Str (short time), v-mil	D149	400(⅛ in.)	500(⅛ in.)	1020(1/32 in.)	425	480	350
Dielectric Constant	D150						
60 cycles		2.65	2.69	2.93	3.06	3.55	3.94
10^6 cycles		2.64	2.68	2.92	3.03	3.41	3.7
Dissipation Factor	D150						
60 cycles		0.0007	0.0007	0.0009	0.0008	0.0019	0.003
10^6 cycles		0.0024	0.0024	0.0015	0.0034	0.0049	0.012
Arc Resistance, sec	D495	75	75	120	122	114	67
HEAT RESISTANCE							
Max Rec Service Temp, F		—	212	—	340	350	500
Deflection Temp, F	D648						
66 psi		230	279	293, 317	358	389	—
264 psi		212	265	282, 310	345	365	525
APPLICABLE PROCESSING METHODS		Injection molding, extrusion, thermoforming, foam molding			njection molding, extrusion, thermoforming		Injection molding
CHEMICAL RESISTANCE		Excellent res to aqueous media such as detergents and weak and strong acids and bases even at elevated temperatures. Many halogenated and aromatic hydrocarbons will soften or partially dissolve these materials			Res to inorganic acids, alkalis and aliphatic hydrocarbons; partially soluble or swells in ketones and aromatic hydrocarbons; soluble in chlorinated hydrocarbons		Res to aqueous acids and bases, fuels, oils, fluorinated solvents. Dissolves in highly polar solvents such as DMF, DMAC, NMP
USES		Automotive dashboards & electrical connectors; appliance & business machine housings, cabinets, consoles & covers; coffee brewers & dispensers; pump & plumbing parts; valves; tape cartridge platforms; coil assemblies; bus bar insulators; switch housings; terminal blocks; tuner bars; light fixtures			Coil bobbins, switches, terminal blocks, battery cases, connectors, circuit carriers, sockets, tube bases, range hardware, coffee maker parts, sight glasses, auto parts, lamp bezels, aircraft ducts, housing and side wall panels, meter housings and components, projector transparencies		Molded parts requiring str and toughness at high temps. Extrusions, coatings, filled compositions for bearing use

[a] Where two values are given, first applies to 20% glass fiber and second to 30%, otherwise same value applies to both. [b] Where two values are given, first applies to 30% glass fiber and second to 40%, otherwise same value applies to both. [c] 10% deformation.

Polybutadienes, Polybutylenes, Polycarbonates and Polyme[...]

Materials	Type →	ASTM	Polybutadienes	Polybutylenes	Polycarbonates (Unfilled)
PHYSICAL PROPERTIES		ASTM			
Specific gravity		D792	1.6-2.0	0.910-0.915	1.19-1.2[…]
Refractive index, n_D		D542	—	150	1.586
Transparency (visible light), %		D1003	Opaque	Translucent	87-89
MECHANICAL PROPERTIES					
Tensile		D638			
Strength, 1000 psi (yld)			5-12	2.2	8.5-9.[…]
Elongation, % (ult)			—	300	10-11
Modulus, 1000 psi		D790	400-1200	26-35	340-45[…]
Flexural					
Strength, 1000 psi			7-21	—	12-1[…]
Modulus, 1000 psi			700-1800	—	310-5[…]
Compressive str, 1000 psi (2% offset)		D695	11-20	—	10-1[…]
Impact strength					
Izod (notched), lb/in.		D256	1-10	No break	4-1[…]
Gardner, ft-lb/in.			1-4	—	—
Hardness					
Rockwell		D785	E30-70	D65-53	M68-[…]
Shore		D785	—	—	—
Bending fatigue str (Woehler), 1000 psi		D671	—	—	1-3
Abrasion res (Taber, CS-17 Wheel, 1000 g wt), mg loss/1000 cycles		D1044	—	—	9-2[…]
Coef of friction (static, 40 psi)		D1894			
Against self			—	—	—
Against steel			—	—	0.3[…]
ELECTRICAL PROPERTIES					
Vol res, ohm-cm		D257	1.5×10^{15}	—	2.5-8.2
Dielec str, v/mil (dry)		D149			
Short time			400-600	—	380
Step-by-step			—	—	—
Dielec constant		D150			
60 Hz			3.3	2.25	3.10
10^6 Hz			3.3	2.25	2.96
Dissip factor		D150			
60 Hz			0.009	0.005	0.0[…]
10^6 Hz			0.003	0.005	0.010
Arc res, sec (tungsten electrodes)		D495	—	—	1[…]
THERMAL PROPERTIES					
Ther cond, Btu/hr/sq ft/°F/in.		C177	—	—	1.3
Coef of therm exp. 10^{-5}/°F		D696	—	—	1.7
Specific heat, Btu/lb/°F			—	—	[…]
Heat deflection temp, F		D648			
At 66 psi			—	215-235	18[…]
At 264 psi			500	120-140	26[…]
Brittleness temp, F		D746	—	−13	[…]
Max temp for cont use (no load), F		—	350-500	225	[…]
CHEMICAL AND ENVIRONMENTAL RES		D570			
Water absorption, %					
In 24 hr			0.10	0.01-0.02	0.[…]
At saturation			—	—	[…]
Weathering			Discolors	Embrittles	D[…]
Acids					
Weak			Resists	Resists	[…]
Strong			Resists	Attacks	[…]
Alkali					
Weak			Resists	Resists	R[…]
Strong			Resists	Resists	[…]
Organic solvents			Resists	—	[…]
Fuels			Resists	—	[…]
Oil and grease			Resists	—	[…]
METHODS OF PROCESSING			Calendering, casting, coating, compression molding, extrusion, injection molding, rotational molding, transfer molding	Casting, compression mldg, extrusion, injection mldg, rotational mldg, thermoforming	Blow sion, inject tation therm[…]

Polyesters—Thermosets

Type →	ASTM	Cast Polyester		Reinforced Polyester Moldings			Pultrusions	
		Rigid	Flexible	High Strength (glass fibers)	Heat and Chemical Resistant (asbestos)	Sheet Molding Compounds, Genera. Purpose	General purpose	High perform-ance
PHYSICAL PROPERTIES								
Specific Gravity	D792	1.12-1.46	1.06-1.25	1.8-12.0	1.5-1.75	1.65-1.80	1.61	1.94
Ther Cond, Btu/hr/sq ft/°F/in.	—	0.10-0.12	—	1.32-1.68	—	—	4	5
Coef of Ther Exp, per °F	D696	$3.9\text{-}5.6 \times 10^{-6}$	—	$13\text{-}19 \times 10^{-6}$	—	—	5×10^{-6}	3×10^{-6}
Specific Heat, Btu/lb/°F	—	0.30-0.55	—	0.25-0.35	—	0.20-0.25	0.28	0.24
Water Absorption (24 hr), %	D570	0.20-0.60	0.12-2.5	0.5-0.75	0.25-0.50	0.15-0.25	0.75	0.75
Transparency (visible light), %	—	—	—	Opaque	Opaque	Opaque	Opaque	Opaque
Refractive Index, n_D	D542	1.53-1.58	1.50-1.57	—	—	—	—	—
MECHANICAL PROPERTIES								
Tensile Strength, 1000 psi	D638	4-10	1-8	5-10	4-6	15-17	20[a]	100[a]
Tensile Modulus, 10^5 psi	D638	1.5-6.5	0.001-0.10	16-20	12-15	15-20	23[a]	60[a]
Elongation, %	D638	1.7-2.6	25-300	0.3-0.5	—	—	—	—
Impact Strength (Izod notched), ft/lb/in.	D256	0.18-0.40	4.0	1-10	0.45-1.0	5-15	18[a]	18[a]
Flexural Strength, 1000 psi	D790	14-18	4-16	6-26	10-13	26-32	30[a]	—
Mod of Elast in Flex, 10^5 psi	D790	1-9	0.001-0.39	15-25	—	15-18	20[a]	—
Compr Strength, 1000 psi	D690	12-37	1-17	20-26	20-25	22-36	—	—
Hardness (Barcol)	D785	35-50	6-40	60-80	40-70	45-60	50[a]	50[a]
ELECTRICAL PROPERTIES								
Volume Resistivity, ohm-cm	D257	10^{13}	10^{12}	1×10^{12} 1×10^{13}	$1 \times 10^{12}\text{-}$ 1×10^{13}	$6.4 \times 10^{15}\text{-}$ 2.2×10^{16}	—	—
Dielectric Str (step-by-step), v/mil	D149	300-400	300-400	200-400	350	400-440	—	—
Dielectric Constant	D150							
60 cycles		2.8-4.4	3.18-7.0	—	—	4.62-5.0	4.5(1)	4.5(1)
10^6 cycles		2.8-4.4	3.7-6.1	—	—	4.55-4.75	—	—
Dissipation Factor	D150							
60 cycles		0.003-0.04	0.01-0.18	—	—	0.0087-0.04	0.03(1)	0.03(1)
10^6 cycles		0.006-0.04	0.02-0.06	—	—	0.0086-0.022	—	—
Arc Resistance, sec	D495	115-135	125-145	130-170	—	130-180	80	80
APPLICABLE PROCESSING METHODS		Casting		Layup molding, laminating, compression molding, transfer molding		Matched metal die compression molding	Pultrusion	
HEAT RESISTANCE								
Max Rec Svc Temp, F	—	250-300	150-250	250-400	300	300	250-300	300
Heat Deflection Temp (264 psi), F	D648	120-400	—	400	375-400	375-400	400	400
CHEMICAL RESISTANCE		Slightly to heavily attacked by strong acids; attacked by strong alkalis, ketones and solvents		Good to exc res to weak acids, organic solvents, weak alkalis; good res to strong acids; poor to fair res to strong alkalis	Good to exc res to weak acids, organic solvents and weak alkalis; good res to strong acids; air to good res to strong alkalis	Good to exc res to weak acids, organic solvents and weak alkalis; good res to strong acids; poor to fair res to strong alkalis	Good to exc res to weak acids, organic solvents, weak alkalies good res to strong acids; poor to fair res to strong alkalies	
USES		Electrical components, buttons, decorative architectural uses	Flooring, tooling encapsulants, buttons and shields	Chairs, housings, covers, trays molded panels, bezels, motor shrouds, electrical parts, fans, helmets			Elec components, corrosion res construction, high-str-to-wt mech parts	

[a] Longitudinal direction.

TABLE A-1 *Continued*

Polyethylenes

Type →	ASTM	Type I—Lower Density (0.910-0.925)			Type II—Medium Density (0.926-0.940)	
		Melt Index 0.3-3.6	Melt Index 6-26	Melt Index 200	Melt Index 20	Melt Index 1.0-1.9
PHYSICAL PROPERTIES						
Specific Gravity	D792	0.910–0.925	0.918–0.925	0.910	0.930	0.930–0.940
Ther Cond, Btu/hr/sq ft/°F/ft	C177	0.19	0.19	0.19	0.19	0.19
Coef of Ther Exp, 10^{-5} per °F	D696	8.9–11.0	8.9–11.0	11.0	8.3–16.7	8.3–16.7
Refractive Index	D542	1.51	1.51	1.51	1.51	1.51
Spec Ht, Btu/lb/°F		0.53–0.55	0.53–0.55	0.53–0.55	0.53–0.55	0.53–0.55
Water Absorption (24 hr), %	D570	<0.01	<0.01	<0.01	<0.01	<0.01
MECHANICAL PROPERTIES						
Mod of Elast in Tension, 10^5 psi	D638	0.21–0.27	0.20–0.24	—	—	—
Ten Str, 1000 psi	D412	1.4–2.5	1.4–2.0	0.9–1.1	2.0	2.3–2.4
Elong, %	D412	500–725	125–675	80–100	200	200–425
Hardness (Shore)	D785	C73, D50–52	C73, D47–53	D45	D55	D55–D56
Impact Str (Izod), ft-lb/in. notch	D256	—	—	—	—	—
Brittleness Temp, F		<—94	<—4	<14	<—148	<—148
Mod of Elast in Flex, 10^3 psi	D747	13–27	12–30	10	35-50	35–50
Shear Str, 1000 psi		1.6–1.85	1.4–1.7	1	—	—
ELECTRICAL PROPERTIES						
Vol Res, ohm-cm	D257	10^{17}–10^{19}	10^{17}–10^{19}	10^{17}–10^{19}	>10^{15}	>10^{15}
Dielec Str (short time), v/mil	D149	480	480	480	480	480
Dielec Const	D150	2.3	2.3	2.3	2.3	2.3
Dissip Factor	D150	<0.0005	<0.0005	<0.0005	<0.0005	<0.0005
APPLICABLE PROCESSING METHODS		Injection molding, extrusion, rotational molding, blow molding, foam molding			Injection molding, extrusion, thermoforming, rotational molding, blow molding	
HEAT RESISTANCE Vicat Softening Point, F		176–201	176–201	—	215-230	220–235
CHEMICAL RESISTANCE		Excellent resistance to acids and alkalis at normal temperature, except oxidizing acids such as nitric, chlorosulfonic and fuming sulfuric. Below 122 F, insoluble in organic solvents; at higher temperatures, soluble to varying degrees in hydrocarbons and halogenated hydrocarbons, but insoluble in more polar liquids. Generally, a higher melt index material has greater solubility				
USES		Injection moldings: kitchen utilityware, toys, process tank liners, closures, packages, sealing rings, battery parts. Blow moldings: squeeze bottles for packaging, containers for drugs. Film: wrapping materials for food, clothes, other items. Wire and cable: high frequency insulation, jacketing. Pipe: chemicals handling, irrigation systems, natural gas transmission				

Type →	ASTM	Type III—Higher Density (0.941-0.965)			High Molecular Weight
		Melt Index 0.2-0.9	Melt Index 0.1-12.0	Melt Index 1.5-15	
PHYSICAL PROPERTIES					
Specific Gravity	D792	0.96	0.950–0.955	0.96	0.94
Ther Cond, Btu/hr/sq ft/°F/ft	C177	0.19	0 19	0.19	0.19
Coef of Ther Exp, 10^{-5} per°F	D696	8.3–16.7	8.3–16.7	8.3–16.7	—
Refractive Index	D542	1.54	1 54	1.54	—
Spec Ht., Btu/lb/°F		0.46–0.55	0.46–0.55	0.46–0.55	—
Water Absorption (24 hr), %	D570	<0.01	<0.01	<0.01	<0.01
Flammability, ipm	D635	1.0	1.0	1.0	1.0
MECHANICAL PROPERTIES					
Mod of Elast in Tension, 10^5 psi	D638	—	—	—	1.0
Ten Str, 1000 psi	D412	4.4	2.9–4.0	4.4	5.4
Ultimate/Elongation, %	D412	700-1000	50-1000	100-700	400
Hardness (Shore)	D785	D68–70	D60–70	D68–70	60–65

Polyethylenes

Type →		Type III—Higher Density (0.941-0.965)			High Molecular Weight
		Melt Index 0.2-0.9	Melt Index 0.1-12.0	Melt Index 1.5-15	
MECHANICAL PROPERTIES (Cont'd)					
Impact Str (Izod), ft-lb/in. notch	D256	4.0–14	0.4–6.0	1.2–2 5	>20
Brittleness Temp, F		−106 to −180	<−76 to <−170	−100 to −180	<−100
Mod of Elast in Flex, 10^3 psi	D747	130–150	90–125	150	75
ELECTRICAL PROPERTIES					
Vol Res, ohm-cm	D257	$>10^{15}$	$>10^{15}$	$>10^{15}$	$>10^{15}$
Dielec Str (short time), v/mil	D149	480	480	480	480
Dielec Const	D150	2.3	2 3	2.3	2 3
Dissip Factor	D150	<0.0005	<0.0005	<0.0005	<0.0005
APPLICABLE PROCESSING METHODS		Injection molding, extrusion, thermoforming, rotational molding, blow molding, foam molding			Extrusion, compression molding, and special injection molding
HEAT RESISTANCE Vicat Softening Point, F		258-266	240-255	250-260	—
CHEMICAL RESISTANCE		Same basic chemical resistance as Types I and II, but better resistance to some specific chemicals			
USES		Refrigerator parts, packaging, structural housing panels, pipe, defroster and heater ducts, sterilizable housewares and hospital equipment, hoops, battery parts, blow molded containers and parts, film wrapping materials, wire and cable insulation, and chemical resistant pipe			

ᵃ Powder. ᵇ Flow molding.

Olefin Copolymers

Type →		EEA (Ethylene Ethyl Acrylate)	EVA (Ethylene Vinyl Acetate)	Ethylene-Butene	Propylene-Ethylene	Ionomer	Poly-allomer
PHYSICAL PROPERTIES	ASTM						
Specific Gravity	D792	0.93	0.94	0.95	0.91	0.94	0.898-0.904
Tensile Impact, ft-lb/in.²	D1822	500	690	—	150	400	—
Tensile Str, 100 psi	D638	2.0	3.6	3.5	4	4	30-43
Izod Impact Str (notched), ft-lb/in.	D256	—	—	0.4	1.1	9-14	1.5
Hardness, Shore D	D785	35	36	65	—	60	—
Elongation (in 1 in.), %	D638	650	650	20	—	450	300-400
Flexural Modulus, psi	D790	—	—	165	140	—	$0.7\text{-}1.3 \times 10^5$
ELECTRICAL PROPERTIES							
Vol Res, 10^{15} ohm-cm	D257	2.4	0.15	—	—	10	$>10^{16}$
Dielec Str (short time), v/mil	D149	550	525	—	—	1000	500-650
Dielec Const, 60 cps	D150	2.8	3.16	—	—	2.4	2.3
Dissip Factor, 60 cps	D150	0.001	0.003	—	—	0.003	>0.0005
THERMAL PROPERTIES							
Heat Deflection Temp (66 psi), F	D648	—	—	—	104	105	122-133ᵃ
Brittleness Temp, F	D746	−155	−148	−35	—	−160	—
Softening Point, Vicat	D1525	147	147	243	—	162	—
CHEMICAL RESISTANCE		Res most weak mineral acids, alkalis; att by chlorinated hydrocarbons, straight-chain paraffinic solvents, benzene		Generally satisfactory res to chemicals; good res to stress corrosion		Res to organic solvents; att by oxidizing acids	Res to strong alk, weak acids, organic solvents; att slowly by acid reagents
USES		Tubing, seals, bushings, dampening pads, rug backing, elec insulation, floor mats, other flexible items		Packaging, appliances, furniture, wire and cable insulation, rigid parts		Skin packaging; coated substrates; clear, flexible parts	Molded parts req toughness and good hinge properties

ᵃ For 264 psi.

TABLE A-1 *Continued*

Polyimides, Poly(amide-imide)

Material →		Polyimides			Poly(amide-imide)	
	ASTM	Unreinforced		Glass Reinf	High Modulus	High Impact
PHYSICAL PROPERTIES						
Specific Gravity	D792	14.3	1.47	1.90	1.41	1.38
Ther Cond, Btu/hr/sq ft/°F/in.	Cenco Fitch	6.78[b]	3.8[b]	3.59[b]	—	—
Coef of Ther Exp, per °F x 10^{-5}	D696	2.5	3.0	0.8	1.9	2.4
Specific Heat, Btu/lb/°F	—	0.31[b,c]	0.28[b,c]	0.27[b,c]	—	—
Refractive Index, n_D	D542	Opaque	Opaque	Opaque	Opaque	Opaque
Water Absorption (24 hr), %	D570	0.47	0.68	0.2	—	—
Coef of Friction	—	—	—	—	—	—
MECHANICAL PROPERTIES						
Tensile Strength, 1000 psi	D638					
Ultimate		—	—	—	—	—
Yield		7.5	5	28	13.3	17.0
Elongation, %	D638					
Ultimate		<1	1.2	<1	5	15
Yield		—	—	—	—	—
Mod of Elast in Tension, 10^5 psi	D638	7.5	5.4	45	—	—
Flex Strength, 1000 psi	D790	11	6.6	56	23.4	27.4
Mod of Elast in Flex, 10^5 psi	D790	7.0	5.0	38.4	7.0	5.2
Impact Str (Izod notched), ft-lb/in.	D638	0.5[d]	0.5[d]	17[d]	1.0	3.0
Compr Strength, 1000 psi	D695	27.4	18.4	42	35	35
Hardness (Rockwell)	D785	99	95	114E	104E	78E
Abrasion Res (Taber, CS-10 wheel), mg/1000 cycles, gm loss	D1044	0.080[a]	0.004[e]	20	—	—
ELECTRICAL PROPERTIES						
Volume Resistivity, ohm-cm	D257	[f]	4×10^{15}	9.2×10^{15}	0.7×10^{15}	0.7×10^{15}
Dielectric Str (short time), v/mil	D149	[f]	310	300	430	440
Dielectric Constant	D150					
60 cycles		[f]	4.12	4.84	—	—
10^6 cycles		[f]	3.96	4.74	3.8	3.9
Dissipation Factor	D150					
60 cycles		[f]	0.003	0.0034	—	—
10^6 cycles		[f]	0.011	0.0055	0.001	0.009
Arc Resistance, sec	D495	[f]	152	50-180	—	—
HEAT RESISTANCE						
Max Rec Service Temp, F	—	500	500	500	—	—
Deflection Temp, F	D648					
66 psi		—	—	—	—	—
264 psi		582	615	660	545	520
APPLICABLE PROCESSING METHODS		Compression molding, laminating, filament winding			Compression molding, laminating, injection molding, foam molding, transfer molding	
CHEMICAL RESISTANCE		Resist polar and nonpolar organic solvents, dilute acids and bases			Resists most acids, alkalis, oils, fuels, greases and alcohol; attacked by organic solvents	
USES		Bushings, valve seats and high temp mechanical parts		High temp uses, such as jet engine components	Bushings, valve seats, grinding wheels and other high temperature mechanical parts	

[a] Extruded sheet. [b] G. E. test. [c] Cal/gm/°C. [d] ASTM D256. [e] In./1000 hr. dry, 10,000 PV. [f] This grade is an electrical conductor.

Polypropylene, Polyphenylene Sulfide, Polyether Sulfone

Material ➤		Polypropylene					Polyphenylene Sulfide		Poly-ether sulfone
		General Purpose	High Impact	Asbestos Filled	Glass Reinforced	Flame Retardant	Standard	40% Glass Reinforced	
PHYSICAL PROPERTIES	ASTM								
Specific Gravity	D792	0.900-0.910	0.900-0.910	1.11-1.36	1.04-1.22	1.2	1.34	1.64	1.37
Ther Cond, Btu/hr/sq ft/ °F/in.	—	1.21-1.36	1.72	—	—	—	2.0	—	—
Coef of Ther Exp, 10^{-6} per °F	D696	3.8-5.8	4.0-5.9	2-3	1.6-2.4	—	3.0	2.2	—
Specific Heat, Btu/lb/°F	—	0.45	0.45-0.48	—	—	—	0.26	—	—
Refractive Index, n_D	D542	Transl-op	Trans-op	Opaque	Opaque	Opaque	Opaque	Opaque	Opaque
Water Ab (24 hr), %	D570	<0.01-0.03	<0.01-0.02	0.02-0.04	0.02-0.05	0.02-0.03	0.2	0.01	0.43
MECHANICAL PROPERTIES									
Tensile Strength, 1000 psi	D638, C								
Maximum		4.8-5.5	—	—	—	—	—	—	—
Yield		4.8-5.2	2.8-4.3	3.3-8.2	6-10	3.6-4.2	11	21	12.2
Elongation, %	—								
Break		30->200	30->200	3-20	2-4	3-15	3	3-9	—
Yield		9-15	7-13	5	—	—	—	—	—
Mod of Elast in Tension, 10^5 psi	D638, B	1.6-2.2	1.3	—	812	1.5-2.4	4.8	11.2	3.54
Flex Yld Strength, 1000 psi	D790, B	6-7	4.1	7.5-9	8-11	—	20	37	18.65
Mod of Elast in Flex, 10^5 psi	D790, B	1.7-2.5	1.0-2.0	3.4-6.5	4-8.2	1.9-6.1	6.0	22.0	3.73
Impact Str (Izod notched), ft-lb/in.	D256	0.4-2.2	1.5-12	0.5-1.5	0.5-2	2.2	3.0	6.0	1.6
Compr Yield Strength, 1000 psi	D695	5.5-6.5	4.4	7.0	6.5-7	—	16	21	—
Hardness (Rockwell)	D785	R80-R100	R28-95	R90-R110	R90-R115	R60-R105	R124	R123	M88
ELECTRICAL PROPERTIES									
Vol Res, ohm-cm	D257	$>10^{17}$	10^{17}	1.5×10^{16}	1.7×10^{16}	4×10^{16}-10^{17}	10^{16}	10^{16}	10^{17}-10^{18}
Diel Str (short time), vpm	D149	650(125 mil)	450-650	450	317-475	485-700	595	490	—
Dielec Const	D150								
60 Cycles		2.20-2.28	2.20-2.28	2.75	2.3-2.5	2.46-2.79	3.11	3.79	3.5
10^6 Cycles		2.23-2.24	2.23-2.27	2.6-3.17	2-2.5	2.45-2.70	3.22	3.88	3.5
Dissip Factor	D150								
60 Cycles		0.0005-0.0007	<0.0016	0.007	0.002	0.0007-0.017	0.0004	0.0037	0.001
10^6 Cycles		0.0002-0.0003	0.0002-0.0003	0.002	0.003	0.0006-0.003	0.0007	0.0041	0.006
Arc Res, sec	D495	125-136	123-140	121-125	73-77	15-40	—	160	—
HEAT RESISTANCE									
Max Rec Svc Temp, F	—	230	—	250	250	205	500	500	—
Deflection Temp, F	D648								
66 psi		205-230	190-235	270-290	275-310	245-280	—	—	—
264 psi		135-140	120-140	170-220	250-300	155	278	425	400
APPLICABLE PROCESSING METHODS		Extrusion, injection mldg, thermoforming, rotational molding, blow mldg, coating, foam mldg					Injection mldg, coating, compression mldg		Injection mldg, extrusion
CHEMICAL RESISTANCE		Res to most acids, alkalis and saline solutions, even at higher temp; res to higher aliphatic solvents and polar substances. Above 175 F, soluble in such aromatic substances as toluene and xylene, and chlorinated hydrocarbons					Exc res to org solv below 375 F. Unaff by strong alkalis or aqueous inorg salt solns		Res most inorg reagents, org chem. Att by conc oxid acids, some org solv, chlor hydrocarbons
USES		Hospital ware, housewares, appliances, radio and TV housings, film fibers	Luggage seating packaging, housings, automotive parts, containers, wire coating	Housings, automobile fan shrouds, covers	Housings, shrouds, cases, panels and mechanical parts	Electrical uses to meet UL requirements, housings and shields	Corrosion resistant pump components, valves and pipe	Pump vanes, valve parts, gaskets, fuel cells and auto parts req chem res at higher temps	Engine fittings, hi-temp elec parts, appliance components

TABLE A-1 *Continued*

Polystyrenes

Type →	ASTM	Polystyrenes				Styrene Acrylonitrile (SAN)	Glass Fiber (30%) Reinforced SAN
		General Purpose	Medium Impact	High Impact	Glass Fiber (30%) Reinf		
PHYSICAL PROPERTIES							
Specific Gravity	D792	1.04	1.04-1.07	1.04-1.07	1.29	1.04-1.07	1.35
Ther Cond, Btu/hr/sq ft/°F/ft	—	0.058–0.090	0.024–0.090	0.024–0.090	0.117	—	—
Coef of Ther Exp, 10^{-5} per °F	D696	3.3–4.8	3.3–4.7	2.2–5.6	1.8	3.6–3.7	1.6
Specific Heat, Btu/lb/°F	—	0.30–0.35	0.30–0.35	0.30–0.35	0.256	0.33	—
Refractive Index, n_D	D542	1.60	Opaque	Opaque	Opaque	1.565–1.569	Opaque
Water Absorption (24 hr), %	D570	0.03-0.2	0.03-0.09	0.05-0.22	0.07	0.20-0.35	0.15
MECHANICAL PROPERTIES							
Tensile Strength, 1000 psi	D638						
Ultimate		5.0-10	6.0	3.3–5.1	14	9.5-12.0	18
Yield		5.0-10	6.0	2.8–5.3	14	—	18
Elongation, %	D638						
Ultimate		1.0-2.3	3.0-40	—	1.1	0.5-3.7	1.4
Yield		1.0-2.3	1.2–3.0	1.5-2.0	1.1	—	1.4
Mod of Elast in Tension, 10^5 psi	D638	4.6–5.0	3.9–4.7	1.50-3.80	12.1	4.0-5.0	17.5
Flex Strength, 1000 psi	D790	10-15	—	—	17	—	22
Mod of Elast in Flex, 10^5 psi	D790	4–5	3.5–5.0	2.3–4.0	12	—	14.5
Impact Str (Izod notched), ft-lb/in.	D638	0.2–0.4	0.5–0.7	0.8–1.8	2.5	0.30–0.45	3.0
Compr Strength, 1000 psi	D695	11.5–16.0	4–9	4–9	19	—	2.3
Hardness (Rockwell)	D785	M72	M47–65	M3–43	M85–95	M80-85	M90–100
Abrasion Res (Taber), mg/1000 cycles	—	—	—	—	164	—	—
ELECTRICAL PROPERTIES							
Volume Resistivity, ohm-cm	D257	$>10^{16}$	$>10^{16}$	$>10^{16}$	3.6×10^{16}	$>10^{16}$	4.4×10^{16}
Dielectric Str (short time), v/mil	D149	>500	>425	300–650	396	400–500	515
Dielectric Constant	D150						
60 cycles		2.45–2.65	2.45–4.75	2.45–4.75	3.1	2.6–3.4	3.5
10^6 cycles		2.45–2.65	2.4–3.8	2.5–4.0	3.0	2.6–3.02	3.4
Dissipation Factor	D150						
60 cycles		0.0001–0.0003	0.0004–0.002	0.0004–0.002	0.005	>0.006	0.005
10^6 cycles		0.0001–0.0005	0.0004–0.002	0.0004–0.002	0.002	0.007-0.010	0.009
Arc Resistance, sec	D495	60-135	20-135	20–100	28	100–150	65
HEAT RESISTANCE							
Max Rec Service Temp, F	—	160–205	125–165	125–165	190–200	175–190	—
Deflection Temp, F	D648						
66 psi		—	—	—	230	—	230
264 psi		220 max	210 max	210 max	220	210–220	220

APPLICABLE PROCESSING METHODS
Injection molding, extrusion, thermoforming, rotational molding, blow molding, foam molding

CHEMICAL RESISTANCE

- General/Medium/High Impact: Res alkalis, salts, low alcohols, glycols and water. Fair res to mineral chemicals and vegetable oils. Not res to aromatic and chlorinated hydrocarbons
- Glass Fiber (30%) Reinf: No effect by weak acids, strong acids; att by oxid acids; no effect by weak alkalis; att slowly by str alkalis; soluble in aromatic and chlor hydrocarbons
- Styrene Acrylonitrile (SAN): Res to alkalis and acids, animal and vegetable oils, soaps, detergents and household chemicals
- Glass Fiber (30%) Reinforced SAN: No effect by weak acids, alkalis, strong acids; att by oxid acids, str alkalis; soluble in ketones, esters, some chlor hydrocarbons

USES

- General Purpose: Thin parts, long flow parts, toys, appliances, containers, film, monofilaments and housewares
- Medium Impact: Radio cabinets, toys, containers, packaging and closures
- High Impact: Containers, cups, lids, large thin wall parts, auto parts, TV cabinets, trays and appliance housings
- Glass Fiber (30%) Reinf: Auto dashboard skeletons, camera housings and frames, tape reels, fan blades
- Styrene Acrylonitrile (SAN): Kitchenware, tumblers, broom bristles, ice buckets, closures, film, containers, lenses, battery cases
- Glass Fiber (30%) Reinforced SAN: Camera housings and frames, auto bezels, electrical components, handles, auto panels

Polyvinyl Chloride and Copolymers

Type ➤		Polyvinyl Chloride, Polyvinyl Chloride Acetate			Vinylidene Chloride[a] Copolymer	Chlorinated Polyvinyl Chloride
		Nonrigid—General	Nonrigid—Electrical	Rigid—Normal Impact		
PHYSICAL PROPERTIES	ASTM					
Specific Gravity	D792	1.20-1.55	1.16-1.40	1.32-1.44	1.68-1.75	1.49-1.58
Ther Cond, Btu/hr/sq ft/°F/ft	D325	0.07-0.10	0.07-0.10	0.07-0.10	0.053	—
Coef of Ther Exp, 10^{-5} per °F	D696	—	—	2.8-3.3	8.78	3.8
Refractive Index	D542	—	—	—	1.60-1.63	—
Spec Ht, Btu/lb/°F	—	—	—	—	0.32	—
Water Absorption (24 hr), %	D570	0.2-1.0	0.40-0.75	0.03-0.40	>0.1	0.02-0.15
Oxygen Index	D2863					53-63
MECHANICAL PROPERTIES						
Mod of Elast in Tension, 10^5 psi	D412	0.004-0.03	0.01-0.03	3.5-4.0[b]	0.7-2.0	—
Ten Str, 1000 psi	D412	1-3.5	2-3.2	5.5-8	4-8, 15-40	—
Elong (in 2 in.), %	D638	200-450	220-360	1-10	15-25, 20-30	—
Hardness (Rockwell)	D785	—	—	R110-120	M50-65	R117-122
Hardness (Shore)	D676	A50-100	A78-100	D70-85	>A95	—
Impact Str (Izod notched), ft-lb/in	D256	Variable	Variable	0.5-10	2-8, 0.053	1.0-3.0
Mod of Elast in Flex, psi	D790	—	—	$3.8\text{-}5.4 \times 10^5$	—	3.8-4.50
100% Modulus, psi	—	600-2800	600-2800	—	—	
Flex Str, 1000 psi	D790	—	—	11-16	15-17, flexible	14.5-17
Compr Str, 1000 psi	D695	—	—	11-12	—	—
Compr Yld Str, 1000 psi	D695	—	—	10-11	75-85	—
Cold Flex Temp, F	D1043	−70 to 0	−7 to +20	—	—	—
Cold Bend Temp, F	—	−40 to −4	−49 to −4	—	—	—
ELECTRICAL PROPERTIES						
Vol Res, ohm-cm	D257	$1\text{-}700 \times 10^{12}$	$4\text{-}300 \times 10^{11}$	$10^{14}\text{-}{>}10^{16}$	$10^{14}\text{-}10^{16}$	10^{15}
Dielec Str (short time), v/mil	D149	—	24-500	725-1400	—	1220-1500
Dielec Const (60 cycles)	D150	5.5-9.1	6.0-8.0	2.3-3.7	3-5	3.08
Dissip Factor (60 cycles)	D150	0.05-0.15	0.08-0.11	0.020-0.03	0.03-0.15	0.019-0.021
Loss Factor (60 cycles)	D150		1.0-1.2	0.030-0.072	—	—
APPLICABLE PROCESSING METHODS		Injection molding, extrusion, thermoforming, blow molding, foam molding, slush molding, calendering			Extrusion, calendering	Calendering, compression molding, extrusion, injection molding, thermoforming
HEAT RESISTANCE						
Max Rec Svc Temp, F	—	150-220	140-220	150-165	170-212	230
Heat Dist Temp, F						
66 psi	D648	—	—	170-185	190-210	215-247
264 psi	D648	—	—	140-170	130-150	202-234
CHEMICAL RESISTANCE		Generally resistant to alkalis and weak acids. Moderately to not resist to strong acids. Not resistant to keotones and esters; aromatic hydrocarbons produce swelling			Excellent to all acids and most common alkalis[c]	Res to acids, alkalis, oil, grease and most organic solvents
USES		Parts made by molding, high speed extrusion, calendering. Blown extruded film. Vacuum cleaner parts, handlebar grips, doll parts, hair curlers, safety goggle cups, grommets, toy tires, garden hose, and protective garments	Parts made by calendering, extrusion. Insulation and jacketing for: communication and low tension power wire and cable, building wiring, appliance and machine tool cords, and switchboard cable	Parts made by calendering, laminating, molding, extrusion. Fume hoods and ducts, storage tanks, chemical piping, plating tanks, phonograph records. Sheets and shapes for decorative panels, other building uses	Extrusions: gasket rods, valve seats, flexible chemical tubing and pipe, tape for wrapping joints, chemical conveyor belts. Moldings: spray-gun handles, acid dippers, parts for rayon producing equipment	Hoods, ducts, exterior bldg components, pipe.

[a] Where two values or ranges are given, they represent unoriented and oriented forms, respectively. [b] Modulus of elasticity in compression. [c] Unaffected by aliphatic and aromatic hydrocarbons, alcohols, esters, etc. [d] Barrel temperature. [e] Stock temperature.

TABLE A-1 *Continued*

Silicones

Type →	ASTM	Glass Fiber Reinforced Silicones	Granular (Silica) Reinforced Silicones	Woven Glass Fabric/ Silicone Laminate
PHYSICAL PROPERTIES	ASTM			
Specific Gravity	D792	1.88	1.86–2.00	1.75–1.8
Ther Cond, Btu/hr/sq ft/°F/ft	—	0.18	0.25–0.5	0.075–0.125
Coef of Ther Exp, per °F x 10^{-5}	D696	3.17–3.23	2.5–5.0	—
Specific Heat, Btu/lb/°F	—	—	—	0.246
Water Absorption (24 hr), %	D570	0.1–0.15	0.08–0.1	0.03–0.5
Transparency (visible light), %	—	Opaque	Opaque	Opaque
Refractive Index, n_D	D542	—	—	—
MECHANICAL PROPERTIES				
Tensile Strength, 1000 psi	D651	6.5	4–6	30–35
Tensile Modulus, 10^5 psi	D651	—	—	28
Elongation, %	D651	<3	<3	—
Impact Str (Izod notched), ft-lb/in.	D256	10	0.34	10–25
Flexural Strength, 1000 psi	D790	16–19	6–10	33–47
Mod of Elast in Flex, 10^5 psi	D790	25	14–17	26–32
Compr Strength, 1000 psi	D690	10–12.5	10.6–17.0	15–24
Hardness (Rockwell)	D785	M87	M71–95	75 (Barcol)
Abrasion Res (Taber)	—	—	—	—
ELECTRICAL PROPERTIES				
Volume Resistivity, ohm-cm (dry)	D257	9×10^{14}	5×10^{14}	$2–5 \times 10^{14}$
Dielectric Str (short time), v/mil	D149	280 (in oil)	380 (in oil)	725
Dielectric Constant	D150			
60 cycles		4.34	4.1–4.5	3.9–4.2
10^6 cycles		4.28	3.4–4.3	3.8–3.97
Dissipation Factor	D150			
60 cycles		0.01	0.002–0.004	0.020
10^6 cycles		0.004	0.001–0.004	0.002
Arc Resistance, sec	D495	240	250–310	225–250
APPLICABLE PROCESSING METHODS		Compression molding, transfer molding and injection molding		Layup molding, laminating
TRANSFER MOLDING	—			
Pressure, 1000 psi		0.5–10	0.150–5	—
Temperature, F		350	310–350	—
Shrinkage, %		0.0005	0.004–0.006	—
HEAT RESISTANCE				
Max Rec Service Temp, F	—	>500	>500	450–500
Heat Deflection Temp (264 psi), F	D648	>900	520–>900	>900
CHEMICAL RESISTANCE		Resistant to aviation gasoline, lubricating oil, and sulphuric and hydrochloric acids. Slightly softened and pitted by sodium hydroxide, except some mineral-filled materials. Should be tested if resistance to ketones, toluene, ethylenes, etc., is required		Satis res to aviation gas, lube oils, 40% sulfuric acid, 5% hydrochloric acid and freon 114. Slightly att by 5% hydrochloric acid. Severely att by many organic solvents
USES		Connector plugs, and other structural electronic parts requiring heat resistance	Electronic component encapsulation, such as transistors, diodes, resistors and capacitors	Special high temp struct or elect parts, such as aircraft radomes and ductwork, thermal and arc barriers, covers and cases for high freq equip

TABLE A-2 RUBBER AND ELASTOMERS
Rubber — Molded, Extruded

	Type ➡	Natural Rubber (Cis-polyisoprene)	Butadiene-Styrene (GR-S)	Synthetic (Polyisoprene)	Butadiene-Acrylonitrile (Nitrile)	Chloroprene (Neoprene)	Butyl (Isobutylene-isoprene)
PHYSICAL PROPERTIES	ASTM						
Specific Gravity	D792	0.93	0.94	0.93	0.98	1.25	0.90
Ther Cond, Btu/hr/ sq ft/°F/ft	C177	0.082	0.143	0.082	0.143	0.112	0.053
Coef of Ther Exp (cubical), 10^{-5} per °F	D696	37	37	—	39	34	32
Electrical Insulation		Good	Good	Good	Fair	Fair	Good
Flame Resistance		Poor	Poor	Poor	Poor	Good	Poor
Min Rec Svc Temp, F		−60	−60	-60	−60	−40	−50
Max Rec Svc Temp, F		180	180	180	300	240	300
MECHANICAL PROPERTIES							
Ten Str, psi							
Pure Gum	D412	2500–3500	200–300	2500-3500	500–900	3000–4000	2500–3000
Black	D412	3500–4500	2500–3500	3500-4500	3000–4500	3000–4000	2500–3000
Elongation, %							
Pure Gum	D412	750–850	400–600	—	300-700	800–900	750–950
Black	D412	550–650	500–600	300–700	300-650	500–600	650–850
Hardness (durometer)		A30–A90	A40–A90	A40–A80	A40–A95	A20–A95	A40–A90
Rebound							
Cold		Excellent	Good	Excellent	Good	Very good	Bad
Hot		Excellent	Good	Excellent	Good	Very good	Very good
Tear Resistance		Excellent	Fair	Excellent	Good	Fair to good	Good
Abrasion Resistance		Excellent	Good to excellent	Excellent	Good to excellent	Good	Good to excellent
CHEMICAL RESISTANCE							
Sunlight Aging		Poor	Poor	Fair	Poor	Very good	Very good
Oxidation		Good	Good	Excellent	Good	Excellent	Excellent
Heat Aging		Good	Very good	Good	Excellent	Excellent	Excellent
Solvents							
Aliphatic Hydrocarbons		Poor	Poor	Poor	Excellent	Good	Poor
Aromatic Hydrocarbons		Poor	Poor	Poor	Good	Fair	Poor
Oxygenated, Alcohols		Good	Good	Good	Good	Very good	Very good
Oil, Gasoline		Poor	Poor	Poor	Excellent	Good	Poor
Animal, Vegetable Oils		Poor to good	Poor to good	—	Excellent	Excellent	Excellent
Acids							
Dilute		Fair to good	Fair to good	Fair to good	Good	Excellent	Excellent
Concentrated		Fair to good	Fair to good	Fair to good	Good	Good	Excellent
Permeability to Gases		Low	Low	Low	Very low	Low	Very low
Water Swell Resistance		Fair	Excellent	Excellent	Excellent	Fair to excellent	Excellent
USES		Penumatic tires and tubes; power transmission belts and conveyor belts; gaskets; mountings; hose; chemical tank linings; printing press platens; sound or shock absorption; seals against air, moisture, sound and dirt	Same as natural rubber	Carburetor diaphragms, self-sealing fuel tanks, aircraft hose, gaskets, gasoline and oil hose, cables, machinery mountings, printing rolls	Wire and cable, belts, hose, extruded goods, coatings, molded and sheet goods, adhesives, automotive gaskets and seals, petroleum and chemical tank linings	Truck and automobile tire inner tubes, curing bags for tire vulcanization and molding, steam hose and diaphragms, flexible electrical insulation, shock, vibration absorption	

TABLE A-2 *Continued*

Rubber — Molded, Extruded

Type →	Polysulfide	Silicone (Polysiloxane)	Urethane (Diisocyanate polyester)	Polyacrylate	Polybutadiene
PHYSICAL PROPERTIES					
Specific Gravity	1.35	1.1–1.6	1.25	1.09	0.91
Ther Cond, Btu/hr/sq ft/°F/ft	—	0.13	—	—	—
Coef of Ther Exp (cubical), 10^{-6} per °F	—	45	—	—	37.5
Electrical Insulation	Fair	Excellent	Fair	Fair	Good
Flame Resistance	Poor	Good	Good	Poor	Poor
Min Rec Svc Temp, F	—60	—178	—65	—20	—150[d]
Max Rec Svc Temp, F	250	600	240	350	200
MECHANICAL PROPERTIES					
Ten Str, psi					
Pure Gum	250-400	600–1300[a]	>5000	250–400	200–1000
Black	>1000	—	—	1000–2500	2000 3000
Elongation, %					
Pure Gum	450–650	100–500[a]	540–750	450–750	400–1000
Black	150-450	—	—	150–450	450–600
Hardness (durometer)	A40–85	A30–90	A35–100	A40–90	A40–90
Rebound					
Cold	Good	Very good	Bad	Fair	Excellent
Hot	Good	Very good	Good	Very good	Excellent
Tear Resistance	Poor	Fair	Good	Fair to good	Fair
Abrasion Resistance	Poor	Poor	Excellent	Good	Excellent
CHEMICAL RESISTANCE					
Sunlight Aging	Very good	Excellent	Excellent[b]	Excellent	Poor
Oxidation	Very good	Excellent	Very good	Excellent	Good
Heat Aging	Fair	Excellent	Excellent[b]	Excellent	Good
Solvents					
Aliphatic Hydrocarbons	Excellent	Fair	Excellent	Excellent	Poor
Aromatic Hydrocarbons	Excellent	Poor	Excellent[c]	Fair to good	Poor
Oxygenated, Alcohols	Very good	Excellent	Poor	Poor	—
Oil, Gasoline	Excellent	Poor	Excellent	Good to excellent	Poor
Animal, Vegetable Oils	Excellent	Excellent	Excellent	Very good	Poor to good
Acids					
Dilute	Good	Very good	Fair	Fair	—
Concentrated	Good	Good	Poor	Fair	—
Permeability to Gaes	Very low	High	Very low	Low	Low
Water Swell Resistance	Excellent	Excellent	Excellent	Poor to fair	Excellent
USES	Seals, gaskets, diaphragms, valve seat disks, flexible mountings, hose in contact with solvents, balloons, boats, life vests and rafts	High and low-temperature electrical insulation, seals, gaskets, diaphragms, ductwork, o-rings	Fork lift truck wheels, airplane tail wheels, back-up wheels for turbine blade grinders, spinning cots for glass fiber, hydraulic accumulators, shoe heels	Oil hose, searchlight gaskets, white or pastel colored goods, and automotive gaskets and o-rings (especially for resistance to extreme pressure lubricants and oils containing sulfur)	Pneumatic tires, shoe heels and soles, gaskets and seals, belting, sponge stocks; often used in blends with other rubbers to impart better resilience, abrasion resistance and low temperature properties

[a] Reinforced with high temperature non-organic fillers.
[b] Discolors, but no change in properties.
[c] For up to 80% aromatics.
[d] Brittle point < —150 F; may stiffen at higher temperatures.

Rubber—Molded, Extruded

Type →	Epichlorohydrin Homopolymer & Copolymer	Fluorosilicone	Ethylene Propylene (EPDM)	Chloro-sulfonated Polyethylene (Hypalon)	Fluorocarbon Elastomers	Propylene Oxides	Styrene-Iso-prene-Styrene & Styrene-Buta-diene-Styrene Block Polymers
PHYSICAL PROPERTIES							
Specific Gravity	1.32-1.49	1.4	0.86	1.11-1.26	1.4-1.95	1.02	0.94-1.15
Ther Cond[a]	—	0.13	—	0.065	0.13	—	0.087
Coef of Ther Exp, $10^{-5}/°F$	—	45	—	27	8.8	—	7.5
Flame Res	Fair	Poor	Poor	Good	Excellent	Poor	Poor
Colorability	Good	Good	Excellent	Excellent	Good	Good	Good
MECHANICAL PROPERTIES							
Hardness (Shore A)	30-95	40-70	30-90	45-95	65-90	40-80	35-90
Tensile Strength, 1000 psi							
Pure Gum	—	1	<1	4	<2	>1	0.7-4.5
Reinforced	2-3	<2	0.8-3.2	1.5-2.5	1.5-3	>2	—
Elongation,%							
Reinforced	320-350	200-400	200-600	250-500	100-450	500-670	350-1350
Resilience	Poor-Exc	Good-Fair	Good	Good	Fair	Very Good	Good
Compression Set Res	Very Good	—	Good	Fair-Good	Good-Exc	Fair	—
Hysteresis Resistance	Good	Good	Good	Good	Good	Very Good	—
Flex Cracking Resistance	Very Good	Good	Good	Good	Good	Very Good	Good
Slow Rate	Very Good	Good	Good	Good	Good	Very Good	Good
Fast Rate	Good	Good	Good	Good	Good	Good	—
Tear Strength	Good	Fair	Poor-Fair	Fair-Good	Poor-Fair	Excellent	Fair-Good
Abrasion Resistance	Fair-Good	Poor	Good	Excellent	Good	Good	Good
ELECTRICAL PROPERTIES							
Dielectric Strength	Fair	Good	Excellent	Excellent	Good	—	Good
Electrical Insulation	Fair	Good	Very Good	Good	Fair-Good	—	Good
THERMAL PROPERTIES							
Service Temperature, F							
Min for Cont Use	-15 to -80	−90	−60	−40	−10	−80	−60 to −80
Max for Cont Use	300	400	<350	<325	<500	<250	150
Low Temp Stiffening, F	-15 to -80	<−100	-20 to -60	−30 to −50	20 to −30	—	−60 to −80
CORROSION RESISTANCE							
Weather	Excellent	Excellent	Excellent	Excellent	Excellent	Very Good	Fair
Oxidation	Very Good	Excellent	Excellent	Excellent	Outstanding	Very Good	Good
Ozone	Good-Exc	Excellent	Excellent	Excellent	Excellent	Very Good	Fair
Radiation	—	Good	Excellent	Fair-Good	Fair-Good	—	Poor
Water	Good	Excellent	Good-Exc	Good	Good	Excellent	Good
Acids	Good	Very Good-Exc	Good-Exc	Excellent	Good-Exc	Good	Good
Alkalis	Good	Very Good	Good-Exc	Excellent	Poor-Good	Very Good-Exc	Good
Aliphatic Hydrocarbons	Excellent	Excellent	Poor	Fair	Excellent	Poor-Fair	Poor
Aromatic Hydrocarbons	Very Good	Excellent	Fair	Poor-Fair	Excellent	Poor-Fair	Poor
Halogenated Hydrocarbons	Good	—	Poor	Poor-Fair	Good	Poor	Poor
Alcohol	Good	—	Good	Very good	Excellent	—	Good
Synthetic Lubricants (diester)	Fair-Good	Excellent	Poor-Fair	Poor	Fair-Good	Fair-Good	Poor
Hydraulic Fluids							
Silicates	Very Good	Excellent	Fair-Good	Good	Good	—	—
Phosphates	Poor-Fair	Excellent	Good-Exc	Poor-Fair	Poor	—	—
USES	Diaphragms, print rolls, belts, oil seals, molded mech goods; gaskets, hose for petroleum handling; low temperature parts	Parts requiring res to high temp solv or oils; seals, gaskets, O-rings	Elec insul and jacketing; footwear, sponge, proofed fabrics, automotive weather stripping, hose, belts, auto, appliance parts; parts req outstanding ozone res and heat res	Flex chemical and petroleum tube and hose, rolls, tank linings; high temp belts; wire and cable; shoe soles and heels; flooring; building products; extruded and molded parts	O-rings, brake seals, shaft seals, gaskets, hose and ducting, connectors, diaphragms carburetor needle tips, lined valves, packings, roll coverings	Electrical insulation, molded mechanical goods	Thermoplastic grades: molded mechanical goods, packaging, sports equip, disposable pharmaceutical items Solution grades: adhesives, coatings, caulking, sealants

[a]Btu/hr/sq ft/°F/ft.

TABLE A-3 CERAMOPLASTICS

Type →	Natural Muscovite	Synthetic Fluor-Phlogopite	Glass-Bonded Mica		Ceramoplastic	
			Compression-Molded	Precision Transfer Molded	Compression-Molded	Precision Transfer Molded
PHYSICAL PROPERTIES						
Specific Gravity	2.6–3.2	2.9	3.0	2.6–3.8	3.0–3.3	2.7–3.8
Ther Cond, Btu/hr/sq ft/°F/ft	0.25–0.36	0.3–0.4	0.24	0.24–0.27	0.29	0.29–0.36
Coef of Ther Exp, per °F	1.8×10^{-5}	1.44–2.70×10^{-5}	0.58×10^{-5}	0.56–0.59×10^{-5}	0.62–0.67×10^{-5}	0.52–0.63×10^{-5}
Spec Heat, Btu/lb/°F	0.20	0.25	0.16	0.13–0.22	0.18	0.15–0.24
Water Absorption	Low	Low	Nil	Nil	Nil	Nil
MECHANICAL PROPERTIES						
Mod of Elast in Tension, psi	20–30×10^6	25×10^6	11.0×10^6	7.5–8.0×10^6	12.0–12.5×10^6	8.0–12.0×10^6
Ten Str, 1000 psi	40–50	45–55	6.0	5.0	5.0–6.0	4.0–5.0
Hardness						
Knoop	90	200	—	—	—	—
Mohs	3–4	3–4	—	—	—	—
Rockwell M	—	—	110–120	115–125	110–120	115–125
Impact Str (Izod), ft-lb/in	—	—	2.0	1.2–1.5	1.2–2.4	1.2–2.0
Flex Str, 1000 psi	—	—	15	12–14	10–15	12
Compr Str, 1000 psi	>150	>150	45	25–40	35–40	25–40
ELECTRICAL PROPERTIES						
Vol Res, ohm-cm	10^{13}–10^{17}	10^{13}–10^{17}	10^{17}	10^{16}–10^{17}	10^{13}–10^{17}	10^{14}–10^{17}
Dielec Str (step by step, ⅛ in.), v/mil	1000–2000	1000–2000	500	350–400	500	270–400
Dielec Const 10^6 cycles	5.4–8.7	6.5	6.7	6.3–9.3	6.8–6.9	6.7–8.8
Dissip Factor 10^6 cycles	0.0001–0.0004	0.0001–0.0003	0.0018	0.0011–0.0025	0.0013–0.0015	0.0015–0.0023
Arc Resistance, sec	High	High	300	200–250	300	250–300
FABRICATING PROPERTIES						
Machinability	Fair, readily punched	Fair, punches more readily than Natural	Good	Good	Good	Good
Moldability	—	—	Good	Good	Good	Good
HEAT RESISTANCE						
Max Endurance Temp, F	1110	1400–1800	700	575–650	700–1550	650–1200
Heat Dist Temp (264 psi), F	—	—	850	690	900–1360	700–1100
Ther Shock Res	—	—	Moderate	Moderate	Moderate	Moderate
CHEMICAL RESISTANCE	Good res to most chemicals and molten materials exc hydrofluoric acid. Oils generally cause delamination	Slightly better chemical res than Natural; high res to oils. High res to high pressure, high temp water	Good res to organic solvents; poor res to strong acids and alkalis			
USES	Furnace peepholes, boiler gage glass, capacitors, tube spacers	Experimental high temp tube spacers and waveguide windows; used to make glass-bonded synthetic mica; with inorganic binder used as potting or embedding compound	Electromechanical devices, high temp insulators (575-850 F) requiring high stability (e.g., computer parts)		Electromechanical devices, high temp insulators (650-1550°F) requiring high stability. Hermetic seals	

TABLE A-4 PLASTICS VS. TIME

Isochronous stress/strain properties

Materials	Strain (%) at room temperature				At 140 F			
	Loading of 1000 psi		2000 psi		1000 psi		2000 psi	
	At 100 hr	1000 hr	100 hr	1000 hr	100 hr	1000 hr	100 hr	1000 hr
ABS								
High impact[a]	0.5	0.6	1.05	1.25	1.70	2.40	4.00	5.85
High modulus[b]	0.4	0.5	0.8	1.0	1.25	1.70	2.65	3.60
Acetals								
Homopolymer[c]	0.35	0.40	0.70	0.85	0.88*	0.96*	1.70*	2.20*
GR homopolymer[d]	0.15	0.18	0.32	0.41	0.31*	0.39*	0.85*	1.20*
Copolymer[e]	0.50	0.64	1.05	1.35	—	1.00	—	—
GR copolymer	—	—	—	—	—	—	—	—
Acrylics								
General purpose[g]	0.25	0.30	0.60	0.68	0.60	0.80	1.20	2.20
High impact	—	—	—	—	—	—	—	—
Fluoroplastics								
PETFE	—	—	—	—	—	—	—	—
GR PETFE	—	—	—	—	—	—	—	—
Noryl phenylene oxides								
General purpose[k]	—	—	0.60	0.85	—	—	—	—
GR[m]	—	—	0.26	0.30	0.35*	0.32**	0.75*	0.70**
Nylons								
Type 6	—	—	—	—	—	—	—	—
Type 6/6[p]	1.00	1.25	2.35	3.00	1.50	1.65	4.00	—
Type 6/6 MR[q]	—	—	0.01	0.01	0.02	0.025	—	—
Type 6/6 GR[r]	0.15	0.18	0.30	0.35	—	—	0.34	0.37
Type 6/12[s]	0.75	1.05	1.75	2.40	1.45	1.60	3.50	—
Type 6/12 GR	—	—	—	—	—	—	—	—
Type 11[u]	—	1.50	—	7.10	—	—	—	—
Polycarbonates								
General purpose[v]	0.30	0.35	0.82	0.90	0.45	0.50	1.05	1.20
High performance	—	—	—	—	—	—	—	—
GR	—	—	—	—	—	—	—	—
Polyethylenes								
High density	—	—	—	—	—	—	—	—
GR	—	—	—	—	—	—	—	—
Polypropylenes								
General purpose[aa]	1.55	1.58	2.5[t]	2.95[t]	3.00[tt]	2.87[o]	—	—
GR[bb]	0.20	0.24	0.42	0.50	0.25	—	0.60[oo]	—
Polyesters								
General purpose[cc]	0.35	0.45	0.75	0.90	2.25*	2.55*	4.60*	5.15*
GR[dd]	0.10	0.11	0.20	0.22	0.18	0.20*	0.35*	0.40*
Polysulfones								
General purpose[ee]	0.30	0.30	0.60	0.65	0.30	0.33	0.65	0.70
GR	—	—	—	—	—	—	—	—
Polyphenylene sulfides								
General purpose	-	—	—	—	—	—	—	—
GR	—	—	—	—	—	—	—	—
Styrenes								
General purpose	—	—	—	—	—	—	—	—
GR	—	—	—	—	—	—	—	—
Acrylonitriles	—	—	—	—	—	—	—	—
GR	—	—	—	—	—	—	—	—
Vinyls								
Type I[nn]	0.26	0.29	—	—	0.65#	1.30#	1.15##	2.40##
Type II[pp]	—	—	0.90	1.30	1.50#	—	5.80#	—

(Courtesy *Materials Engineering*, Van Nostrand Reinhold Co.)

TABLE A-4A   CREEP PROPERTIES

Materials (A)	Apparent modulus after 1000 hr									
	At 500 psi		1000 psi		2000 psi		3000 psi		5000 psi	
	At RT	140 F	RT	140 F	RT	140 F	RT	140 F	RT	140 F
ABS										
High impact[n] (2)	—	—	210	—	250	—	—	—	—	—
Heat resistant[b] (2)	—	—	245	—	210	—	—	—	—	—
Acetals										
Homopolymer[c] (4)	250	120	250	120	210	100	—	—	—	—
GR homopolymer[d] (4)	650	300	550	275	500	190	—	—	775	—
Copolymer[e] (2)	270	70*	270	45**	—	—	220#	—	—	—
GR copolymer[f] (2)	—	—	—	—	—	—	1150#	—	750	—
Acrylics										
General purpose[g] (4)	—	175##	245	—	—	—	—	—	—	—
High impact	—	—	—	—	—	—	—	—	—	—
Noryl phenylene oxides										
General purpose[k] (4)	—	—	245[t]	90[tt]	230	89[tt]	230	—	—	—
GR self-exting[m] (4)	—	—	—	—	1050	970[tt]	980	860[tt]	—	—
Nylons										
Type 6/6[n] (4)	90	60	85	55	65	25	—	—	—	—
Type 6/6 GR[p] (4)	—	—	—	—	—	—	585[x]	480#	625	410*
Type 6/6 MR[q]	—	—	—	—	210	—	—	—	—	—
Type 6/12[r] (4)	—	—	—	—	75	45	—	—	—	—
Type 6/12 GR[s] (4)	—	—	—	—	—	—	—	—	510	—
Polycarbonates										
General purpose[t] (4)	—	190[xx]	—	178[xx]	—	225[o]	310	—	—	—
High performance[u] (4)	—	—	—	—	—	—	430	—	—	—
GR[v] (4)	—	—	—	—	—	—	760	950	1050	830*
Polyesters										
General purpose	—	—	—	—	—	—	—	—	—	—
GR	—	—	—	—	—	—	—	—	—	—
Polyethylenes										
High density[y]	55[oo]	—	40[oo]	—	—	—	—	—	—	—
GR[z] (4)	—	—	420[t]	—	—	—	—	—	—	—
Polypropylenes										
General purpose[aa]	65[oo]	30[oo]	60[oo]	—	—	—	—	—	—	—
GR[bb]	—	—	—	—	825	—	—	—	—	—
Polysulfones										
General purpose[cc] (4)	—	—	—	—	—	—	325	310	—	—
GR[dd] (4)	—	—	—	—	—	—	—	—	1075	795*
Polyphenylene sulfides										
General purpose[ee] (4)	—	—	—	—	—	—	550#	—	—	—
GR[ff] (4)	—	—	—	—	—	—	2100#	—	1475	795*
Styrenes										
General purpose	—	—	—	—	—	—	—	—	—	—
GR[hh] (4)	—	—	—	—	—	—	—	—	1000	490
Acrylonitrile[ii]	—	—	—	—	—	—	340	—	—	—
GR[jj] (4)	—	—	—	—	—	—	920#	—	1280	750
Vinyls										
Type I[mm] (2)	320	45°	325	25°	—	—	—	—	—	—
Type II[nn] (2)	160	—	250	—	215	—	—	—	—	—

(Courtesy *Materials Engineering*, Van Nostrand Reinhold Co.)

TABLE A-5 SPE GUIDE FOR UNIFORM TESTING AND REPORTING OF IDENTIFICATION PROPERTIES OF PLASTICS MATERIALS
(Updated in 1972 and issued by Engineering Properties and Structure Division)

Property	ASTM test method[a]	Recommendations[b]	U.S. customary units	SI units	X[c]	s[c]
Apparent density	D1895					
Free flowing	Method A	—	—	g/cu cm		
Non-pouring	Method B	—	—	g/cu cm		
Bulk factor	D1895	—	Dimensionless			
Specific gravity	D792 Method A or B	—	Dimensionless			
Density	D1505	—	—	G/cu cm		
Mold shrinkage	D955	—	in./in.	mm/mm		
Flow temperature Rossi-Peakes	D569 Procedure A	Used to classify cellulose acetate and cellulose acetate butyrate compounds	°C	°C		
Melt-flow rate, thermoplastics	D1238	State condition employed	—	g/10 min		
Molding index	D731	—	1 bf (at close time), sec	kgf (at close time), s		
Dielectric constant; dissipation factor at 60 Hz, 1 kHz, 1 MHz	D150	⅛-in.-thick specimens	Dimensionless			
Volume resistivity 1 min at 500 v	D257	⅛-in.-thick specimens	—	ohm-cm		
Arc resistance high voltage, low current	D495 Stainless steel electrodes	⅛-in.-thick specimens	sec	s		

Dielectric strength
  Short time
  Step by step
	D149
	Sec. 6.1.1
	Sec. 6.1.3
		⅛-in.-thick specimens	v/mil	v/mm		

Tensile strength	D638	⅛-in.-thick type 1 specimens; state testing speed used	psi State whether at yield or break	Pa		
Elongation	D638	⅛-in.-thick type I specimens; state testing speed used	% State whether at yield or break	%		
Modulus of elasticity, tensile	D638	⅛-in.-thick type I specimens; state testing speed used	psi	Pa		
Flexural strength	D790	State procedure employed	psi State whether at yield, at rupture, or 5% strain	Pa		
Tangent modulus of elasticity, flexural	D790	State procedure employed	psi	Pa		
Compressive strength	D695	—	psi State whether at yield, at failure, or at given strain	Pa		
Modulus of elasticity, compressive	D695	—	psi	Pa		
Impact resistance, Izod	D256 Method A	Specimen width, as tested, ⅛ in.	ft-1 bf/in. of notch	joule/m of notch		
Notch sensitivity	D256 Method D	Specimen width, as tested, ⅛ in.	ft-1 bf/in. (notch)-in. (rad)	joule/m (notch)-m (rad)		
Hardness, durometer	D2240	—	Scale reading; state durometer type and time under pressure			
Hardness, Rockwell	D785 Procedure A	—	Scale reading; state scale			
Haze	D1003	⅛-in.-thick specimens; state method (within D1003) used	%	%		
Luminous transmittance	D1003	⅛-in.-thick specimens; state method (within D1003) used	%	%		
Index of refraction	D542	⅛-in.-thick specimens; state method (within D542) used	Dimensionless			

Water absorption
  24 hr immersion
  Long-term immersion
	D570
	Sec. 6.1
	Sec. 6.4
		—	%	%		

Brittleness temperature	D746	—	°C at 50% failure point	°C at 50% failure point		
Coefficient of linear thermal expansion	D696	—	°C⁻¹	°C⁻¹		

TABLE A-5 *Continued*

Property	ASTM test method[a]	Recommendations[b]	U.S. customary units	SI units	X[c]	s[c]
Deflection temperature	D648	¼-in.-thick specimens	°C at 264 and 66 psi fibers stress	°C at 18.5 and 4.6 kgf/sq cm stress		
Vicat softening point	D1525	—	°C	°C		
Flammability	D635	⅛-in.-thick specimens	in./min (if burning) in., sec (if SE)	cm/min (if burning) cm, s (if SE)		
Oxygen index	D2863	—	%	%		
Deformation under load	D621	State temperature and stress	%	%		
Rigid plastics	Method A					
Non-rigid plastics	Method B					
Dynamic mechanical properties	D2236	Report values over range of temperature of show plot	Dimensionless			
Logarithmic decrement						
Elastic shear modulus			psi	Pa		
Creep	D2990	⅛-in.-thick type II specimens; report stress to produce 1% creep strain in 1000 hr	psi	Pa		
Creep rupture	D2990	⅛-in.-thick type II specimens; report stress to produce rupture in 1000 hr	psi	Pa		

[a] Use the latest revision of any ASTM method referenced.

[b] Applicable material specification should be the primary guide to test specimen dimensions and methods of preparing and conditioning specimens. (See Part 26 of the Annual Book of ASTM Standards.) If specifications are lacking, test method recommendations should be followed. In their absence, specimens should be prepared by compression molding, fully positive. Conditioning in accordince with Procedure A of ASTM Methods D618 is usually appropriate for materials which absorb little moisture. Hygroscopic plastics are routinely tested in the dry or "as molded" state; specimens conditioned to an "equilibrium moisture content" (at 23°C and 50% RH) may exhibit properties more closely related to those which obtain in use.

[c] Characterization of a given grade of material should be based on tests of representative samples from at least five batches. The arithmetic average "X" and the estimated standard deviations "s" for each property of that material would be computed from the average values for all batches evaluated.

Note: The reporting of all significant test parameters is strongly recommended. (See the Report Section of each test method used.) All departures from standard or recommended procedures should be noted clearly on the report or data sheet. The use of international SI units is encouraged.

(Courtesy Society of Plastics Engineers)

TABLE A-6 HOW TO IDENTIFY SOME PLASTICS

A preliminary guide that will help to identify many of the basic types of plastics using simple techniques and readily available tools. Naturally, these tests should be used only for tentative identification because some complex plastic compounds require a rigorous analysis for identification.

To initially determine whether a material is termoset or thermoplastic, heat a stirring rod (to about 500°F) and press it against the sample. If the sample softens, the material is thermoplastic; if not, it is probably thermosetting.

Next, hold the sample to the edge of a flame until it ignites. (Hold in the flame for about 10 sec. if no flame is produced immediately.) If the material burns, note the color of the flame, the nature of the smoke, the presence of soot in the air and if in burning the sample drips. Next, extinguish the flame and cautiously smell the fumes. (In identifying odor, a known sample is most helpful for comparison.). Finally, check your observations against the known characteristics of each plastic given below. Once you have made a tentative identification it usually is desirable to make one additional test to confirm the results of the original identification.

Materials	No Flame	Burns, But Extinguishes on Removal of Flame Source			Continues to Burn After Removal of Flame Source				Remarks
	Odor	Odor	Color of Flame	Drips	Odor	Color of Flame	Drips	Speed of Burning	
THERMOPLASTICS									
ABS	—	Acrid[a]	Yellow, blue edges[a]	No[a]	Acrid	Yellow, blue edges	Yes	Slow	Black smoke with soot in air
Acetals	—	—	—	—	Formaldehyde	Blue, no smoke	Yes	Slow	—
Acrylics	—	—	—	—	Fruity	Blue, yellow tip	No (cast) Yes (mold-ed)	Slow	Flame, may spurt if rubber modified
Cellulosics									
Acetate	—	Vinegar[c]	Yellow with sparks[a]	No[a]	Vinegar	Yellow	Yes	Slow	Flame may spark
Acetate Butyrate	—	—	—	—	Rancid butter	Blue, yellow tip	Yes	Slow	Flame may spark
Ethyl Cellulose	—	—	—	—	Burnt sugar	Yellow, blue edges	Yes	Rapid	—
Nitrate	—	—	—	—	Camphor	White	No	Rapid	—
Propionate	—	—	—	—	Burnt sugar	Blue, yellow tip	Yes	Rapid	—
Chlorinated Polyether	—	[b]	Green, yellow tip	No	—	—	—	—	Black smoke with soot in air
Fluorocarbons									
FEP	Faint odor of burnt hair	—	—	—	—	—	—	—	Deforms; no combustion, but drips
PTFE	Faint odor of burnt hair	—	—	—	—	—	—	—	Deforms; does not drip
CTFE	Faint odor of acetic acid	—	—	—	—	—	—	—	Deforms; no combustion, but drips
PVF$_2$	Acidic	—	—	—	—	—	—	—	Deforms
Nylons									
Type 6	—	Burnt wool	Blue, yellow tip	Yes	—	—	—	—	—
Type 6/6	—	Burnt wool or hair	Blue, yellow tip	Yes	—	—	—	—	More rigid than type 6 nylon
Phenoxies	—	Acrid[c]	Yellow[c]	No[c]	Acrid[d]	Yellow[d]	Yes[d]	Slow[d]	Black smoke with soot in air
Polycarbonates	—	Faint, sweet aromatic ester	Orange	Yes	—	—	—	—	Black smoke with soot in air
Polyethylenes	—	—	—	—	Paraffin	Blue, yellow tip	Yes	Slow	Floats in water

TABLE A-6 *Continued*

Materials	No Flame	Burns, But Extinguishes on Removal of Flame Source			Continues to Burn After Removal of Flame Source				Remarks
	Odor	Odor	Color of Flame	Drips	Odor	Color of Flame	Drips	Speed of Burning	
THERMOPLASTICS									
Polyphenylene Oxides (PPO)	—	Phenol	Yellow–orange	No	—	—	—	—	Flame spurts; very difficult to ignite
Modified Grade	—	Phenol	Yellow–orange	No	—	—	—	—	Flame spurts; difficult to ignite; soot in air
Polyimides	[b]	—	—	—	—	—	—	—	Chars; mat'l very rigid
Polypropylenes	—	Acrid[a]	Yellow[a]	No[a]	Sweet	Blue, yellow tip	Yes	Slow	Floats in water; more difficult to scratch than polyethylene
Polystyrenes	—	—	—	—	Illuminating gas	Yellow	Yes	Rapid	Dense black smoke with soot in air
Polysulfones	—	[b]	Orange	Yes	—	—	—	—	Black smoke
Polyurethanes	—	—	—	—	[b]	Yellow	No	Slow	Black smoke
Vinyls									
Flexible	—	Hydrochloric acid	Yellow with green spurts	No	—	—	—	—	Chars, melts
Rigid	—	Hydrochloric acid	Yellow with green spurts	No	—	—	—	—	Chars, melts
Polyblends									
ABS/Polycarbonate	—	—	—	—	[b]	Yellow, blue edges	No	—	Black smoke with soot in air
ABS/PVC	—	Acrid	Yellow, blue edges	No	—	—	—	—	Black smoke with soot in air
PVC/Acrylic	—	Fruity	Blue, yellow tip	No	—	—	—	—	—
THERMOSETS									
Alkyds	—	—	—	—	—	—	—	—	—
Diallyl Phthalates	—	—	—	—	Phenolic	Yellow	No	Slow	Black smoke, cracks
Diglycol Carbonate	—	—	—	—	Acrid	Yellow	No	Slow	Black smoke with soot
Epoxies	—	—	—	—	Phenol	Black smoke	No	Slow	Black smoke with soot in air
Malamines	Formaldehyde and fish	—	—	—	—	—	—	—	—
Phenolics	Formaldehyde and phenol[c]	Phenol and wood or paper[d]	Yellow[d]	No	—	—	—	—	May crack
Polyesters	—	Hydrochloric acid[a]	Yellow[a]	No[a]	[b]	Yellow, blue edges	No	Slow	Cracks and breaks
Silicones	[b]	—	—	—	—	—	—	—	Deforms
Ureas	Formaldehyde	—	—	—	—	—	—	—	—

[a] Flame retardant grade. [b] Nondescript. [c] Inorganic filler. [d] Organic filler.

(Courtesy *Materials in Design Engineering*)

INDEX